Xenobiotics in Aquatic Animals

Mohd Ashraf Rather • Adnan Amin •
Younis Ahmad Hajam • Ankur Jamwal •
Irfan Ahmad
Editors

Xenobiotics in Aquatic Animals

Reproductive and Developmental Impacts

Springer

Editors
Mohd Ashraf Rather
Division of Fish Genetics and Biotechnology
Sher-e-Kashmir University of Agricultural Sciences and Technology, Kashmir
Rangil, Ganderbal, Jammu and Kashmir, India

Adnan Amin
Division of Aquatic Environmental Management
Sher-e-Kashmir University of Agricultural Sciences and Technology, Kashmir
Rangil, Ganderbal, Jammu and Kashmir, India

Younis Ahmad Hajam
Department of Life Sciences and Allied Health Sciences
Sant Baba Bhag Singh University
Padhiana, Jalandhar, Punjab, India

Ankur Jamwal
Azim Premji University
Bengaluru, Karnataka, India

Irfan Ahmad
Division of Fish Genetics and Biotechnology
Sher-e-Kashmir University of Agricultural Sciences and Technology, Kashmir
Rangil, Ganderbal, Jammu and Kashmir, India

ISBN 978-981-99-1213-1 ISBN 978-981-99-1214-8 (eBook)
https://doi.org/10.1007/978-981-99-1214-8

This Springer imprint is published by the registered company Springer Nature Singapore Pte Ltd.
The registered company address is: 152 Beach Road, #21-01/04 Gateway East, Singapore 189721, Singapore

Dedicated
To
The Earth, the lovely planet that we know of

Foreword 1

Xenobiotic pollution of the environment has become a serious problem on a global front. Due to high toxicity, protracted persistence, and restricted biodegradability, many xenobiotic chemicals have a negative influence on the environment, especially the aquatic environment. Different types of xenobiotics, which are credited with saving millions of lives in recent decades, have emerged as a new category of an environmental hazard. These substances can affect the natural flora and fauna in ways that are both acute and chronic. These pollutants can be found in a variety of water sources, including groundwater, surface waters, seawater, wastewater treatment plant influents and effluents, soils, and sludges.

Looking into the threat posed by hazardous chemicals from the perspective of species vulnerability, researchers have the opportunity to take steps to reduce those risks, such as managing habitat, the ocean, freshwater, and fisheries. To manage the vulnerability of aquatic ecosystems to toxic chemicals, more research is needed to develop biomarkers of vulnerability, identify the most vulnerable life stages and populations of the aquatic ecosystem, and comprehend the linkage between pathogens, hazardous chemicals, nutritional status, and environmental changes on a global scale.

The exposure of aquatic animals to xenobiotics and the demand for chemical therapeutic agents for aquatic animals will both rise as the aquaculture sector is in an escalating stage. The book *Xenobiotics in Aquatic Animals: Reproductive and Developmental Impacts* has been the outcome of the contributions made by different national and international experts working on xenobiotics in aquatic systems. I sincerely believe the book would greatly help towards enhancing the understanding of readers including scientists, teachers, students, researchers, policymakers, and all those who are interested in working in this particular area.

General (Fisheries Science), ICAR,
New Delhi, India
J. K. Jena

Foreword 2

The idea that environmental risks are to blame for the reproductive and developmental problems aquatic animals experience has been brought to light most convincingly by several studies undertaken over the last few decades. Recent years have seen a significant increase in the number of chemicals discharged into our environment, especially aquatic ecosystems. Each year, thousands of tons of synthetic chemical substances are generated; these compounds have already been approved for use in industry and agriculture. Their use has been continuing unabated, despite increasing awareness of their longstanding impacts.

In order to identify, regulate, and potentially intervene in xenobiotic exposure and its impacts on ecosystems and people, a significant amount of research has been conducted recently to understand better how xenobiotics behave in aquatic organisms. As a resource for human needs, the ecosystem's general health and safety should be our concern in this predicament.

In this context, I am happy to see this compilation of xenobiotic stressors on aquatic ecosystems that significantly impact aquatic organisms and dependent communities. This book, titled *Xenobiotics in Aquatic Animals: Reproductive and Developmental Impacts*, focuses on several significant and recently updated aspects of various xenobiotics entering aquatic ecosystems, including their effects on reproductive physiology, developmental biology, breeding biology, hormonal imbalance, aquatic ecology, and pollution to aquatic ecosystems. Readers, including scientists, teachers, students, researchers, policymakers, or anyone interested in this field, can benefit from the book's unique mix of various stressors on aquatic organisms, under a single cover.

Aquaculture Program, Asian Institute of Technology (AIT), Bangkok, Thailand

Krishna R. Salin

Preface

India has an exclusive economic zone that is 2.02 million square kilometres in size, a continental shelf that is 0.506 million square kilometres, and a coastline that is around 8129 km long. In 2019–20, India produced 14.16 million metric tonnes of fish, keeping its position as the second largest fish producer in the world. Fish production has increased by an average of 7.53% annually over the last 5 years, demonstrating the industry's phenomenal expansion. There are more native freshwater fish species in India than in any other nation in continental Asia, making it one of the world's key hotspots for biodiversity. The cold-water fisheries support 258 species of freshwater fish from 21 families and 76 genera. Snow-trout, brown trout, Chinese carps, and rainbow trout are among the 258 cold-water fish species both native and foreign reported from Indian uplands that are commercially significant. Following shrimp, which makes up 15% of all trade in terms of value, trout and salmon are the second most valuable fish species traded worldwide. More than 6% (five million tonnes) of the world's aquaculture production is made up of trout and salmon (SOFIA FAO 2020).

The entire world is searching for a sustainable source of food security. Aquaculture is steadfastly upholding expectations. The dominance of fish and fisheries products in terms of growth rate and future potential to provide nutritional security is evident from the food production sector scenario of today. According to "The State of World Fisheries and Aquaculture," aquaculture generated almost 46% of the total 178.5 million tonnes of fish produced in 2020. Around 88% of the total fish production was used for direct human consumption (SOFIA 2020). India's aquaculture industry has grown significantly over the past 10 years, and the country is now prepared to expand by embracing cutting-edge aquaculture technologies and species diversification to make the most of its vast potential water resources.

Many efforts have been made recently to comprehend how xenobiotics operate in aquatic species in order to recognise, control, and maybe intervene in xenobiotic exposure and its effects on ecosystems and people. In this case, our primary concerns should be the ecosystem's overall health and safety as well as its capacity to meet human requirements.

The focus of this book, ***Reproductive and Developmental Impact of Xenobiotics in Aquatic Animals***, is on a number of significant and recently updated aspects of various xenobiotics entering aquatic ecosystems, such as their effects on reproductive physiology, developmental biology, breeding biology, hormonal imbalance, aquatic ecology, and pollution to aquatic ecosystems.

The unique grouping of numerous stressors on aquatic organisms under a single title has benefits for readers, including scientists, teachers, students, and researchers.

Rangil, Ganderbal, Jammu and Kashmir, India — Mohd Ashraf Rather
Rangil, Ganderbal, Jammu and Kashmir, India — Adnan Amin
Padhiana, Jalandhar, Punjab, India — Younis Ahmad Hajam
Bengaluru, Karnataka, India — Ankur Jamwal
Rangil, Ganderbal, Jammu and Kashmir, India — Irfan Ahmad

Acknowledgements

The capacity to finish this book has been a gift from the Almighty, and we want to start by thanking Him for that. To all the authors who contributed chapters to this book, we would like to express our profound gratitude and respect. Thank you to everyone who helped with the publication of this book; it gives us great joy. Our effort would not have been possible if they had not continued to support and encourage us. All authors have our sincere gratitude forever.

We are very much thankful to Dr. Joykrushna Jena, Deputy Director General (Fisheries Science), and Dr. Krishna R. Salin, Director, Associate Professor, Asian Institute of Technology (AIT) Bangkok, Thailand, for their compliment, praise, and blessing to the entire team of young researchers for their contributions to this book.

We appreciate Dean Faculty of Fisheries Rangil Ganderbal, at Sher-e-Kashmir University of Agricultural Sciences and Technology in Kashmir, for their unwavering support and encouragement.

We greatly appreciate the support of our students, co-workers, and family members.

We will also take this opportunity to express our sincere gratitude to Springer Publishing Company for publishing this work. Thank a lot.

Last but not least, we would like to bow down before the Almighty, who has given us the bravery, passion, and strength to finish this effort.

Dated: 30/12/2022.

Mohd Ashraf Rather
Adnan Amin
Younis Ahmad Hajam
Ankur Jamwal
Irfan Ahmad

Contents

Editors and Contributors

About the Editors

Mohd Ashraf Rather works as an Assistant Professor cum Junior Scientist in the Division of Fish Genetics and Biotechnology at the Faculty of Fisheries, Rangil, Ganderbal, Sher-e-Kashmir University of Agricultural Sciences and Technology—Kashmir. He graduated from the College of Fisheries, Ratnagiri, Maharashtra, and obtained master's and doctoral degrees in Fisheries Biotechnology from the Division of Genetics and Biotechnology, Central Institute of Fisheries Education (CIFE), Mumbai. His area of research interest is reproductive physiology and molecular endocrinology of fish proteins. His major research contribution is the identification, characterization, and expression profiling of more than 25 reproductive genes in fish including kisspeptin genes in Indian major carps. He has more than 8 years of experience of working on molecular endocrinology of fish.

Adnan Amin is presently working as Assistant Professor Cum Junior Scientist in the Division of Aquatic Environmental Management at the Faculty of Fisheries, Rangil, Ganderbal, Sher-e-Kashmir University of Agricultural Sciences and Technology—Kashmir. He is an undergraduate from the College of Fishery Science, Muthukur, Nellore SVVU, Tirupati, Andhra Pradesh. He did his master's from the College of Fisheries, Mangalore, in the Department of Aquatic Environment Management. He completed his doctoral degree from the College of Fisheries, Ratnagiri, in the Department of Aquatic Environment Management. His area of research interest is aquatic toxicology and limnology. His major research contribution is heavy metal and pesticide toxicity in aquatic organisms, and impact of climate change in aquatic biodiversity.

Younis Ahmad Hajam holds a doctorate in Zoology from Guru Ghasidas Vishwavidyalaya (A Central University), Bilaspur, Chhattisgarh, India. He is currently working as Assistant Professor in the Department of Life Sciences and Allied Health Sciences, Sant Baba Bhag Singh University, Padhiana, Jalandhar, Punjab,

India. Formerly Dr. Younis Ahmad had worked as assistant Professor and Head of the Department of Zoology, Career Point University, Hamirpur, Himachal Pradesh, India. He is teaching undergraduate and postgraduate and doctoral students. Dr. Younis Ahmad has published 33 research papers in leading national and international journals (such as *Life Sciences*, Impact factor 5.037, and *Neuroendocrinology* 4.96) and Toxicology Reports, 25 book chapters, and four national proceedings. Dr. Younis Ahmad has published four books as editor in Taylor and Francis, etc.

Ankur Jamwal is an assistant professor at Azim Premji University, Bengaluru, India. Dr. Jamwal has over a decade of experience as an aquatic biologist and ecotoxicologist and has published over twenty research articles in international journals of repute. In addition to studying the effects of stress on the physiology of aquatic animals, Dr. Jamwal is also interested in the risk assessment of consumption of food cultivated using water that may be contaminated with toxic elements and compounds. He applies the concepts of aquatic physiology to developing sustainable aquaculture practices. Dr. Jamwal has received multiple awards, including the Best Fisheries Graduate of India—2010 and ICAR International Fellow. He is a passionate speaker and loves to spread awareness on climate change and the conservation of aquatic ecosystems.

Irfan Ahmad works as a Professor cum Chief Scientist in the Division of Fish Genetics and Biotechnology at the Faculty of Fisheries, Rangil, Ganderbal, Sher-e-Kashmir University of Agricultural Sciences and Technology—Kashmir. He obtained his master's and doctoral degrees in Zoology from Kashmir University. He was the first student in Zoology from Jammu and Kashmir to clear National Eligibility Test (CSIR/UGC) in Zoology. His area of research interest is molecular markers, pathogenic organisms in fish and aquatic environment. He is also the recipient of first prize from Wiley Blackwell (International Society of Zoological Sciences, China). Dr. Ahmad is the recipient of various International Travel Scholarships from World Fisheries Congress, SERB, GoI, Sabin Vaccine Institute, and Gates Foundation. He has travelled to more than eight countries to present his research in various international congress, symposiums, and conferences.

Contributors

Adnan Abubakr Division of Aquatic Environmental Management, Faculty of Fisheries, Rangil, SKUAST-Kashmir, Srinagar, Jammu and Kashmir, India

Deepak Agarwal Tamil Nadu Dr. J. Jayalalithaa Fisheries University—IFPGS, OMR Campus, Chennai, India

Fayaz Ahmad Advanced Research Laboratory, Department of Zoology, University of Kashmir, Srinagar, Jammu and Kashmir, India

Irfan Ahmad Division of Fish Genetics and Biotechnology, Sher-e-Kashmir University of Agricultural Sciences and Technology, Kashmir, Rangil, Ganderbal, Jammu and Kashmir, India

Ishtiyaq Ahmad Faculty of Fisheries Rangil, SKUAST-Kashmir, Srinagar, India

Qurat ul Ain Department of Chemistry, Government College Women University, Faisalabad, Pakistan

Sadanand Dangari Aksahy Nitte Centre for Science Education and Research, NUCSER, Nitte (Deemed to be University), Mangaluru, Karnataka, India

Adnan Amin Division of Aquatic Environmental Management, Sher-e-Kashmir University of Agricultural Sciences and Technology, Kashmir, Rangil, Ganderbal, Jammu and Kashmir, India

Saima Andleeb Division of Fish Genetics and Biotechnology, Faculty of Fisheries, Sher-e-Kashmir University of Agricultural Sciences and Technology, Kashmir, Rangil, Ganderbal, Jammu and Kashmir, India

Ihtesham Arshad Department of Biotechnology, Faculty of Life Sciences, University of Okara, Okara, Pakistan

Priyanka Ashwath Nitte Centre for Science Education and Research, NUCSER, Nitte (Deemed to be University), Mangaluru, Karnataka, India

Oyas Asimi Division of Fish Nutrition and Biochemistry, Faculty of Fisheries, Sher-e-Kashmir University of Agricultural Sciences and Technology, Kashmir, Rangil, Ganderbal, Jammu and Kashmir, India

Sana Aziz Department of Zoology, Wildlife and Fisheries, University of Agriculture, Faisalabad, Pakistan

Sajad H. Baba Division of Agri Economics and Hotibusiness Management, Sher-e-Kashmir University of Agricultural Sciences and Technology, Kashmir, Shalimar, Jammu and Kashmir, India

Muddasir Sharief Banday Department of Clinical Pharmacology, Sher-e-Kashmir Institute of Medical Sciences, Soura, Srinagar, Jammu and Kashmir, India

Indrashis Bhattacharya Department of Zoology, The Central University of Kerala, Kasaragod, Kerala, India

Pradyut Biswas Department of Aquaculture, College of Fisheries, Lembucherra, Central Agricultural University, Imphal, Manipur, India

Hafsa Farooq Chashoo Division of Aquatic Environmental Management, Faculty of Fisheries, Rangil, SKUAST-Kashmir, Srinagar, Jammu and Kashmir, India

Reshmi Debbarma Department of Aquaculture, College of Fisheries, Lembucherra, Central Agricultural University, Imphal, Manipur, India

Menakshi Dhar Department of Zoology, University of Jammu, Jammu, Jammu and Kashmir, India

Melissa K. Driessnack Puyallup Research and Extension Center, Washington State University, Puyallup, WA, USA

Akhila D S Nitte Centre for Science Education and Research, NUCSER, Nitte (Deemed to be University), Mangaluru, Karnataka, India

Anees Fatima Department of Botany, Govt College for Women, Cluster University-Kashmir, Srinagar, Jammu and Kashmir, India

Shams M. Galib Department of Biosciences, University of Durham, Durham, UK

Udai Ram Gurjar DRPCAU-Krishi Vigyan Kendra, East Champaran, Bihar, India

Mehak Hafeez Division of Fish Genetics and Biotechnology, Faculty of Fisheries, Sher-e-Kashmir University of Agricultural Sciences and Technology, Kashmir, Rangil, Ganderbal, Jammu and Kashmir, India

Younis Ahmad Hajam Department of Life Sciences and Allied Health Sciences, Sant Baba Bhag Singh University, Padhiana, Jalandhar, Punjab, India

Mouada Hanane VTRS Laboratory, Department of Process Engineering, University Center of Tipaza, El Oued, Algeria

Farah Hanief Division of Aquatic Environmental Management, Faculty of Fisheries, Rangil, SKUAST-Kashmir, Srinagar, Jammu and Kashmir, India

Muzammal Hoque ICAR-Central Institute of Fisheries Education, Mumbai, India

Ishraq Hussain Division of Vety Biochemistry, FVSc&AH, Sher-e-Kashmir University of Agricultural Sciences and Technology, Kashmir, Shuhama, Jammu and Kashmir, India

Mir Shabir Hussain Wildlife Research Lab, Department of Zoology, University of Kashmir, Srinagar, India

S. Abuthagir Iburahim ICAR-Central Institute of Fisheries Education, Mumbai, India

Rehana Iqbal Institute of Zoology, Bahauddin Zakariya University, Multan, Pakistan

Iqram ul Haq Division of Wildlife Sciences, SKAUST-Kashmir, Srinagar, India

Saman Iram Institute of Zoology, Bahauddin Zakariya University, Multan, Pakistan

Ankur Jamwal Azim Premji University, Bengaluru, Karnataka, India

Raman Jasrotia Department of Zoology, University of Jammu, Jammu, Jammu and Kashmir, India

Mohammad Amjad Kamal Institutes for Systems Genetics, Frontiers Science Center for Disease-Related Molecular Network, West China Hospital, Sichuan University, Chengdu, China
King Fahd Medical Research Center, King Abdulaziz University, Jeddah, Saudi Arabia
Department of Pharmacy, Faculty of Allied Health Sciences, Daffodil International University, Daffodil Smart City, Bangladesh
Novel Global Community Educational Foundation, Hebersham, NSW, Australia

Sutanu Karmakar Faculty of Fishery Sciences, West Bengal University of Animal and Fishery Sciences, Kolkata, India

Sachin Kothiyal Department of Zoology, HNB Garhwal University, Pauri Garhwal, Uttarakhand, India

J. Mori Krinal Postgraduate Institute of Fisheries Education and Research, Kamdhenu University, Himmatnagar, India

Gulshan Kumar College of Fisheries Sciences, Gumla, Birsa Agricultural University, Ranchi, Jharkhand, India

Rajesh Kumar Department of Biosciences, Himachal Pradesh University, Shimla, Himachal Pradesh, India

Sujit Kumar Postgraduate Institute of Fisheries Education and Research, Kamdhenu University, Himmatnagar, India

Jham Lal Department of Aquaculture, College of Fisheries, Lembucherra, Central Agricultural University, Imphal, Manipur, India

Seema Langer Department of Zoology, University of Jammu, Jammu, Jammu and Kashmir, India

Fariha Latif Institute of Pure and Applied Biology, Bahauddin Zakariya University, Multan, Pakistan
Institute of Zoology, Bahauddin Zakariya University, Multan, Pakistan

Monisa Mehboob Malik Division of Aquatic Environmental Management, Sher-e-Kashmir University of Agricultural Sciences and Technology, Kashmir, Faculty of Fisheries Rangil, Ganderbal, Jammu and Kashmir, India

Kavitha Guladahalli Manjunatha Nitte Centre for Science Education and Research, NUCSER, Nitte (Deemed to be University), Mangaluru, Karnataka, India

Nafee Maqbool Advanced Research Laboratory, Department of Zoology, University of Kashmir, Srinagar, Jammu and Kashmir, India

Umer Aziz Mir Advanced Research Laboratory, Department of Zoology, University of Kashmir, Srinagar, India

Nuzaiba P. Muhammad Centurion University of Technology and Management, Paralakhemundi, Odisha, India

Mumaiza Mumraiz Department of Zoology, Wildlife and Fisheries, University of Agriculture, Faisalabad, Pakistan

Bejawada Chanikya Naidu ICAR-Central Institute of Fisheries Education, Mumbai, India

Maria Nazir Institute of Zoology, Bahauddin Zakariya University, Multan, Pakistan

T. Nirmal Centurion University of Technology and Management, Paralakhemundi, Odisha, India

Nevil Pinto ICAR-CIFE, Versova, Andheri (W), Mumbai, Maharashtra, India

Arya Prabhakaran ICAR-Central Institute of Fisheries Education, Versova, Andheri (w), Mumbai, Maharashtra, India

Gora Shiva Prasad Faculty of Fishery Sciences, West Bengal University of Animal and Fishery Sciences, Kolkata, India

Premlata Division Zoology, Department of Biosciences, Career Point University, Hamirpur, Himachal Pradesh, India

Shriya Purohit Department of Zoology, HNB Garhwal University, Pauri Garhwal, Uttarakhand, India

Gopika Radhakrishnan Department of Mathematics and Natural Science, Institute of Marine Research, University of Bergen, Bergen, Norway

Seema Rai Department of Zoology, Guru Ghasidas Vishwavidyalya (A Central University), Bilaspur, Chhattisgarh, India

Raghvendra Rao Department of Biotechnology, Alvas PG College, Moodibidre, Karnataka, India

Sumaira Rashid Division of Fish Genetics and Biotechnology, Faculty of Fisheries, Sher-e-Kashmir University of Agricultural Sciences and Technology, Kashmir, Rangil, Ganderbal, Jammu and Kashmir, India

Mohd Ashraf Rather Division of Fish Genetics and Biotechnology, Sher-e-Kashmir University of Agricultural Sciences and Technology, Kashmir, Rangil, Ganderbal, Jammu and Kashmir, India

A. Rathipriya Tamil Nadu Dr. J. Jayalalithaa Fisheries University—IFPGS, OMR Campus, Chennai, India

Nakeer Razak Division of Fish Genetics and Biotechnology, Faculty of Fisheries, Sher-e-Kashmir University of Agricultural Sciences and Technology, Kashmir, Rangil, Ganderbal, Jammu and Kashmir, India

DeviPrasadh Reddy Krishi Vigyan Kendra, Dr. Y.S.R. Horticultural University, Venkataramannagudem, Andhra Pradesh, India

Sabeehah Rehman Centre of Research for Development, University of Kashmir, Srinagar, India

Sangram Keshari Rout Faculty of Fishery Sciences, West Bengal University of Animal and Fishery Sciences, Kolkata, India

Arfa Safder Institute of Molecular Biology and Biotechnology, The University of Lahore, Lahore, Punjab, Pakistan

M. D. Sahana ICAR-Central Institute of Fisheries Education, Mumbai, India

Muhammad Sarfraz Ali Department of Zoology, Wildlife and Fisheries, University of Agriculture, Faisalabad, Pakistan

A. Sathiyanarayanan ICAR-CIFE, Versova, Andheri (W), Mumbai, Maharashtra, India

Azra Shah Division of Fish Genetics and Biotechnology, Faculty of Fisheries, Sher-e-Kashmir University of Agricultural Sciences and Technology, Kashmir, Rangil, Ganderbal, Jammu and Kashmir, India

Ishteyaq Majeed Shah Advanced Research Laboratory, Department of Zoology, University of Kashmir, Srinagar, Jammu and Kashmir, India

Muhammad Shakeel Institute of Zoology, Bahauddin Zakariya University, Multan, Pakistan

Partigya Sharma Department of Zoology, HNB Garhwal University, Pauri Garhwal, Uttarakhand, India

Soibam Khogen Singh Department of Aquaculture, College of Fisheries, Lembucherra, Central Agricultural University, Imphal, Manipur, India

E. Suresh Tamil Nadu Dr. J. Jayalalithaa Fisheries University—IFPGS, OMR Campus, Chennai, India

Nurul Suwartiningsih Laboratory of Ecology and Systematics, Study Program of Biology, Ahmad Dahlan University, Yogyakarta, Indonesia

Suman Takar The Neotia University, Kolkata, West Bengal, India

Tincy Varghese ICAR-Central Institute of Fisheries Education, Versova, Andheri (w), Mumbai, Maharashtra, India

G. Ad Viral Postgraduate Institute of Fisheries Education and Research, Kamdhenu University, Himmatnagar, India

Rajeshwari Vittal Nitte Centre for Science Education and Research, NUCSER, Nitte (Deemed to be University), Mangaluru, Karnataka, India

Rasy Fayaz Choh Wani Advanced Research Laboratory, Department of Zoology, University of Kashmir, Srinagar, India

K. A. Martin Xavier Fisheries Resource Harvest and Post-Harvest Management Division, ICAR-Central Institute of Fisheries Education, Mumbai, Maharashtra, India

B. S. Yashwanth ICAR-CIFE, Versova, Andheri (W), Mumbai, Maharashtra, India

Waqas Yousaf Department of Botany, Institute of Molecular Biology and Biotechnology (IMBB), The University of Lahore, Lahore, Punjab, Pakistan

Imran Zafar Department of Bioinformatics and Computational Biology, Virtual University, Lahore, Punjab, Pakistan

Abbreviations

2-NA	2-Naphthylamine
4-ABP	4-Aminobiphenyl
AhR	Aryl hydrocarbon receptor
ALL	Acute lymphocytic leukaemia
ANLL	Acute nonlymphocytic leukaemia
ASCR	Association of Swiss Cancer Registries
ASL	Angiosarcoma of the liver
ATSDR	Agency for Toxic Substances and Disease Registry
b.p.	Boiling point
BCME	Bis(chloromethyl)ether
BP	Benzo pyrene
BSF	Benzene-soluble fraction
BSM	Benzene-soluble materials
BW	Body weight
CA	Chromosomal aberration
CAS	Chemical Abstracts Service (of the American Chemical Society)
CI	Confidence interval
CMME	Chloromethyl methyl ether
CNS	Central nervous system
CPTV	Coal-tar pitch volatiles
CSF	Cyclohexane-soluble fraction
CYP	Cytochrome P450
DCM	Dichloromethane
DEHP	Diethylhexylphthalate
DHBMA	1,2-Dihydroxybutyl mercapturic acid
DLC	Dioxin-like compound
DMSO	Dimethyl sulfoxide
EH	Epoxide hydrolase
EPA	Environmental Protection Agency

EU	European Union
FISH	Fluorescence in situ hybridization
GST	Glutathione S-transferase
HCC	Hepatocellular carcinoma
HPRT	Hypoxanthine-guanine phosphoribosyltransferase
HRR	Hazard rate ratio
IH	Industrial hygiene
IRR	Incidence rate ratio
ISCO	International Standard Classification of Occupations
ISIC	International Standard Industrial Classification
MBS	Morpholinomercaptobenzothiazole
MBT	2-Mercaptobenzothiazole
MDS	Myelodysplastic syndromes
mEH	Microsomal epoxide hydrolase
MGP	Manufactured gas plant residues
MHBMA	Monohydroxy-3-butenyl mercapturic acid
MHBVal	N-(2-hydroxy-3-butenyl)valine
MN	Micro nucleus
MOCA	4,4′-Methylene-bis-(2-chloroaniline)
MPD	Myeloproliferative disorder
MT	Metallothionein
NAT	N-acetyltransferase
NDBA	N-nitrosodibutylamine
NDEA	N-nitrosodiethylamine
NDMA	N-nitrosodimethylamine
NDPA	N-nitrosodiphenylamine
NG	Not given
NIOSH	National Institute for Occupational Safety and Health (USA)
NMor	N-nitrosomorpholine
NP	Nanoparticle
NPIP	N-nitrosopiperidine
NTP	National Toxicology Program
OR	Odds ratio
PAH	Polycyclic aromatic hydrocarbons
PARP	Poly(ADP-ribose) polymerase
PBNA	Phenyl-β-naphthylamine
PCB	Polychlorobiphenyl
PCDD	Polychlorinated dibenzo-para-dioxins
PCP	Pentachlorophenol
PCR	Polymerase chain reaction
PeCDF	2,3,4,7,8-Pentachlorodibenzofuran
PKC	Protein kinase C
POG	Paediatric oncology group
ppb	Parts per billion

ppm	Parts per million
ppt	Parts per trillion
PVC	Polyvinyl chloride
ROS	Reactive oxygen species
RR	Relative risk
SCE	Sister chromatid exchange
SES	Socioeconomic status
TCDD	Tetrachlorodibenzo-p-dioxin
TCE	Trichloroethylene
TEF	Toxicity equivalence factor
TWA	Time-weighted average
USA	United States of America
VC	Vinyl chloride
VCM	Vinyl chloride monomer
VOC	Volatile organic compound
wk	Week
yr	Year
ZDEC	Zinc-diethyldithiocarbamate

Pesticide and Xenobiotic Metabolism in Aquatic Organisms

Akhila D S, Priyanka Ashwath, Kavitha Guladahalli Manjunatha, Sadanand Dangari Aksahy, Raghvendra Rao, DeviPrasadh Reddy, and Rajeshwari Vittal

1 Introduction

Scientific, technological, and industrial revolutions have permitted humans to overutilize the resources, thereby creating an imbalance in the natural ecosystem (Sikandar et al. 2013). Enormous amounts of toxic effluents let out as a result of industrial processes have caused widespread contamination of the ecosystem. It was reported that nitrated and halogenated hydrocarbons are few among the major contaminants (Jain et al. 2011). Various fertilizers, insecticides, and herbicides have been employed in agricultural activities. In addition, industry-based synthetic compounds like dyes, pharmaceuticals, hydraulics, pigments, agrochemicals, halogenated compounds, and fire retardants have been extensively used (Reineke and Knackmuss 1988). Due to the inevitable uses in veterinary and anthropoid medications, pharmaceutical wastes have developed as a significant cause of prolonged environmental condition (Gani et al. 2021). The chemicals thus let out into the surrounding environments are believed to possess specific modes of action and thus impart a certain number of hazards on the aquatic flora and fauna, in comparison to the other chemical substances. These chemical compounds persist within the environment making them a potential agent causing health hazards and posing toxic effects on the surrounding niche. Xenobiotics can have a range of impacts, such as

Akhila D S · P. Ashwath · K. G. Manjunatha · S. D. Aksahy · R. Vittal (✉)
Nitte (Deemed to be University), Nitte University Centre for Science Education and Research, NUCSER, Deralakatte, Mangalore, Karnataka, India
e-mail: rajeshwari.vittal@nitte.edu.in

R. Rao
Department of Biotechnology, Alvas PG College, Moodibidre, Karnataka, India

D. Reddy
Krishi Vigyan Kendra, Dr. Y.S.R. Horticultural University, Venkataramannagudem, Andhra Pradesh, India

M. A. Rather et al. (eds.), *Xenobiotics in Aquatic Animals*,
https://doi.org/10.1007/978-981-99-1214-8_1

immunological reactions, medicine toxicity, and climate change. Xenobiotics are substances that are either nonbiodegradable or only partially biodegradable, which may result in sluggish biotransformation and persistence in the setting for an extended period of time (Dar et al. 2020).

With this point of view, in this chapter, we sought to discuss dormant pollutants (pesticides and xenobiotics), their metabolism, and their harmful effects on aquatic life.

2 Pesticides and Xenobiotics

Pesticides are a collective term used to represent all the compounds including herbicides, insecticides, and fungicides that are applied to regulate pests or undesirable organisms. Water resources contaminated with pesticides are found to affect both humans and the ecosystem. Pesticides have sought to be as probable mutagens, as they are capable of triggering deviations in the DNA. The World Health Organization has estimated that around 1,000,000 human beings have been affected by acute poising as a result of toxicant contact; in addition, annually, a death rate ranging between 0.4% and 1.9% has been recorded. It has been reported that constant exposure to pesticides in the medium and long term resulted in various syndromes and tumors including the nervous system disorder. Accustomed application of agrochemicals such as pesticides, soil conditioners, acidifying agents, chemicals involved in animal husbandry (hormones and antibiotics), and fertilizers is popular. Since the beginning of the industrial era in 1950, the use of pesticides has had a severe impact on the environment niche. It has been extensively employed as a pest control agent wherein monocular cultivation is involved. Few demerits are still involved, despite the development of chemistry in the field of pesticide formulations; majorly, the pesticides perturb the predator-prey interactions, thus causing an imbalance in the biodiversity. Furthermore, the pesticides can cause exceptional health concerns. Though the usage of a few chemical compounds is limited/controlled by the agricultural sector, it is said that agriculture is one of the areas which purposefully discharges chemicals into the surrounding niche leading to the adverse effects (WHO 2020; Warra and Prasad 2020). Of the manufactured compounds, the usage of the pesticides by the agricultural sector has been recorded the highest (Sharma et al. 2019; Laxmi et al. 2019). It is often difficult to distinguish among the pesticide effect and environmental effects on the ecosystem because the industrial effluents are let out into the surroundings to a greater extent accidently or intentionally. On the contrary, significant evidence suggests that the usage of pesticides in the agricultural field has had a considerable impact on the water quality causing major influence and concerns on the environment (FAO 1990; Warra and Prasad 2020). Despite the fact that the amount of pesticide practice is relatively extensive, it is plausible that substantial use of chemicals is associated only with few pesticide products. Of the million tons of pesticides produced, 29.5% insecticides, 47.5% herbicides, 5.5% other pesticides, and 17.5% fungicides are the categories extensively in application

(De et al. 2014). Howbeit, the use of pesticides is in a range of low to zero in subsistence and conventional farming in countries like Asia and Africa. However, the environment, water quality, and health hazards are associated with its inappropriate usage.

2.1 Classification of Pesticides

Pesticide is a collective term that describes diverse groups of insecticides, herbicides, rodenticides, garden chemicals, fungicides, and household disinfectants employed to both protect and destroy pests (Mohapatra et al. 2021). Each pesticide groups differ in its physical and chemical properties. On that account, it is preferable to classify them based on their properties. As recommended by Drum (1980), the widely used method of classification of pesticide groups depends on their (a) chemical structure, (b) mode of entry, and (c) mode of action and type of organisms they target (Yadav et al. 2015). Based on their source, chemical pesticides have been classified into four types: pyrethroid pesticides, organophosphate, carbamate, and organochlorine (Yadav and Devi 2017). Another class of pesticides are biopesticides, occurring naturally or are naturally derived from living organisms such as bacteria, plants, and fungi (Mehrotra et al. 2017). Microbial pesticides, biochemical pesticides, and incorporated protectants are the three major groups of biopesticides.

2.2 Pesticide Categories

2.2.1 Chemical Structure of Pesticides

The most general and applicable method of classifying insecticides is based on their chemical description and the chemical composition of the active ingredients. The chemical classification of pesticides delivers the efficacy and physical and chemical properties of special pesticides (Fig. 1).

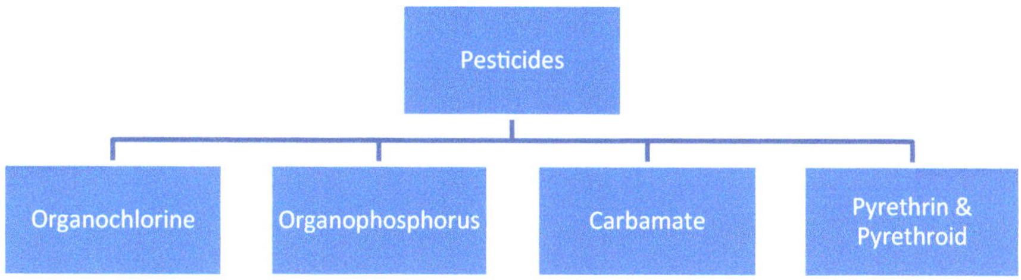

Fig. 1 Classification of Pesticides

2.2.2 Organochlorine Pesticides (OCPs)

Chemicals belonging to this group are very stable and persistent in the environment and have the potential of accumulating in the adipose tissue (Lee et al. 2017). It is said that in humans, these chemicals/their metabolites largely work on the central nervous system by altering their electrophysiological properties and altering enzymatic nerve membranes and changing the flow kinetics of K+ and Na+ through the nerve cell membrane and may cause seizures from apnea and acute poisoning death (Zaffar et al. 2016). Based on their structure, the organochlorines are categorized into five categories: (a) hexachlorocyclohexane (HCH), such as lindane, (b) DDT and its analogues including DDT and dichlorodiphenyldichloroethylene (DDE), (c) dichlorodiphenyldichloroethane (DDD), (d) mirex and chlordecone, and (e) cyclodienes including aldrin, dieldrin, endrin, heptachlor, chlordane, and endosulfan (Singh and Singh 2017). Higher concentrations of most of the OCPs generally result in acute toxicity and death under natural circumstances and may be gradually causing chronic illness. The persistent nature of the OCPs and their lipophilic nature might lead to long-term storage in the adipose tissue, following their release into the circulatory system (Kumar et al. 2013). This process takes a longer duration from the initial time of exposure to the onset of effects; it is said that DDT may remain in the human body for 50 years and more.

Organophosphate Pesticides

Organophosphate pesticides are ester derivatives of phosphoric acid. These esters work on the central nervous system by blocking acetylcholine. This enzyme is responsible for maintaining the levels of acetyl cholinesterase, which disturbs the nerve impulse by the phosphorylation of the serene OH group in the active site of the enzyme (Lionetto et al. 2011; Laxmi et al. 2020). Headache, dizziness, loss of reactions, nausea, convulsions, cramps, and ultimately coma and death are the intoxication symptoms of organophosphate pesticides.

Atropine is said to be the definitive treatment for organophosphate poisoning, which competes with acetylcholine at the muscarinic receptors. The recommended initial dose for adults is 2–5 mg IV, and for children, a dosage of 0.05 mg/kg IV is recommended. In case the patient does not respond to the initial doses, the dosage is doubled every 3–5 min until the respiratory secretions have been cleared and there are no signs of bronchoconstriction (van Heel and Hachimi-Idrissi 2011). The atropine is given as a continuous infusion or in bolus in patients with severe poisoning for a couple of days until the patient shows any signs of improvement.

Carbamate Pesticides

Some organic ester compounds which are the derivatives of dimethyl N-methyl carbamic acid are applied as fungicides, herbicides, nematicides, and insecticides, collectively termed as carbamates. Propoxur, pyridostigmine, molinate, carbaryl, thiobencarb, methiocarb, and disulfiram (Antabuse) are being widely used in dogs and cats (Hassaan and El Nemr 2020). The toxicity of the carbamate compounds depends on their molecular structure, and generally, they have short duration in comparison to that of the organochlorines and organophosphates; organochlorines are said to inhibit acetyl cholinesterase. Carbamates are short-lived, and thus precautions must be taken while administering the atropine (Hernández et al. 2013). Symptoms caused by acute poisoning of carbamate insecticides and organophosphate are often common and severe. The symptoms of poisoning develop in different organs as follows: cough, increased secretions, pulmonary edema, bronchial tree (dyspnea, wheezing,), cardiovascular effects (hypotension and bradycardia), abdominal cramps, gastrointestinal manifestations (diarrhea, vomiting, nausea, and incontinence), glandular stimulation (lacrimation, increased salivation, and sweating), eye problems (miosis, and blurred vision), compromised motor activity (cramps, fasciculation, muscle twitching, weakness, depression of respiratory and circulatory), bladder dysfunction (incontinence and frequency), sympathetic dysfunction (tachycardia, hypertension, and pallor), central nervous system effects (generalized weakness, drowsiness, restlessness, tremor, emotional lability, confusion, slurred speech, Cheyne-Stokes respiration , areflexia, ataxia, convulsion, hypothermia, and coma), nicotinic receptor stimulation (including sympathetic and motor neurons), bronchoconstriction, and cyanosis (Muhammad et al. 2017).

Pyrethroid Pesticides

Pyrethroids are natural insecticides that are the derivatives of pyrethrum extracts from the flowering chrysanthemum, commonly called pyrethrin. These chemicals act on the central nervous system causing fluctuations in the dynamics of sodium cation channels on the membrane of the nerve cells, leading to an increased opening time of the sodium channels. In both insects and vertebrates, the sodium cation stream extends across the membrane, and the neuronal hyperexcitation experienced could be the result of all these symptoms put together (Clark and Symington 2011). Since there is an increased demand for the usage of pyrethroids, and also a shortage of the essential oils required for the production of natural organic pyrethrum, researchers have opted for synthetic pyrethroids. Majority of the pyrethroid insecticides exhibit lower toxicity toward mammals and birds while higher rates of toxicity to arthropods, as they require very low doses for the effects to show up (Lengai et al. 2020). When applied directly to water, these are highly toxic to the fishes and act rapidly against chewing insects. Even though majority of the pyrethroid insecticides are absorbed by the insect pests, they are not very effective in penetrating the soil in order to kill the underground pests, as they adhere tightly to the organic matter and soil. Pyrethroids are employed as active substance in the production of several products such as topical mosquito repellents, pet shampoos, human head lice treatments, pet sprays, and insecticide sprays in farms and homes (Hassaan and El Nemr 2020).

Pesticides in Water Resources

Water is considered as one of the most predominant natural resources, essential for all the living creatures. The condition of surface water and soil is a matter of concern in many developing countries due to the contaminants reaching the surface water and soil over the past few years.

It is mandatory for the pesticide users to clearly understand the cycle of pesticide contamination of both surface and ground water (Fig. 2). In order to reduce the pesticide off-site movement, it is utmost important for the pesticides users to implicate safe practices when introducing pesticides into the surface and ground water. The groundwater contamination by the pesticides is hard to be eliminated due to the pollutant concentration, in comparison to other environments, which is worrisome (Sjerps et al. 2019). The entry of pesticides into the water source is by rundown (flowing) and leaching (filtering), of which both modes are related to the earth's hydrological cycle. While taking into consideration the utilization of water for surface flow, the pesticides in the municipal wastewater are suitable for hydrological model. Some of the factors affecting the quality of water are usage of active ingredients in the formulation of pesticides, degradable compounds as a result of microbial/photochemical or chemical degradation of the active compounds, adhesives, buffers, preservatives, and emulsifier/wetting agents used as additive mixture along with the active ingredients.

Fate and Effects of Pesticides in Aquatic Ecosystems

Employing pesticides in various sectors contaminates the aquatic niche via several ways, namely, leaching, spray drift and, runoff, and this may cause harmful effects on the terrestrial and aquatic life (Van den Brink 2013; Wijngaarden et al. 2005). Fishes are said to be directly affected by the pesticide usage; the small fishes are affected severely in comparison to the larger ones. Indirect toxic effects of pesticides on fishes include decrease in the fish's food sources (plankton and algae), deterioration in the quality of the aquatic habitat, and changes in their food pattern. The reduced abundance of primary producers as a result of pesticide application significantly decreases the primary and secondary consumers.

Organochlorine pesticides severely affect the primary consumers such as zooplankton. Microcrustaceans are also being affected by the employment of various insecticides. It can be noted that the interferences used in the pesticide formulation might further increase the toxic effects of the pesticides; however, these are not a part of the active ingredients (Singh et al. 2016). For instance, TFM (3-trifluoromethyl-4-

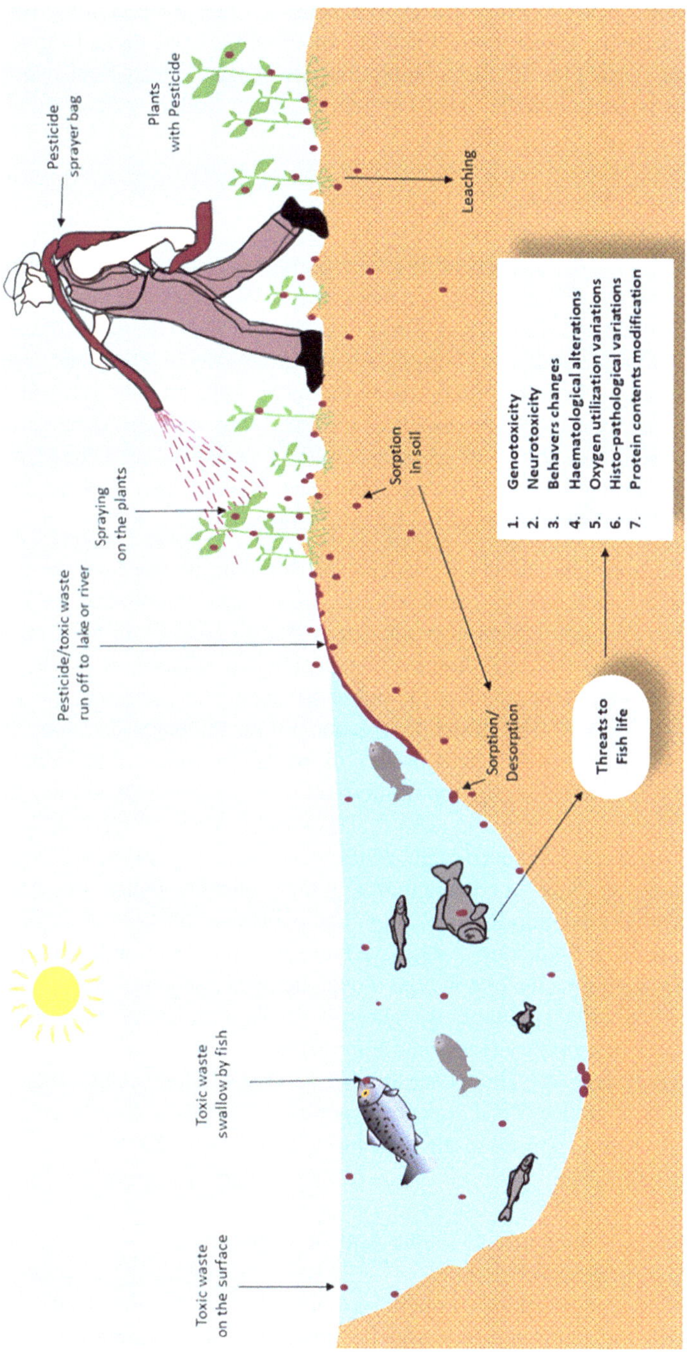

Fig. 2 Effect of pesticide on aquatic environment

nitro phenol), a lamprey pesticide, has found its application in the tributaries of the Great Lakes for years together in Egypt, in order to control the lamprey (Gllderhus and Johnson 1980). The environmental fate of TFM has been studied for years, and reports have stated that the formulation consists of additional potent interferences that have an impact on the hormonal system of fishes, leading to liver diseases.

Direct Effects of Pesticides

The aquatic ecosystem consists of diverse groups of organisms such as microorganisms, invertebrates, plants, fishes, and amphibians. Pesticides can impart both direct and indirect effects on these living forms. The physiological action of the particular pesticides within the organisms brings about the direct effects. The biotic community is distinguished by the interactions among the species like predation or competition; indirect effects involve the consequences mediated through these communications.

Mostly, the direct effects of chemicals on any organism are dependent on the concentration or the dosage of the pesticide used. Nevertheless, few factors impact the magnitude and occurrence of the adverse effects.

Exposed life stage: The effects of pesticides in different life stages of the organisms vary, for instance, younger invertebrates, fishes, and amphibians are more susceptible to the toxic effects than adult ones (Babalola and Van Wyk 2018).

Duration of exposure: Longer duration of exposure to pesticides has potent effects. It can be noted that biomagnification indirectly delays the temporal effects caused by pesticide usage. The pesticide organochlorine biomagnifies along the food web, resulting in higher magnitudes of pesticide concentration/lipid weight within the organisms, at the top of the food web pyramid (Tooker and Pearsons 2021).

Additional stressors: Enhanced effect could be visualized when an additional single compound was found in conjunction with the pesticides. Other stressors include food scarcity, predation, UV radiation, and parasitism (Hooper et al. 2013).

Population density: Reduction of negative intraspecific interaction, toxicant-induced effects in high-density populations could be seen, although there remains an altered population structure, the toxicant displays delay in development and age-dependent mortality (Liess and Foit 2010).

History of the community: Higher sensitivity or acquired tolerance/pollution-induced community tolerance (PICT) may be experienced, in the case of the communities previously being exposed to the toxicants (Schmitt et al. 2005).

Indirect Effects of the Pesticide on the Aquatic Community

In the ecosystem, interactions within the species and with the other species and their respective abiotic environment could be observed. These interactions could be altered as a result of the direct effects of the pesticides, causing the secondary or indirect effects on the related species which are not affected directly. Listed ecological relationships may result in the indirect effects through the propagating direct effects, namely, predation, which comprises the parasite-host, predator-prey, and herbivore-plant relationships; competition, which is intraspecific/interspecific; and species-habitat relation, that is, commensalism or mutualism.

As mentioned earlier, the primary producers are usually at a higher risk of being adversely affected by herbicide usage. Thus, a reduction in the population of the primary producers eventually leads to a decrease in the herbivore populations, as a consequence of habitat loss or limitations in the food source. In a study reporting the effects of the herbicide (atrazine) on the freshwater communities, for instance, in artificial ponds, the reproduction rate of the zooplankton (e.g., *Daphnia pulex*, *Simocephalus serrulatus*) gradually decreased as a result of the reduction in the phytoplankton biomass. Similarly, it was observed that due to the reduction of periphyton and as a loss of macrophyte habitat (*Typha latifolia*, *Chara* sp.), there was diminished biomass of the amphibian tadpole (*Rana catesbeiana*) (DeNoyelles et al. 1982, 2020). While in the nutrient-dense ecosystems, the indirect effect pronounced on the herbivores as a result of reduction in the food sources is often less common. Effects may be propagated to the subsequent higher tropic levels, when there is a reduction in the herbivore source, for instance, the predators preying on the herbivores. The reduction in the populations of the macroinvertebrates (e.g., *Chironomidae* spp.) and zooplanktons ensued owing to herbicide-induced reductions and resulted in the loss of food sources like the primary producers and habitat depletion, ultimately leading to the total biomass decrease of bluegill sunfish (*Lepomis macrochirus*) (DeNoyelles et al. 2020). The ecological chain effect indicates the bottom-up indirect effects due to the primary producers (lowest trophic level) bearing alterations on the higher trophic levels. These indirect effects being observed are a consequence of predatory ecological relationships, that is, elimination of sensitive primary competitors by the pesticides, thereby promoting competitive relationship among the primary consumers making them tolerant. For instance, the herbicide linuron increased the algal bloom *Chlamydomonas* sp., due to the decreased population of the macrophyte *Elodea nuttallii* (Van den Brink et al. 1997).

Photosynthesis is considered a significant part of the ecosystem, influencing the water quality. The outcome of the herbicide usage is lowered pH value of water during day time and lowered concentration of dissolved oxygen (DO). As recorded by a study, usage of linuron (50–150 μg/L) resulted in lowered pH of about 40% and 25% and reduced the DO (Cuppen et al. 1997). Acute mortality of the macrophages as a result of herbicide usage further has been said to reduce the concentration of DO and pH (Brock et al. 2000). This decrease in the water quality can have detrimental effects on the sensitive invertebrate species and results in indirect effects as a result of species-habitat relationships. Decreased populations of *Copepoda* and Cladocera were characterized to lowered DO levels (20%) in comparison to the controls (Thompson et al. 1993).

The usage of insecticides adversely affects the zooplanktons and invertebrate species. The impact of pesticides on the freshwater ecosystems is the reduction of the invertebrate prey. The significantly decreased populations of macroinvertebrates such as dipterans and ephemeropterans (mayflies) and also two zooplankton groups *Cyclopidae* and *Daphniidae*, followed by treatment with methyl parathion, resulted in the mean weights of the rainbow trout (*Salmo gairdneri*) (Crossland 1984). Reduction of *Copepoda* (*Isopoda Cladocera*) and few invertebrate communities and the eradication of insecta and amphipoda due to the usage of the insecticide chlorpyrifos resulted in an increase (two to threefold) in the periphyton chlorophyll-a, along with blooms of *Oscillatoria* sp. (Brock et al. 1992). In a study conducted on chlorpyrifos, herbivorous rotifers, and freshwater bivalve mollusks (Sphaeriidae), an increase was found in their population, as a result of reduced competition for food sources among the sensitive invertebrates (Brock et al. 1992). Identical study was performed to check the exposure effects of pesticide on the increased tolerant species and decreased sensitive species; in this scenario the effects on the ecosystem are ambiguous. As there is observed increased pH and DO concentration, due to the increase in the population of the primary producers, the decomposition of the zooplankton and the dead invertebrates by the bacteria or the fungi decreases the water quality (Brock et al. 2000). A study conducted on the French streams illustrated that a three to fivefold reduction in the invertebrate fauna inhibited the leaf-litter decomposition due to the application of the insecticides. The decomposition of the leaf-litter decomposition describes a significant energy source in the stream ecosystems, and the reduction in the process could negatively affect the river sections up to several kilometers downstream. These alterations put together indirectly affect the flora and fauna of the aquatic ecosystem in the long term (Schäfer et al. 2007).

Lethal concentration 50 (LC50), a common toxicity dosage measurement, is employed since not all animals of a species perish at the same dose (some are more tolerant than others). This is the demonstration of a pesticide that, in a certain time frame, often between 24 and 96 h kills 50% of the test population of animals.

LC50s and hazard ratings ranging from minor to extremely hazardous for frequently used fungicides, insecticides, and herbicides are mentioned in the Table 1. For instance, the pesticide permethrin's 24 h LC50 for rainbow trout is 12.5 ppb. This indicates the pesticide's extreme toxicity to trout, as half of the fish exposed to 12.5 ppb of permethrin died within 24 h (Maurya et al. 2019).

A pesticide's biological availability (bioavailability), bioconcentration, biomagnification, and environmental persistence all affect how much fish and other aquatic creatures are exposed to it. The quantity of pesticide in the environment that is accessible to fish and other wildlife is referred to as bioavailability (Maurya et al. 2019; Adnan et al. 2016). Some insecticides degrade quickly after being used. Some are less available because they cling firmly to stream bottoms or soil particles floating in the water column. Some are less accessible to aquatic life because they swiftly dissolve in water or rapidly volatilize into the air. The accumulation of pesticides at levels higher than those in the water or soil where they were sprayed is known as bioconcentration. Some fish may have concentrations of some pesticides ten million times higher than those found in the water in their bodily tissues and organs (particularly lipids) (Katagi 2010; Adnan and Indulkar 2017).

Table 1 Acute toxicity (LC 50) of pesticides against fish

Name of the chemical	Fish species	Duration exposure/dose	References
DDT	Rainbow trout	96 h–8.7 μg/L	Maurya et al. (2019)
Cypermethrin	Nile tilapia (*Oreochromis niloticus*)	0.082 ppm	Yuniari et al. (2016)
Akton	Channel catfish	400 μg/L	Maurya et al. (2019)
Acephate	Feathed fish	1000 μg/L	
Alachlor	Rainbow trout	2.4 μg/L	
Dieldrin	African catfish (*Clarias gariepinus*)	0.056 (0.006–0.144) mg/L	Islam et al. (2021)
Endosulfan	Channel catfish	1.5 μg/L	Maurya et al. (2019)
Endosulfan	Snakehead (*Channa striatus*)	0.8–4.8 ng/g	Ibrahim (2007)
Endosulfan	Javanese carp (*Puntius gonionotus*)	1.1–4.8 ng/g	
Endosulfan	Gourami (*Trichogaster* sp.)	0.4–3.9 ng/g	
Endosulfan	Climbing perch (*Anabas testudineus*)	0.4–4.2 ng/g	
Malathion	*Labeo rohita*	15 μg/L	Maurya et al. (2019)
Malathion	*Heteropneustes fossilis*	0.98 ppm	
Methyl parathion	*Catla catla*	4.8 ppm	
Roger	*Pontius stigma*	7.1 and 7.8 ppm	
Endosulfan	African catfish (*Clarias gariepinus*)	0.004 (0.001–0.01) mg/L	Islam et al. (2021)
Heptachlor	African catfish (*Clarias gariepinus*)	0.006 mg/L	
Dieldrin	Catfish (*Arius* sp.)	0.02–0.50 ng/g	Ibrahim (2007)
Dieldrin	Blood cockle (*Anadara granosa*)	0.01–0.70 ng/g	
Dieldrin	Mullet (*Valamugil* sp.)	0.02–0.8 ng/g	
Heptachlor	Catfish (*Arius* sp.)	0.3–8.2	
Heptachlor	Blood cockle (*Anadara granosa*)	0.27–3.54	
Heptachlor	Mullet (*Valamugil* sp.)	0.1–5.2 ng/g	
Chlorpyrifos	*C. striata*	4.521 ppm	Suja and Williams (2018)

2.3 *Pesticide Metabolism in Aquatic Animals*

The chemicals released by industrial effluents, agricultural runoff, petroleum refining, home sewage, etc. affect aquatic life by causing water pollution. The adverse effects in aquatic life are primarily reducing the dissolved oxygen, direct toxicity to animals, and reduced taste of the meat. These pollutants seriously damage the physical and biological process of the animals. Sometimes, they also affect the embryonic stages. Knowledge on pesticide toxicity and metabolism in aquatic environment is essential for better understanding and application of such chemicals. The most commonly used fungicide is copper sulfate ($CuSO_4$) as fungicide in agriculture and paraquat (PQ, 1,1′-dimethyl-4,4′-bipyridinium-dichloride), a common herbicide.

2.3.1 Paraquat

A paraquat is a nonselective, broad-spectrum herbicide that affects the photosynthetic system (PQ, 1,1′-dimethyl-4,4′-bipyridinium-dichloride). PQ is frequently used in agriculture to manage weeds and kills practically all plants after spraying within a few days. PQ easily dissolves in water to its fullest extent. Invading the plant, the positively charged PQ ion impacts the harmful effect by obstructing photosynthesis. Paraquat is a potent electron acceptor, which limits the amount of NADPH 84 formed by competitively inhibiting ferredoxin reduction by removing electrons from photosystem I. PQ ion is converted to a water-soluble free radical during this process and is comparatively stable (Nemcsok and Benedeczky 1995).

2.3.2 Copper Sulfate

Initially, copper sulfate was used as a herbicide to control weeds in wheat fields and as a caustic against wheat mildew. Its use spreads after the discovery of the so-called Bordeaux combination, wherein the initial application took place. Given that metals may form chelates, copper sulfate functions as a fungicide. The electron negativity of the cations has a big impact on this process' stability of metal binding, as well as that of metal chelates and sulfide. The following metal cations lose relative effectiveness as fungicides: Ag, Hg, Cu, Cd, Cr, Ni, Pb, Co, Zn, Fe, and Ca. The extent of the fungicidal effects is probably determined by the strength of the covalent or coordinational binding of the metal complexes attached to cell walls. When Cu2+ ions penetrate cells, they form complexes with the thiol and amino groups therein, preferentially inhibiting key enzymes and other proteins. This is the basis for the fungicidal activity of copper-containing medications (Nemcsok and Benedeczky 1995).

2.3.3 Diquat

Diquat is a broad-spectrum herbicide that can be used to reduce algae and weeds that are submerged, but it is not particularly effective against weeds that are emerging. Diquat-treated water for livestock consumption, agricultural irrigation, or drinking must be contained for 14 days according to the legislation before use. There are no limitations on fishing; however, for swimming, a waiting period of 1 day should be observed. After 10 days, diquat is seldom ever found in treated water. Even when a herbicide is applied that is not specifically hazardous to fish, fish deaths might nevertheless happen following application. Because of the massive amounts of decaying water weeds destroyed by the herbicide, which break down and lower oxygen levels, fish indirectly suffocate rather than poisoned by the herbicide. To provide fish the flexibility to relocate to untreated, oxygen-rich sections of the pond or lake, treat no more than half (or less) of the lake at a time when applying herbicides. Utilize herbicides in the spring when dissolved oxygen levels are greater and water temperatures are lower than in the summer. At lower temperatures, some herbicides are less harmful. Apply in the early spring when there are fewer weeds to break down and they are tiny and less well-established (Helfrich et al. 2009).

2.4 *Bioconcentration of Pesticides*

A specific instance of bioaccumulation is bioconcentration. The definition of bioconcentration is the uptake and retention of a chemical from just water. The use of other sources is not taken into account. The majority of laboratory investigations on the absorption by aquatic creatures expose the organisms to water containing the target chemical(s) in solution.

The bioconcentration factor (BcF) quantifies the magnitude of bioconcentration. The BcF is the ratio of a chemical's concentration in an organism's tissues to its concentration in solution in the water to which it was exposed at equilibrium. The uptake rate constant or uptake clearance to the release rate constant ratio can also be used to calculate the BcF:

$$\mathrm{BcF} = C_{\mathrm{t}}/C_{\mathrm{W}} = k_1/k_2 \tag{1}$$

where C_t and C_w represent the chemical's equilibrium concentrations in tissues and water, respectively; k_1 denotes uptake clearance; and k_2 denotes release rate constant. Units for uptake clearance are mass of chemical/mass of tissue/mass of time, which translates to time^{-1} (Spacie and Hamelink 1982). BcF is unitless since the release rate constant likewise has units of time^{-1}.

Pesticides are primarily bioconcentrated toward aquatic species through passive diffusion through the gastrointestinal system, epithelial tissues, and gills. According to Miyamoto et al. (1990), Connell (1988), Barron (1990), and Landrum and Fisher (1999), the physicochemical properties of the chemicals, the surrounding environmental conditions, and the physiological disposition of each organism involved are what essentially influence bioconcentration. Since compounds must first pass a diffusion barrier, like mucus and biological membranes, to reach circulation fluids, the relative solubilities of such molecules in water and n-octanol may act as a stand-in for lipids. Another important factor that could repeat the processes of diffusion and partitioning in tissues that contain lipids is the size of the molecules. The propensity to bioconcentrate is influenced by both lipid solubility and molecular size (Fig. 3).

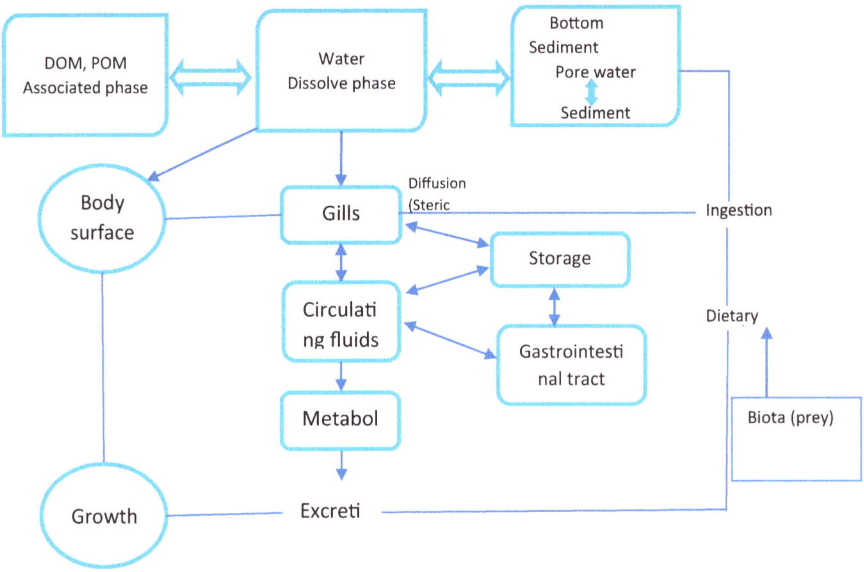

Fig. 3 Schematic representation of bioconcentration and bioaccumulation of pesticide in aquatic organisms

For a number of chemical class combinations with fish, the classic hydrophobicity model has demonstrated a strong correlation between the n-octanol/water partition coefficient (K_{ow}) or water solubility (WS) and BcF. Equation (2) describes this relationship, where a and b are constants (Mackay 1982; Ellgehausen et al. 1980; Veith et al. 1979; Neely et al. 1974):

$$\log \mathrm{BCF} = a \log K_{ow}(\mathrm{or\,WS}) + b. \tag{2}$$

The association between physicochemical traits and BcF has been studied for a range of aquatic creatures, as indicated in Table 1. When many aquatic species or chemical classes are taken into consideration, lower values are seen in the coefficient of correlation (r^2) between the two approaches (Zaroogian et al. 1985; Axelman et al. 1995; Mailhot 1987; Hawker and Connell 1986). However, the correlation coefficient (r^2) is typically greater than 0.7–0.8. A range of urea herbicides were tested on *Chlorella fusca*, and Manthey et al. (1993) found a moderate association; metabolism and hydrophobicity both played a part in this investigation.

The link between fish BcF and K_{ow} was initially discussed by Mackay (1982). For a chemical involved in passive uptake from water, the biological membrane, which is primarily composed of lipid bilayers, acts as the principal barrier. An example of this is the gill epithelium. Steric characteristics like molecule size and shape can occasionally play a significant role in passive absorption (Barron 1990; Landrum and Fisher 1999). Sometimes a chemical's isomorphism affects bioconcentration. There are four isomers (α–δ) of hexachlorocyclohexane (HCH), and the lindane-30

Table 2 Mode of action of some herbicides

Sl. no	Active ingredient	Class (mode of action)	Source
1	Glyphosate	Amino acid synthesis Inhibitors	Vonk and Kraak (2020)
2	EPTC	Seedling growth inhibitors	
3	2,4-D	Growth regulators (interfere with plant hormones)	
4	Atrazine	Inhibitors of photosynthesis	
5	Sethoxydim	Lipid synthesis inhibitors	
6	Paraquat	Cell membrane disrupters	
7	Clomazone	Inhibitors of protective pigments	

isomer is the most common. BCFs in clams vary according to isomer as follows: $\delta > \alpha > \beta \approx \gamma$. However, the order of the elimination procedure was different: $\gamma > \alpha \gg \delta \approx \beta$ (Yamato et al. 1983). Similar results in the mussel Mytilus edulis (Ernst 1979) and blue-green alga *Anabaena sp.* (Mathur and Saxena 1986) suggested the presence of yet another separate order; these changes imply that the bioconcentration of these isomers varies depending on the species. According to Moore et al. (1977), there were species-specific changes in the absorption of chlordane-24 isomers, and the predominant difference between the isomers in *Mysis relicta* was due to the enantioselective metabolism of the (or trans)-isomer (Warner and Wong 2006). Additionally, flucythrinate's enantioselective bioconcentration in oysters has been studied (Schimmel et al. 1983). Endosulfan-27 has two isomers, α (or I) and β (or II), that differ in the structure of the cycloheptyl moiety. According to studies on crayfish *Procambarus clarkii* (Naqvi and Newton 1990) and the *Daphnia magna* (DeLorenzo et al. 2002), the β-isomer bioconcentrates more than the α-isomer, most likely because it undergoes less metabolic conversion to the equivalent sulphate. However, it was noted that algae had a higher bioconcentration of the α-isomer (Narayana Rao and Lal 1987; DeLorenzo et al. 2002).

The main factors influencing how quickly a chemical gets bioconcentrated into an organism from water are absorption and removal mechanisms. In aquatic creatures such as mollusks, algae, crustaceans, and insects (other than fish), the pesticide bioconcentration factor and elimination clearance periods (CL_{50}) are provided in Tables 2, 3, 4, and 5. Data on the larvae, nymphs, and inhabitants of the aquatic environment are provided for *Insecta* members. Exposures to pesticides are carried out utilizing static or flow-through systems. The hydrolytic stability and water solubility of the chemical determine which system is employed. The 287 compounds that the EPI Suite program determined to have log K_{ow} values between 3 and 7 have log BcF values that are usually between 0 and 6 for such xenobiotics (USEPA 2008). BcF levels among studied species greatly vary, even for a single chemical. Log BcF values for the organochlorine pesticides mentioned in Table 2 typically range from 3 to 5, with some species showing variation within a factor of two. Endosulfan, hexachlorobenzene, and DDT showed a higher variance. High log K_{ow} values (5.73 and 6.91) for the latter two pesticides are likely caused by the individual species'

Table 3 Correlation of BcF with physicochemical properties for pesticides

Compound	Species	n[a]	Range[b]	Equation (log BcF=)[c]	r^{2} [d]	References
Organochlorines and pesticides	Mollusks (four species)	34	4.0–7.8	0.844 log K_{ow} – 1.235k	0.69	Hawker and Connell (1986)
Herbicides	*Mytilus edulis*	4	2–6	−0.58 log $WS_{(ppb)}$ + 4.5	0.92	Watanabe et al. (1985)
Insecticides		6	3.4–6.0	0.66 log K_{ow} – 0.05	0.96	Zaroogian et al. (1985)
Pesticides		16	1.7–6.2	0.843 log K_{ow} – 0.808	0.91	Geyer et al. (1982)
Organochlorines and pesticides	*Crassostrea virginica*	17	3.9–6.5	0.72 log K_{ow} + 0.41	0.29	Zaroogian et al. (1985)
Organochlorines	*Daphnia pulex*	22	1.8–6.2	0.898 log K_{ow} – 1.315, k	0.93	Hawker and Connell (1986)
Pesticides	*Daphnia magna*	52	0.9–6.7	0.850 log K_{ow} – 1.10	0.91	Geyer et al. (1982)
Pesticides	*Lemna minor*	10	0.3–6.6	0.49 log K_{ow} + 0.0562	0.91	Lockhart et al. (1983)
Pesticides and organics urea herbicides	*Chlorella fusca*	41	0.6–6.4	0.681 log K_{ow} + 0.164	0.81	Govers et al. (1984)
Urea herbicides		15	1.5–4.3	0.53 log k_{w} + 0.99	0.56	Manthey et al. (1993)
Pesticides	*Scenedesmus acutus*	8	1.7–6.4	0.70 log K_{ow} – 0.26	0.93	Ellgehausen et al. (1980)
Organochlorines and pesticides	*Selenastrum capricornutum*	5	4.1–7.1	0.28 log K_{ow} + 2.6	0.64	Mailhot (1987)

na not available

[a]Total number of chemicals used to derive the equation

[b]Range of physicochemical properties (WS, K_{ow}, k_{w}) in a logarithm unit

[c]*BcF* bioconcentration factor on a wet weight basis, *k* kinetic value, *WS* water solubility with a unit in the parentheses, K_{ow} 1-octanol-water partition coefficient, k_{w} HPLC capacity factor, $3\chi cv$ third-order molecular connectivity index, Σ *Fi* empirical correction factor

[d]Coefficient of correlation

varied lipid levels. The crayfish *Procambarus clarkii* may more significantly metabolize xenobiotics to their sulfate forms, which results in significantly lower BcF values, by resemblance to the metabolism of *D. magna* (DeLorenzo et al. 2002). Aldrin has an extremely high hydrophobicity (log $K_{ow} = 6.5$); however, due to its oxidation to dieldrin, it is rapidly eliminated in ostracods (Kawatski and Schmulbach 1972).

Some pesticides seem to be quite species-specific in their ability to be eliminated. In polychaetes as opposed to bivalves and isopods, the CL50 values for lindane-30 were longer and had a higher bioconcentration (Thybaud and Caquet 1991; Thybaud and Le Bras 1988; Ernst 1979). Pentachlorophenol, DDT, and its metabolite DDE, which are more hydrophobic insecticides, showed significant differences in

Table 4 A summary of organochlorine pesticides bioconcentration on aquatic organisms

Pesticide (log K_{ow})[a]	Species[b]	Experimental conditions[c] (concentration and duration)	log BcF[d]	CL_{50}[e]	References
Aldrin (6.5)	B: *Anabaena cylindrica*	1 ppb, 23 °C, 7 days/na	3.1∗	na	Schauberger and Wildman (1977)
	Aulosira fertilissima	0.1–1 ppm, 29 °C, 2 days/na	2.3–2.6	na	Dhanaraj et al. (1989)
	Anacystis nidulans	1 ppb, 23 °C, 7 days/na	3.0∗	na	Schauberger and Wildman (1977)
Dieldrin (5.2–5.4)	Os: *Chlamydotheca arcuata*	8.4 ppb, 22 °C, 4 days/2 days	3.9∗	1 day	Kawatski and Schmulbach (1972)
	W: *Daphnia magna*	0.02 ppb, 21 °C, 3 days/na	5.1*	na	Johnson et al. (1971)
	I: *Hexagenia bilineata*	0.02 ppb, 21 °C, 3 days/na	4.5	na	Johnson et al. (1971)
	Chironomus sp.	0.02 ppb, 21 °C, 3 days/na	4.4	na	Johnson et al. (1971)
	C: *Rangia cuneata*	0.55 ppb, 25 °C, 72 h/na	2.9–3.3	na	Petrocelli et al. (1973)
	Corbicula manilensis	0.89 ppb, na, 72 days/na	3.5	na	Hartley and Johnston (1983)
	Mu: Lampsilis siliquoidea	0.57 ppb, 20 °C,3 weeks/3 weeks	3.0–3.1	4.7 days	Bedford and Zabik (1973)
	Sphaerium corneum	2.4–2.6 ppb, 10–19 °C, 1 day/na	2.8	na	Boryslawskyj et al. (1988)
	Oy: Crassostrea virginica	0.5 and 9 ppb, 23 °C, 7 days/na	3.3–3.5	na	Mason and Rowe (1976)
	G: Scenedesmus obliquus	1–20 ppb, 25 °C, 36 h/na	3.1∗	na	Reinert (1972)
	B: Anabaena cylindrica	1 ppb, 23 °C, 7 days/na	2.3∗	na	Schauberger and Wildman (1977)
	Aulosira fertilissima	0.1–1 ppm, 27 °C, 5 days/na	1.4–2.5∗	na	Kumar et al. (1988)
	Anacystis nidulans	1 ppb, 23 °C, 7 days/na	2.7∗	na	Schauberger and Wildman (1977)

(continued)

Table 4 (continued)

Pesticide (log K_{ow})[a]	Species[b]	Experimental conditions[c] (concentration and duration)	log BcF[d]	CL_{50}[e]	References
	Nostoc muscorum	1 ppb, 23 °C, 7 days/na	3.3∗	na	Schauberger and Wildman (1977)
	W: Daphnia magna	2–13 ppb, 21 °C, 6 days/na	4.1∗	na	Reinert (1972)
	Os: Chlamydotheca arcuata	6.6 ppb, 22 °C, 4 days/2 days	3.4∗	1 day	Kawatski and Schmulbach (1972)
Photodieldrin (4.13)	*G: Ankistrodesmus amelloides*	0.72 ppb, na, 2 days/na	1.6–1.7∗	na	Neudorf and Khan (1975)
	W: Daphnia pulex	3.3 ppb, na, 1 day/4 days	1.9∗	4 days	Khan et al. (1975)
Chlordane (6.1–6.22)	*C: Corbicula manilensis*	0.3–0.4 ppb, na, 72 days/na	3.7 (α)	na	Hartley and Johnston (1983)
	Mu: Anodonta piscinalis	1.5 ppb, 10 °C, 20 days/na	3.6 (γ) 3.0	na	Sabaliūnas et al. (1998)
	G: Scenedesmus quadricauda	0.1–100 ppb, na, 1 day/na	3.7–4.1 (α) 3.8–4.2 (γ)	na	Glooschenko et al. (1979)
	Ankistrodesmus amelloides	0.72 ppb, 25 °C, 1 day/na	3.7∗	1 day	Moore et al. (1977)
	Mc: Hydrilla verticillata	5 ppb, 25 °C, 6 days/na	3.0	na	Hinman and Klaine (1992)
	W: Daphnia pulex	0.5 ppb, 25 °C, 1 day/3 days	4.4∗	1 day	Moore et al. (1977)
	I: Chironomus decorus	0.7–1.4 ppt, 20 °C, 50 days/na	2.3–2.5 (α)	na	Harkey and Klaine (1992)
Endrin (5.2–5.4)	*Oy: Crassostrea virginica*	0.1–50 ppb, 23 °C, 7 days/na	3.2–3.4	na	Mason and Rowe (1976)
	Mu: Mytilus edulis	na, 15 °C, 7 days/na	3.4	na	Donkin et al. (1997)
	I: Pteronarcys dorsata	0.03–0.15 ppb, 15 °C, 28 days/na	2.9	na	Anderson and DeFoe (1980)
	G: Scenedesmus quadricauda	1 ppm, 25 °C, 7 days/na	2.2	na	Vance and Drummond (1969)

Endosulfan (3.83)	*Oy: Crassostrea madrasensis*	0.1–1.4 ppb, 28 °C, 10 days/na	1.0–1.9	na	Rajendran and Venugopalan (1991)
	C: Katelysia opima	0.1–1.4 ppb, 28 °C, 10 days/na	1.0–1.8	na	Rajendran and Venugopalan (1991)
	Mu: Anodonta piscinalis	1.5 ppb, 10 °C, 20 days/na	2.8	na	Sabaliūnas et al. (1998)
	B: Anabaena sp. ARM310	0.1–1 ppm, 29 °C, 2 days/na	2.7–3.7	na	Narayana Rao and Lal (1987)
	G: Selenastrum capricornutum	0.1 ppm, 25 °C, 16 h/na	3.4	na	DeLorenzo et al. (2002)
	Cr: Procambarus clarkii	0.1 ppm, 8 weeks/8 weeks	– 0.7 (α) 0.3 (β)	na	Naqvi and Newton (1990)
	W: Daphnia magna	0.1 ppm, 25 °C, 24 h/na	3.5	na	DeLorenzo et al. (2002)
Lindane (3.72–4.14)	*C: Corbicula manilensis*	0.3 ppb, na, 72 days/na	3.4	na	Hartley and Johnston (1983)
	Mu: Mytilus edulis	0.9–1.8 ppb, 22 °C, 8 days/na 2–5 ppb, 10 °C, 8 days/18 days na, 15 °C, 7 days/na	2.4–2.6 2.1 2.5	na 22 h na	Renberg et al. (1985), Ernst (1979) and Donkin et al. (1997)
	Sn: Lymnaea palustris	6–600 ppb,20 °C, 10 days/7 days	1.6–1.7	0.7 h	Thybaud and Caquet (1991)
	P: Lanice conchilega	2–5 ppb, 10 °C, 8 days/18 days	3.1	4.7 days	Ernst (1979)
	Is: Asellus aquaticus	2 ppb, 18 °C, 5 days/3 days	1.7	1–2 days	Thybaud and Le Bras (1988)
	G: Chlorella pyrenoidosa	0.01–1 ppm, 20 °C, 6 days/na	2.6–3.1	na	Hansen (1979)
Hexachlorohexane (4.52)	*C: Venerupis japonica*	1–2 ppb, na, 10 days/4 days	2.2 (α), 2.1 (β) 2.1 (γ) 2.4 (δ)	na	Yamato et al. (1983)
	Mc: Hydrilla verticillata	0.12 ppm, 25 °C, 6 days/na	1.6	na	Hinman and Klaine (1992)

(continued)

Table 4 (continued)

Pesticide (log K_{ow})[a]	Species[b]	Experimental conditions[c] (concentration and duration)	log BcF[d]	CL_{50}[e]	References
α-isomer	*G: Chlorella pyrenoidosa*	0.01–0.8 ppm, 28 °C, 3 h/na	2.2–2.4	na	Canton et al. (1975)
	Chlamydomonas sp.	0.1–0.8 ppm, na, 2–3 h/na	3.4*	na	Canton et al. (1977)
	Dunaliella sp.	0.1–0.8 ppm, na, 2–3 h/na	3.2*	na	Canton et al. (1977)
	W: Daphnia magna	0.01–0.8 ppm, 20 °C, 2 days/na	1.8–2.5	na	Canton et al. (1975)
Hexachlorobenzene (5.73)	*C: Corbicula manilensis*	0.59 ppb, na, 72 days/na	3.4	na	Hartley and Johnston (1983)
	Mu: Mytilus edulis	0.5 ppb, 10 °C, 103 h/na	>3.0	na	Bauer et al. (1989)
	Sn: Lymnaea palustris	0.5–5 ppb, na, 21 days/na	5.0–5.4	na	Baturo and Lagadic (1996)
	O: Lumbriculus variegates	0.7–4 ppb, 20 °C, 28–44 days/na	3.7–4.3	na	Schuytema et al. (1988)
	A: Hyalella azteca	1.2–7.7 ppb, 20 °C, 28 days/na	4.1–4.4	na	Schuytema et al. (1988)
	Gammarus lacustris	0.3–5 ppb, 20 °C, 28 days/na	4.3–4.5	na	Nebeker et al. (1989)
Hexachlorobenzene (5.73)	*G: Scenedesmus* sp.	0.08 ppm, 5–39 °C, 2 days/na	1.6–2.2*	na	Koelmans and Jiménez (1994)
	Mc: Myriophyllum spicatum	na, 21 °C, 21 days/133	3.0	8.5 days	Gobas et al. (1991)
	I: Chironomus decorus	0.1 ppb, 20 °C, 2 /na	2.9	na	Knezovich and Harrison (1988)
Toxaphene (5.78)	*Oy: Crassostrea virginica*	5–46 ppb, 28 °C, 4 days/na	3.9–4.2	na	Schimmel et al. (1977)
	S: Penaeus duorarum	0.8–4 ppb, 26 °C, 4 days/na	2.6–2.9	na	Schimmel et al. (1977)
	Palamonetes pugio	3–10 ppb, 21 °C, 4 days/na	2.9–3.1	na	Schimmel et al. (1977)

Pentachlorophenol (5.12)	*C: Corbicula fluminea*	30–100 ppb, 20 °C, 72 h/72 h	1.9–2.0	1 day	Basack et al. (1997)
	Mu: Mytilus edulis	2–5 ppb, 10 °C, 8 days/18 days	2.6	3.2 days	Ernst (1979)
	Dreissena polymorpha	3 ppb–11 ppm, 10–25 °C, 6 h/7 days	2.6–3.2	0.6 ° Cday	Fisher et al. (1999)
	Anodonta anatina	7–14 ppb, 13 °C, 1 day/na	2.2–2.5	na	Mäkelä and Oikari (1990)
	Oy: Crassostrea gigas	0.8 ppm, 15 °C, 5 h/13 h	1.4–1.7	<0.5 day	Shofer and Tjeerdema (1993)
	Ab: Haliotis rufescens	1.2 ppm, 14 °C, 5 h/13 h	1.2–1.3	<0.5 day	Tjeerdema and Crosby (1992)
	Haliotis fulgens	0.5 ppm, 15 °C, 5 h/13 h	1.6–1.8	<0.5 day	Shofer and Tjeerdema (1993)
	P: Lanice conchilega	2–5 ppb, 10 °C, 8 days/18 days	3.6	21 days	Ernst (1979)
	Mc: Eichhornia crassipes	0.5 ppm, 26 °C, 2 days/na	2.1 (leaf) 2.2 (root)	na	Roy and Hänninen (1994)
	A: Gammarus pulex	70 ppb, 16 °C, 3 days/3 days	1.7	0.4 day	Ashauer et al. (2006)
	Hyalella azteca	132 ppb, 21 °C, 1 day/3 days	2.1	3.6 h	Nuutinen et al. (2003)
	Pontoporeia hoyi	na, 4 °C, 6 h/7 days	3.0	8.8 days	Landrum and Dupuis (1990)
	S: Mysis relicta	na, 4 °C, 6 h/7 days	2.1	15 days	Landrum and Dupuis (1990)
	I: Chironomus riparius	0.9 ppb, 20 °C, 16 h/36 h	2.7	15 h	Lydy et al. (1994)
	W: Daphnia magna	20 ppb, 20 °C, 1 day/na	2.2–2.8	na	Kukkonen and Oikari (1988)
DDT (6.91)	*Mu: Anodonta grandis*	0.62 ppb, 20 °C,3 weeks/4 weeks	3.4	13.6 days	Bedford and Zabik (1973)
	Sn: Vivipara heliciformis	5–50 ppb, na, 22 days/19 days	1.9–2.5	<1 day	Yadav et al. (1978)
	C: Indonaia caerulea	5–50 ppb, na, 19 days/11 days	2.5–2.8	4–5 days	Pillai et al. (1980)

(continued)

Table 4 (continued)

Pesticide (log K_{ow})[a]	Species[b]	Experimental conditions[c] (concentration and duration)	log BcF[d]	CL_{50}[e]	References
	G: Selenastrum capricornutum	16 ppb, 22 °C, 2 h/na	4.6–5.5	na	Halling-Sørensen et al. (2000)
	Chlorella vulgaris	10 ppb, 23 °C, 2 days/2 days	2.6	17 h	Kikuchi et al. (1984)
	Scenedesmus obliquus	1 ppm, 26 °C, 7 days/na	2.8	na	Gregory et al. (1969)
	B: Anacystis nidulans	1 ppm, 26 °C, 7 days/na	2.9	na	Gregory et al. (1969)
	D: Nitzschia closterium	10 ppb, 23 °C, 2 days/2 days	4.9	8.4 days	Kikuchi et al. (1984)
	Cylindrotheca closterium	0.1 ppm, na, 21 days/na	2.3	na	Keil and Priester (1969)
	E: Euglena gracilis	1 ppm, 26 °C, 7 days/na	2.0	na	Gregory et al. (1969)
	W: Daphnia magna	8–16 ppb, 21 °C, 26 h/na	4.2–4.4	na	Crosby and Tucker (1971)
	A: Gammarus fasciatus	0.08 ppb, 21 °C, 3 days/na	4.3	na	Johnson et al. (1971)
	Hyalella azteca	0.02–0.12 ppb, 20 °C, 2 days/8 days	4.4–4.6	3–4 days	Lotufo et al. (2000)
	Diporeia sp.	0.2–3.3 ppb, 4 °C, 2 d/90 d	5.3–5.5	3–5 month	Lotufo et al. (2000)
	Cr: Orconectes nais	0.1 ppb, 21 °C, 3 days/na	3.4	na	Johnson et al. (1971)
	S: Palaemonetes kadiakensis	0.08 ppb, 21 °C, 3 days/na	3.7	na	Johnson et al. (1971)
	Artemia nauplii	0.5–1 ppb, 20 °C, 1 day/na	2.4	na	Wang and Simpson (1996)
	I: Chironomus sp.	0.05 ppb, 21 °C, 3 days/na	4.7	na	Johnson et al. (1971)
	Ephemera danica	0.76 ppb, 14 °C, 9 days/na	2.6–3.9	na	Södergren and Svensson (1973)
	Ci: Blepharisma intermedium	1 ppm, 24 °C, 10 days/na	4.8	na	Saxena et al. (1982)

Dicofol (5.02)	*A: Hyalella azteca*	1.6–1.8 ppm, 15 °C, 28 days/na	3.8–4.1	na	Saxena et al. (1982)
DDE (6.51)	*W: Daphnia pulex*	0.11 ppt, 5–25 °C, 1 day/na	3.7–4.3	na	Nawaz and Kirk (1996)
	S: Artemia nauplii	0.5–1 ppb, 20 °C, 1 day/na	1.6–1.7	na	Wang and Simpson (1996)
	I: Chironomus tentans	0.1–1 ppb, 21 °C, 30 days/na	4.1–4.3	na	Derr and Zabik (1972)
	A: Hyalella azteca	1.12 ppb, 20 °C, 2 days/8 days	4.6	2.3 days	Lotufo et al. (2000)
	Diporeia sp.	2.3–20 ppb, 4 °C, 2 days/90 days	5.5–5.7	144 days	Lotufo et al. (2000)
Methoxychlor (5.08)	*Mu: Mytilus edulis*	9–12 ppb, 22 °C, 21 days/na	4.1	na	Renberg et al. (1985)
	Sn: Physa integra	0.42–4.2 ppb, 15 °C, 28 days/na	3.8	na	Anderson and DeFoe (1980)
	G: Chlorella pyrenoidosa	8–50 ppb, 15 °C, 200 h/na	3.9*	na	Paris and Lewis (1976)
	I: Pteronarcys dorsata	0.15–4.2 ppb, 15 °C, 28 days/na	2.8	na	Anderson and DeFoe (1980)

[a]Evaluated by EPI Suite (USEPA 2008) or experimental data therein

[b]Designation of species: *A* amphipod, *Ab* abalone, *B* blue-green algae, *C* clam. *Ci* ciliate, *Cr* crayfish, *D* diatom, *E* Euglenophyta, *G* green algae, *I* aquatic insect, *Is* isopod, *Mc* macrophyte, *Mu* mussel, *O* oligochaete, *Os* ostracod, *Oy* oyster, *P* polychaete, *S* shrimp, *Sn* snail, *W* water flea

[c]Application of pesticide to water, presence (y) or absence (n) of sediment in the system, exposure condition [static(s)/flow through (fl)], concentration, temperature (°C), periods of exposure/elimination

[d]Experimentally obtained bioconcentration factor based on the overlying water concentration in the whole body (or each tissue). The asterisk means dry-weight basis

[e]50% clearance time

Table 5 A summary of organophosphorus pesticides bioconcentration in aquatic organisms

Pesticide (log K_{ow})[a]	Species[b]	Experimental conditions[c] (concentration and duration)	log BCF[d]	CL_{50}[e]	References
Parathion (3.83)	G: *Scenedesmus obliquus*	1 ppm, 26 °C, 7 days/na	1.9∗	na	Gregory et al. (1969)
	B: *Anacystis nidulans*	1 ppm, 26 °C, 7 days/na	1.7∗	na	Gregory et al. (1969)
	E: *Euglena gracilis*	1 ppm, 26 °C, 7 days/na	1.8∗	na	Gregory et al. (1969)
	A: *Hyalella azteca*	2.3 ppb, 22 °C, 1 day/3 days	1.1	0.7hr	Nuutinen et al. (2003)
Fenitrothion (3.30)	Mu: *Mytilus edulis*	0.2–13 ppb, 15 °C, 14 days/28 days	1.9–2.1	1–2 days	McLeese et al. (1979)
	C: *Mya arenaria*	0.2–13 ppb, 15°C, 14 days/28 days	1.3–2.5	1–2 days	McLeese et al. (1979)
	Anodonta cataracta	0.8 ppb, 12 °C, 14 days/28 days	1.0	2 days	McLeese et al. (1979)
	Sn: *Cipangopaludina japonica*	0.1 ppm, 25 °C, 3 days/1–3 days	1.3	0.4 days	Takimoto et al. (1987a)
	Physa acuta	0.1 ppm, 25 °C, 3 days/1–3 days	1.7	0.4 days	Takimoto et al. (1987a)
	G: *Chlamydomonas reinhardtii*	1 ppm, 20 °C, 1 day/na	2.5∗	na	Kent and Currie (1995)
	Chlorella vulgaris	10 ppb, 23 °C, 2 days/2 days	1.6∗	0.9 h	Kikuchi et al. (1984)
	Scenedesmus quadricauda	1–50 ppb, 25 °C, 1 day/na	1.0–1.5∗	na	Guanzon Jr et al. (1996)
	B: *Anabaena flos-aquae*	10 ppb, 23 °C, 2 days/2 days	1.7∗	2.6 h	Kikuchi et al. (1984)
	Aulosira fertilissima	1–10 ppm, 29 °C, 5 days/na	2.0–2.9∗	na	Lal et al. (1987)
	Microcystis aeruginosa	1–50 ppb, 25 °C, 1 day/na	1.5–2.4∗	na	Guanzon Jr et al. (1996)
	D: *Aulacoseira granulata*	1–50 ppb, 20 °C, 1 day/na	1.1–1.7∗	na	Guanzon Jr et al. (1996)
	Nitzschia closterium	10 ppb, 23 °C, 2 days/2 days	2.0∗	0.9 h	Kikuchi et al. (1984)
	Mc: *Lemna minor*	3.9 ppb, 25 °C, 5 days/25 days	1.3	1.3 days	Lockhart et al. (1984)
	W: *Daphnia pulex*	1 ppb, 18 °C, 1 day/1 days	1.9	5 h	Takimoto et al. (1987b)
	S: *Palaemon paucidens*	1 ppb, 25 °C, 3 days/1 days	0.8	1.5 h	Takimoto et al. (1987b)

(continued)

Table 5 (continued)

Pesticide (log K_{ow})[a]	Species[b]	Experimental conditions[c] (concentration and duration)	log BCF[d]	CL_{50}[e]	References
Fenitrothion (3.30)	S: *Penaeus japonicus*	0.5 ppb, 25 °C, 1 day/na	2.1	na	Kobayashi et al. (1985)
		0.3 ppb, 25 °C, 12 h/72 h	2.1	6 h	Kobayashi et al. (1990)
	Artemia salina	0.1 ppm, 25 °C, 3 days/na	3.7	na	Kashiwada et al. (1995)
	Cr: *Procambarus clarkii*	20 ppb, 22 °C, 2 days/21 days	2.1	na	Escartín and Porte (1996)
	Co: *Sinocalanus tenellus*	0.1 ppm, 25 °C, 3 days/na	3.6	na	Kashiwada et al. (1995)
	R: *Brachionus plicatilis*	0.1 ppm, 25 °C, 3 days/na	3.3	na	Kashiwada et al. (1995)
Fenitrooxon (1.69)	S: *Penaeus japonicus*	3 ppb, 25 °C, 12 h/72 h	0.5	12 h	Kobayashi et al. (1990)
Chlorothion (3.45)	Sn: *Lymnaea stagnalis*	27 ppb, 20 °C, 10 days/na	1.5	2 h	Legierse et al. (1998)
Fenthion (4.84)	C: *Marcia hiantina*	0.02–0.2 ppm, 28 °C,15 days/15 days	1.0–2.3	na	Sathe et al. (2005)
Chlorpyrifos (4.7) C	Mu: *Mytilus edulis*	1–3.2 ppm, 18 °C, 24–38 days/38 days	2.4–2.7	5 days	Serrano et al. (1997a)
	Mytilus galloprovincialis	1 ppm, 18 °C, 35 days/na	2.6	na	Serrano et al. (1997b)
	Oy: *Crassostrea virginica*	0.7 ppb, 22 °C, 28 days/ 14 days	2.8	2.5 days	Woodburn et al. (2003)
		1–56 ppm, 18 °C, 4 days/na	0.5	na	Serrano et al. (1995)
	B: *Anabaena* sp. ARM310 *Aulosira fertilissima*	1–10 ppm, 29 °C, 5 days/na 1–10 ppm, 29 °C, 5 days/na	0.8–2.8∗ 1.7-2.6∗	na	Lal et al. (1987)
	S: *Artemia parthenogenetica*	0.5–100 ppb, 20 °C, 2 days/na	3.0–3.8	na	Varo et al. (2000)
	Is: *Asellus aquaticus*	0.1 ppb, 12 °C, 3 days/3 days	3.2	1.5 days	Ashauer et al. (2006)
	A: *Gammarus pulex*	0.7–5 ppb, 12–25 °C, 23 days/na	5.4	1–4 days	Montañés et al. (1995)
	I: *Hydropsyche* sp.	3 ppb, 20 °C, 6 h/na	1.6	na	Tang and Siegfried (1996)
	Stenacron spp.	3 ppb, 20 °C, 6 h/na	1.1	na	Tang and Siegfried (1996)

(continued)

Table 5 (continued)

Pesticide (log K_{ow})[a]	Species[b]	Experimental conditions[c] (concentration and duration)	log BCF[d]	CL_{50}[e]	References
Diazinon (3.81)	Sn: *Cipangopaludina malleata*	10 ppb, 20 °C, 7 days/na	0.8	na	Kanazawa (1978)
	Cr: *Procambarus clarkii*	10 ppb, 20 °C, 7 days/na	0.7	na	Kanazawa (1978)
	S: *Artemia salina*	1 ppm, 25 °C, 3 days/na	2.6	na	Kashiwada et al. (1995)
	Co: *Sinocalanus tenellus*	1 ppm, 25 °C, 3 days/na	3.5	na	Kashiwada et al. (1995)
	R: *Brachionus plicatilis*	1 ppm, 25 °C, 3 days/na	2.7	na	Kashiwada et al. (1995)
Pyridaphenthion (3.2)	G: *Chlorella saccharophila*	10 ppm, 22 °C, 7 days/na	1.4	na	Jonsson et al. (2001)
Demeton-*S*-methyl (1.02)	Mc: *Spirodela oligorrhiza*	1 ppm, 22 °C, 8 days/na	1.1	na	Gao et al. (2000)
	Myriophyllum aquaticum	1 ppm, 22 °C, 8 days/na	–1.6	na	Gao et al. (2000)
	Elodea canadensis	1 ppm, 22 °C, 8 days/na	–0.4	na	Gao et al. (2000)
Phorate (3.56)	B: *Anabaena* sp. ARM310	0.1–1 ppm, 29 °C, 2 days/na	0.5–1.1*	na	Dhanaraj et al. (1989)
	Aulosira fertilissima	0.1–1 ppm, 29 °C, 2 days/na	0.9–1.1*	na	Dhanaraj et al. (1989)
Dimethoate (0.78)	Mu: *Mytilus galloprovincialis*	3–56 ppm, 18 °C, 4 days/na	–0.5	na	Serrano et al. (1995)
	B: *Anabaena* sp. ARM310	0.1–1 ppm, 27 °C, 5 days/na	0.3–1.9*	na	Kumar et al. (1988)
	Aulosira fertilissima	0.1–1 ppm, 27 °C, 5 days/na	0.0–2.1*	na	Kumar et al. (1988)
Malathion (2.36)	B: *Anabaena* sp. ARM310	0.1–1 ppm, 29 °C, 2 days/na	1.9–2.5*	na	Narayana Rao and Lal (1987)
	Aulosira fertilissima	0.1–1 ppm, 29 °C, 2 days/na	2.0–2.3*	na	Narayana Rao and Lal (1987)
	Mc: *Spirodela oligorrhiza*	1 ppm, 22 °C, 8 days/na	1.4	na	Gao et al. (2000)
	Myriophyllum aquaticum	1 ppm, 22 °C, 8 days/na	0.5	na	Gao et al. (2000)
	Elodea canadensis	1 ppm, 22 °C, 8 days/na	0.08	na	Gao et al. (2000)
	Co: *Sinocalanus tenellus*	15 ppb, 20 °C, 3 days/na	3.7	na	Kashiwada et al. (1995)
	S: *Artemia salina*	0.3 ppm, 25 °C, 3 days/na	2.3	na	Kashiwada et al. (1995)

(continued)

Table 5 (continued)

Pesticide (log K_{ow})[a]	Species[b]	Experimental conditions[c] (concentration and duration)	log BCF[d]	CL_{50}[e]	References
Malathion (2.36)	R: *Brachionus plicatilis*	0.3 ppm, 25 °C, 3 days/na	4.6	na	Kashiwada et al. (1995)
Methidathion (2.20)	Mu: *Mytilus galloprovincialis*	1 ppm, 18 °C, 35 days/na	2.3	na	Serrano et al. (1997b)
	C: *Venus gallina*	3–56 ppm, 18 °C, 4 days/na	–0.03	na	Serrano et al. (1995)
	W: *Daphnia magna*	1.2 ppb, 20 °C, 2 days/na	0.3	na	Julin and Sanders (1977)
	I: *Chironomus plumosus*	1.2 ppb, 20 °C, 2 days/na	0.8	na	Julin and Sanders (1977)
	Cr: *Orconectes nais*	1.2 ppb, 20 °C, 2 days/na	0.8	na	Julin and Sanders (1977)
Dichlorvos (1.47)	Mu: *Mytilus edulis*	15 °C, 3–7 days/na	0.04	na	Donkin et al. (1997)
Chlorfenvinphos (3.8)	Mu: *Mytilus galloprovincialis*	1 ppm, 18 °C, 35 days/na	2.4	na	Serrano et al. (1997b)
Crufomate (3.42)	Mc: *Spirodela oligorrhiza*	1 ppm, 22 °C, 8 days/na	0.4	na	Gao et al. (2000)
	Myriophyllum aquaticum	1 ppm, 22 °C, 8 days/na	0.4	na	Gao et al. (2000)
	Elodea canadensis	1 ppm, 22 °C, 8 days/na	–1.3	na	Gao et al. (2000)

[a]Evaluated by EPI Suite (USEPA 2008) or experimental data therein

[b]Designation of species: *A* amphipod, *Ab* abalone, *B* blue-green algae, *C* clam. *Ci* ciliate, *Cr* crayfish, *D* diatom, *E* Euglenophyta, *G* green algae, *I* aquatic insect, *Is* isopod, *Mc* macrophyte, *Mu* mussel, *O* oligochaete, *Os* ostracod, *Oy* oyster, *P* polychaete, *S* shrimp, *Sn* snail, *W* water flea

[c]Application of pesticide to water, presence (y) or absence (n) of sediment in the system, exposure condition [static(s)/flow through (fl)], concentration, temperature (°C), periods of exposure/elimination

[d]Experimentally obtained bioconcentration factor based on the overlying water concentration in the whole body (or each tissue). The asterisk means dry-weight basis

[e]50% clearance time

elimination. Compared to polychaetes and shrimp, which require 2–3 weeks for elimination (Landrum and Dupuis 1990: Ernst 1979), mollusks swiftly eliminated, which has a CL_{50} of less than 1 day (Tjeerdema and Crosby 1992; Shofer and Tjeerdema 1993). Significant variations in CL_{50} (by 100 days) were found for the amphipods (Lotufo et al. 2000). Organophosphorus insecticides with intermediate hydrophobicity were reported to have lower BcF values, often between 1 and

2 (Table 3). In comparison to other aquatic animals, the BcF values of chlorpyrifos are 1–2 orders of magnitude higher in the isopods *Artemia* sp. and *Asellus aquaticus* (Varo et al. 2000; Montañés et al. 1995). In shrimp and copepods, fenitrothion, diazinon, and malathion were found to have a comparable effect (Kashiwada et al. 1995). In these species, higher bioconcentration rates probably result from less metabolic activity. Organophosphorus insecticides have a relatively quick clearance rate (CL_{50} values are fewer than 4 days). The log BcF values of the pyrethroid insecticides shown in Table 4 are 2–4, which is nearly an order of magnitude less than what the regression equations in Table 1 would indicate. Toxaphene and hexachlorobenzene are two organochlorine insecticides with log K_{ow} values about 6. Organochlorine insecticides are often believed to be resistant to metabolism, despite the fact that they demonstrate log BcF values of 4–6 in mollusks, which are close to predicted levels. However, these values are higher than those observed for pyrethroid. The terminal residues of the pyrethroids are therefore more hydrophilic than their parent compounds, indicating that they are likely to be metabolized through ester cleavage and hydroxylation.

Other than the pesticides described above, bioconcentration values for pesticides are shown in Table 5. Carbamates, triazines, ionizable acids, ureas, esters, and amides have BcF values less than 1000. In addition to having these characteristics, many pesticides also feature chemical groups that make hydrogen bonding easier and easier to digest. Such groups often produce BcF values below those predicted by the regression equations in Table 1. In comparison, the amphipod *Pontoporeia hoyi*'s log BCF value of carbaryl (log $K_{ow} = 1.85$) is 4.3, which is significantly higher than the similar value (2.2) in *Mysis relicta*. The very sluggish elimination and decreased metabolic activity account for the disparity (Landrum and Dupuis 1990). It may be because of its impact on algal development that the ionizable salicylic acid demonstrated a surprisingly high bioconcentration (log BCF = 3) in green algae (Wang and Lay 1989). These results show that the chemical class being tested and the type of aquatic life have a substantial impact on the BcF values for pesticides.

2.5 Bioaccumulation of Pesticides

Bioaccumulation, or the buildup of a chemical in an organism in relation to its level in the surrounding environment, is a significant environmental hazard. Thus, a common and growing method for determining the chemical state of aquatic ecosystems is to measure chemical concentrations in biota. Aquatic species can take in chemicals by eating contaminated prey or silt, or they can take them directly from the water by using their respiratory organs (such their gills). Therefore, in any kinetic investigation of bioaccumulation, each of these exposure pathways should be considered. A substance that is not yet digested and is dissolved in interstitial water or sediment is the most basic model for bioaccumulation. Bioaccumulation factor values and biosediment accumulation factors of different pesticides were reported in Tables 6 and 7.

While individual pesticides are bad for the ecology, pesticide mixtures increase their toxicity. There is a dearth of information regarding how particular herbicides from these families interact with aquatic life. As a result, it might be challenging to forecast whether or not a given combination of drugs will increase toxicity (Deneer 2000). The degree of bioaccumulation of several pesticides in fish is influenced by water solubility and the polarity of the pesticides. The bioaccumulation of a pesticide chemical in fish is inversely correlated with its water solubility. As the pesticide becomes more soluble in water, the amount of bioaccumulation decreases. As a result, water solubility is a crucial factor in reducing pesticide dynamics in aquatic ecosystems (Haque et al. 1977). Pesticide desorption or elimination rates seem to vary depending on the species. The rate of elimination reactions and rate of absorption determine the quantity of pesticides in a specific species (Matsumura 1977). To identify and map the spread of pesticide residues in the aquatic ecosystem, significant effort has been made. A thorough residue analysis program has revealed numerous harmful effects connected to organochlorine pesticides and metabolites of the organophosphates (Livingston and de La Cruz 1977). Pesticides that enter aquatic environments have the potential to induce undesirable ecological loss in the form of disease and aquatic animal mortality. As a result, aquatic bacteria, vertebrates like fish and water birds, and invertebrates like frogs, muscles, turtles, and prawns all experience deterioration. Because they are a component of natural food chains and other animals rely on them for food, the hazardous compounds present in pesticides have an effect on these aquatic species (Lakhani 2015).

In aquatic environments including estuaries and rivers, pesticides can have an impact on the microorganisms through spills, agricultural runoff, and drift. Both the structure and the function of microbial communities can be harmed by pesticide toxicity. Pesticides can be bioaccumulated in the ecosystem or digested by microorganisms. Different mechanisms of pesticide toxicity in microorganisms exist depending on the chemical and the microbial species exposed to it (DeLorenzo et al. 2001).

D. magna has been proven to be susceptible to pesticide contamination in bioassays in which *Parachromis dovii* and *Daphnia magna* were exposed to contaminated water taken from the field (Diepens et al. 2014). Shrimp farming is one of the significant sources of seafood that could be damaged by environmental or accidental exposure to neonicotinoid insecticides like imidacloprid. Neonicotinoids influence how insects' neural systems function. In a study on adult black tiger shrimp (*Penaeus monodon*), stress enzyme activity was assessed for both acute and chronic imidacloprid effects in the abdomen, head, gills, and hepatopancreas. This demonstrated an increase in these biomarkers' activity, and the enzymatic activity was positively linked with imidacloprid accumulation in the tissue. The effects had dose- and time-dependent variations in how various tissues reacted. Based on an elevated response in each of the aforementioned biomarkers during routine monitoring, imidacloprid seems to serve as an ambient chemical stressor for adult black tiger shrimp (*Penaeus monodon*) (Butcherine et al. 2022). On primary cell cultures of the

Table 6 A summary of bioconcentration studies performed on pyrethroid insecticides in aquatic organisms

Pesticide (log K_{ow})[a]	Species[b]	Experimental conditions[c] (concentration and duration)	log BCF[d]	CL_{50}[e]	References
Allethrin (4.78)	*Mu: Anodonta piscinalis*	1.5 ppb, 10 °C, 20 days/na	2.3	na	Sabaliünas et al. (1998)
Permethrin (6.5)	*Oy: Crassostrea virginica*	28 °C, 31 days/40 days	3.3	<1 week	Schimmel et al. (1983)
	Sn: Helisoma trivolvis	16 ppb, 15 °C, 28 days/na	2.9	na	Spehar et al. (1983)
	B: Anabaena sp. ARM310	0.1 ppm, 27 °C, 5 days/na	1.8–2.9	na	Kumar et al. (1988)
	Aulosira fertilissima	0.1 ppm, 27 °C, 5 days/na	1.6–3.4	na	Kumar et al. (1988)
	Mc: Lemna minor	2.3–3 ppb, 25 °C, 5 days/15 days	2.3 (trans), 2.2 (cis)	18 days	Lockhart et al. (1984)
	I: Hydropsyche sp.	3 ppb, 20 °C, 6 hr/na	1.5	na	Tang and Siegfried (1996)
	Stenacrn sp.	3 ppb, 20 °C, 6 hr/na	1.4	na	Tang and Siegfried (1996)
	Pteronarcys dorsata	0.03–0.4 ppb, 15 °C, 4 days/na	2.3	na	Anderson (1982)
Lambda-cyhalothrin (6.8–7.0)	*I: Chironomus riparius*	0.2 ppb, 23 °C, 4 days/na	3.2–3.3	na	Hamer et al. (1999)
		0.2 ppb, 23 °C, 4 days/na	3.1–3.5	na	Hamer et al. (1999)
Fenvalerate (6.2)	*Mu: Anodonta piscinalis*	1.5 ppb, 10 °C, 20 days/na	2.5	na	Sabaliünas et al. (1998)
	Oy: Crassostrea virginica	1 ppb, 29 °C, 28 days/37 days	3.7	<1 week	Schimmel et al. (1983)
	Sn: Helisoma trivolvis	0.02–0.8 ppb, 15 °C, 28 days/na	2.6–3.1	na	Anderson (1982)
	G: Chlamydomonas reinhardtii	16 ppb, 20 °C, 1 h/na	2.3	na	Day and Kaushik (1987)

(continued)

Table 6 (continued)

Pesticide (log K_{ow})[a]	Species[b]	Experimental conditions[c] (concentration and duration)	log BCF[d]	CL_{50}[e]	References
	W: Daphnia galeata	0.1–0.5 ppb, 20 °C, 2 days/na	4.0–4.6	na	Day and Kaushik (1987)
Flucythrinate (6.2)	*Mu: Mytilus edulis*	0.04–0.4 ppm, 15 °C, 3–7 days/na	2.9–3.6	na	Donkin et al. (1997)
	Oy: Crassostrea virginica	1 ppb, 29 °C, 28 days/52 days	3.4	10 days	Schimmel et al. (1983)

na not available
[a]Evaluated by EPI Suite (USEPA 2008) or experimental data therein
[b]Designation of species: *A* amphipod, *Ab* abalone, *B* blue-green algae, *C* clam, *Ci* ciliate, *Cr* crayfish, *D* diatom, *E* Euglenophyta, *G* green algae, *I* aquatic insect, *Is* isopod, *Mc* macrophyte, *Mu* mussel, *O* oligochaete, *Os* ostracod, *Oy* oyster, *P* polychaete, *S* shrimp, *Sn* snail, *W* water flea
[c]Application of pesticide to water, presence (y) or absence (n) of sediment in the system, exposure condition [static(s)/flow through (fl)], concentration, temperature (°C), periods of exposure/elimination
[d]50% clearance time

gonad, mantle, digestive gland tissues, and gill of *Unio* sp., the cytotoxic effects of a neonicotinoid, a pyrethroid insecticide, acetamiprid, and flumethrin were investigated. This investigation demonstrated that flumethrin was more cytotoxic to all tested cells than acetamiprid (Arslan et al. 2021). Daphnia was shown to have a greater risk quotient for the majority of organophosphate pesticides than fish or algae (Sumon et al. 2018).

The simplest model for bioaccumulation is a nonmetabolized chemical dissolved in soil or interstitial water. When evaluating dissipation from the body of aquatic creatures that have consumed polluted sediments, both gastrointestinal clearance and metabolism of the contaminating substances should be taken into account. In order to evaluate the k_U, k_M, k_{PE}, and k_{ME} values (wherein k is the first-order rate constant; U, uptake; M, metabolism; P, parent molecule; E, elimination), Schuler et al. (2003) employed a three-compartment model to investigate the bioaccumulation of sediment-spiked benzo[α]pyrene in worms, amphipods, and larval midges. With alterations that were equivalent to those in kPE and kME values, only the larval midge demonstrated substantial metabolism (75% after 3 days).

In aquatic species, the elimination of particular compounds can occasionally result in biphasic profiles (Egeler et al. 1997; Muir et al. 1982; Landrum and Scavia 1983; Shaw and Connell 1987; Richter and Nagel 2007). Such profiles are the result of a number of mechanisms, such as decreased metabolite clearance or bound residue formation. Additionally, because stomach contents can leave the body either rapidly or slowly, they might affect how chemical elimination curves seem.

According to research by Bartlett et al. (2004), *Hyalella azteca* held in a cage above the sediment at an early stage of exposure exhibited higher body residues of

Table 7 A summary of bioconcentration studies performed on other pesticides and other chemicals in aquatic organisms

Pesticide (log K_{ow})[a]	Species[b]	Experimental conditions[c] (concentration and duration)	log BCF[d]	CL_{50}[e]	References
Diuron (2.68)	G: *Chlorella fusca*	0.16 ppm, 22 °C, 0.5 h/na	3.4	na	Manthey et al. (1993)
Isoproturon (2.87)	G: *Chlorella fusca*	0.14 ppm, 22 °C, 0.5 h/na	2.6	na	Manthey et al. (1993)
	Mc: *Lemna minor*	53 ppb, 23 °C, 21 days/na	1.2	na	Böttcher and Schroll (2007)
	Elodea densa	2 ppb, 21 °C, 22 days/na	1.8	na	Feurtet-Mazel et al. (1996)
Diflubenzuron (3.88)	G: *Scenedesmus subspicatus*	0.2 ppm, na, 7 days/ 4 days	2.9–3.6	3 days	Yu-yun et al. (1993)
CCU (3.84)	G: *Scenedesmus subspicatus*	0.2 ppm, na, 7 days/ 4 days	2.5–3.8	1 days	Yu-yun et al. (1993)
Chlorsulfuron (2.0)	G: *Chlorella fusca*	7.2 ppm, 22 °C, 2 h/na	0.9(pH6), 1.7(pH 5)	Qna	Fahl et al. (1995)
Metsulfuron-methyl (2.2)	G: *Chlorella fusca*	7.6 ppm, 22 °C, 2 h/ na	0.0 (pH 6) 1.2(pH 5)	na	Fahl et al. (1995)
Atrazine (2.61)	Sn: *Lymnaea palustris*	5–125 ppb, na, 21 days/na	0.6–0.9	na	Baturo and Lagadic (1996)
	G: *Scenedesmus acutus*	11ppb, 30 °C, 8 h/na	1.5	na	Böhm and Müller (1976)
	Chlorella sp.	40 ppb, 20 °C, 1 day/na	2.2*	na	Tang et al. (1998)
	Pediastrum sp.	40 ppb, 20 °C, 1 day/na	2.5*	na	Tang et al. (1998)
	D: *Cyclotella gamma*	40 ppb, 20 °C, 1 day/na	1.6*	na	Tang et al. (1998)
	Synedra acus	40 ppb, 20 °C, 1 day/na	1.6*	na	Tang et al. (1998)
	Mc: *Hydrilla verticillata*	57ppb, 25 °C, 6 days/na	0!.98	na	Hinman and Klaine (1992)
	Periphyton community	0.27 ppm, 10 °C, 1 day/na	2.3–2.5	na	Nikkilä et al. (2001)
	W: *Daphnia magna*	23 ppb, 20 °C, 1 day/na	0.6–0.7	na	Nikkilä et al. (2001)
	Daphnia pulicaria	0.17–0.3 ppm, 8–20 °C, 7–8 days/ na	0.3–1.0	na	Heisig-Gunkel and Gunkel (1982)

(continued)

Table 7 (continued)

Pesticide (log K_{ow})[a]	Species[b]	Experimental conditions[c] (concentration and duration)	log BCF[d]	CL_{50}[e]	References
Cyanazine (2.22)	*G: Scenedesmus quadricauda*	10–40 ppb, 24 °C, 22 days/na	2.3–4.1	na	Aly et al. (1984)
Terbutryn (3.74)	*I: Chironomus tentans*	14 ppb, 25 °C, 4 h/3 days	1.7	<8 h	Muir et al. (1982)
	A: Gammarus fossarum	0.67 ppm, 15 °C, 2 days/2 days	1.1	10 h	Richter and Nagel (2007)
	Is: Asellus aquaticus	0.67 ppm, 15 °C, 2 days/2 days	1.5	7 h	Richter and Nagel (2007)
Fluridone (3.16)	*Mc: Potamogeton pectinatus*	1 ppm, 20 °C, 2 weeks/na	2.0 (shoot),1.6 (root)	na	Marquis et al. (1981)
	I: Chironomus tentans	7.8 ppb, 25 °C, 4 h/3 days	2.1	<8 hr	Muir et al. (1982)
Fipronil (4.0)	*W: Daphnia pulex*	0.4 ppb–1 ppm, na, 2 days/na	1.4	na	Chaton et al. (2002)
	Os: Eucypris virens	0.4 ppb–1 ppm, na, 2 days/na	1.0	na	Chaton et al. (2002)
	I: Chironomus annularius	0.4 ppb–1 ppm, na, 2 days/na	1.8	na	Chaton et al. (2002)
Tetradifon (4.61)	*G: Nannochloropsis oculata*	1.5–4.4 ppm, 22 °C, 4 days/na	2.0	na	Ferrando et al. (1996)
	W: Daphnia magna	1.5–4.4 ppm, 22 °C, 4 days/na	1.7	na	Ferrando et al. (1996)

na not available
[a]Evaluated by EPI Suite (USEPA 2008) or experimental data therein
[b]Designation of species: *A* amphipod, *Ab* abalone, *B* blue-green algae, *C* clam, *Ci* ciliate, *Cr* crayfish, *D* diatom, *E* Euglenophyta, *G* green algae, *I* aquatic insect, *Is* isopod, *Mc* macrophyte, *Mu* mussel, *O* oligochaete, *Os* ostracod, *Oy* oyster, *P* polychaete, *S* shrimp, *Sn* snail, *W* water flea
[c]Application of pesticide to water, presence (y) or absence (n) of sediment in the system, exposure condition [static(s)/flow through (fl)], concentration, temperature (°C), periods of exposure/elimination
[d]50% clearance time

tributyltin than those kept in a water-sediment system with added sediment. If all processes involved in uptake and dissipation are considered, the kinetic and fugacity models may be able to satisfactorily describe the bioaccumulation profiles of a chemical in aquatic species (i.e., uptake via gills, ingestion of food or sediment, metabolism, and elimination). Gobas et al. (1991) developed the fugacity model for fish and recommended feeding rates and transport variables that lead to gastrointestinal absorption in the water and lipid phases. Using monitoring data on sediment, crayfish, mussels, caddisfly larvae, and gammarus, the fugacity of PCB congeners in biota (fB) was theoretically estimated for the detritivorous and filter-feeding benthic invertebrates (Morrison et al. 1996).

The biosediment accumulation factor (BASF) values, which were calculated using the connection between $f_B/f = 0.62 \times \text{BASF}$, were found to be 0.5–12.4. The ratio of f_B to the fugacity in sediment (f_S) ranged from 0.3 to 7.7. The f_B/f_S values demonstrated a propensity to follow a parabolic relationship with log K_{ow} and statistically outperformed the equilibrium partition strategy in predicting bioaccumulation. The percentage of organic carbon or fat in the meal that was lost during digestion, together with the velocity of food absorption, was the most sensitive factor. Additionally, the fugacity model was utilized to estimate bioaccumulation at various trophic levels (Hendriks et al. 2001) (Table 8). In various aquatic creatures that interact via a food web, this model was only marginally successful in forecasting the absorption process ($r^2 = 0.39$), but it was more successful in predicting elimination ($r^2 = 0.70$). For the chemicals maintaining log K_{ow} values of 2–7, the projected bioaccumulation values were usually consistent with those that had been observed. Thomann and Komlos (1999) successfully implemented a kinetic model to analyze the low BSAF values (0.01–0.1) observed in crayfish exposed to different polycyclic aromatic hydrocarbons (PAH) in a creek (Table 9).

They found that for log K_{ow} values of PAHs above roughly 5–6, the proportional contribution of the food exposure pathway toward water steadily increased, with its magnitude being underlined by alkyl substitution in the ring structures. Aquatic animals have a tendency to bioaccumulate these toxins along the food chain, making compliance with current rules and developing mitigation measures essential.

3 Xenobiotics

To begin with, the term xenobiotic is of Greek origin from the word *xenos* meaning strange or foreign and *bio* which means life. These are chemicals that exhibit abnormal structural characteristics (Fetzner 2002). The aberrant presence of any substance for that matter at higher concentrations could also be termed as xenobiotics. At the same time, natural substances, which found its way into animals or humans, could also be defined as xenobiotic. Banjoko (2014) suggested the term xenobiotics as the biological and physiological effects of exogenous substances either synthetic or natural on the cells, tissues, or the organs of organisms. The xenobiotic sources that are caused by humans include industrial, domestic, pharmaceutical, agricultural, and transportation sources (Essumang 2013). Based on the physiological and biological effects of exogenous substances, whether natural or synthetic (drugs, chemicals), on the cells, tissues, or organs of animals, Banjoko (2014) introduced the term "xenobiotic."

Numerous xenobiotics are possibly hazardous to the organisms which are exposed to them within the environment niche. Nevertheless, the bioavailability of such chemical substances depends on the attribute of the chemical, organism, and the

Table 8 A summary of bioaccumulation studies of selected pesticides in aquatic organisms

Pesticide	Species[a]	Experimental conditions[b] (concentration and duration)	log BAF[c]	CL_{50}[d]	References
Photo-dieldrin	*W: Daphnia pulex*	1 ppb, na, 36 h/7 days	1.2	4 days	Khan et al. (1975)
Lindane α-isomer of (31)	*O: Tubifex tubifex*	0.74 ppm, 8 °C, 79 days/ 84 days	0.88	<5 days	Oliver (1987)
	W: Daphnia magna	10 ppb, 20 °C, 2 days/na	–0.5	na	Canton et al. (1975)
Hexachlorobenzene	*O: Tubifex tubifex*	0.9 ppm, 8 °C, 79 days/ 84 days	0.49	24 days	Oliver (1987)
Pentachlorophenol	*O: Lumbriculus variegatus*	0.4 ppm, 20 °C, 14 days/ na	1.46	na	Nikkilä et al. (2003)
DDT	*O: Tubifex tubifex*	0.1 ppm, 8 °C, 79 days/ 84 days	–0.3	53 days	Oliver (1987)
DDE	*O: Tubifex tubifex*	0.29 ppm, 8 °C, 79 days/ 84 days	0.63	80 days	Oliver (1987)
Terbutryn	*Is: Asellus aquaticus*	na, 15 °C, 2 days/1 days	–0.37	4.3 hr	Richter and Nagel (2007)
	A: Gammarus fossarum	na, 15 °C, 2 days/3 days	–1.4	2.9 hr	Richter and Nagel (2007)
Bentazone	*O: Lumbriculus variegatus*	20 °C, 10 days/na	–0.3 to 0.6	na	Mäenpää et al. (2003)
Pendimethalin	*O: Lumbriculus variegatus*	20 °C, 10 days/na	–1.0–0.9	na	Mäenpää et al. (2003)
Ioxynil	*O: Lumbriculus variegatus*	20 °C, 10 days/na	0.4–1.7	na	Mäenpää et al. (2003)

na not available
[a]Evaluated by EPI Suite (USEPA 2008) or experimental data therein
[b]Application of pesticide to sediment (s) or food (f), exposure condition [static(s)/flow through (fl)], concentration, temperature (°C), periods of exposure/elimination
[c]Bioaccumulation factor based on the sediment or food concentration in the whole body
[d]50% clearance time

environment. The bioaccumulation of the chemical residue within the organism decides the toxicity of the xenobiotic (Mäenpää 2007). It is said that the long-term effects of the xenobiotics in the environment might be contained up to months to years within the environment.

Advanced technologies to regulate the trace polar compounds have aided in providing new perceptions on the removal of xenobiotics. Initially, in the USA, pharmaceutical products (0.8–2 μg/L) were recorded in the treated wastewater

Table 9 Biota-sediment accumulation factors (BSAF) of pesticides and chemicals in aquatic organisms

Pesticide	Species[a]	f_{bb}	Source[c]	f_{ocd}	BSAF	References
Pyrene	*O: Lumbriculus variegates*	NA	Lake X 2	1.4–3.0	0.42–0.59	Leppänen and Kukkonen (2006)
Benzo[a]pyrene	*P: Nereis diversicolor*	1.2–1.3	Marine	0.6	0.028	Driscoll and McElroy (1996)
	Leitoscolopos fragilis	1.4–1.7	Marine	0.6	1.40	Driscoll and McElroy (1996)
Lindane	*O: Tubifex tubifex*	2.9–5.2	Artificial soil	2.0	2.81–2.93	Egeler et al. (1997)
Hexachlorobenzene	*O: Tubifex tubifex*	2.9–5.2	Artificial soil	2.0	3.13–4.96	Egeler et al. (1997)
DDT	*A: Hyalella azteca*	0.7–1.8	Lake	0.4–0.6	0.44–2.08	Lotufo et al. (2001a)
	Leptocheirus plumulosus	1.9 ±0.2	Marine	1.78	2.51 ±0.54	Lotufo et al. (2001b)
DDE	*O: Lumbriculus variegates*	1.1–2.3	Lake X 2	1.3–7.9	2.00–5.31	You et al. (2006)
TCBP	*O: Lumbriculus variegates*	1.0 ±0.3	Creek	0.29	0.05–0.33	Leppänen et al. (2003)
	A: Hyalella azteca	NA	Creek	0.29	1.73–3.85	Leppänen et al. (2003)
	I: Chironomus tentans	NA	Creek	0.29	0.47–1.33	Leppänen et al. (2003)
Chlorpyrifos	*O: Lumbriculus variegates*	1.5–2.3	Lake X 2	1.3–7.9	1.49–6.54	You et al. (2006)
Azinphos-methyl	*Co: Amphiascus tenuiremis*	2.2	Estuarine	3.85	26.8	Klosterhaus et al. (2003)
	Microarthridion littorale	5.6	Estuarine	3.85	2.2	Klosterhaus et al. (2003)
Permethrin	*O: Lumbriculus variegates*	1.5–2.6	Lake × 2	1.3–7.9	2.17–3.79	You et al. (2006)
Cypermethrin	*W: Daphnia magna*	NA	River × 3	1–13	0.08–0.31	Maund et al. (2002)
	I: Chironomus tentans	NA	River × 3	1–13	0.08–0.63	Maund et al. (2002)
Bentazone	*O: Lumbriculus variegates*	1.23	Lake × 4	0.54–24	2.9–14.9	Mäenpää et al. (2003)
Pendimethalin	*O: Lumbriculus variegates*	1.23	Lake × 4	0.54–24	2.2–5.5	Mäenpää et al. (2003)

na not available, *TCBP* 3,4,3′,4′-tetrachlorobiphenyl
[a]Evaluated by EPI Suite (USEPA 2008) or experimental data therein
[b]Wet weight fraction of lipid in an organism
[c]Source of sediment or soil
[d]Organic carbon %

(Garrison et al. 1976). Following which, clofibric acid (1 μg/L) was reported in the UK, which was found in the rivers (Richardson and Bowron 1985). Ibuprofen and naproxen concentrations were identified in the Canadian wastewaters by Rogers. Significant reduction in the population of the Indian and Asian white-backed vultures nesting in the Keoladeo National Park from North-Western India was an impact of the diclofenac (pain killer) accumulation, a pain killer applied in the veterinary sector to treat cattle. Several toiletries and drugs, phthalates, insect repellents, and steroids were reported by the Geological Survey Department of the USA (Embrandiri et al. 2016). Although, there was traces of the xenobiotic concentrations, effects of the chronic exposure were uncertain. Increase in the usage of bulk drugs lately, has been associated as a significant source of environmental pollution consisting of active pharmaceutical compounds in definitive locations (Gunnarsson et al. 2009; Fick et al. 2010). In addition, there is a cause of concern globally on the residues (pharmaceutical) found in the surface water which can pose negative effects on the aquatic organisms. Thereby, there is challenging to develop, an unambiguous strategy to prioritize drugs on which major focus is on the environmental research (Fick et al. 2010).

3.1 Common Xenobiotic Compounds Based on Its Course of Action

Xenobiotics target specified molecular and metabolic pathways in animals and humans in the ecosystem. Howbeit, the xenobiotics when introduced into the environment might impart effects on some pathways in animals, exhibiting identical target biomolecules, cells, tissues, or organs. The present ecotoxicological effects of pharmaceuticals confront basically acute toxicity and are extensively centered on aquatic organisms. The impact of environmental variables such as pH on its toxicity has been rarely or has not been investigated. The majority of the studies have been directed toward the acidic pharmaceuticals that induce various toxicities subjected to speciation at distinct ambient pH. Furthermore, till date, limited research has been conducted on the effects of drug metabolites. Discussed below are the frequently used xenobiotic compounds that are predicted to pose environmental distress.

3.1.1 Analgesics and Nonsteroidal Anti-inflammatory Drugs (NSAIDs)

Extensively used nonsteroidal anti-inflammatory drugs (NSAIDs) are diclofenac, ibuprofen, and naproxen, and their metabolites like the carboxy-ibuprofen and hydroxyl-ibuprofen could be traced in the water sewage and surface. In the USA, the levels of NSAID in the sewage system exceeded to 1 μg/L, and its concentration in the effluents of the conventional sewage plants (biological treatment and mechanical clarification) exceeded to 0.1 μg/L (Gross et al. 2004). The deacylated form of acetylsalicylic acid, which is believed to be the more active form, has been determined in majority of the municipal wastewaters to levels up to 4.1 μg/L, 13 μg/L, or even 59.6 μg/L, respectively (Embrandiri et al. 2016.). Acetaminophen (paracetamol), which is similar to the acetylsalicylic acid, has been recovered from the sewage-treated water. However, it was reported that in 24% of samples from US streams, acetaminophen (up to 10 μg/L, median 0.11 μg/L) was spotted. In addition, in several countries, analgesic codeine (median 0.01 μg/L) was recurrently detected in 7% of samples. Furthermore, diclofenac was also commonly identified in the wastewater and in minute amounts in the surface water. Ibuprofen and their metabolites (0.1–20 μg/L) were detected in all the seawater and sewage samples in Norway (Wiegel et al. 2004). In the stream water samples, ibuprofen (1 μg/L, median 0.2 μg/L) was traced in high concentrations (Kolpin et al. 2002). Besides, many other NSAID compounds have also been traced in the surface, sewage, and drinking water samples.

3.1.2 Mode of Action

Pharmaceutical and xenobiotic substances have an impact on traditional sewage treatment facilities and may hinder biological processes like nitrification. The oxidation of ammonium to nitrite, the initial stage in the nitrification process, is sensitive to the presence of xenobiotic chemicals. Under uncontrolled circumstances, xenobiotics can completely stop the biological nitrogen process by inhibiting the first stages of nitrification (Essumang et al. 2009). It has been determined that many medication combinations are exposed in streams and rivers. Diphenhydramines that are antidiabetic and antihistamine have been proven to significantly disturb the biofilm community, which is crucial to the ecology. In biofilms, which are microbial aggregations, cells that are frequently encased in extracellular polymeric substances (EPS) matrix cling to one another or to a surface. The vital food source for invertebrates, which in turn feed fish and other large animals, is biofilms. Therefore, animals in the stream food web including insects and fish may be affected by diphenylamines' impacts on biofilm (Rosi-Marshall 2013). Antidepressant use causes some shellfish to begin spawning early, upsetting the aquatic balance. Additionally, it was discovered that fluoxetine and propranolol negatively affected zooplankton and benthic species (Hoffman et al. 2005). The development of biofilms by bacteria is caused by a variety of factors, including cellular recognition of specific or nonspecific attachment sites, nutritional signals, or exposure of planktonic cells to subinhibitory concentrations of antibiotics (Hoffman et al. 2005; Karatan and Watnick 2009). When exposed to tributyltin, female marine snails displayed masculinization (imposex) (TBT). Due to imposex, local populations of the dog whelk (*Nucella lapillus*), a "species of predatory sea snail," have declined or gone extinct all across the world, especially in coastal regions all over Europe and the North Sea. Diverse fish species exposed to effluents have been negatively impacted by EDCs (endocrine disruptors), leading to reproductive issues. Similar effects have also been seen in turtles (Cleuvers 2003; Le Page et al. 2011).

Most cleaning products contain the broad-spectrum antimicrobial ingredient triclosan (TCS), which works to stop bacterial, fungal, and mildew growth. Triclosan is released into water streams by leaking sewers, sewage overflows, and domestic wastewater. The ongoing use of these antibiotics causes the development of resistant bacteria, which could reduce the effectiveness of crucial antibiotics (Drury et al. 2013).

Xenobiotic substances that are released into surface water may leak into groundwater, although this practice is currently severely prohibited since it could compromise the ecological integrity of aquatic ecosystems. Important biological markers of xenobiotic contamination include some aquatic creatures (Fent et al. 2006). Xenobiotic substances can enter the environment as metabolites or in their original forms. Humans may process xenobiotic chemicals through consumption, excretion, and wastewater disposal (Singh et al. 2016). In typical sewage treatment facilities, some xenobiotic chemicals are nonbiodegradable and discharged with treated runoff, which could contaminate aquatic systems like rivers, lakes, and estuaries (Embrandiri et al. 2016). The most significant and crucial characteristics of xenobiotics are their high production, environmental persistence, and biological impacts. Concerns have been raised around the world due to studies showing an increase in the number of xenobiotic chemicals discovered in aquatic systems (surface water) (Embrandiri et al. 2016). Animals in the food chain, such as fish and insects, are impacted by xenobiotics (Rosi-Marshall 2013).

3.1.3 NSAIDs (Nonsteroidal Anti-inflammatory Drugs)

The NSAIDs (nonsteroidal anti-inflammatory drugs) are usually employed to relieve fever and treat pain and inflammation, and under few circumstances, they are employed in the long-term treatment of rheumatic diseases. NSAIDs act by inhibiting either one isoforms of the cyclooxygenase enzyme (COX-1 and COX-2), either reversibly or irreversibly, which are involved in the synthesis of various prostaglandins from arachidonic acid. The classical NSAIDs inhibit the COX-1 and COX-2 at different degrees, while the new NSAIDs act more specifically on the COX-2, which is an inducible form, responsible for various inflammatory reactions. The selectivity of the drugs is mainly due to the differences in binding site size, while the former NSAIDs work in nonspecific fashion (Szewczuk et al. 2004).

3.1.4 Blood Lipid Regulators

Habitually described pharmaceutical in the observational studies is the clofibric acid, an active metabolite and is commonly employed to regulate blood lipids like etofibrate, etofylline, and clofibrate. These chemicals have been detected in copious amounts in surface waters, wastewaters, and seawaters, specifically at a quite higher concentration in the groundwater (4 μg/L) and drinking water (0.07–0.27 μg/L), respectively. Gemfibrozil and bezafibrate, known as lipid-lowering agents, have been found in extreme concentrations (up to 4.6 and 0.79 μg/L) in surface water and wastewater, respectively (Kolpin et al. 2002). Additionally, auxiliary drugs acting as metabolites of fenofibrate such as fenofibric, gemfibrozil, and clofibric acid have also been traced in surface water and sewage water (Heberer 2002).

Two types of antilipidemic drugs are fibrates and statins, which are employed in lowering the triglycerides (fibrates) and cholesterol concentration (statins and fibrates) in the blood plasma. These drugs are often frequently targeted in the aquatic environment. Statins (inhibitors of cholesterol synthesis) are said to act by the inhibition of 3-hydroxymethylglutaril coenzyme A reductase (HMG-CoA), which is important to convert HMG-CoA to mevalonate. Reports have demonstrated that the statins have exhibited effects on the biosynthesis (in vitro) and on the mandibular organ of lobsters and also on the juvenile hormone synthesis in insects. Alteration in the gene transcription, which encode for the proteins that control the metabolism of lipoprotein, in addition to activating the enzyme lipoprotein lipase, whose main role is controlling the protein metabolism, is one of the major effects of the fibrates. The enzyme lipoprotein lipase has a role to play in the conversion of very-low-density lipoprotein (VLDL) to high-density lipoproteins (HDL), thereby decreasing the concentration of the plasma triglyceride. Moreover, it stimulates the uptake by converting the acetyl CoA derivatives and catabolism by the beta-oxidation pathways. A combination of these processes leads to the reduction of the triglyceride synthesis and fatty acids and thus decreases the VLDL production. Studies have demonstrated that chronic exposure to fibrates caused hepatic damages; this could be due to the mitochondrial oxidative phosphorylation inhibition. At the same time, it was observed that the fibrates in rodents caused massive proliferation of peroxisomes. There was a strong co-relation established between hepatocarcinogenicity and fibrate exposure in rodents, while this wasn't demonstrated in humans (Cajaraville et al. 2003). These demonstrations enhance the interest to focus on the ecotoxicological impact of therapeutic applications of these drugs.

3.1.5 Beta-Blockers

Various beta-blockers like metoprolol, propranolol, and bisoprolol were found in the wastewater of concentrations 2.2 μg/L, 0.59 μg/L, and 2.9 μg/L, respectively. Few other beta-blockers, betaxolol (0.028 μg/L) and nadolol, in the surface waters were detected at its lowest concentration (Ternes 1998). The presence of metoprolol, bisoprolol, and propranolol was detected in the surface water, in addition to the detection of sotalol in groundwater (Sacher et al. 2001).

They act by inhibiting the beta-adrenergic receptors. Generally, these are employed in the treatment of hypertension and preventing heart attacks in high-risk patients. Some of the functions of the adrenergic system are bronchodilation, vasodilatation of blood vessels, oxygen supply, and heartbeat regulation. Moreover, it is necessary for the metabolism of lipids and carbohydrates in case of starvation. Beta-blockers could selectively hinder one or more β-receptor types based on the requirement. For example, these chemicals are involved in the treatment of hypertension by preventing the cardiac arrests, as the β-2 blocker subtype is not present in the heart.

Unlike the metoprolol which lack the ability to stabilize cell membranes, the beta-blocker 9 propranolol, a beta-1-adrenoceptor antagonist, exhibits those properties (Doggrell 1990). Negative effects of these beta-blockers are majorly disturbed peripheral circulations and bronchoconstriction. These work by passing the blood-brain barrier, to act on the central nervous system, due to their lipophilicity (Heberer 2002). Ractopamine and clenbuterol have the role of β-agonist in mammals; however, they showed different reactions in rainbow trouts. The difference in the functions may be due to the difference in their function and structures and varied affinity with β-blockers and mechanisms triggered by these drugs.

3.1.6 Neuroactive Compounds (Antiepileptics and Antidepressants)

The antiepileptic carbamazepine was most commonly detected at the highest concentration in the wastewater (up to 6.3 μg/L) (Ternes 1998) and in lesser concentrations in other media (Heberer 2002). In all the effluent samples, carbamazepine was traced on the Canadian sewage treatment plant (STP) (2.3 μg/L). In addition, it was also reported to be traced on all the samples of German river Elbe and streams (Wiegel et al. 2004), which exceeded a concentration of 1 μg/L in the surface waters (Ternes 1998; Heberer 2002) and also was detected in the groundwater (Sacher et al. 2001). Carbamazepine was detected at average concentrations of 20.9 ng/mg in the STP. In Germany, diazepam was reported to be present in 8 out of 20 treatments, at lower concentrations (0.04 μg/L) (Ternes 1998), whereas it was recorded at a concentration of 0.66 μg/L (van der Ven et al. 2004) in Belgium. Fluoxetine, an antidepressant, was found in the US streams and Canadian effluent samples at a median concentration of 0.012 μg/L (Kolpin et al. 2002). Besides, primidone (0.6 μg/L), an antiepileptic drug, was reported in the sewage (Heberer 2002) samples.

3.1.7 Mode of Action

The overall neuronal activity will be suppressed by the antiepileptic drugs. This could be brought about either by enhancing the inhibitory effects of the neurotransmitter (GABA) by binding to the site exactly which corresponds to the gamma subunit of the corresponding receptor (member of benzodiazepine family/diazepam). One more way is by blocking the voltage-dependent sodium channels of excitatory neurons (e.g., carbamazepine). Serotonin uptake is inhibited by fluoxetine (antidepressant). Serotonin is a neurotransmitter which interferes with the food intake, sexual behavior, and neuronal and hormonal mechanisms. Norfluoxetine, fluoxetine, desmethylsertraline, and sertraline have been found accumulated from the wild fish samples in the, reflecting on the bioaccumulation potential (Brooks et al. 2005).

3.1.8 Various Other Compounds

The effluents of the surface waters and sewage treatment plants have found to be contaminated by drugs comprising of cotinine and caffeine (a nicotine metabolite). Caffeine was found detectable in higher levels (6.0 μg/L (median 0.1 μg/L (Kolpin et al. 2002); this can act as an anthropogenic marker in aquatic systems because of its ubiquity in the groundwater, seawater, and surface water (Wiegel et al. 2004). In the streams of the USA, the antacids ranitidine and cimetidine were found to occur at the respective concentrations (0.58 and 0.01 μg/L) (Kolpin et al. 2002). In the surface water (0.49 μg/L), groundwater, and municipal wastewater (15 μg/L), iopamidol was found to be detected. Metformin, an antidiabetic compound (5%), was found in the stream water samples, with the estimated levels of 0.11 μg/L (Kolpin et al. 2002). Bronchodilators such as salbutamol and β2-sympathomimetic terbutaline were also observed in the sewage waste waters, however not exceeding 0.2 μg/L concentration (Ternes 1998).

Ranitidine and cimetidine compounds hinder the histamine receptor type 2 in the gastric system, hence inhibiting the antacid (acid secretion) and thus used in the treatment of gastric ulcer. Metformin is an antidiabetic agent; however, the mechanism of action isn't fully understood. It has been studied that this drug increases the cellular glucose usage, thus inhibiting gluconeogenesis. Metformin is said to act on the insulin receptor by directly stimulating the insulin receptor or indirectly via inhibition of tyrosine phosphatase (Holland et al. 2004).

3.1.9 Effects of Xenobiotics on Ecosystem

It has been recorded that more than 13 million deaths and 24% of the world diseases are a cause of environmental pollution and exposures to various contaminates, which could be indeed avoided. As of today, traceable amounts of pharmaceutical preparations (metabolite/parent drug) are found in both food and water sources (Banjoko 2014). Severe consequences could be predicted as a result of medications for humans and animals, which extends beyond the conventional medical care. Healthcare sectors are one of the significant causes of active pharmaceutical ingredients (API) let out from medication residues and a prime cause of environmental pollution.

3.1.10 Effects on Aquatic Ecosystem

The aquatic organisms could be an important biological indicator of pollution. A comprehensive study was conducted on the occurrence, ultimate fate of the pharmaceuticals in the aquatic environment, and the mechanism of action of various pharmaceuticals and expanded chronic and acute ecotoxicological effects on organisms (Fent et al. 2006).

Pharmaceuticals are said to be the most frequently released effluents into the environment either in their metabolite or the original form. The main pathway in the humans includes ingestion, excretion, and disposal through the wastewater. The largest source of human pharmaceuticals is wastewater.

Wastewaters let out from manufacturers, hospitals, and landfill leachates might contain significant concentrations of pharmaceuticals. The nondegradable pharmaceuticals in the sewage treatment plant (STP) are released into the treated effluents which ultimately results in the contamination of drinking water, groundwater, estuaries, rivers, and lakes. Agricultural sector contamination is a possibility when sewage water is let into the farms and fields. Moreover, the drugs used in the veterinary sector enter the waterways as a course of surface application for agricultural farming purposes, and runoff affects the fish farming. Effluents let out from the pharmaceuticals have high environmental significance and sometimes have high production volume and increased environmental persistence particularly after the long-term exposure. It has been noted that the pharmaceutical concentrations traced in surface waters globally are a matter of concern, mostly with respect to the aquatic flora and fauna. Thereby, its is a huge task in initiating a strategy to prioritize drugs on which majority of the environmental research focus must rely upon. Among the aquatic life, most often shared drug targets with humans are the fishes. Very little information is known regarding the long-term effect of drugs in aquatic organisms. Diclofenac is said to interfere with the organ histology and gene expression of fishes while exposed at a concentration of 1 μg/L (Cuklev et al. 2012). In India, surface water collected from 27 locations of the rivers in southern India (Tami and Kaveri) exposed the presence of nonsteroidal anti-inflammatory drugs (NSAIDs) such as acetylsalicylic acid, naproxen, ketoprofen, and ibuprofen. This alarming situation imparts direct toxicity in the case of all the consumers of the water (Shanmugam et al. 2014). In a similar scenario, the effluents let out from a treatment plant in Hyderabad, India, was reported to be the reason for the deleterious effects on aquatic organisms. Embryo toxicity assay carried out revealed that smaller concentrations (0.2%) of effluents hindered the growth of tadpole by 40%; however, the growth rate of zebra fish (*Danio rerio*) was not impeded. Regardless, the study focused on fishes; meanwhile, it also shed light on how the aquatic vertebrates are probably affected as a result of effluent exposures and the substances responsible in causing toxic effects at the threshold dilutions (Shanmugam et al. 2014). Rivers and streams have been identified as sources often exposed to various drug combinations. Significant disruption was caused to the biofilm community, an important part of the ecosystem as a result of usage of antihistamine diphenhydramines and antidiabetics. In addition, the biofilms also serve as a major source of food for the invertebrates that in return are fed by larger animals of the aquatic lives.

Thus, the diphenylamines could affect biofilms and therefore have repercussion for animals in stream food webs like the fishes and insects (Rosi-Marshall 2013). Disruption of the aquatic equilibrium by early activation of spawning in some shellfish might be one of the effects of antidepressant usage.

Moreover, deleterious effects were posed as a result of fluoxetine and propranolol usage on the benthic organisms and zooplanktons (Hoffman et al. 2005). Aspects such as nutritional cues, nonspecific attachment sites, cellular recognition of specific attachment sites, and planktonic cells exposed to antibiotics at the subinhibitory concentrations lead to the formation of biofilms by microbes (Hoffman et al. 2005; Karatan and Watnick 2009). Exposure of female marine snails to tributyltin (TBT) resulted in masculinization (imposex). The decline or extinction of local populations of the dog whelk (*Nucella lapillus*), a predatory sea snail species in the coastal areas and all over the North Sea and Europe, is due to imposex. One of the best examples of reproductive impairment causing population decline is the DDE (dichlorodiphenyldichloroethylene)-induced eggshell thinning in birds in North America and Europe. Ovotestis in male western gulls is a result of gradual DDT complex (dichlorodiphenyltrichloroethane) exposure. A variety of fishes, exposed to effluents (EDC-endocrine disruptors) have negative effects on their reproductive system. Similar effects have been demonstrated in turtles (Cleuvers 2003; Le Page et al. 2011). Broad-spectrum antimicrobial compounds (Triclosan-TCS) have been active ingredients in the cleaning products for preventing the growth of mildew, bacteria, and fungi. This is reported to enter the domestic wastewater, sewage overflows, water streams, and leaking sewerage, causing adverse effects on the aquatic life forms.

The removal of these contaminants is a severe environmental problem because there are several xenobiotic chemicals present in typical sewage treatment systems, each of which has its own effects. Xenobiotics are persistent in the environment and difficult to break down, for example, trichloroethylene (TCE) and polycyclic aromatic hydrocarbons (PAHs).

Due to these xenobiotics' unique chemical characteristics, they accumulate in the environment. As a result, these xenobiotics exhibit traits of toxicity and accumulation in the environment and have an impact on both the natural world and human life. Xenobiotic pollutants can have an impact on the climate and human health and are typically found in biological systems, agricultural runoff, and water and wastewater sources (Fatta-Kassinos et al. 2011). Common xenobiotic receptors exist in traditional sewage treatment plants and must be treated with municipal wastewater before being released into aquatic systems. The presence of trace metals, xenobiotic substances, and synthetic organic chemicals such as PAHs, phthalates, and pesticides in water bodies has been documented (Essumang and Ankrah 2010).

Pharmaceutical and xenobiotic substances have an impact on traditional sewage treatment facilities and may hinder biological processes like nitrification. The oxidation of ammonium to nitrite, the initial stage in the nitrification process, is sensitive to the presence of xenobiotic chemicals. Under uncontrolled circumstances, xenobiotics can completely stop the biological nitrogen process by inhibiting the first stages of nitrification (Essumang et al. 2009). It has been determined that many medication combinations are exposed in streams and rivers. Diphenhydramines that are antidiabetic and antihistamine have been proven to significantly disturb the biofilm community, which is crucial to the ecology. In biofilms, which are microbial aggregations, cells that are frequently encased in an extracellular polymeric substance (EPS) matrix cling to one another or to a surface. The vital food source for invertebrates, which in turn feed fish and other large animals, is biofilms. Therefore, animals in the stream food web including insects and fish may be affected by diphenylamines' impacts on biofilm (Rosi-Marshall 2013). Antidepressant use causes some shellfish to begin spawning early, upsetting the aquatic balance. Additionally, it was discovered that fluoxetine and propranolol negatively affected zooplankton and benthic species (Hoffman et al. 2005). The development of biofilms by bacteria is caused by a variety of factors, including cellular recognition of specific or nonspecific attachment sites, nutritional signals, or exposure of planktonic cells to subinhibitory concentrations of antibiotics (Hoffman et al. 2005; Karatan and Watnick 2009). When exposed to tributyltin, female marine snails displayed masculinization (imposex) (TBT). Due to imposex, local populations of the dog whelk (*Nucella lapillus*), a species of predatory sea snail, have declined or gone extinct all across the world, especially in coastal regions all over Europe and the North Sea. Diverse fish species exposed to effluents have been negatively impacted by EDCs (endocrine disruptors), leading to reproductive issues. Similar effects have also been seen in turtles (Cleuvers 2003; Le Page et al. 2011). Most cleaning products contain the broad-spectrum antimicrobial ingredient triclosan (TCS), which works to stop bacterial, fungal, and mildew growth. Triclosan is released into water streams by leaking sewers, sewage overflows, and domestic wastewater. The ongoing use of these antibiotics causes the development of resistant bacteria, which could reduce the effectiveness of crucial antibiotics (Drury et al. 2013).

Xenobiotic substances that are released into surface water may leak into groundwater, although this practice is currently severely prohibited since it could compromise the ecological integrity of aquatic ecosystems. Important biological markers of xenobiotic contamination include some aquatic creatures (Fent et al. 2006). Xenobiotic substances can enter the environment as metabolites or in their original forms. Humans may process xenobiotic chemicals through consumption, excretion, and wastewater disposal (Singh et al. 2016). In typical sewage treatment facilities, some xenobiotic chemicals are nonbiodegradable and discharged with treated runoff, which could contaminate aquatic systems like rivers, lakes, and estuaries (Embrandiri et al. 2016). The most significant and crucial characteristics of xenobiotics are their high production, environmental persistence, and biological impacts. Concerns have been raised around the world due to studies showing an increase in the number of xenobiotic chemicals discovered in aquatic systems (surface water) (Embrandiri et al. 2016). Animals in the food chain, such as fish and insects, are impacted by xenobiotics (Rosi-Marshall 2013).

4 Bioconcentration and Bioaccumulation of Xenobiotics

Bioaccumulation of persistent hydrophobic xenobiotics in aquatic species can occur through a variety of methods, including bioconcentration, ingestion, and biomagnification. Even if the subchronic, chronic, or acute consequences are not obvious, bioaccumulation should be considered a hazard criterion in and of itself, because certain harmful effects may not be noticed until later in life.

Bioconcentration refers to the absorption and retention of a substance in an organism solely through respiration from aquatic ecosystems or terrestrial ones. The bioconcentration factor (BCF) is defined as the concentration of a chemical in an organism divided by the concentration of the same chemical in the environment or a component of the environment (e.g., water). The BCF is mostly used to predict the degree of accumulation of an organic contaminant in water by organisms. For terrestrial animals, food is usually the primary source of many xenobiotic compounds, and if the rate of intake is constant, a steady state is eventually formed. Chemicals can enter the organism primarily through three different entryways such as gills, skin, and digestive system. The chemicals are dispersed throughout the various tissues once they have entered the body. The distribution process is influenced by the characteristics of the absorbed compounds, which may have a strong affinity for specific biomolecules like membrane lipids, blood proteins, structural proteins, or storage lipids. As a consequence, they tend to accumulate in organs that are abundant in these biomolecules (Da Cuña et al. 2020; Hou et al. 2017; Grech et al. 2016).

The process by which pesticides enter organisms directly from water through the gills or through epithelial tissues is known as bioconcentration. In the contrary, bioaccumulation includes the effect of dietary assimilation through food consumption or ingestion of bottom sediments. Organic hydrophobic compounds, including PCBs, are predominantly bioaccumulated in the lipid body component of the organisms. The discovery of DDT (dichlorodiphenyltrichloroethane) and methyl mercury residues in fish, fish-eating birds, and other species in the 1960s brought the phenomena of bioaccumulation to the attention of the general public.

Intake of a chemical and its concentration in the organism by all possible means, including contact, respiration, and ingestion, is called bioaccumulation. Bioaccumulation of organic hydrophobic chemicals, such as PCBs, mainly happen in the lipid body fraction of the organisms. The xenobiotic compounds dissolved/suspended in the medium, as well as ingested food and sediment residues, are taken up. Bioaccumulation factor needs to be incorporated.

In the case of a predator, the majority of xenobiotic compounds are consumed through the ingestion of prey that has previously concentrated on a xenobiotic (Fig. 4).

4.1 Biomagnification of Xenobiotics

Biomagnification is the process of transferring xenobiotic chemicals from food to an organism, resulting in larger quantities than the source. The term biomagnification includes the full process of bioconcentration and bioaccumulation. Furthermore, it considers the slow increase in chemical concentration in the tissues of organisms as it moves through the food chain. This is widely assumed to be a common occurrence

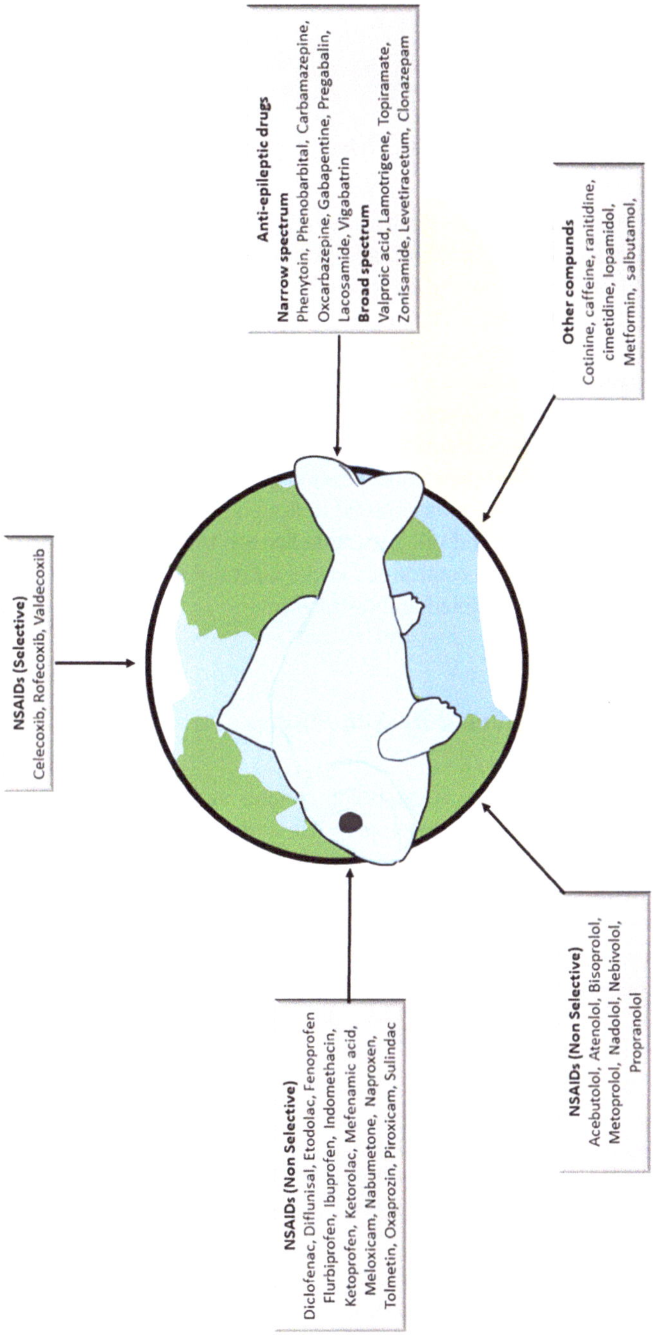

Fig. 4 Types of xenobiotics

in marine food webs. Only organic mercury displays biomagnification in studies on metals, while the majority of metals are controlled and eliminated and do not biomagnify.

4.2 *Xenobiotic Compounds: Other Effects*

High toxicity: Many xenobiotics are harmful to bacteria, lower eukaryotes, and even humans, such as halogenated and aromatic hydrocarbons. They may cause numerous skin issues and impair reproductive capacity at low dosages.

Cancer-causing substances: The majority of xenobiotic chemicals include carcinogens. They experience significant bioaccumulation and biomagnification. The cells become resistant to antibodies as a result of this transmission in the food chain. They have an extremely high risk of developing cancer-like diseases.

High resistivity to the environment: Many xenobiotics are recalcitrant and remain in the environment, causing their concentration to rise with time.

Coral bleaching: When coral is exposed to high quantities of various xenobiotics such as copper, herbicides, and oil, zooxanthellae are lost.

Since high xenobiotic concentrations cause zooxanthellae loss, bleaching from such sources is typically confined and/or transient.

4.3 *Marine Life Affected by Xenobiotics*

Xenobiotics have a deleterious impact on a number of marine creatures' metabolic processes, particularly those of developing fish embryos, which results in morphological and functional defects, stunted growth, and eventual death. Fish have also been known to have altered body shape, bodily abnormalities, and delayed hatching, and some even die (Arya et al. 2019). The fact that dyes and paints impede sunlight and obstruct gas exchange makes them xenobiotics even when they are present in minute concentrations (Rápó et al. 2021). The main sources of xenobiotic contamination in marine life are pesticides and herbicides. Organophosphorus, nitrophenols, morpholine, synthetic pyrethroids, and carbamates are just a few examples of the chemicals that are frequently employed in agriculture and daily life. These chemicals later find their way into various water bodies, such as the sea and ocean. Marine life and invertebrates are seriously threatened by insecticides like cypermethrin (Tornero and Hanke 2016). The bioaccumulation of xenobiotics causes the ingested substances to change into a variety of metabolites. In organs such as the liver, gills, and kidneys, metabolic processes occur very vigorously (Gomez et al. 2010). Consequently, typically simpler-to-expel molecules are created, yet there are also known instances in which the metabolic byproducts prove to be more toxic and accumulative than the initial chemicals (La Farre et al. 2008). Excretion is the final stage of interaction of chemicals with the body. In this scenario, elimination occurs predominantly through the gills and through fecal egestion, and the primary routes are identical to absorption (Arnot and Gobas 2006). Additionally, a pseudo-elimination process that involves dilution of chemical substances in developing tissues should be taken into account (Segner 2015). The bioaccumulation phenomenon, which denotes the accumulation of ambient chemicals in living things' tissues, is caused by all of the abovementioned processes.

4.4 Food-Based Xenobiotic Uptake

The gastrointestinal system easily absorbs the xenobiotics drawn in through the integument and gill surface using similar diffusion and transport mechanisms. Due to prolonged interaction between the food and membranes, lipophilic xenobiotics ingested are easily absorbed. In unionized form, weak acids and bases are absorbed. While the intestinal pH favors the absorption of neutral or weakly basic xenobiotics, the stomach pH favors the diffusion of weak acids.

To fully understand how chemicals enter the body, what happens to them inside, and how they are excreted, it is worthwhile to trace the processes of uptake, distribution, metabolism, and excretion.

Due to the resistivity of xenobiotics, they are highly resistive and complex. There is a need to control the spread of these compounds in the food chain and prevent their further magnification. For the removal and detoxification of toxins from the environment, the microbial bioremediation technique has recently emerged as the best alternative. Synthetic biology is addressing xenobiotic and related compound decontamination and remediation solutions in the environment. It has been discovered that understanding existing metabolic pathways is a prerequisite for removing xenobiotic compounds.

4.5 Xenobiotic Metabolism in Aquatic Animals

Xenobiotics are chemicals which are foreign to an organism's normal metabolism (Brodie et al. 2002). Most commonly, xenobiotics are referred to carcinogens, drugs and various compounds artificially introduced into environments. Majority of xenobiotics are toxic in nature, and an organism tries to overcome the toxic effect by modifying the chemical structure by set of interconnected reactions called metabolic pathways. These pathways are mediated by a variety of enzymes.

The xenobiotics are ultimately excreted from the body. The excretion can happen in two ways: (1) excretion in unchanged state and (2) metabolized endogenously and then excreted (Johnson et al. 2012). Xenobiotics may be water soluble or lipid soluble. Water-soluble compounds may be eliminated from the body in the unchanged state, whereas the lipophilic or lipid soluble compounds must be metabolized in order to make it more polar and water soluble and can be excreted from the body easily (Schenkman 1999).

Metabolism of xenobiotics can result in (1) activation and (2) detoxification. In case of activation, the metabolism of xenobiotics can result in increasing the toxicity, whereas in the case of detoxification, metabolism results in decreasing toxicity. There are two types of metabolism: Phase I and Phase II.

Phase I: This type of metabolism is also called functionalization reaction. The enzymes involved in this phase are responsible for oxidation, reduction, hydration, and hydrolysis, or they introduce functional group like -OH, -COOH, etc. to xenobiotics so that enzymes of Phase II metabolism attach large polar moieties such as glutathione, sulfate, amino acid, etc. The metabolites resulting from Phase I metabolism are more reactive chemically than the parent compound. This metabolism ultimately develops metabolites which is suitable for undergoing Phase II metabolism (Livingstone 1998).

Phase II: Alternatively, this type of metabolism is called conjugative reaction. In this metabolism, activated derivatives produced in Phase I metabolism are conjugated with polar moieties like glutathione, amino acids, etc. to produce water-soluble derivatives which can be easily excreted (Livingstone 1998).

4.5.1 Enzymes Involved in Phase I Metabolism

As mentioned earlier, Phase I involves oxidation, reduction, hydrolysis, and hydroxylation. Different enzymes participate in each reaction:

1. **Oxidation:** Oxidation is performed mainly by three categories, namely, (1) cytochrome P-450 monooxygenase or mixed function oxidase (MFO), (2) microsomal flavin-containing monooxygenase (MFMO), and (3) other oxidative enzymes.

   (a) *Cytochrome P-450 monooxygenase or Mixed Function Oxidase*

   The name MFO is coined because it catalyzes reactions in which each of two atoms of O_2 is utilized for different purposes in a reaction. These enzymes oxidize two different substrates. The MFO converts the lipophilic substrates (RH in the following equation) into a metabolite which is more hydrophilic than RH (ROH in the following equation) as follows:

   $$NADPH + H^+ + O_2 = RH \xrightarrow{MFO} NADP^+ + H_2O + ROH$$

   Herein, reaction oxidation of RH and NADPH is performed by each of the atoms of the oxygen. This family of enzymes has the ability to convert a wide variety of substrates (i.e., xenobiotics) such as insecticides, carcinogens, and environmental pollutants to more polar compounds that can be very easily excreted into the environment (Cederbaum 2015).

(b) *Microsomal Flavin-Containing Monooxygenase (MFMO)*

Flavin-containing monooxygenases (FMOs) and cytochrome P450 are two categories of proteins of microsomal origin. They add molecular oxygen to lipophilic compounds and convert them into water-soluble compound to ensure its rapid excretion. FMOs are responsible for oxygenation of nucleophilic S, N, O, and Se atoms of a wide range of substrates such as thiols, amines, amides, sulfides, etc. (Eswaramoorthy et al. 2006).

(c) *Other Oxidative Enzymes*

There are many oxidative enzymes like alcohol dehydrogenase, aldehyde dehydrogenase, and aldehyde oxidase involved in degradation of xenobiotics.

2. **Reduction:** The enzymes which are involved in this mechanism of detoxification of xenobiotics belong to cytochrome P 450 and P 450 reductase. These systems play important role in the metabolism of both endogenous and exogenous compounds including insecticides (Jing et al. 2018). There are many substrates for these enzymes which include epoxides, azo and nitro-compounds, halogenated compounds, and heterocycles.
3. **Hydrolysis:** It is a reaction involving addition of water as a result of which toxic substances splits into two smaller molecules. During the course of hydrolysis, hydroxyl group is added to one of the fragments, and hydrogen atom is added to another. A broad range of esterases such as carboxyl esterase, amidases, and phosphatases participate in this reaction (Arand et al. 2005). The substrates for these enzymes are amides, esters, hydrazides, and carbonates.
4. **Hydration:** Hydration is the process of combining with water. The enzymes epoxide hydratase (epoxide hydrase or hydrolase) are mainly involved in hydration. The substrates of this enzyme are epoxides.

4.5.2 Enzymes Involved in Phase II Metabolism

These enzymes facilitate reactions such as glucouronidation, glycosidation, sulfation, etc.

1. **Glucouronidation:** It is a very important Phase II metabolic pathway. This process involves metabolism of parent compound by an enzyme called UDP-glucuronosyltransferases (also called as UGTs) into negatively charged hydrophilic glucuronides that require efflux transporters for excretion out of the cell. Therefore, removal of xenobiotics via glucouronidation in metabolically active cell requires (1) UGT enzymes for production of glucuronides from the parent compound and (2) efflux transporters for excretion of glucuronides (Yang et al. 2017). Substrates for this reaction include alcohols, carboxylic acids, amides, thiols, sulfonamides, and phenols.
2. **Glycosidation:** This is a process of addition of sugars to small organic molecules mediated by a superfamily of enzyme called UDP-glycosyltransferases. The enzyme catalyzes the transfer of glucuronic acid to a wide variety of exogenous and endogenous lipophilic substrates (Meech et al. 2019).

3. **Sulfation**: It is a reaction involving addition of SO_3 group. It is mediated by cytosolic enzyme sulfotransferases. The substrates for this enzyme are phenols, alcohols, amines, and thiols.

Toxicity of Xenobiotics in Aquatic Ecosystems

A broad range of chemicals are used in many industrial and household activities. It has been observed that these chemicals are known to disturb the normal physiology and endocrinology of living organisms. The xenobiotics are mainly known to cause three major problems: (1) neurophysiological, (2) reproductive, and (3) behavioral. These effects are interrelated i.e., neurophysiological changes cause behavioral changes and behavioral changes affect the reproduction. The effect of xenobiotic on the target organism or community depends on concentration of compound and time of exposure. The toxic effect of xenobiotics can be acute or chronic. In case of acute toxicity, the effect is very rapid, clearly defined, whereas the induction of chronic toxicity requires long exposure to low doses (Zaki and Hamaam 2014).

Xenobiotics are negatively affecting a plethora of metabolic processes in aquatic animals particularly in developing fish embryos, which causes abnormal and retarded growth leading to death, resulting in functional and morphological abnormalities. In addition, several studies recorded the abnormalities like altered body shape, body abnormalities, and delay in hatching in fishes (Arya and Haq 2019). The xenobiotics like dyes and paints restrict the penetration of light and inhibit the gas exchange (Abdelkader et al. 2011). For marine life the pesticides and herbicides are the major sources of xenobiotic pollution. Chemicals like nitrophenols, organophosphorus, morpholine, and synthetic pyrethroids used as agricultural chemicals reached the various waterbodies including the sea. Pesticides like β-cypermethrin causes severe problem to marine life and invertebrates (Zhang et al. 2011).

5 Conclusion

At present, majority of the researchers have focused on the impact of pesticides and xenobiotics on the climate change and their impact on the environment, including the aquatic life forms as well as the human health. Both the chemical forms have adversely affected the climate (salinity and temperature). The increase in temperatures could enhance their metabolism and degradation. In addition, these changes could also facilitate the contaminants to get into the ground and surface waters, thereby affecting the aquatic life forms. These chemical contaminants are transferred into the food webs and chains, thereby affecting all the organisms at the tropic levels as well as the nutrient cycle. In order to mitigate the combined effects of the pollutants on the climate, humans, and aquatic life forms, enhanced awareness to the society must be conveyed regarding the effects of the chemical pollutants and their impact on the environment. The increase in the salinity and temperature, linked to the change in climate, could probably affect the distribution of toxicity of the pollutants employed and increase its persistence in the aquatic ecosystems. Currently, many researchers have been focusing on ways to minimize the pollutant effects on the change in climate and the ecosystems, in addition to making it sustainable for its usage and also enhancing its sustainability toward the environment.

References

Abdelkader E, Nadjia L, Ahmed B (2011) Degradation study of phenazin neutral red from aqueous suspension by paper sludge. J Chem Eng Process Technol 2:109–114

Adnan A, Indulkar ST (2017) Chronic effect of Imidacloprid pesticide, on carcass composition of common carp fingerlings. *Int J Agric Sci Res* 7(2):343–348

Adnan A, Indulkar ST, Pai R (2016) Sublethal effect of Buprofezin pesticide on carcass composition of *Cyprinus carpio communis* fingerlings. J Exp Zool India 20(1):269–272

Aly OA, Shehata SA, Farag H (1984) Uptake and accumulation of selected herbicides by the freshwater alga Scenedesmus. Arch Environ Contam Toxicol 13(6):701–705

Anderson RL (1982) Toxicity of fenvalerate and permethrin to several nontarget aquatic invertebrates. Environ Entomol 11(6):1251–1257

Anderson RL, DeFoe DL (1980) Toxicity and bioaccumulation of endrin and methoxychlor in aquatic invertebrates and fish. Environ Pollut Ser A Ecol Biol 22(2):111–121

Arand M, Cronin A, Adamska M, Oesch F (2005) Epoxide hydrolases: structure, function, mechanism, and assay. Methods Enzymol 400:569–588. https://doi.org/10.1016/S0076-6879(05)00032-7

Arnot JA, Gobas FA (2006) A review of bioconcentration factor (BCF) and bioaccumulation factor (BAF) assessments for organic chemicals in aquatic organisms. Environ Rev 14(4):257–297

Arslan P, Yurdakok-Dikmen B, Kuzukiran O, Ozeren SC, Filazı A (2021) Effects of acetamiprid and flumethrin on Unio sp. primary cells. Biologia 76(4):1359–1365

Arya P, Haq SA (2019) Effects of xenobiotics and their biodegradation in marine life. In: Smart bioremediation technologies. Academic Press, Cambridge, MA, pp 63–81

Arya M, Patil S, Shruti N (2019) Xenobiotics and their adverse impact on living organisms. J Curr Microbiol 23(7):348–357

Ashauer R, Boxall A, Brown C (2006) Uptake and elimination of chlorpyrifos and pentachlorophenol into the freshwater amphipod Gammarus pulex. Arch Environ Contam Toxicol 51(4): 542–548

Axelman J, Broman D, Näf C, Pettersen H (1995) Compound dependence of the relationship log Kow and log BCFL. Environ Sci Pollut Res 2(1):33–36

Babalola OO, Van Wyk JH (2018) Comparative early life stage toxicity of the African clawed frog, Xenopus laevis following exposure to selected herbicide formulations applied to eradicate alien plants in South Africa. Arch Environ Contam Toxicol 75(1):8–16

Banjoko B (2014) Environmental pharmacology—an overview. In: Pharmacology and therapeutics. Intech, Rijeka

Barron MG (1990) Bioconcentration. Will water-borne organic chemicals accumulate in aquatic animals? Environ Sci Technol 24(11):1612–1618

Bartlett AJ, Borgmann U, Dixon DGE, Atchelor SPB, Maguire RJ (2004) Tributyltin uptake and depuration in Hyalella azteca: implications for experimental design. Environ Toxicol Chem 23(2):426–434

Basack SB, Oneto ML, Verrengia Guerrero NR, Kesten EM (1997) Accumulation and elimination of pentachlorophenol in the freshwater bivalve Corbicula fluminea. Bull Environ Contam Toxicol 58(3):497–503

Baturo W, Lagadic L (1996) Benzo [a] pyrene hydroxylase and glutathione S-transferase activities as biomarkers in Lymnaea palustris (Mollusca, Gastropoda) exposed to atrazine and hexachlorobenzene in freshwater mesocosms. Environ Toxicol Chem 15(5):771–781

Bauer I, Weigelt S, Ernst W (1989) Biotransformation of hexachlorobenzene in the blue mussel (Mytilus edulis). Chemosphere 19(10-11):1701–1707

Bedford JW, Zabik MJ (1973) Bioactive compounds in the aquatic environment: uptake and loss of DDT and dieldrin by freshwater mussels. Arch Environ Contam Toxicol 1(2):97–111

Boryslawskyj M, Garrood T, Stanger M, Pearson T, Woodhead D (1988) Role of lipid/water partitioning and membrane composition in the uptake of organochlorine pesticides into a freshwater mussel. Mar Environ Res 24(1–4):57–61

Böttcher T, Schroll R (2007) The fate of isoproturon in a freshwater microcosm with Lemna minor as a model organism. Chemosphere 66(4):684–689

Brock TC, Crum SJH, Van Wijngaarden R, Budde BJ, Tijink J, Zuppelli A, Leeuwangh P (1992) Fate and effects of the insecticide Dursban® 4E in indoor Elodea-dominated and macrophyte-free freshwater model ecosystems: I. Fate and primary effects of the active ingredient chlorpyrifos. Arch Environ Contam Toxicol 23(1):69–84

Brock TCM, Lahr J, Van den Brink PJ (2000) Ecological risks of pesticides in freshwater ecosystems; Part 1: herbicides (No. 88). Alterra, Wageningen

Brodie ED, Ridenhour BJ, Brodie ED (2002) The evolutionary response of predators to dangerous prey: hotspots and coldspots in the geographic mosaic of coevolution between garter snakes and newts. Evolution 56(10):2067–2082

Brooks BW, Chambliss CK, Stanley JK, Ramirez A, Banks KE, Johnson RD, Lewis RJ (2005) Determination of select antidepressants in fish from an effluent-dominated stream. Environ Toxicol Chem 24(2):464–469

Butcherine P, Kelaher BP, Benkendorff K (2022) Assessment of acetylcholinesterase, catalase, and glutathione S-transferase as biomarkers for imidacloprid exposure in penaeid shrimp. Aquat Toxicol 242:106050

Cajaraville MP, Cancio I, Ibabe A, Orbea A (2003) Peroxisome proliferation as a biomarker in environmental pollution assessment. Microsc Res Tech 61(2):191–202

Canton JH, Greve PA, Slooff W, Van Esch GJ (1975) Toxicity, accumulation and elimination studies of α-hexachlorocyclohexane (α-HCH) with freshwater organisms of different trophic levels. Water Res 9(12):1163–1169

Canton JH, Vanesch GJ, Greve PA, Vanhellemond ABAM (1977) Accumulation and elimination of gamma-hexachlorocyclohexane (gamma-HCH) by the marine algae Chlamydomonas and Dunaliella. Water Res:111–115

Cederbaum AI (2015) Molecular mechanisms of the microsomal mixed function oxidases and biological and pathological implications. Redox Biol 4:60–73. https://doi.org/10.1016/j.redox.2014.11.008

Chaton PF, Ravanel P, Tissut M, Meyran JC (2002) Toxicity and bioaccumulation of fipronil in the nontarget arthropodan fauna associated with subalpine mosquito breeding sites. Ecotoxicol Environ Saf 52(1):8–12

Clark JM, Symington SB (2011) Advances in the mode of action of pyrethroids. Top Curr Chem 314:49–72

Cleuvers M (2003) Aquatic ecotoxicity of pharmaceuticals including the assessment of combination effects. Toxicol Lett 142(3):185–194

Connell DW (1988) Bioaccumulation behavior of persistent organic chemicals with aquatic organisms. Rev Environ Contam Toxicol 102:117–154

Crosby DG, Tucker RK (1971) Accumulation of DDT by Daphnia magna. Environ Sci Technol 5(8):714–716

Crossland NO (1984) Fate and biological effects of methyl paration in outdoor ponds and laboratory aquaria: II. Effects. Ecotoxicol Environ Saf 8(5):482–495

Cuklev F, Gunnarsson L, Cvijovic M, Kristiansson E, Rutgersson C, Björlenius B, Larsson DJ (2012) Global hepatic gene expression in rainbow trout exposed to sewage effluents: a comparison of different sewage treatment technologies. Sci Total Environ 427:106–114

Cuppen JG, Van den Brink PJ, Van der Woude H, Zwaardemaker N, Brock TC (1997) Sensitivity of macrophyte-dominated freshwater microcosms to chronic levels of the herbicide linuron. Ecotoxicol Environ Saf 38(1):25–35

Da Cuña RH, Lo Nostro FL, Shimabukuro V, Ondarza PM, Miglioranza KSB (2020) Bioaccumulation and distribution behavior of endosulfan on a cichlid fish: differences between exposure to the active ingredient and a commercial formulation. Environ Toxicol Chem 39 (3):604–611

Dar SA, Chatterjee A, Rather MA, Chetia D, Srivastava PP, Gupta S (2020) Identification, functional characterization and expression profiling of cytochrome p450 1A (CYP1A) gene in Labeo rohita against emamectin benzoate. Int J Biol Macromol 158:1268–1278

Day K, Kaushik NK (1987) The adsorption of fenvalerate to laboratory glassware and the alga Chlamydomonas reinhardii, and its effect on uptake of the pesticide by Daphnia galeata mendotae. Aquat Toxicol 10(2-3):131–142

De A, Bose R, Kumar A, Mozumdar S (2014) Targeted delivery of pesticides using biodegradable polymeric nanoparticles. Springer, New Delhi, pp 5–6

DeLorenzo ME, Scott GI, Ross PE (2001) Toxicity of pesticides to aquatic microorganisms: a review. Environ Toxicol Chem 20(1):84–98

DeLorenzo ME, Taylor LA, Lund SA, Pennington PL, Strozier ED, Fulton MH (2002) Toxicity and bioconcentration potential of the agricultural pesticide endosulfan in phytoplankton and zooplankton. Arch Environ Contam Toxicol 42(2):173–181

Deneer JW (2000) Toxicity of mixtures of pesticides in aquatic systems. Pest Manag Sci 56(6): 516–520

DeNoyelles F, Kettle WD, Sinn DE (1982) The responses of plankton communities in experimental ponds to atrazine, the most heavily used pesticide in the United States. Ecology 63(5): 1285–1293

DeNoyelles F, Dewey SL, Huggins DG, Kettle WD (2020) Aquatic mesocosms in ecological effects testing: detecting direct and indirect effects of pesticides. In: Aquatic mesocosm studies in ecological risk assessment. CRC, Boca Raton, FL, pp 577–603

Derr SK, Zabik MJ (1972) Biologically active compounds in the aquatic environment: The uptake and distribution of [1, 1-dichloro-2, 2-bis (p-chlorophenyl) ethylene], DDE by Chironomus tentans Fabricius (Diptera: Chironomidae). Trans Am Fish Soc 101(2):323–329

Dhanaraj PS, Kumar S, Lal R (1989) Bioconcentration and metabolism of aldrin and phorate by the blue-green algae Anabaena (ARM 310) and Aulosira fertilissima (ARM 68). Agr Ecosyst Environ 25(2–3):187–193

Diepens NJ, Pfennig S, Van den Brink PJ, Gunnarsson JS, Ruepert C, Castillo L (2014) Effect of pesticides used in banana and pineapple plantations on aquatic ecosystems in Costa Rica. J Environ Biol 35(sp. issue):73–84

Doggrell SA (1990) The membrane stabilizing and beta 1-adrenoceptor blocking activity of (+)-and (-)-propranolol on the rat left atria. Gen Pharmacol 21(5):677–680

Donkin P, Widdows J, Evans SV, Staff FJ, Yan T (1997) Effect of neurotoxic pesticides on the feeding rate of marine mussels (Mytilus edulis). Pestic Sci 49(2):196–209

Driscoll SK, McElroy AE (1996) Bioaccumulation and metabolism of benzo [a] pyrene in three species of polychaete worms. Environ Toxicol Chem 15(8):1401–1410

Drum C (1980) Soil chemistry of pesticides. PPG Industries, Pittsburgh, PA

Drury B, Rosi-Marshall E, Kelly JJ (2013) Wastewater treatment effluent reduces the abundance and diversity of benthic bacterial communities in urban and suburban rivers. Appl Environ Microbiol 79(6):1897–1905

Egeler P, Römbke J, Meller M, Knacker T, Franke C, Studinger G, Nagel R (1997) Bioaccumulation of lindane and hexachlorobenzene by tubificid sludgeworms (Oligochaeta) under standardised laboratory conditions. Chemosphere 35(4):835–852

Ellgehausen H, Guth JA, Esser HO (1980) Factors determining the bioaccumulation potential of pesticides in the individual compartments of aquatic food chains. Ecotoxicol Environ Saf 4(2): 134–157

Embrandiri A, Kiyasudeen SK, Rupani PF, Ibrahim MH (2016) Environmental xenobiotics and its effects on natural ecosystem. In: Plant responses to xenobiotics. Springer, Singapore, pp 1–18

Ernst W (1979) Factors affecting the evaluation of chemicals in laboratory experiments using marine organisms. Ecotox Environ Saf 3:90–98

Escartín E, Porte C (1996) Bioaccumulation, metabolism, and biochemical effects of the organophosphorus pesticide fenitrothion in Procambarus clarkii. Environ Toxicol Chem 15(6): 915–920

Essumang DK (2013) Environmental xenobiotics: PAHs in soil (heavy metals), indoor air and water environment, case studies of Ghana and Denmark

Essumang DK, Ankrah DA (2010) Distribution, levels, and risk assessment of polycyclic aromatic hydrocarbons (PAHs) in some water bodies along the coastal belt of Ghana. Sci World J TSW Environ 10(972):985

Essumang DK, Adokoh C, Afriyie J, Mensah E (2009) Source assessment and analysis of polycyclic aromatic hydrocarbon (PAH's) in the Oblogo waste disposal sites and some water bodies in and around the Accra Metropolis of Ghana

Eswaramoorthy S, Bonanno JB, Burley SK, Swaminathan S (2006) Mechanism of action of a flavin-containing monooxygenase. Proc Natl Acad Sci U S A 103(26):9832–9837. https://doi.org/10.1073/pnas.0602398103

Fahl GM, Kreft L, Altenburger R, Faust M, Boedeker W, Grimme LH (1995) pH-dependent sorption, bioconcentration and algal toxicity of sulfonylurea herbicides. Aquat Toxicol 31(2): 175–187

FAO (1990). http://www.fao.org/3/w2598e/w2598e07.htm

Fatta-Kassinos D, Kalavrouziotis IK, Koukoulakis PH, Vasquez MI (2011) The risks associated with wastewater reuse and xenobiotics in the agroecological environment. Sci Total Environ 409(19):3555–3563

Fent K, Weston AA, Caminada D (2006) Ecotoxicology of human pharmaceuticals. Aquat Toxicol 76(2):122–159

Ferrando MD, Sancho E, Andreu-Moliner E (1996) Accumulation of tetradifon in an algae (Nannochloris oculata) and the cladoceran, Daphnia magna. Bull Environ Contam Toxicol 57(1):139–145

Fetzner S (2002) Biodegradation of xenobiotics. Biotechnology 10:215–246

Feurtet-Mazel A, Grollier T, Grouselle M, Ribeyre F, Boudou A (1996) Experimental study of bioaccumulation and effects of the herbicide isoproturon on freshwater rooted macrophytes (Elodea densa and Ludwigia natans). Chemosphere 32(8):1499–1512

Fick J, Lindberg RH, Tysklind M, Larsson DJ (2010) Predicted critical environmental concentrations for 500 pharmaceuticals. Regul Toxicol Pharmacol 58(3):516–523

Fisher SW, Hwang H, Atanasoff M, Landrum PF (1999) Lethal body residues for pentachlorophenol in zebra mussels (Dreissena polymorpha) under varying conditions of temperature and pH. Ecotoxicol Environ Saf 43(3):274–283

Gani KM, Ali M, Dubey M, Kazmi AA, Kumari S, Bux F (2021) Transport of emerging contaminants from agricultural soil to groundwater. In: Sustainable agriculture reviews, vol 50. Springer, Cham, pp 261–281

Gao J, Garrison AW, Hoehamer C, Mazur CS, Wolfe NL (2000) Uptake and phytotransformation of organophosphorus pesticides by axenically cultivated aquatic plants. J Agric Food Chem 48(12):6114–6120

Garrison AW, Pope JD, Allen FR (1976) GC-MS analysis of organic compounds in domestic wastewater. In: Keith LH (ed) Identification and analysis of organic pollutants in water. Ann Arbor Science, Michigan, pp 517–556

Geyer H, Sheehan P, Kotzias D, Freitag D, Korte F (1982) Prediction of ecotoxicological behaviour of chemicals: relationship between physico-chemical properties and bioaccumulation of organic chemicals in the mussel Mytilus edulis. Chemosphere 11(11):1121–1134
Gllderhus PA, Johnson BGH (1980) Effects of sea lamprey (Petromyzon marinus) control in the Great Lakes on aquatic plants, invertebrates, and amphibians. Can J Fish Aquat Sci 37(11): 1895–1905
Glooschenko V, Holdrinet M, Lott JN, Frank R (1979) Bioconcentration of chlordane by the green alga Scenedesmus quadricauda. Bull Environ Contam Toxicol 21(1):515–520
Gobas FA, McNeil EJ, Lovett-Doust L, Haffner GD (1991) Bioconcentration of chlorinated aromatic hydrocarbons in aquatic macrophytes. Environ Sci Technol 25(5):924–929
Gomez CF, Constantine L, Huggett DB (2010) The influence of gill and liver metabolism on the predicted bioconcentration of three pharmaceuticals in fish. Chemosphere 81(10):1189–1195
Govers H, Ruepert C, Aiking H (1984) Quantitative structure-activity relationships for polycyclic aromatic hydrocarbons: correlation between molecular connectivity, physico-chemical properties, bioconcentration and toxicity in Daphnia pulex. Chemosphere 13(2):227–236
Grech A, Quignot N, Brochot C, Dorne J, Bois FY, Beaudouin R (2016, May) Development and application of generic toxicokinetic models in fish to environmental risk assessment of chemicals. In 26. SETAC Europe annual meeting. SETAC
Gregory WW Jr, Reed JK, Priester LE Jr (1969) Accumulation of parathion and DDT by some algae and protozoa. J Protozool 16(1):69–71
Gross B, Montgomery-Brown J, Naumann A, Reinhard M (2004) Occurrence and fate of pharmaceuticals and alkylphenol ethoxylate metabolites in an effluent-dominated river and wetland. Environ Toxicol Chem 23(9):2074–2083
Guanzon NG Jr, Fukuda M, Nakahara H (1996) Accumulation of agricultural pesticides by three freshwater microalgae. Fisheries Sci 62(5):690–697
Gunnarsson L, Kristiansson E, Rutgersson C, Sturve J, Fick J, Förlin L, Larsson DJ (2009) Pharmaceutical industry effluent diluted 1: 500 affects global gene expression, cytochrome P450 1A activity, and plasma phosphate in fish. Environ Toxicol Chem 28(12):2639–2647
Halling-Sørensen B, Nyholm N, Kusk KO, Jacobsson E (2000) Influence of nitrogen status on the bioconcentration of hydrophobic organic compounds to Selenastrum capricornutum. Ecotoxicol Environ Saf 45(1):33–42
Hamer MJ, Goggin UM, Muller K, Maund SJ (1999) Bioavailability of lambda-cyhalothrin to Chironomus riparius in sediment-water and water-only systems. Aquat Ecosyst Health Manage 2(4):403–412
Hansen PD (1979) Experiments on the accumulation of lindane (γ-BHC) by the primary producers Chlorella spec. and Chlorella pyrenoidosa. Arch Environ Contam Toxicol 8(6):721–731
Haque R, Kearney PC, Freed VH (1977) Dynamics of pesticides in aquatic environments. In: Pesticides in aquatic environments. Springer, Boston, MA, pp 39–52
Harkey GA, Klaine SJ (1992) Bioconcentration of trans-chlordane by the midge, Chironomus decorus. Chemosphere 24(12):1911–1919
Hartley DM, Johnston JB (1983) Use of the freshwater clam Corbicula manilensis as a monitor for organochlorine pesticides. Bull Environ Contam Toxicol 31(1):33–40
Hassaan MA, El Nemr A (2020) Pesticides pollution: Classifications, human health impact, extraction and treatment techniques. Egypt J Aquat Res 46(3):207–220
Hawker DW, Connell DW (1986) Bioconcentration of lipophilic compounds by some aquatic organisms. Ecotoxicol Environ Saf 11(2):184–197
Heberer T (2002) Occurrence, fate, and removal of pharmaceutical residues in the aquatic environment: a review of recent research data. Toxicol Lett 131(1–2):5–17
Heisig-Gunkel G, Gunkel G (1982) Distribution of a herbicide (atrazine, s-triazine) in Daphnia pulicaria: a new approach to determination. Arch Hydrobiol 59(4):359–376
Helfrich LA, Weigmann DL, Stinson ER (2009) Pesticides and aquatic animals: a guide to reducing impacts on aquatic systems. Communications and Marketing, College of Agriculture and Life Sciences, Virginia Polytechnic Institute and State University, Publication, Blacksburg, VA, p 24

Hendriks AJ, van der Linde A, Cornelissen G, Sijm DT (2001) The power of size. 1. Rate constants and equilibrium ratios for accumulation of organic substances related to octanol-water partition ratio and species weight. Environ Toxicol Chem 20(7):1399–1420

Hernández AF, Parrón T, Tsatsakis AM, Requena M, Alarcón R, López-Guarnido O (2013) Toxic effects of pesticide mixtures at a molecular level: their relevance to human health. Toxicology 307:136–145

Hinman ML, Klaine SJ (1992) Uptake and translocation of selected organic pesticides by the rooted aquatic plant Hydrilla verticillata Royle. Environ Sci Technol 26(3):609–613

Hoffman LR, D'Argenio DA, MacCoss MJ, Zhang Z, Jones RA, Miller SI (2005) Aminoglycoside antibiotics induce bacterial biofilm formation. Nature 436(7054):1171–1175

Holland W, Morrison T, Chang Y, Wiernsperger N, Stith BJ (2004) Metformin (Glucophage) inhibits tyrosine phosphatase activity to stimulate the insulin receptor tyrosine kinase. Biochem Pharmacol 67(11):2081–2091

Hooper MJ, Ankley GT, Cristol DA, Maryoung LA, Noyes PD, Pinkerton KE (2013) Interactions between chemical and climate stressors: A role for mechanistic toxicology in assessing climate change risks. Environ Toxicol Chem 32(1):32–48

Hou R, Liu C, Gao X, Xu Y, Zha J, Wang Z (2017) Accumulation and distribution of organophosphate flame retardants (PFRs) and their di-alkyl phosphates (DAPs) metabolites in different freshwater fish from locations around Beijing, China. Environ Pollut 229:548–556

Ibrahim MS (2007) Persistent organic pollutants in Malaysia. Dev Environ Sci 7:629–655. https://doi.org/10.1016/S1474-8177(07)07014-3

Islam MA, Amin SMN, Brown CL, Juraimi AS, Uddin MK, Arshad A (2021) Determination of median lethal concentration (LC50) for Endosulfan, Heptachlor and Dieldrin Pesticides to African Catfish, Clarias gariepinus and their impact on its behavioral patterns and histopathological responses. Toxics 9(12):340. https://doi.org/10.3390/toxics9120340

Jain PK, Gupta VK, Gaur RK, Lowry M, Jaroli DP, Chauhan UK (2011) Bioremediation of petroleum oil contaminated soil and water. Res J Environ Toxicol 5(1):1

Jing TX, Tan Y, Ding BY, Dou W, Wei DD, Wang JJ (2018) NADPH-Cytochrome P450 Reductase mediates the resistance of Aphis (Toxoptera) citricidus (Kirkaldy) to Abamectin. Front Physiol 9:986

Johnson BT, Saunders CR, Sanders HO, Campbell RS (1971) Biological magnification and degradation of DDT and aldrin by freshwater invertebrates. J Fisheries Board Canada 28(5): 705–709

Johnson CH, Patterson AD, Idle JR, Gonzalez FJ (2012) Xenobiotic metabolomics: major impact on the metabolome. Annu Rev Pharmacol Toxicol 52:37–56. https://doi.org/10.1146/annurev-pharmtox-010611-134748

Jonsson CM, Paraiba LC, Mendoza MT, Sabater C, Carrasco JM (2001) Bioconcentration of the insecticide pyridaphenthion by the green algae Chlorella saccharophila. Chemosphere 43(3): 321–325

Julin AM, Sanders HO (1977) Toxicity and accumulation of the insecticide imidan in freshwater invertebrates and fishes. Trans Am Fish Soc 106(4):386–392

Kanazawa J (1978) Bioconcentration ratio of diazinon by freshwater fish and snail. Bull Environ Contam Toxicol 20(1):613–617

Karatan E, Watnick P (2009) Signals, regulatory networks, and materials that build and break bacterial biofilms. Microbiol Mol Biol Rev 73(2):310–347

Kashiwada S, Mochida K, OzoE Y, Nakamura T (1995) Contribution of Zooplankton to disappearance of organophosphorus insecticides in environmental water. J Pestic Sci 20:503–512

Katagi T (2010) Bioconcentration, bioaccumulation, and metabolism of pesticides in aquatic organisms. In: Whitacre D (ed) Reviews of environmental contamination and toxicology, vol 204. Springer, New York, NY. https://doi.org/10.1007/978-1-4419-1440-8_1

Kawatski JA, Schmulbach JC (1972) Uptake and elimination of 14C-aldrin and 14C-dieldrin by the ostracod Chlamydotheca arcuata (Sars). Int J Environ Anal Chem 1(4):283–291

Keil JE, Priester LE (1969) DDT uptake and metabolism by a marine diatom. Bull Environ Contam Toxicol 4(3):169–173

Kent RA, Currie D (1995) Predicting algal sensitivity to a pesticide stress. Environ Toxicol Chem 14(6):983–991

Khan HM, Neudorf S, Khan MAQ (1975) Absorption and elimination of photodieldrin by Daphnia and goldfish. Bull Environ Contam Toxicol 13(5):582–587

Kikuchi R, Yasutaniya T, Takimoto Y, Yamada H, Miyamoto J (1984) Accumulation and metabolism of fenitrothion in three species of algae [concerning ecotoxicology]. J Pestic Sci 9:331

Klosterhaus SL, DiPinto LM, Chandler GT (2003) A comparative assessment of azinphosmethyl bioaccumulation and toxicity in two estuarine meiobenthic harpacticoid copepods. Environ Toxicol Chem 22(12):2960–2968

Knezovich JP, Harrison FL (1988) The bioavailability of sediment-sorbed chlorobenzenes to larvae of the midge, Chironomus decorus. Ecotoxicol Environ Saf 15(2):226–241

Kobayashi K, Nakamura Y, Imada N (1985) Metabolism of an organophosphorus insecticide, fenitrothion, in tiger shrimp Penaeus japonicus. Nippon Suisan Gakkaishi 51(4):599–603

Kobayashi K, Rompas RM, Oshima Y, Imada N (1990) A comparative study on the toxicity, absorption and depuration of fenitrothion and its oxon in Japanese tiger shrimp. Nippon Suisan Gakkaishi 56(6):923–928

Koelmans AA, Jiménez CS (1994) Temperature dependency of chlorobenzene bioaccumulation in phytoplankton. Chemosphere 28(12):2041–2048

Kolpin DW, Furlong ET, Meyer MT, Thurman EM, Zaugg SD, Barber LB, Buxton HT (2002) Pharmaceuticals, hormones, and other organic wastewater contaminants in US streams, 1999–2000: a national reconnaissance. Environ Sci Technol 36(6):1202–1211

Kukkonen J, Oikari A (1988) Sulphate conjugation is the main route of pentachlorophenol metabolism in Daphnia magna. Comp Biochem Physiol C Comp Pharmacol 91(2):465–468

Kumar S, Lal R, Bhatnagar P (1988) Uptake of dieldrin, dimethoate and permethrin by cyanobacteria, Anabaena sp. and Aulosira fertilissima. Environ Pollut 54(1):55–61

Kumar S, Sharma AK, Rawat SS, Jain DK, Ghosh S (2013) Use of pesticides in agriculture and livestock animals and its impact on environment of India. Asian J Environ Sci 8(1):51–57

La Farre M, Pérez S, Kantiani L, Barceló D (2008) Fate and toxicity of emerging pollutants, their metabolites and transformation products in the aquatic environment. TrAC Trends Anal Chem 27(11):991–1007

Lakhani L (2015) How to reduce impact of pesticides in aquatic environment. Soc Iss Environ Probl 3(9):29–38

Lal S, Lal R, Saxena DM (1987) Bioconcentration and metabolism of DDT, fenitrothion and chlorpyrifos by the blue-green algae Anabaena sp. and Aulosira fertilissima. Environ Pollut 46(3):187–196

Landrum PF, Dupuis WS (1990) Toxicity and toxicokinetics of pentachlorophenol and carbaryl to Pontoporeia hoyi and Mysis relicta. ASTM Spec Tech Publ 1096:278–289

Landrum PF, Fisher SW (1999) Influence of lipids on the bioaccumulation and trophic transfer of organic contaminants in aquatic organisms. In: Lipids in freshwater ecosystems. Springer, New York, NY, pp 203–234

Landrum PF, Scavia D (1983) Influence of sediment on anthracene uptake, depuration, and biotransformation by the amphipod Hyalella azteca. Can J Fish Aquat Sci 40(3):298–305

Laxmi B, Madhavi K, Adnan A, Chamundeswari Devi B, Dhanapal K, Ramana TV (2019) Sub lethal effects of dichlorvos on physiological parameters in fingerlings of *Cyprinus carpio*. Int J Curr Microbiol App Sci 8(8):372–377

Laxmi B, Madhavi K, Dhanapal K, Adnan A, Sudhakar O, Jesintha N (2020) Biochemical responses of fingerling *Cyprinus carpio* (Linnaeus, 1758) exposed to sub-lethal concentrations of Dichlorvos. IntJ Curr Microbiol App Sci 9(10):3250–3257

Le Page Y, Vosges M, Servili A, Brion F, Kah O (2011) Neuroendocrine effects of endocrine disruptors in teleost fish. J Toxicol Environ Health Pt B 14(5-7):370–386

Lee YM, Kim KS, Jacobs DR Jr, Lee DH (2017) Persistent organic pollutants in adipose tissue should be considered in obesity research. Obes Rev 18(2):129–139

Legierse KC, Sijm DT, van Leeuwen CJ, Seinen W, Hermens JL (1998) Bioconcentration kinetics of chlorobenzenes and the organophosphorus pesticide chlorthion in the pond snail Lymnaea stagnalis—a comparison with the guppy Poecilia reticulata. Aquat Toxicol 41(4):301–323

Lengai GM, Muthomi JW, Mbega ER (2020) Phytochemical activity and role of botanical pesticides in pest management for sustainable agricultural crop production. Sci Afr 7:e00239

Leppänen MT, Kukkonen JV (2006) Evaluating the role of desorption in bioavailability of sediment-associated contaminants using oligochaetes, semipermeable membrane devices and Tenax extraction. Environ Pollut 140(1):150–163

Leppänen MT, Landrum PF, Kukkonen JV, Greenberg MS, Burton GA Jr, Robinson SD, Gossiaux DC (2003) Investigating the role of desorption on the bioavailability of sediment-associated 3, 4, 3′, 4′-tetrachlorobiphenyl in benthic invertebrates. Environ Toxicol Chem Int J 22(12):2861–2871

Liess M, Foit K (2010) Intraspecific competition delays recovery of population structure. Aquat Toxicol 97(1):15–22

Lionetto MG, Caricato R, Calisi A, Schettino T (2011) Acetylcholinesterase inhibition as a relevant biomarker in environmental biomonitoring: new insights and perspectives. In: Ecotoxicology around the globe, pp 87–115

Livingston RJ, de La Cruz AA (1977) Review of current literature concerning the acute and chronic effects of pesticides on aquatic organisms. Crit Rev Environ Sci Technol 7(4):325–351

Livingstone DR (1998) The fate of organic xenobiotics in aquatic ecosystems: quantitative and qualitative differences in biotransformation by invertebrates and fish. Comp Biochem Physiol A 120:43–49

Lockhart WL, Billeck BN, de March BG, Muir DC (1983) Uptake and toxicity of organic compounds: studies with an aquatic macrophyte (Lemna minor). In: Aquatic Toxicology and Hazard Assessment: Sixth Symposium. ASTM International, West Conshohocken, PA

Lockhart WL, Metner DA, Billeck BN, Rawn GP, Muir DCG (1984) Bioaccumulation of some forestry pesticides in fish and aquatic plants. In: Garner WY, Harvey J Jr (eds) Chemical and biological controls in forestry. ACS symposium series 238. American Chemical Society, Washington, DC, pp 297–315

Lotufo GR, Landrum PF, Gedeon ML, Tigue EA, Herche LR (2000) Comparative toxicity and toxicokinetics of DDT and its major metabolites in freshwater amphipods. Environ Toxicol Chem 19(2):368–379

Lotufo GR, Farrar JD, Duke BM, Bridges TS (2001a) DDT toxicity and critical body residue in the amphipod Leptocheirus plumulosus in exposures to spiked sediment. Arch Environ Contam Toxicol 41(2):142–150

Lotufo GR, Landrum PF, Gedeon ML (2001b) Toxicity and bioaccumulation of DDT in freshwater amphipods in exposures to spiked sediments. Environ Toxicol Chem 20(4):810–825

Lydy MJ, Hayton WL, Staubus AE, Fisher SW (1994) Bioconcentration of 5, 5′, 6-trichlorobiphenyl and pentachlorophenol in the midge, Chironomus riparius, as measured by a pharmacokinetic model. Arch Environ Contam Toxicol 26(2):251–256

Mackay D (1982) Correlation of bioconcentration factors. Environ Sci Technol 16(5):274–278

Mäenpää K (2007) The toxicity of xenobiotics in an aquatic environment: connecting body residues with adverse effects. Joensuun yliopisto

Mäenpää KA, Sormunen AJ, Kukkonen JV (2003) Bioaccumulation and toxicity of sediment associated herbicides (ioxynil, pendimethalin, and bentazone) in Lumbriculus variegatus (Oligochaeta) and Chironomus riparius (Insecta). Ecotoxicol Environ Saf 56(3):398–410

Mailhot H (1987) Prediction of algal bioaccumulation and uptake rate of nine organic compounds by ten physicochemical properties. Environ Sci Technol 21(10):1009–1013

Mäkelä P, Oikari AO (1990) Uptake and body distribution of chlorinated phenolics in the freshwater mussel, Anodonta anatina L. Ecotoxicol Environ Saf 20(3):354–362

Manthey M, Faust M, Smolka S, Grimme LH (1993) Herbicide bioconcentration in algae: Studies on lipophilicity-sorption-activity relationships (LSAR) with Chlorella fusca. Sci Total Environ 134:453–459

Marquis LY, Comes RD, Yang CP (1981) Absorption and translocation of fluridone and glyphosate in submersed vascular plants. Weed Sci 29(2):229–236

Mason JW, Rowe DR (1976) The accumulation and loss of dieldrin and endrin in the eastern oyster. Arch Environ Contam Toxicol 4(1):349–360

Mathur R, Saxena DM (1986) Effect of hexachlorocyclohexane (HCH) isomers on growth of, and their accumulation in, the blue-green alga, Anabaena Sp.(ARM 310). J Environ Biol 7:239

Matsumura F (1977) Absorption, accumulation, and elimination of pesticides by aquatic organisms. In: Pesticides in aquatic environments. Springer, Boston, MA, pp 77–105

Maund SJ, Hamer MJ, Lane MC, Farrelly E, Rapley JH, Goggin UM, Gentle WE (2002) Partitioning, bioavailability, and toxicity of the pyrethroid insecticide cypermethrin in sediments. Environ Toxicol Chem 21(1):9–15

Maurya PK, Malik DS, Sharma A (2019) Impacts of pesticide application on aquatic environments and fish diversity. In: Kumar V, Kumar R, Singh J, Kumar P (eds) Contaminants in agriculture and environment: health risks and remediation, vol 1. Agro Environ Media, Haridwar, pp 111–128. https://doi.org/10.26832/AESA-2019-CAE-0162-09

McLeese DW, Zitko V, Sergeant DB (1979) Uptake and excretion of fenitrothion by clams and mussels. Bull Environ Contam Toxicol 22(1):800–806

Meech R, Hu DG, McKinnon RA, Mubarokah SN, Haines AZ, Nair PC, Rowland A, Mackenzie PI (2019) The UDP-glycosyltransferase (UGT) superfamily: new members, new functions, and novel paradigms. Physiol Rev 99(2):1153–1222. https://doi.org/10.1152/physrev.00058.2017

Mehrotra S, Kumar S, Zahid M, Garg M (2017) Biopesticides. In: Principles and applications of environmental biotechnology for a sustainable future. Springer, Singapore, pp 273–292

Miyamoto J, Mikami N, Takimoto Y (1990) The fate of pesticides in aquatic ecosystems. Prog Pest Biochem Toxicol 7:123–147

Mohapatra S, Kumar R, Sundaray JK, Patnaik ST, Mishra CSK, Rather MA (2021) Structural damage in liver, gonads, and reduction in spawning performance and alteration in the haematological parameter of Anabas testudineus by glyphosate-a herbicide. Aquacult Res *52*(3):1150–1159

Montañés JC, Van Hattum B, Deneer J (1995) Bioconcentration of chlorpyrifos by the freshwater isopod Asellus aquaticus (L.) in outdoor experimental ditches. Environ Pollut 88(2):137–146

Moore R, Toro E, Stanton M, Khan MAQ (1977) Absorption and elimination of 14C-alpha- and gamma-chlordane by a freshwater alga, daphnid, and goldfish. Arch Environ Contam Toxicol 6(1):411–420

Morrison HA, Gobas FA, Lazar R, Haffner GD (1996) Development and verification of a bioaccumulation model for organic contaminants in benthic invertebrates. Environ Sci Technol 30(11):3377–3384

Muhammad G, Rashid I, Firyal S (2017) Practical aspects of treatment of organophosphate and carbamate insecticide poisoning in animals. Matrix Sci Pharma 1(1):10–11

Muir DCG, Grift NP, Townsend BE, Metner DA, Lockhart WL (1982) Comparison of the uptake and bioconcentration of fluridone and terbutryn by rainbow trout and Chironomus tentans in sediment and water systems. Arch Environ Contam Toxicol 11(5):595–602

Naqvi SM, Newton DJ (1990) Bioaccumulation of endosulfan (Thiodan R insecticide) in the tissues of Louisiana crayfish, Procambarus clarkii. J Environ Sci Health Pt B 25(4):511–526

Narayana Rao VS, Lal R (1987) Uptake and metabolism of insecticides by blue-green algae Anabaena and Aulosira fertilissima. Microbios Lett 36(143-144):143–147

Nawaz S, Kirk K (1996) Temperature effects on bioconcentration of DDE by Daphnia. Freshw Biol 35(1):173–178

Nebeker AV, Griffis WL, Wise CM, Hopkins E, Barbitta JA (1989) Survival, reproduction and bioconcentration in invertebrates and fish exposed to hexachlorobenzene. Environ Toxicol Chem 8(7):601–611

Neely WB, Branson DR, Blau GE (1974) Partition coefficient to measure bioconcentration potential of organic chemicals in fish. Environ Sci Technol 8(13):1113–1115

Nemcsok J, Benedeczky I (1995) Pesticide metabolism and the adverse effects of metabolites on fishes. In: Hochachka PW, Mommsen TP (eds) Biochemistry and molecular biology of fishes, vol 59. Elsevier Science, Amsterdam, pp 313–348

Neudorf S, Khan MAQ (1975) Pick-up and metabolism of DDT, dieldrin and photodieldrin by a fresh water alga (Ankistrodesmus amalloides) and a Microcrustacean(Daphnia pulex). Bull Environ Contam Toxicol 13(4):443–450

Nikkilä A, Paulsson M, Almgren K, Blanck H, Kukkonen JV (2001) Atrazine uptake, elimination, and bioconcentration by periphyton communities and Daphnia magna: effects of dissolved organic carbon. Environ Toxicol Chem 20(5):1003–1011

Nikkilä A, Halme A, Kukkonen JV (2003) Toxicokinetics, toxicity and lethal body residues of two chlorophenols in the oligochaete worm, Lumbriculus variegatus, in different sediments. Chemosphere 51(1):35–46

Nuutinen S, Landrum PF, Schuler LJ, Kukkonen JV, Lydy MJ (2003) Toxicokinetics of organic contaminants in Hyalella azteca. Arch Environ Contam Toxicol 44(4):0467–0475

Oliver BG (1987) Biouptake of chlorinated hydrocarbons from laboratory-spiked and field sediments by oligochaete worms. Environ Sci Technol 21(8):785–790

Paris DF, Lewis DL (1976) Accumulation of methoxychlor by microorganisms isolated from aqueous systems. Bull Environ Contam Toxicol 15(1):24–32

Petrocelli SR, Hanks AR, Anderson J (1973) Uptake and accumulation of an organochlorine insecticide (dieldrin) by an estuarine mollusc, Rangia cuneata. Bull Environ Contam Toxicol 10:315

Pillai MK, Mittal PK, Agarwal HC (1980) Bioaccumulation, metabolism & elimination of DDT by the fresh water clam Indonaia caerulea (Lea). Indian J Exp Biol 18(12):1439–1442

Rajendran N, Venugopalan VK (1991) Bioconcentration of endosulfan in different body tissues of estuarine organisms under sublethal exposure. Bull Environ Contam Toxicol 46(1):151

Rápó E, Tonk S (2021) Factors affecting synthetic dye adsorption; desorption studies: a review of results from the last five years (2017–2021). Molecules 26(17):5419

Reineke W, Knackmuss HJ (1988) Microbial degradation of haloaromatics. Annu Rev Microbiol 42(1):263–287

Reinert RE (1972) Accumulation of dieldrin in an alga (Scenedesmus obliquus), Daphnia magna, and the guppy (Poecilia reticulata). J Fisheries Board Canada 29(10):1413–1418

Renberg L, Tarkpea M, Lindén E (1985) The use of the bivalve Mytilus edulis as a test organism for bioconcentration studies: I. Designing a continuous-flow system and its application to some organochlorine compounds. Ecotoxicol Environ Saf 9(2):171–178

Richardson ML, Bowron JM (1985) The fate of pharmaceutical chemicals in the aquatic environment. J Pharm Pharmacol 37(1):1–12

Richter S, Nagel R (2007) Bioconcentration, biomagnification and metabolism of 14C-terbutryn and 14C-benzo [a] pyrene in Gammarus fossarum and Asellus aquaticus. Chemosphere 66(4): 603–610

Rosi-Marshall E (2013) Streams stressed by pharmaceutical pollution. https://environmentalchange.nd.edu/news-events/

Roy S, Hänninen O (1994) Pentachlorophenol: uptake/elimination kinetics and metabolism in an aquatic plant, Eichhornia crassipes. Environ Toxicol Chem 13(5):763–773

Sabaliūnas D, Lazutka J, Sabaliūniene I, Södergren A (1998) Use of semipermeable membrane devices for studying effects of organic pollutants: comparison of pesticide uptake by semipermeable membrane devices and mussels. Environ Toxicol Chem 17(9):1815–1824

Sacher F, Lange FT, Brauch HJ, Blankenhorn I (2001) Pharmaceuticals in groundwaters: analytical methods and results of a monitoring program in Baden-Württemberg, Germany. J Chromatogr A 938(1-2):199–210

Sathe MC, Shrotri RV, Raghu K, Murthy NBK (2005) Observations on accumulation and depuration of fenthion in various tissues of marine edible clam, Marcia hiantina (Lamarck). J Ecophysiol Occup Health 5(1):33–36

Saxena DM, Lal R, Reddy BVP (1982) DDT uptake and metabolism in Blepharisma intermedium. Acta Protozool 21(2):173

Segner H (2015) In vitro methodologies in ecotoxicological hazard assessment: the case of bioaccumulation testing for fish. Altern Lab Anim 43(2):P14–P16

Schäfer RB, Caquet T, Siimes K, Mueller R, Lagadic L, Liess M (2007) Effects of pesticides on community structure and ecosystem functions in agricultural streams of three biogeographical regions in Europe. Sci Total Environ 382(2-3):272–285

Schauberger CW, Wildman RB (1977) Accumulation of aldrin and dieldrin by blue-green algae and related effects on photosynthetic pigments. Bull Environ Contam Toxicol 17(5):534–541

Schenkman JB (1999) The fate of xenobiotics in the body. In: Arinç E, Schenkman JB, Hodgson E (eds) Molecular and applied aspects of oxidative drug metabolizing enzymes. NATO ASI series, vol 303. Springer, Boston, MA. https://doi.org/10.1007/978-1-4615-4855-3_1

Schimmel SC, Patrick JM, Forester J (1977) Uptake and toxicity of toxaphene in several estuarine organisms. Arch Environ Contam Toxicol 5(1):353–367

Schimmel SC, Garnas RL, Patrick JM Jr, Moore JC (1983) Acute toxicity, bioconcentration, and persistence of AC 222,705, benthiocarb, chlorpyrifos, fenvalerate, methyl parathion, and permethrin in the estuarine environment. J Agric Food Chem 31(1):104–113

Schmitt H, Haapakangas H, van Beelen P (2005) Effects of antibiotics on soil microorganisms: time and nutrients influence pollution-induced community tolerance. Soil Biol Biochem 37(10): 1882–1892

Schuler LJ, Wheeler M, Bailer AJ, Lydy MJ (2003) Toxicokinetics of sediment-sorbed benzo [a] pyrene and hexachlorobiphenyl using the freshwater invertebrates Hyalella azteca, Chironomus tentans, and Lumbriculus variegatus. Environ Toxicol Chem 22(2):439–449

Schuytema GS, Krawczyk DF, Grlffis WL, Nebeker AV, Robideaux ML, Brownawell BJ, Westall JC (1988) Comparative uptake of hexachlorobenzene by fathead minnows, amphipods and oligochaete worms from water and sediment. Environ Toxicol Chem 7(12):1035–1045

Serrano R, Hernandez F, Pena JB, Dosda V, Canales J (1995) Toxicity and bioconcentration of selected organophosphorus pesticides in Mytilus galloprovincialis and Venus gallina. Arch Environ Contam Toxicol 29(3):284–290

Serrano R, Hernández F, López FJ, Pena JB (1997a) Bioconcentration and depuration of chlorpyrifos in the marine mollusc Mytilus edulis. Arch Environ Contam Toxicol 33(1):47–52

Serrano R, Lopez FJ, Hernandez F, Pena JB (1997b) Bioconcentration of chlorpyrifos, chlorfenvinphos, and methidathion in Mytilus galloprovincialis. Bull Environ Contam Toxicol 59(6):968–975

Shanmugam G, Sampath S, Selvaraj KK, Larsson DG, Ramaswamy BR (2014) Non-steroidal anti-inflammatory drugs in Indian rivers. Environ Sci Pollut Res 21(2):921–931

Sharma A, Kumar V, Shahzad B, Tanveer M, Sidhu GPS, Handa N, Thukral AK et al (2019) Worldwide pesticide usage and its impacts on ecosystem. SN Applied Sciences 1(11):1–16

Shaw GR, Connell DW (1987) Comparative kinetics for bioaccumulation of polychlorinated biphenyls by the polychaete (Capitella capitata) and fish (Mugil cephalus). Ecotoxicol Environ Saf 13(1):84–91

Shofer SL, Tjeerdema RS (1993) Comparative disposition and biotransformation of pentachlorophenol in the oyster (Crassostrea gigas) and abalone (Haliotis fulgens). Pestic Biochem Physiol 46(2):85–95

Sikandar A, Shehzadi K, Arshad Q, Munir K (2013) Phytoremediation: an analytical technique for the assessment of biodegradation of organic xenobiotic pollutants: a review. Int J Sci Res 4(2): 2250–2253

Singh T, Singh DK (2017) Phytoremediation of organochlorine pesticides: Concept, method, and recent developments. Int J Phytoremediation 19(9):834–843

Singh Z, Kaur J, Kaur R, Hundal SS (2016) Toxic effects of organochlorine pesticides: a review. Am J Biol Sci 4(3):11

Sjerps RM, Kooij PJ, van Loon A, Van Wezel AP (2019) Occurrence of pesticides in Dutch drinking water sources. Chemosphere 235:510–518

Södergren A, Svensson B (1973) Uptake and accumulation of DDT and PCB by Ephemera danica (Ephemeroptera) in continuous-flow systems. Bull Environ Contam Toxicol 9:345–350

Spacie A, Hamelink JL (1982) Alternative models for describing the bioconcentration of organics in fish. Environ Toxicol Chem 1(4):309–320

Spehar RL, Tanner DK, Nordling BR (1983) Toxicity of the synthetic pyrethroids, permethrin and AC 222, 705 and their accumulation in early life stages of fathead minnows and snails. Aquat Toxicol 3(2):171–182

Suja S, Williams SE (2018) Determination of lethal concentration (LC 50) of Channa striata. Int J Sci Res 7(8):603–605

Sumon KA, Rashid H, Peeters ET, Bosma RH, Van den Brink PJ (2018) Environmental monitoring and risk assessment of organophosphate pesticides in aquatic ecosystems of north-west Bangladesh. Chemosphere 206:92–100

Szewczuk LM, Forti L, Stivala LA, Penning TM (2004) Resveratrol is a peroxidase-mediated inactivator of COX-1 but not COX-2: a mechanistic approach to the design of COX-1 selective agents. J Biol Chem 279(21):22727–22737

Takimoto Y, Ohshima M, Miyamoto J (1987a) Comparative metabolism of fenitrothion in aquatic organisms: II. Metabolism in the freshwater snails, Cipangopaludina japonica and Physa acuta. Ecotoxicol Environ Saf 13(1):118–125

Takimoto Y, Ohshima M, Miyamoto J (1987b) Comparative metabolism of fenitrothion in aquatic organisms: III. Metabolism in the crustaceans, Daphnia pulex and Palaemon paucidens. Ecotoxicol Environ Saf 13(1):126–134

Tang JX, Siegfried BD (1996) Bioconcentration and uptake of a pyrethroid and organophosphate insecticide by selected aquatic insects. Bull Environ Contam Toxicol 57(6):993–998

Tang J, Hoagland KD, Siegfried BD (1998) Uptake and bioconcentration of atrazine by selected freshwater algae. Environ Toxicol Chem 17(6):1085–1090

Ternes TA (1998) Occurrence of drugs in German sewage treatment plants and rivers. Water Res 32(11):3245–3260

Thomann RV, Komlos J (1999) Model of biota-sediment accumulation factor for polycyclic aromatic hydrocarbons. Environ Toxicol Chem 18(5):1060–1068

Thompson DG, Holmes SB, Wainio-Keizer K, MacDonald L, Solomon KR (1993) Impact of hexazinone and metsulfuron methyl on the zooplankton community of a boreal forest lake. Environ Toxicol Chem 12(9):1709–1717

Thybaud E, Caquet T (1991) Uptake and elimination of lindane by Lymnaea palustris (Mollusca: Gastropoda): a pharmacokinetic approach. Ecotoxicol Environ Saf 21(3):365–376

Thybaud E, Le Bras S (1988) Absorption and elimination of lindane by Asellus aquaticus (Crustacea, Isopoda). Bull Environ Contam Toxicol 40(5):731–735

Tjeerdema RS, Crosby DG (1992) Disposition and biotransformation of pentachlorophenol in the red abalone (Haliotis rufescens). Xenobiotica 22(6):681–690

Tooker JF, Pearsons KA (2021) Newer characters, same story: neonicotinoid insecticides disrupt food webs through direct and indirect effects. Curr Opin Insect Sci 46:50–56

Tornero V, Hanke G (2016) Chemical contaminants entering the marine environment from sea-based sources: a review with a focus on European seas. Mar Pollut Bull 112(1–2):17–38

USEPA (2008) Estimation Programs Interface SuiteTM for Microsoft® Windows, v3.20. United States Environmental Protection Agency, Washington, DC, USA

Van den Brink PJ (2013) Assessing aquatic population and community-level risks of pesticides. Environ Toxicol Chem 32(5):972–973

Van den Brink PJ, Hartgers EM, Fettweis U, Crum SJ, Van Donk E, Brock TC (1997) Sensitivity of macrophyte-dominated freshwater microcosms to chronic levels of the herbicide linuron. Ecotoxicol Environ Saf 38(1):13–24 van der Ven K, Van Dongen W, Maes BU, Esmans EL, Blust R, De Coen WM (2004) Determination of diazepam in aquatic samples by capillary liquid chromatography–electrospray tandem mass spectrometry. Chemosphere 57(8):967–973

van Heel W, Hachimi-Idrissi S (2011) Accidental organophosphate insecticide intoxication in children: a reminder. Int J Emerg Med 4(1):1–4

Vance BD, Drummond W (1969) Biological concentration of pesticides by algae. J AWWA 61: 360–362

Varo I, Serrano R, Pitarch E, Amat F, Lopez FJ, Navarro JC (2000) Toxicity and bioconcentration of chlorpyrifos in aquatic organisms: Artemia parthenogenetica (Crustacea), Gambusia affinis, and Aphanius iberus (Pisces). Bull Environ Contam Toxicol 65(5):623–630

Veith GD, DeFoe DL, Bergstedt BV (1979) Measuring and estimating the bioconcentration factor of chemicals in fish. J Fisheries Board Canada 36(9):1040–1048

Vonk JA, Kraak MHS (2020) Herbicide exposure and toxicity to aquatic primary producers. In: de Voogt P (ed) Reviews of environmental contamination and toxicology, vol 250, pp 119–171. https://doi.org/10.1007/398_2020_48

Wang WH, Lay JP (1989) Fate and effects of salicylic acid compounds in freshwater systems. Ecotoxicol Environ Saf 17(3):308–316

Wang JS, Simpson KL (1996) Accumulation and depuration of DDTs in the food chain from Artemia to brook trout (Salvelinus fontinalis). Bull Environ Contam Toxicol 56(6):888–895

Warner NA, Wong CS (2006) The freshwater invertebrate Mysis relicta can eliminate chiral organochlorine compounds enantioselectively. Environ Sci Technol 40(13):4158–4164

Warra AA, Prasad MNV (2020) African perspective of chemical usage in agriculture and horticulture—their impact on human health and environment. In: Agrochemicals detection, treatment and remediation. Butterworth-Heinemann, Oxford, pp 401–436

Watanabe S, Watanabe S, Ito K (1985) Accumulation and excretion of herbicides in various tissues of mussel. Food Hyg Saf Sci 26(5):496–499

Wiegel S, Aulinger A, Brockmeyer R, Harms H, Löffler J, Reincke H, Wanke A et al (2004) Pharmaceuticals in the river Elbe and its tributaries. Chemosphere 57(2):107–126

Wijngaarden R, Van PA, Brock T, Brink PJ (2005) Threshold levels for effects of insecticides in freshwater ecosystems: a review. Ecotoxicology 14(3):355–380

Woodburn KB, Hansen SC, Roth GA, Strauss K (2003) The bioconcentration and metabolism of chlorpyrifos by the eastern oyster, Crassostrea virginica. Environ Toxicol Chem 22(2):276–284

World Health Organization (2020) The WHO recommended classification of pesticides by hazard and guidelines to classification 2019. World Health Organization

Yadav IC, Devi NL (2017) Pesticides classification and its impact on human and environment. Environ Sci Eng 6:140–158

Yadav DV, Agarwal HC, Pillai MKK (1978) Uptake, metabolism and excretion of DDT by the fresh water snail, Vivipara heliciformis. Bull Environ Contam Toxicol 15:300–306

Yadav IC, Devi NL, Syed JH, Cheng Z, Li J, Zhang G, Jones KC (2015) Current status of persistent organic pesticides residues in air, water, and soil, and their possible effect on neighboring countries: a comprehensive review of India. Sci Total Environ 511:123–137

Yamato Y, Kiyonaga M, Watanabe T (1983) Comparative bioaccumulation and elimination of HCH isomers in short-necked clam (Venerupis japonica) and guppy (Poecilia reticulata). Bull Environ Contam Toxicol 31(3):352–359

Yang G, Ge S, Singh R, Basu S, Shatzer K, Zen M, Liu J, Tu Y, Zhang C, Wei J, Shi J, Zhu L, Liu Z, Wang Y, Gao S, Hu M (2017) Glucuronidation: driving factors and their impact on glucuronide disposition. Drug Metab Rev 49(2):105–138. https://doi.org/10.1080/03602532.2017.1293682

You J, Landrum PF, Lydy MJ (2006) Comparison of chemical approaches for assessing bioavailability of sediment-associated contaminants. Environ Sci Technol 40(20):6348–6353

Yuniari SH, Hertika AMS, Leksono AS (2016) Lethal concentration 50 (LC50—96 hours) Nile Tilapia (Oreochromis niloticus) exposed Cypermethrin-based pesticide. JExp Life Sci 6(2): 58–62

Yu-yun T, Thumm W, Jobelius-Korte M, Attar A, Freitag D, Kettrup A (1993) Fate of two phenylbenzoylurea insecticides in an algae culture system (Scenedesmus subspicatus). Chemosphere 26(5):955–962

Zaffar H, Irshad U, Pervez A, Naqvi TA (2016) Mode of action, toxicity and biodegradation of organochlorinated pesticides: a mini overview. J Appl Environ Biol Sci 6(8):1–6

Zaki MS, Hamaam AMM (2014) Xenobiotics as stressful factors in aquatic system (in fish). Life Sci J 11(4):188–197

Zaroogian GE, Johnson M, Heltshe JF (1985) Estimation of bioconcentration in marine species using structure-activity models. Environ Toxicol Chem 4(1):3–12

Zhang C, Wang S, Yan Y (2011) Isomerization and biodegradation of beta-cypermethrin by Pseudomonas aeruginosa CH7 with biosurfactant production. Bioresour Technol 102:7139–7146

Endocrine-Disrupting Activity of Xenobiotics in Aquatic Animals

Prem Lata, Younis Ahmad Hajam, Rajesh Kumar, and Seema Rai

1 Introduction

From the last decade, there are several evidences which state that with the increase in anthropogenic activities, some compounds are introduced in the ecosystem which affect the endocrine system of vertebrates (Oberdörster and Cheek 2001). These compounds are known as "endocrine-disrupting chemicals" (EDCs). The US Environmental Protection Agency defined them as "exogenous compounds" that interfere with the physiological functions and actions of endogenous hormones associated with homeostasis maintenance and regulate the developmental processes (Kavlock et al. 1996). Recent studies extended this definition and stated that EDCs may be natural or synthetic as they have the potential to disrupt the hormonal and homeostatic systems that regulate the communicating pathways of organisms with respect to the environment (Diamanti-Kandarakis et al. 2009). A consensus was conducted by La Merrill et al. (2020), in which they stated that the significant features of EDCs are their interaction with or activation potential of hormone receptors, hormone-mediated signaling in cells, transport of hormones across the plasma membranes, distribution of hormones or level of hormone in circulatory system, metabolism of

P. Lata
Division Zoology, Department of Biosciences, Career Point University, Hamirpur, Himachal Pradesh, India

Y. A. Hajam (✉)
Department of Life Sciences and Allied Health Sciences, Sant Baba Bhag Singh University, Jalandhar, Punjab, India

R. Kumar
Department of Biosciences, Himachal Pradesh University, Shimla, Himachal Pradesh, India

S. Rai
Department of Zoology, Guru Ghasidas Vishwavidyalya (A Central University), Bilaspur, Chhattisgarh, India

M. A. Rather et al. (eds.), *Xenobiotics in Aquatic Animals*,
https://doi.org/10.1007/978-981-99-1214-8_2

hormone or its clearance, etc., which shows hormone-producing cell effects and epigenetic alteration induction in hormone-responsive cells. However, the effects mostly in in vivo conditions and having natural correlation become very tough to recognize. In this there are some EDCs that comprise plasticizers, byproducts of industries, dioxins, drugs, pesticides, etc. These EDCs have created major concerns due to their adverse effects on wildlife and humans (Sharpe and Skakkebaek 1993; Maiti et al. 2019). From the literature survey, it has been reported that EDCs induce various reproductive and developmental disorders in males and females (Crain et al. 2008), breast cancer (Darbre 2006), prostate cancer (Mahajan et al. 2006), and obesity (Grun et al. 2006).

Exogenous molecule endocrine disruptors that change the endocrine glands function in biological system inducing adverse effects on the health of a population or an individual. It is a well-known fact that EDs aggravate the hormonally regulated physiological activities, i.e., growth, development, and reproduction (Sundaray et al. 2021). These EDs work through different mechanisms, like either they bind to hormone receptor or alter the endogenous hormones level or modulate the expression of genomic network. This in turn stimulates/inhibits the downstream cellular and molecular cascades, hence affects the normal indices and functions. Currently, various in vitro assays have been developed comprehensively by different agencies to identify EDs, among them are organic molecules having low molecular weights. In contrast, in vivo assays for the identification of ED are very sensitive and provide biological eco-relevant results than in vitro assays. In addition, endocrine disruptors intermingle with other pollutant families that produce harmful effects when they enter into the aquatic system. Endocrine disruptor comprises pesticides, hormonal mimics, heavy metals, polychlorinated biphenyls, organic solvents, flame retardants, surfactants, pharmaceuticals, etc. Moreover, certain endocrine disruptors are naturally occurring, which are synthesized by natural organisms such as fungi or plants (Liu et al. 2010). On the other hand, other endocrine disruptors are synthesized by degradation of industrial chemicals (polycyclic aromatic hydrocarbons). It has been studied that in wildlife fauna, EDs induce irreversible reproductive dysfunctions reported by the workers (Tubbs and McDonough 2018), which includes feminization of males and infecundity (Gimeno et al. 1998). Besides these reproductive abnormalities, they also cause disorders of other endocrine systems in different axes, by inducing antagonistic or agonistic impacts upon binding to hormone receptors. In addition, some of them are adverse (seen in wild animals/experimental organisms), might also occur in humans when exposed to certain concentration, and cause endocrine dysfunctions. Endocrine-disrupting chemicals (EDCs) are structurally different groups of chemical compounds that are having potential to damage human health, wildlife, fisheries, and their future generations due to their interaction with the endocrine system (Hansen 1998). Generally, endocrine-disrupting chemicals are utilized in different sectors like agriculture and industry which includes PCBs, organochlorine pesticides, plasticizers, and surfactants. There are different well-known EDCs such as estrogenics, impacting the reproductive system due to the lipophilicity and persistent nature of the xenobiotic estrogen and their

Table 1 Commonly detected in aquatic organisms

S. No.	Name of endocrine disruptor (EDs)	Physiological effect	References
1.	Bisphenol A (BPA)	Significantly increases the morphological alteration, nuclear changes, and DNA damage potential of BPA in RBCs	Sharma and Chadha 2021
2.	Dioxins	"Blue sac" toxicity in fish larvae Reproductive toxicity in adults	King-Heiden et al. 2012
3.	Perchlorate	Hepatotoxicity, thyroid hormone which promotes growth, reproduction, embryonic development, and metamorphosis initiation	Power et al. 2001
4.	Perfluoroalkyl and polyfluoroalkyl substances (PFAS)	Decreases ovarian reserve and endogenous hormone synthesis by activation of peroxisome proliferator-activated receptors, deteriorates gap junction intercellular communication between oocyte and granulosa cells, decreases the level of thyroid hormone, antagonizing ovarian enzyme activities associated with ovarian steroidogenesis, or suppresses the kisspeptin signaling in the hypothalamus	Ding et al. 2020
5.	Phthalates	Reproductive, endocrine, immune, genotoxic, and nephrotoxic effects	Munshi et al. 2013
6.	Phytoestrogens	Both phytoestrogens genistein and daidzein show estrogenic effect in Nile tilapia larvae sexual differentiation and in turn impair the masculinization	El-Sayed et al. 2012
7.	Polybrominated diphenyl ethers (PBDE)	Liver, thyroid gland, and possibly developing reproductive organs	Kuriyama et al. 2005
8.	Polychlorinated biphenyls (PCBs)	Polychlorinated biphenyls, in the liver, GI system, skin, blood, and reproductive and nervous system which shows toxic effects	Longnecker et al. 1997
9.	Triclosan	Sinusoidal dilation, congestion, vacuolization, hepatocellular degeneration, and necrosis	Arman 2021

produced metabolites which get accumulated and biomagnified (Adami et al. 1995), environmental estrogen, anti-estrogen, etc. All these endocrine disruptors function same as steroid sex hormones. Herein, estrogen and androgen are bind with the hormone receptor complex, control the cellular level of signing pathways (environmental estrogens and androgens), and change or block the anti-estrogens and anti-androgens reported by various researchers (Fossi and Marsili 2003) (Table 1; Fig. 1).

On our planet has approximately 4500 species of mammals, and aquatic mammals comprise a small percentage of this number; however, they contribute significantly for the maintenance of ecological balance in the marine and freshwater

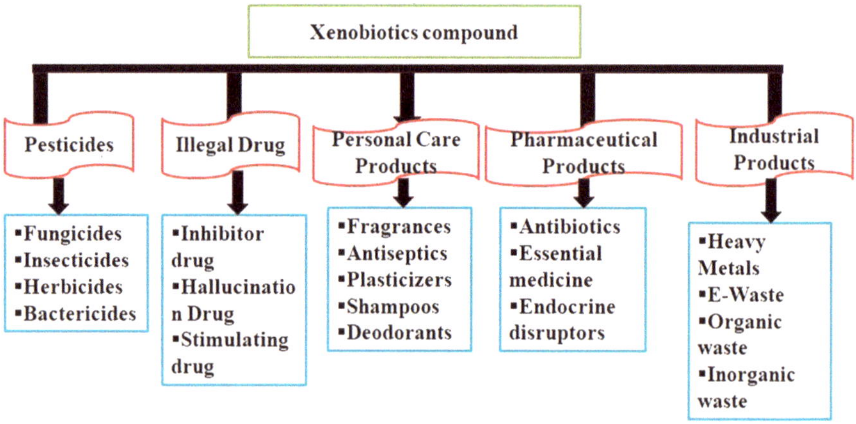

Fig. 1 Figure showing different xenobiotic compounds

ecosystem. There are many aquatic species (mammalian) such as pinnipeds and odontocete cetaceans. There are top predators that help control communities.

However, reduction or disappearance in some areas may considerably alter the structure of community. From the last decades, different studies reported that various mammalian species are prone to the toxicological effects due to the xenobiotic compounds (Bernhoft et al. 2000; UNEP 2001; Dar et al. 2020), with huge endocrine-disrupting chemical class. There are some mammals in the aquatic system that used to eat fish due to their position in the food chain. According to the food web of both the aquatic and marine, they inhibit the higher agriculture and industry areas as per their reproductive biology reported by different workers in their studies (Hajam et al. 2021; Gillesby and Zacharewski 1998). These EDCs becomes bioaccumulated and biomagnified in the aquatic food chain such as organochlorines (Adami et al. 1995).

The exposure of aquatic animals on EDs and their effects has been studied very deeply especially in fishes. A number of EDs have been identified, whereas some of them has been also tested to have a combination of or both endosulfan and flutamide as an insecticide and nonsteroidal anti-androgenic drug, respectively, which impaired the ovarian and testicular growth in catfish (Rajakumar et al. 2012; Chakrabarty et al. 2012).The different researchers reported in their studies the fenvalerate and triadimefon combinatorial disruptive effects in Chinese rare minnow, *Gobiocypris rarus*, in embryonic development (Wu et al. 2018).

Fishes are the most abundant and are most diverse among the vertebrates. They dominate the water bodies in the whole world due to their different adaptations like morphological, physiological, etc. Their diversity is predicted through their large population of living species. Till date, 21,700 species have been described and presently increase up to 28,000 (Nelson 1984). Fishes occupy unusual group of habitats ranging from streams, deserts springs, open oceans, cold mountains lakes, and other aquatic environments (Moyle and Cech 2000). Due to the alterations in environmental conditions, fishes have the ability to adopt various life-history strategies to resolve the reproduction problems (Thorpe 1989). Fishes are a rich source of proteins and lipids, which are used for humans and domestic animals and establish an economically essential fisheries and aquaculture.

2 Xenobiotics-Induced Toxic Manifestations

Nowadays, synthetic chemicals are a big problem. They are primarily utilized in the different household activities and in industrial activity. These activities cause disturbance in physiological as well as endocrinological functions in organisms. The three major impacts of xenobiotics on the organisms, i.e., neurophysiological, reproductive, and behavioral, are discussed. These impacts associate with each other: changes in nervous systems can affect the behavior; behavioral changes can affect the reproductive functions. A compound is not able to induce its effect on a target organism or a community. The effect of compounds depends on its concentration and the duration of treatment given to the organism. The exposure time and effect may range from acute to chronic. Acute toxicity includes exposure of any toxicant for a short period of time, which is well defined, sometimes fatal and infrequently reversible. Chronic toxicity occurs when organism is exposed toward any toxicant and could finally lead to death. Xenobiotics become fatal when it led to death or enough to cause death, through direct target on organism. On the other hand, when the toxicant concentration was less, it is sublethal, thereby reducing the physiological and behavioral processes. The radioactive pollution may also affect the ecosystem irreversibly (www.lenntech.com). In addition, there was loss of some species due to pollution on freshwater; however, some species get advantage with this. In general, there occurs decreases in diversity, but a number of individual species does not change; this change in balance occurs due to predation, competition, and cycling of materials. Due to the complex nature of pollution, the adverse effect on aquatic life depends on the feature of pollutants. Combined effect of two or more pollutants may promote synergetic or additive and antagonistic interaction. Combined effect of zinc and cadmium on fish is an example of additive interaction. Calcium (Ca) is antagonistic toward Pb, Zn, and, Al, whereas copper showed highest additive effect with Cl, Zn, Cd, Ar, and, Hg and also decreases cyanide toxicity (Rani et al. 2022; Hajam et al. 2022; Adnan Amin et al. 2016, 2017; Sruthisree et al. 2015; Adnan Amin et al. 2015). The toxic effect of phenol and ammonia on mayfly *Baetis rhodani* at low concentration is additive; however, at higher concentration it shows more additive effect.

3 Stress

Stress is one more well-known example that suppresses the immune system and decreases the resistance against the diseases. To determine the stress level on immune system, many experiments were conducted on different species of fish.

Herein, cortisol and adrenaline and other proteins such as heat shock proteins and plasma glucose levels were determined in addition to innate and adaptive immune indicator and impacts on disease resistance. There are some factors associated with the aquaculture, i.e., handling transport and high density, which leads to stress, thereby requiring significant attention. Prolonged exposure toward the stressors usually suppresses the immune system, resulting in decreased immunity against diseases in fishes. Initial stimulation include limited time exposure to handle stress of Atlantic salmon and rainbow trout up to the basal level has been generally observed during the induction of chronic stress. Likewise, fishes in aquaculture try to adopt confinement and show decreased stress response in comparison to wild type.

Direct exposure to stress hormones or neuropeptides of fishes under in vivo/ in vitro studies suppresses the functioning of immune system. For instance, in Atlantic salmon, Ig-positive lymphocytes showed that cortisol administration and mitogenic activation downregulation inhibited these cortisol incubation. In contrary, it has been found that endorphin stimulates leukocyte phagocytic activity in rainbow trout and carp fish kidneys under in vitro conditions. Stress affects the immune system and resistance disease by affecting the factors from growth to larva health. Due to this broad range of effects, husbandry practices in aquaculture are trying to avoid stress through the maintenance of environmental conditions, rearing at a particular density, and regulated sexual maturity. Xenoestrogens has the ability to copy or act as a natural estrogen hormone. This group of compounds also includes synthetic steroids used as contraceptive pill (Pelissero et al. 1993). In addition, there are some pesticides such as DDTs, HCHs (Wester and Canton 1986; Palmer and Palmer 1995; Donohoe and Curtis 1996), alkylphenolpolyethoxylates, APEs (Soto et al. 1991; Jobling and Sumpter 1993; White et al. 1994; Jobling et al. 1996; Arukwe et al. 1997a, b), plasticizers, polychlorinated biphenyls, polychlorinated biphenols (McLachlan 1985), and other natural chemicals like phytoestrogens and mycoestrogens (Pelissero et al. 1991a, b). In some cases, the estrogenic activity of these compounds has been found by chance (Soto et al. 1991), and in fishes, estrogenic activities like reduction of zona radiate proteins and vitelogenesis have been reported by different workers (Sumpter and Jobling 1995; Arukwe et al. 1997b). Alkylphenolpolythoxylates constitute an essential class of nonionic surfactant. These are mostly used as detergents and emulsifiers. They were also utilized in plastic products in industries and agriculture (Ahel et al. 1994). Alkylphenols (APs) are produced during the breakdown of APEs. Some reports suggested that APs, one of the essential alkylphenolpolyethoxylate metabolites, because of resistance increased toxicity, biodegradation, etc. in aquatic organisms as reported by different researchers (Ahel et al. 1994). A multifaceted microbe-induced degradation pattern, characterized by the production of different metabolic products showing higher toxic manifestations than the parent compound, has been developed for alkylphenolpolyethoxylates (Ekelund et al. 1990).

4 Effects of Xenoestrogens

Agonist is a substance that binds with receptor to activate its potential to turn on the receptor found in classical pharmacology. ER and E2 are agonist examples. In comparison, it blocks the signals emitted by the agonist reported by the researchers (Nimrod and Benson 1996). It can function through nonreceptor-aided mechanisms. Xenobiotics have the potential to copy the natural estrogen action known as xenoestrogens reported by the workers (Colborn and Clement 1992).

The action mechanism of xenoestrogens is that they are having higher affinity to the ER and starts the action of target tissues, like natural estrogens. Other compounds are also having potential to bind with the receptor; however, they do not induce estrogenic activities and hence block the binding sites on which natural estrogens bind, as reported by different workers (Safe 1995; Safe and Krishnan 1995; Ahlborg et al. 1995). Incorporation of vitellogenin is important, which plays an important role for the development of oocytes during recrudescence of ovary. A possible indirect method for the evaluation of changed hepatic vitellogenin synthesis when fishes are exposed toward xenobiotics is decreased or elevated GSI. These changes can be measured directly in plasma, hepatic, and ovarian vitellogenin levels (Kime 1995).

5 Effect of Xenobiotics on Fish Reproduction

Different studies reported that xenobiotics induces reproductive abnormalities in aquatic organisms (Guillette Jr et al. 1995). Different methods and indices are used for the evaluation of reproductive success in different fish species. These parameters include decreased hatching in *Platichthys flesus* and *Clupea harengus* having PCB association in eggs (Hansen et al. 1985). Higher mortality of eggs in Lake *Salvelinus alpinus* in connection with increased polychlorinated biphenyls and DDT in eggs (Monod 1985) decreased fertilization success and viable hatch in *Platichthys stellatus* from polluted areas of San Francisco Bay (Spies and Rice 1988). Common defects observed during different studies include blister proliferation during initial and late embryos, failure to close the blastopore, and deformity in the notochord. However, considerable association has been observed for malformations of dab and concentrations of p,p'-DDE residues. In addition, deformities in chromosomes and embryo seen in North Sea area fish have been studied. Researchers found the positive correlation in polychlorinated biphenyls, DDT, and DDE in gonads and liver of whiting fish with increases in malformation rates in the areas near the coast of Netherlands and the Rhine River Estuary.

6 Effect of EDCs in Aquatic Mammals

Higher levels of both PCB and DDT as EDCs have been reported in various aquatic mammals. In addition, the physicochemical properties of both polychlorinated dibenzo-p-dioxins and dibenzofurans are less transportable. It has been reported that these contaminants have a wide range of antiestrogenic activities observed under in vivo and in vitro conditions.

The concentration-dependent impacts of various PCDD and PCDF congeners show the antiestrogenic activity of aryl hydrocarbon (Ah)-responsive MCF-7 human breast cancer cell lines. For the PCDDs and PCDFs, the antiestrogenic potential was found in the order of 2,3,7,8-tetrachlorodibenzo-*p*-dioxin > 2,3,7,8-tetrachlorodibenzofuran > 2,3,4,7,8 pentachlorodibenzofuran > 1,2,3,7,9-pentachlorodibenzofuran > 1,3,6,8-tetrachlorodibenzofuran (Haynes et al. 1999). The top predators are consistently exposed to large quantity of pollutants via biomagnifications of compounds obtained from contaminated prey (Vonier et al. 1996). Pinnipeds and cetaceans comparatively have higher quantity of blubber for insulation that steadily retains significant levels of highly persistent organochlorine insecticides, PCBs, dioxins, and other associated compounds (Muir et al. 1996). The predictable TEQ levels in the blubber of some cetacean species, such as the northern right whale dolphin (*Lissodelphis borealis*) and Pacific white-sided dolphin (*Lagenorhynchus obliquidens*) from the northern North Pacific; Dall's porpoise (*Phocoenoides dalli*) from the Japan Sea; striped dolphin (*Stenella coeruleoalba*) off Sanriku and Fraser's dolphin (*Lagenodelphis hosei*) off Kii Peninsula, Japan; and hump-backed dolphin (*Sousa chinensis*) and finless porpoise (*Neophocaena phocaenoides*) from Hong Kong, increased the concentrations related with immunosuppression in harbor seals (*Phoca vitulina*) (Wren 1991). Various studies were published with regard to PCDD and PCDF concentrations in aquatic mammals such as Baikal seal from lake Baikal, harp seal from Greenland sea (Berg et al. 2001; Harding et al. 1999), false killer whale and Risso's dolphin from British Columbia, harbor porpoises from British Columbia and central California (Mazet et al. 2001), and dugong and sea lion from St. Lawrence River estuary, Great Barrier Reef, and Argentina (Jarman et al. 1996; Roos et al. 2001; Mason and Macdonald 1993). There are various studies that reported the association between OCs and their respective reproductive toxicity which have been collected from semi-field studies on both seals and mustelids (Tanabe et al. 1988; Wiig et al. 1998).

7 Freshwater Mammals

In mustelids, *Mustela vison* and *Lontra canadensis* have decreased in these species area owing to the increased percentage of contaminated fish in their diet as reported by different workers (Skaare et al. 2001). In laboratory trials it was demonstrated that among these species, mink are sensitive to OCs (specifically polychlorinated biphenyls and dioxins). Studies reported that from a comparison of mink population from 1982 to 1987, contaminated countries bordering Lake Erie showed decreases than that of other countries, signifying the chemical effect on the mink population (Skaare et al. 2001).

Previous studies reported that Ontario also suggested that mink harvest is having decreased potential higher PCB exposed areas than areas having lower exposure of PCBs. Evidences also reported that effect of endocrine-disruptive chemicals on the otters collected from four New York State countries adjacent to Lake Ontario and the St. Lawrence River. The harvest data collected from four countries revealed that between 1960 and early 1970, otter harvest continued constantly and then elevated. Elevated harvest was found constant with better water quality of Lake Ontario in the past 15 years. Harding (Sandau et al. 2000) evaluated the contamination caused by chlorinated hydrocarbons on mink and river otters in Columbia and Fraser River systems of northwestern North America, in association with their morphological conditions. In the river system, various residues of organochlorine pesticides, PCBs, dibenzo-p-dioxins, and dibenzofurans were evaluated. Concentration of contaminants was found comparatively low as documented in other North American populations, even though the concentrations were found greater than those noted during the previous survey in these regional populations. Even though some certain individual animals having higher reproductive dysfunctions does not reveal higher concentrations of contaminants, the correlation was found considerably negative between total polychlorinated biphenol levels and baculum length in juvenile mink. The ranch-reared mink is used as a model organism for the experimental trials to examine the toxic potency of two petroleum products when exposed to sea otters. Females were exposed to bunker C fuel oil in their diet which considerably decreased the reproductive success; however, their offspring were exposed to petroleum products in utero or during nursing. The reproductive performance was decreased due to the consumption of contaminated food or early oil habitats of *Enhydra lutris* (Reijnders 1980).

Moreover, during 1960–1980, the European otter population decreased dramatically, and exposure to polychlorinated biphenols caused impairments in the reproductive functions, which has been considered the main cause of reduction ((Reijnders 1990). A study was conducted in East Anglia and England rivers on the concentration of polychlorinated biphenyls and otter population and also organochloride residue. In addition, the comparison of population and the concentration of contamination was conducted. First, on the bases on the index value of the population that was stable and, second, as "no effect level" for all peoples (ICES CM 1992). About 44% of the samples shows "level of concern." Contamination caused by polychlorinated biphenyls is hence considered as a major factor affecting otter populations (Minh et al. 2000).

8 Bisphenol A (BPA)

Bisphenol A is broadly used in different products, such as baby bottles to thermal paper for credit card receipts (Liao and Kannan 2011; Vandenberg et al. 2009). BPA is the highest among the produced chemicals, having a production rate of greater than six million pounds per year, and more than about a hundred tons are discharged into the environment every year (Vandenberg et al. 2009; Repossi et al. 2016; Corrales et al. 2015). Bisphenol A makes its entry into the aquatic ecosystem from either sewage treatment wastewater or natural process of degradation (Gatidou et al. 2007; Wintgens et al. 2003). It has been reported that in China, the concentration of BPA is as high in wastewater, rivers, and seawater surfaces, i.e., 370 mg/L, 3.92 mg/L, and 0.19 mg/L, respectively (Huang et al. 2012). Therefore, this increased levels of BPA in different environment especially in aquatic ecosystems, including fishes.

BPA is broadly present in aquatic ecosystems; when aquatic animals are exposed to these chemical compounds, they affect the energy metabolism, individual cortisol stress response, growth, musculature, and motor behavior and hence affects the overall growth and development in fishes (Birceanu et al. 2015; Nagato et al. 2016; Wang et al. 2013; Birceanu et al. 2015). In addition, Bisphenol A is a well-known endocrine-disrupting chemical with harmful impacts on reproduction and sexuality. Cabaton et al. (2011) found that BPA exposure causes reduction in fertility and fecundity and induces puberty at early age in female offspring (Durando et al. 2007); on the other hand, in male offspring, bisphenol A reduces the anogenital distance (Miao et al. 2011). Various studies conducted on fishes have found that BPA affects the reproductive functions of female fish. Sohoni et al. (2001) found that BPA exposure for 164 days slows down the development of gonads and reduces ovulation in female *Pimephales promelas* fathead minnows. In *Oryzias melastigma* BPA exposure slows the oocyte development and decreases estrogen concentrations (Huang et al. 2017). In male fish, exposure of BPA induces adverse effects on the development of gonads such as reduced production of androgens, reduced sperm counts and their mortality in the epididymis, and sperm cell apoptosis and damages (Chen et al. 2015; Zhang et al. 2016; Karnam et al. 2015; Meeker et al. 2010; Zhuang et al. 2015; Wintgens et al. 2003; Xie et al. 2016).

In teleost fish, reproductive functioning depends on various factors. The outmost essential is the interaction of genes associated with sex determination and synthesis of sex steroid hormones (Dar et al. 2012). Distinctive variations have been recognized for the sex determination/differentiation of genes and synthesis of hormones in male and female. Likewise, ovarian differentiation is controlled by genes like CYP19a1a and foxl2 (Piferrer and Guiguen 2008; Guiguen et al. 2010). There are some genes that are harmful as well for the testis development such as dmrt, cyp11b, and dax1 (Marchand et al. 2000; Kusakabe et al. 2002). Moreover, earlier studies revealed that BPA induces its effects on the hypothalamic-pituitary-gonadal (HPG) axis for the regulation of reproduction (Molina et al. 2018). An earlier study demonstrated that for the gender, sustained androgen sex hormones play a significant role as reported by various researchers (Wang et al. 2017). In teleosts, synthesis of steroids is also regulated by HPG axis.

9 Toxic Effects of Bisphenol A on Goldfish Gonad Development and the Possible Pathway of BPA Disturbance in Female and Male Fish Reproduction

It has been studied that BPA exposure decreases the maturation of germ cells in males; however, this process cannot be restarted once BPA exposure has stopped in male goldfish. Histoarchitecture of testis revealed that after an exposure of 50 mg/L and 500 mg/L BPA for 30 days, the STs and SCs disappeared, and the GSI in the 50 mg/L and 500 mg/L BPA-exposed goldfish was found considerably reduced in comparison to the normal fishes. These modulations in spermatogenesis were observed in goldfish exposed to BPA (Golshan et al. 2015; Sohoni et al. 2001). These changes in STs, SCs, and GSI following the exposure of BPA might be due to the disruption caused by BPA on gametogenesis. It has been studied that STs, SCs, and GSI decreased in testis even after the withdrawal of BPA exposure (50 mg/L and 500 mg/L). It has been observed that BPA (1 mg/L, 500 mg/L) exposure of female goldfish causes changes in histological in ovaries from oocytes with CAOs to oocytes with PGs. BPA toxicity alters GSI and ovarian weight which indicates that BPA treatment suppresses the maturation of oocytes. BPA exposure causes alteration in steroidogenesis, by decreasing the mRNA level of CYP11b and 11-KT level in circulatory serum. Both CYP11b and 11-KT are key regulatory factors for spermatogenesis (Fenske and Segner 2004). Moreover, BPA exposure also decreases the maturation potency of testis, which suggests that BPA inhibits the formation of 11-KT and interrupts the spermatogenesis.

Biosynthesis of steroid hormones is regulated by HPG axis in all organisms and in fishes as well. The HPG axis maintains the reproductive function by coordinating with actions of gonadotrophin-releasing hormone and two gonadotrophins (LH and FSH hormones). The toxic effect of BPA exposure was evaluated by assessing the testicular apoptosis and proliferation. Testicular apoptosis and proliferation are assessed via TUNEL and PCNA assays. Various physiological activities including maintenance of approximate germ cells to somatic cells ratio, removal of germ cells (abnormal), and maintenance of sperm production in testis rely on apoptosis and cellular proliferation (Prisco et al. 2003; Shukla et al. 2012). It has been studied that BPA exposure leads to apoptosis of spermatids. It also blocks the spermatocytes proliferation. In rodents, treatment of BPA at higher doses increases the rate of programmed death of germ cells and Leydig cells (Eo and Lim 2008; Li et al. 2009; Xie et al. 2016). In fish, CYP 11b is expressed in Leydig cells in normal testes reported by the workers (Sun et al. 2014). Therefore, due to the apoptosis in Leydig cells, the level of 11-KT decreases. BPA exposure decreases the maturation ability in testis due to apoptosis in Leydig cells and results in the reduction in androgen levels that interrupts spermatogenesis in testis. Moreover, in goldfish, BPA exposure reduces the 11-KT level which is correlated with a decrease in AR level; however, the era, erbI, and erb II do not show any significant decrease.

BPA is a weak estrogen receptor, but it helps activate androgen receptors as reported by various researchers (Kinch et al. 2015; Wetherill et al. 2007). This indicates that BPA has the ability to modulate AR signaling and in turn reduces the level of androgen in goldfish. In female goldfishes, BPA exposure has no significant changes in CYP19a1a and foxl2 mRNA level. CYP19ala plays a significant role in synthesis of E2 and FOX12 and is also important for ovarian differentiation (Guiguen et al. 2010). It has been reported that BPA exposure does not alter the level of estrogen level and the sex-specific gene expression in females. Moreover, BPA exposure via the TUNEL and PCNA assays does not show any positive signs in the ovarian cells; however, histoarchitecture observations revealed that BPA exposure affects the maturation of ovaries. BPA treatment also induces adverse effect on the maturation of ovaries by directly acting on the ovary. This BPA-affected ovarian maturation has been confirmed by studying the levels of sgnrh, fshb, and IHb which were found to be reduced in goldfish after its exposure to BPA 50 mg/L and 500 mg/L. GnRH helps to activate the anterior pituitary. The anterior pituitary secretes LH and FSH hormones and helps in ovary maturation (Molina et al. 2018). Fernández et al. (2010) found that medium and higher doses of BPA alter the HPG axis by blocking the maturation of ovary indirectly through its impact on the HPG axis. Moreover, BPA exposure also increases the expression of AR in the brain; this indicates that BP might be regulating the reproductive axis in female fish through the modulation of AR expression. Hence, BPA disrupts the gonad maturation in both genders. It occurs due to HPG axis modulation that further causes apoptosis in germ cell and Leydig cells.

10 Effect of Endocrine-Disrupting Chemicals on Fish Gamete

10.1 Effects of EDC Exposure on Ovarian Physiology

Endocrine-disrupting chemicals affect the physiology and morphology of the ovary, and oocyte maturation and steroidogenesis change the endocrine and reproductive functions. They occur by suppressing actin of androgen hormones that modulates androgen hormone production (Sonnenschein and Soto 1998). EDC has endocrine-disruptive activity hormones binding with the receptor and specific site. Moreover, the receptors associated with signaling either suppresses or changes serum binding protein in blood plasm.

10.2 Effect of EDCs on Morphology of the Ovary and Oocyte Growth, Maturation, and Fertility

Exposure of bisphenol A (BPA) given to *Gobiocypris rarus* induced changes in ovarian morphology which includes the presence of a large number of premature oocytes and also various atretic follicles. In addition, exposure of dichloro-6-nitrophenol to *Gobiocypris rarus* induces harmful effects in both genders.

Gonads increase the deformed number in female fish with degenerating velogenic oocytes. It has been reported that this degenerating of velogenic oocytes indicates the defect in the follicle process. Catfish and *Clarias gariepinus* were exposed to this and various changes have been observed. It has been studied that exposure to ethinylestradiol and diethylstilbestrol during the early stages of development, up until 50 days post-hatch, can have adverse effects. These changes are observed till adulthood, such as morphological deformities (stunted growth, spinal curvature, and yolk sac fluid retention). The various researchers reported in their studies the rudimentary ovaries with development of precocious oocyte and follicular atresia as well (Sridevi et al. 2015). In *Danio rerio* (zebrafish), chronic exposure to EE2 (25 ng/1) and to di-2-ethylhexyl-phthalate (0.02 μg, 0.2 μg, 2 μg, 20 μg, and 40 μg/I) impaired the production of ovulation and embryo. The ability of the EE2 and di-2-ethylhexyl-phthalate (DEHP) to modulate vitellogenesis is suggested in ovarian observation. It induces a previtellogenic oocyte reduction that is associated with velogenic increases. The different researchers reported that when fishes were crossed after exposure, the number of embryos was 1% which clearly showed fecundity impairment (Carnevali et al. 2018). At present, the same study was conducted with different concentrations (such as 0.42, 4.2, 42, 420, and 4200 μg/L) of DiNP. Di-isononyl-phthalate was utilized in place of di-2-ethylhexyl-phthalate (DEHP). They analyzed the number of reproductive parameters such as size frequency, follicle number, and macromolecular composition of vitellogenin oocytes. Histological analysis showed that in fish gonads, the number of vitellogenic oocytes becomes lesser in number having lower concentrations of DiNP. In addition, the number of mature oocytes gets decreased when exposed to higher doses such as 420 and 4200 μg/L.

Further, the sample was collected from exposed fish in which lipid decrease, phosphate decrease, and protein were measured. FTIR was used to measure these components. The observation may conclude that DiNP affects the uptake of vitellogenin and also stops the changes that occur in the lipid composition. In addition, results affect the physiology of female reproductive system. Wang et al. (2015) reported the same study exposure from 2 hour postfertilization to sexual maturity toward tris1(3-dichloro-2-propyl) phosphate. The maturation of oocytes occurs with an increase in mature oocytes promoted by it as reported by various researchers. Wang et al. (2015) reported high malformation rate in first generation. It was observed that developmental toxicity is caused when parental exposure occurs to TDCPP in offspring (Wang et al. 2015). In *Oryzias latipes* 3,3-diindolylmethane inhibits the increase of VTG storage and reduces the activities of the cathepsin enzyme in growing oocytes. Various researchers have reported that a decrease in eggshell proteins results in inhibited maturation and reduced fecundity of primary oocyte. This reduction is caused by a lower level of eggshell proteins (Chen et al. 2017).

In a 3-month trout, exposure toward 5 µg/L BPA occurs. It proceeded at the "spawning season," which further affected percentage of ovulated egg. After 3 weeks, the ovulation starts with any effect on egg quality as reported by different workers in their studies (Lahnsteiner et al. 2005). Giari et al. (2016) studied the atretic oocytes (with disorganization of ooplasm) present in carp. This carp was treated with 2 mg/L PFOA (perfluorooctanoic acid) for about 56 days. In a 21-day study of reproduction, "fathead minnow" reported that various histopathological conditions were demonstrated such as atretic follicles, oogonia, intestinal connective tissue proliferation, etc., and the accumulation of both adipose tissue and interstitial observed as reported by the researchers (Margiotta-Casaluci et al. 2013).

In *Oreochromis niloticus*, a diuron action is a urea herbicide substitute. It has anti-androgenic effects. Three metabolites were demonstrated in it. Boscolo Pereira and coworkers reported an increase in the gonadosomatic index due to exposure to pollutants. In addition, the higher number of vitellogenic oocytes is found in the ovary, and the decrease of germinative cells is observed. It showed estrogenic action of metabolites and can accelerate the ovarian development of this species (Boscolo Pereira et al. 2016).

10.3 EDC Effects on Molecular Markers Linked to Steroidogenesis, Ovarian Growth, and Maturation

G. rarus exposure to BPA with different concentrations showed adverse effects of steroidogenic enzyme expression at lower concentrations, i.e., 5–15 µg/L. On the other hand, it increases the ovarian transcription genes such as steroidogenic and sexual steroid receptor genes at a high concentration up to 50 µg/L. Zhang et al. reported that mostly the changes or variation in the expression of the steroidogenic gene were due to changes in "et" and "ar" transcripts. Both can affect hormone signaling and nr5als expression (Zhang et al. 2014a).

Zhang et al. (2014b) reported gdf9 and bmp15 mRNA alteration. It has a significant role in weight gain and abnormal development of the ovary (Zhang et al. 2014b). In catfish, EE2 and DES exposure shows a significant role in the alteration of the rate-limiting enzyme of estrogen biosynthesis (Sridevi et al. 2015). In EE2 exposure at 50 and 150dph, the transcript level of star and P450sec increased, a significant increase was observed in "cyp17, 3b-hsd, 17b-hsd1 from 50 to 350 dph," and decrease in the cyp17 transcript levels and others remains unaltered (Sridevi et al. 2015).

In zebrafish, the fecundity gets decreased when zebrafish are exposed chronically to EE2 and DEHP, associated with BMP15 when its level increases. It results in a reduction of one hour or more in ptgs2 expression. Carnevali et al. (2018) reported the DEHP inhibitory action in in vitro when the germinal vesicle breaks down

(Carnevali et al. 2018). Further, the DiNP exposure for the same species was recognized that resulted in gene modulation involved in steroidogenesis but in a nonmonotonic manner. In different concentrations, i.e., 0.42 and 420 μg/L, there were greater differences demonstrated in star and cyp11a1 transcript levels; the lowest dose might affect the fshr transcript. Exposure to DiNP at a concentration of approximately 420 μg/L has been found to affect the levels of Esr1 and esr2b, and results in a reduction in the esr1a transcript level. These effects were observed with significant differences compared to exposure at 0.4 mg/L and at higher concentrations. Pgrmc1 transcript levels did not change exposure to DiNP. 1hcgr and pgrmc2 levels of transcript reduction were measured, and the bmp15 transcript level increased only 0.42 μg/L concentration.

Various researchers reported that DINP lower concentration interferes with both oocyte and steroidogenesis growth. On the other hand, oocyte maturation is impaired with a high concentration of DINP (Santangeli et al. 2017). This concluded that there is a need for an alternative to DEHP as reported by different workers (Forner-Piquer et al. 2017).

Furthermore, concerning estrogenic effects, in fathead minnow, EDCs may have anti-androgenic effects. This occurs when it was treated with dutasteride. It is a product same as 5-alpha-reductase inhibitor. It is an enzyme that helps in the conversion of testosterone into dihydrotestosterone. In fish, dutasteride inhibits or impairs reproductive functions with an anti-androgenic action reported by various workers (Margiotta-Casaluci et al. 2013). In in vitro conditions, the post-vitellogenic oocyte of trout exposure to prochloraz (imidazole fungicide), which is utilized in agriculture, affects the process of oocyte maturation. In rainbow trout, prochloraz with or without LH can alter GVBD in post-vitellogenic follicles. Alone, prochloraz can affect the maturation of the oocyte, stimulating the transcription of the gene. This transcription is responsible for the recruitment of pre-ovulatory follicular phase and its differentiation as well. Among these, some are triggered by LH. It shows a synergistic cooperation with prochloraz (inducing GVBD) as reported by various researchers (Rime et al. 2010).

Blair et al. (2000) reported that there are xenobiotics that also affect the morphology of the ovary, maturation of the oocyte, and steroidogenesis that follow the classical genomic pathway (Blair et al. 2000). After some time, the conflict of interest focused on nongenomic mechanisms that were activated by xenobiotics (Thomas and Doughty 2004). The various rapid estrogenic responses have a key role of seven pass-membrane receptors and GPR30. In zebrafish, the low concentration of BPA and alkyl phenols is responsible for the maturation of disrupted oocytes by nongenomic mechanisms that further activate Gper pathway (Fitzgerald et al. 2015).

10.4 *Effects of EDC Exposure on Male Gonadal Physiology and Functionality*

Linhart et al. (2008) reported that the fertility of males can be checked by determining the sperm mortality rate, the velocity of spermatozoa, and the duration of motility. In vertebrates, spermatogenesis, secondary sexual features, and behavior regulated by androgen are reported by different researchers (Martyniuk and Denslow 2012). Kime (1999) reported that the delay in maturation, changes in male sex cell types, low semen quality, and estrogen and testosterone level changes were due to the exposure to EDC. The significant effects showed at lower concentrations, but its technical and biological differences show difficulty in EDCs classification concerning toxicity. It exhibits more toxicity than organic EDCs; in addition, heavy metals can impair the functioning of spermatozoa (Popek et al. 2006). Endocrine disruptive chemical exposure has been observed on testis morphology and spermatogenesis in medaka at a concentration of 0–8.5 μg/L di-indolylmethane (DIM), which is an antifouling agent that produced eggshell proteins and vitellogenin in testis (Chen et al. 2017). In zebrafish, DEHP concentration (0.2 and 20 μg/L) leading to chronic exposure has an impairing ability to induce a mitotic arrest at the time of spermatogenesis that reduces to 90% production of the embryo. The workers reported that these changes were demonstrated with the fragmentation of spermatozoa DNA (Corradetti et al. 2013).

Exposure to triclocarban and mercury at certain concentrations can lead to severe histological lesions in the testis of spermatozoa. Herein, the size of the sperm and its maturation were decreased in comparison with other fishes that are exposed individually as reported by different workers (Wang et al. 2016). Yin et al. (2017) reposted that the decreased concentration of sperm and spermatogenesis impairment occur when exposure occurs to DES and FLU, also affecting both meiotic and apoptotic processes.

In *Cyprinus carpio* (male carp), a delay in spermatogenesis was demonstrated when exposed to 2 mg/L PFOA for 56 days, where spermatocytes and spermatogonia dominated the germ cells.

Giari et al. (2016) reported that in testis, proliferation of Sertoli cells and interstitial tissue increased. About 14-day exposure to BPA concentrations caused alteration of the lobular structure and spermatogenic cysts with free spermatozoa as reported by the workers (Mandich et al. 2007). BPA exposure of goldfish induced hepatic vitellogenin increase, and gonadosomatic index and hepatosomatic index were not affected by acting as testicular spermatogenesis causing alteration of sperm maturing (Hatef et al. 2012a, c).

11 Spermatozoa Production, Morphology, Motility, and Velocity

Alavi et al. (2012a, b) reported that a dark field microscope was utilized to demonstrate the alteration of morphological sperm (including flagellum damage and beating) with the help of stroboscopic illumination and a phase contrast microscope having a high-speed video camera. They used CASA (computer-assisted sperm analysis) to check the motility and velocity of sperm because CASA is mainly based upon high video recording approaches (Alavi et al. 2012a). There are some parameters of sperm velocity that were affected by ECDs.

The curvilinear velocity and the straight-line velocity of spermatozoa decreased due to cadmium, mercury, bisphenol A, zinc, and tributyltin exposure in fish species. The various researchers reported in their study that when species are exposed to these EDCs, sperm shows a specific response (Hatef et al. (2013). Lahnsteiner and workers studied that exposure to cadmium and copper of different species like African catfish, burbot, and European chub showed a decrease in motility of sperm immediately. In addition, in S trutta fario, there was no such evidence found in response to similar metals (Lahnsteiner et al. 2004). However, exposure to metals and compounds (zinc, mercury, lead, nickel, 2,4-dichlorophenol, cyclohexane) showed effects (spermatozoa motility) in seabass, American catfish, burbot, and European chub as reported by various workers (Abascal et al. 2007; Lahnsteiner et al. 2004). Acosta et al. (2016), reported that in *D. rerio*, cadmium exposure also showed a negative effect on motility and curvilinear, straight line, and average path velocity, which may change in animal fertility rate (Acosta et al. 2016). Similarly, in medaka, there was no effect on its motility when exposed to nonylphenol as reported by the researchers (Hara et al. 2007). In the croaker, there are various pollutants such as estrogenic and non-estrogenic as per environmental requirements. The sperm motility was induced by blocking 17,20-beta,21-trihydroxy-4-pregnen-3-one and physiologically mediated by nongenomic steroid action as reported by the researchers (Thomas and Doughty 2004).

All the shreds of evidence indicate that with EDC recognition mode of action the variability is found in the physiology and biochemistry of spermatozoa and seminal plasma. In Perca fluviatilis, a decrease in sperm velocity was found when incubated for 3 h with Hg (mercury). This experiment was conducted to identify the mercury chloride effects on the sperm physiology and its structure and to find out the action mechanism of spermatozoa. After the experiment, a promising result was found. Mercury chloride acts on different sets of sperm PM, axoneme, and mid-piece. The different researchers reported that due to mercury chloride exposure, the morphological alterations increased, and stored ATP decreased (Hatef et al. 2011).

The in vitro sperm treatment was conducted in the same species with BPA. It showed that the velocity and motility of sperm decreased. In addition, the flagella structure gets changed straight to the c-shape as reported by the researchers (Hatef et al. 2010). The instant exposure of mercury in sea bass, i.e., 1 mg/L mercury for 5 min and 100 mg/L, showed an effect on velocity of sperm, but no effect occurred in the case of lead and copper (Abascal et al. 2007). With TBT, the common carp velocity was affected after the 1-day incubation period and graded mercury and zinc concentration (Rurangwa et al. 2002). In S. trutta fario, there was an observed impairment of spermatogenesis at BPA concentration during spawning period and pre-spawning as well.

At the start of the spawning season, at low BPA concentration velocity, sperm density motility reduced, and BPA reduces the swimming velocity at the middle of the spawning season. The motility rate reduced with a 2.4 μg/L concentration. A high concentration led to a complete impairment of spermatogenesis. Lahnsteiner et al. (2005) found that only one out of eight fish produce low-quality semen. In goldfish, having a long exposure to BPA, e.g., 20–30 days, the mortality of sperm decreased at 15, 30, 60, and 90 s post activation. Similarly, in short-term exposure to BPA, e.g., 10 days, the velocity is affected at 30, 60, and 90 s post-activation. All of this are due to hormonal imbalances as reported by the researchers (Hatef et al. 2012a). Further, in the same species, small flagella length was measured in spermatozoa with the help of contrast microscopy having a high concentration of mercury (Van Look and Kime 2003).

In addition, seminal plasma was utilized in rainbow trout to check or confirm the exposure of mercury and cadmium on sperm and showed an immediate effect on the motility of sperm after dilution at 4 h of exposure. There was no effect on sperm control after treatment up to 24 h. This shows that seminal plasma has an antiprotection role against the toxic effect of Hg on motility as reported by different workers (Dietrich et al. 2010).

Different researchers (Migliarini et al. 2005) reported that for cadmium acetate (1–10 mg/L) exposure for 48 h, a significant increase of HSP 70, mtt, and casp3 mRNA was demonstrated in testis, and the toxicity of the metal was highlighted in testis physiology. Same as the sperm exposure was demonstrated in starlet to NP and DES that induced ROS in spermatozoa. It causes the impairment of sperm, thereby decreasing intact sperm cell percentage (Shaliutina et al. 2017). The in vitro toxicological test system was developed for animal testing as the alternating source. About seven heavy metals were tested on zebrafish sperm parameter of motility, and progressive motility, VCL, and linearity were observed as a result, in which progressive motility was very sensitive. This can be utilized for the quick and accurate bioindicator of heavy metal load as reported by the researchers (Kollár et al. 2018).

12 Effects of EDCs on Steroidogenesis and Hormone Synthesis

In zebrafish, alterations can be observed in the morphology of testis. They occur when zebrafish are exposed to triclocarban and mercury associated with cyp19a, 3betaHSD, cyp17, and 17betaHSD as reported by the researchers (Wang et al. 2016). The exposure to BPA determined nonmonotonic effects in *G. rarus* on mRNA expression of cyp11a1, 3beta-HSD, cyp17a1, and cyp19ala at 15 μg/L concentration. Several mRNAs were codified for testicular steroidogenic enzyme when expression and androgen increased. Liu et al. (2014a) reported that the effect could be recruited through nonsignificant downregulation of cyp17c1 mRNA expression. The analysis performed at mRNA levels reported that the testicular steroidogenesis BPA-mediated actions may involve the signaling of sex steroid hormones, gonadotrophic receptor pathway, forkhead protein, etc. (Wang et al. 2012; Zhang et al. 2013). About 25 ng/L EE2 exposure demonstrated the same steroidogenic genes affected through BPA. This showed negative feedback on EE2 on FSH (Liu et al. 2012).

In goldfish, DEHP's disruptive activity was demonstrated. Herein, there were 30 days of exposure to the DEHP that help reduce 11-KT and LH levels and resulted in a decreased sperm quality due to the contaminated testicular hormone levels; moreover, there was no effect of DEHP on GnRH and Kiss-2/Gpr54 system, which does not exhibit estrogenic activities (Golshan et al. 2015). However, in the same condition, goldfish were exposed to VZ (normal vinclozolin) concentration. VZ can show potential effects by inhibiting 11-KT biosynthesis directly reported by different researchers (Hatef et al. 2012b). Chronic exposure was also demonstrated by it to BPA at different concentrations, i.e., 0.6, 4.5, and 11 μg/L. It decreased testosterone and 11-KT levels significantly and impaired the mortality and velocity of sperm (Hatef et al. 2012a). In spermatogenesis, the plasmatic 11-KT reduction was demonstrated. The deregulation of genes such as a1dh1a2, cyp26a1, nonos1, sycp3 bax, and bcl2 occurs in meiosis for signal regulation. There are some genes for orchestrating gonadal steroidogenesis (Yin et al. 2017). The plasma testosterone concentration increased due to di-*n*-butyl phthalate exposure, and decreased spiggin levels in males were observed. Therefore, DBP acts as an endocrine disruptor at a significant concentration in three-spined sticklebacks. Aoki et al. (2011)) suggested that DBP is similar to the anti-androgenic in fish as well.

Newborn *Poecilia reticulata* was exposed to a sublethal dose of NP (nonylphenol) for 3 months. This was described as the hepatic vitellogenin anomalous with a decrease in GSI (gonadosomatic index). Herein, a male does not approach a female with a wearied behavior. It suggests that nonylphenol has estrogenic potency good for disrupting reproduction as reported by various researchers (Cardinali et al. 2004).

12.1 Effects of EDCs on Sex Reversal

In different studies, various researchers reported that, in fish, at the time of critical developmental stages, exposure to EDCs can alter sex phenotype, impair the development of gonads, and discontinue in the reproductive process (Fenske and Segner 2004; Schäfers et al. 2007; Scholz and Klüver 2009; Zhu et al. 2016). In teleosts, the genetic and environmental processes play a significant role in sex determination which showed it an excellent experiment for this study in teleost (Guiguen et al.

2010). In differentiation in the fish ovary, estrogen plays a significant role that is susceptible to many endocrine disruptors and estrogenic mediators as well. Paul-Prasanth and coworkers analyzed the molecular and cellular mechanism of DES, involved in the differentiation of the ovary in medaka (*Oryzias latipes*) embryo (Paul-Prasanth et al. 2011). The short-term exposure to DES, from 0 to 8 pdf, resulted in inhibition of gene expression coded for GSDF, SCP3, and 42SP43. Similarly, 28 days of exposure causes a decrease in gene expression and the development of ovaries. This ovary highlights the estrogenic properties of pollutants. Generally, a pathological condition is considered when an oocyte is present in the testis of gonochoristic species. In Rutilus rutilus, generally high intersex was observed in wastewater treatment plants downstream in English rivers (Jobling et al. 1998, 2002). With aromatase upregulation and intersex individuals, in juvenile and male fish, vitellogenin and zona radiate proteins were also described that show xenoestrogenic contamination. Furthermore, more molecular markers identified feminization responses and intersex conditions within species, in which ovarian RNA showed expression of 5SrRNA in oocytes. It helps in the utilization of such RNA to find the xenoestrogenicity in field conditions (Ortiz-Zarragoitia et al. 2014). In largemouth bass, intersex was described in the form of testicular oocytes in 5 years in surface water Delmarva Peninsula, USA, as reported by the researchers (Yonkos et al. 2014). There was a permanent disruption resulting from EDCs. Maradonna et al. 2004 reported that it primarily depends upon nature, concentration, and exposure to the chemical. *G. niger*, when exposed to alkylphenols for 48 h, shows a significant increase in protein synthesis and vtg mRNA, but there were no histological changes found in testis. Some studies showed the ability of zebrafish to recover from estrogenic exposure as reported by different workers in their studies (Baumann et al. 2014a, b). The permanent disrupters' sexual development of zebrafish formed due to estrogen-antagonist fadrozole or in some cases EE2, proceeded by inducing masculinization and severe pathological testicular alterations.

Histopathology of gonad demonstrated ovarian atresia and IPFD (intestinal proteinaceous fluid deposits). In addition, Luzio et al. (2016) reported that EE2 was responsible to induce gonadal changes and showed reversible recovery as well. In carp, intersex phenotype was observed with exposure to BPA. About 1000 μg/L BPA was present in the male testis. Few previtellogenic oocytes were found within the testicular tissue (Mandich et al. 2007).

12.2 Effects of EDCs on Epigenetic Process

In 1942, researchers started to talk about epigenetics, because Conrad Waddington reported that it is the interaction of genes with their environment which brings phenotype into being. Epigenetic modifications are defined as reversible modification as well as heritable chromatin chemical modification, with no significant changes in the underlying DNA sequences. Epigenetics showed a significant role in different cellular processes through silencing from growth, metabolism, differentiation, and gene expression regulation by enhancing specific gene as well. Esteller (2007)) reported that no communication occurs in gene transcription and environment in epigenetic mechanisms. There are various epigenetic modifications such as DNA methylation and histone modification (Labbé et al. 2017). Various researchers reported that in EDC response, deregulation of microRNA implication was found in different species of vertebrates (Cameron et al. 2016). Similarly, in *D. rerio* and *Carassius auratus*, it was reported that fluoxetine exposure causes miRNA expression changes and represents a new biomarker group exposure to toxicants. There were increases in knowledge after exposure to chronic EDCs in miRNA transcriptional variation.

For a long time, the effects of EDCs were determined on epigenetic mechanisms. There are some examples such as HBCD (hexa-bromocyclododecane) and E2 (17-beta oestradiol) that were demonstrated in *Gasterosteus aculeatus* on global DNA methylation in gonads, which further showed increases in GGH (global genomic hypermethylation) in both gonads; however, male gonads showed significant increases than that of female gonads as reported by different researchers (Aniagu et al. 2008). Skinner (2014) reported the ability of EDCs to induce ETI (Skinner 2014).

There were some compounds that have the ability to affect germline directly, i.e., dioxin, PCBs, and phthalates. Embryo formation occurs by the fusion of oocytes and spermatozoa. The gamete formation occurs in puberty stage when the embryo developed in mature organism. The various researchers studied that this was the time when epigenetic modification plays a significant role in the cell differentiation at the time of development of embryo and gametogenesis process as well (Labbé et al. 2017).

In this way, it was reported that in zebrafish, their reproductive physiology was impaired when exposed to BPA (5 μl/L). Santangeli and workers reported that in mature follicles, signal downregulation in growth and oocyte maturation acts with apoptosis promotion (Santangeli et al. 2016), due to change in chromatin structure with some modifications of histone. It was observed that BPA shows negative effects on the female reproductive system. It may occur due to the high ability to deregulate the mechanisms. The interference of bisphenol A reported with histone modification in the ovary of zebrafish. It leads to decreases in 1hcgr mRNA levels that further affect the global methylation, resulting in interference in dnmt expression as reported by various researchers (Santangeli et al. 2016; Santangeli et al. 2017). There are some examples that show the upregulation of cyp19a1a mRNA when exposed with BPA (15 μg/L) for 7 days in *G. rarus*. On the other hand, the exposure of BPA for 35 days resulted in aromatase mRNA downregulation. Researchers have reported an inverse correlation between cyp19a1 DNA methylation and its mRNA levels, but only in the ovary. It was observed that cyp191a1 transcriptional control occurs due to DNA methylation (Liu et al. 2014b). BPA helps in DNA methylation changes in cyp17a1 and cyp11a1 5′ flanking regions at CpG loci. The direct and inverse correlation was demonstrated in CpG methylation and cyp11a1 mRNA levels and in CpG methylation or cyp17a1 mRNA levels as reported by various researchers (Zhang et al. 2017). BPA also caused hypermethylation in testis that decreases the

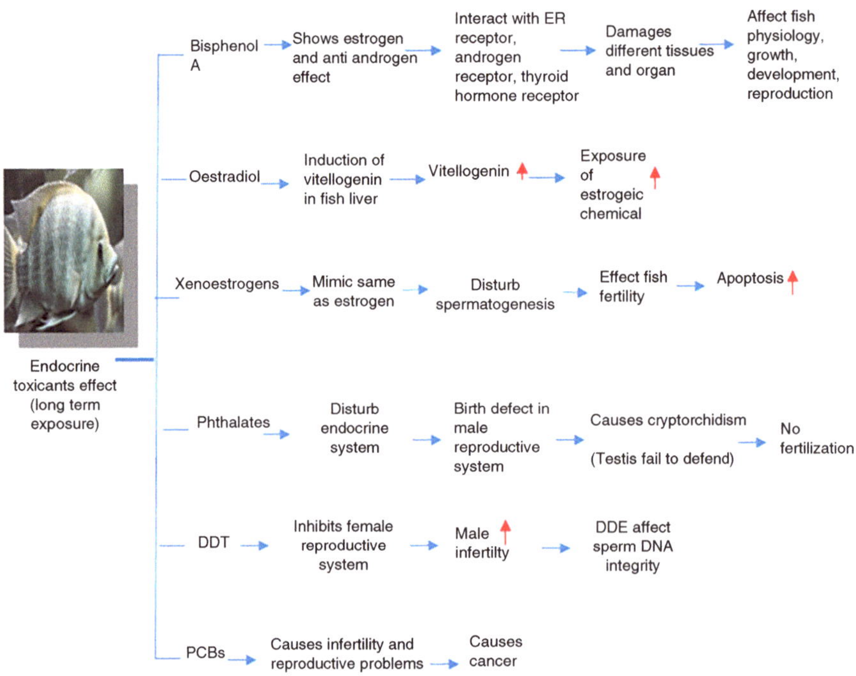

Fig. 2 The mechanism of endocrine toxicant effect after a long-term exposure

TETs after a week of exposure. The more concentration of BPA, the more the global DNA methylation occurs. This is because of the upregulation of DNMTs owing to glutathione synthesis (Yuan et al. 2016). On the other hand, exposure to BPA at a concentration of 1 mg/L has been found to result in a decrease in dnmt1 expression, leading to down-regulation of global methylation (Laing et al. 2016). All these results showed that a single pollutant can induce the epigenetic effects, may be time dependent, and differ in vertebrates. A field survey also demonstrated the different levels of contamination in the same species at different regions in southwest of France. The fish sample from both contaminated part and from clean site showed high aromatase and FSHr levels in DNA methylation levels in genes encoded for these. It can be suggested that gonadal gene methylation level increased due to the pollution that plays a significant role in growth and differentiation of oocytes. Pierron and coworkers reported that chronic pollution affects the animals' reproductive capacity and the offspring may inherit such epigenetic marks (Pierron et al. 2014) (Fig. 2).

References

Abascal FJ, Cosson J, Fauvel C (2007) Characterization of sperm motility in sea bass: the effect of heavy metals and physicochemical variables on sperm motility. J Fish Biol 70(2):509–522

Acosta IB, Junior ASV, Silva EF, Cardoso TF, Caldas JS, Jardim RD, Corcini CD (2016) Effects of exposure to cadmium in sperm cells of zebrafish, Danio rerio. Toxicol Rep 3:696–700
Adami HO, Lipworth L, Titus-Ernstoff L, Hsieh CC, Hanberg A, Ahlborg U, Baron J, Trichopoulos D (1995) Organochlorine compounds and estrogen-related cancers in women. Cancer Causes Control 6(6):551–566
Adnan Amin AT, Naik R, Nayak H, Priyadarshini N (2016) Physiological responses of rohu (Labeo rohita) exposed to sub-lethal concentrations of Lead nitrate. Biochem Cell Arch 17(1):271–274
Adnan Amin AT, Naik R, Priyadarshini N (2015) Sub lethal effect of arsenic on oxygen consumption and behaviour of rohu, Labeo rohita. Asian Journal of Biochemical and Pharmaceutical Research 2(5):43–47
Adnan Amin AT, Naik R, Priyadarshini N, Nayak H, Sruthi Sree C (2017) Toxic effect of heavy metal Lead on oxygen consumption of rohu (Labeo rohita) fingerlings. Biochem Cell Arch 17(1):225–228
Ahel M, Giger W, Schaffner C (1994) Behaviour of alkylphenol polyethoxylate surfactants in the aquatic environment—II. Occurrence and transformation in rivers. Water Res 28(5):1143–1152
Ahlborg UG, Lipworth L, Titus-Ernstoff L, Hsieh CC, Hanberg A, Baron J, Trichopoulos D, Adami HO (1995) Organochlorine compounds in relation to breast cancer, endometrial cancer, and endometriosis: an assessment of the biological and epidemiological evidence. Crit Rev Toxicol 25(6):463–531
Alavi SMH, Hatef A, Pšenička M, Kašpar V, Boryshpolets S, Dzyuba B, Cosson J, Bondarenko V, Rodina M, Gela D, Linhart O (2012a) Sperm biology and control of reproduction in sturgeon: (II) sperm morphology, acrosome reaction, motility and cryopreservation. Rev Fish Biol Fish 22(4):861–886
Alavi SMH, Rodina M, Gela D, Linhart O (2012b) Sperm biology and control of reproduction in sturgeon:(I) testicular development, sperm maturation and seminal plasma characteristics. Rev Fish Biol Fish 22(3):695–717
Aniagu SO, Williams TD, Allen Y, Katsiadaki I, Chipman JK (2008) Global genomic methylation levels in the liver and gonads of the three-spine stickleback (Gasterosteus aculeatus) after exposure to hexabromocyclododecane and 17-β oestradiol. Environ Int 34(3):310–317
Aoki KA, Harris CA, Katsiadaki I, Sumpter JP (2011) Evidence suggesting that di-n-butyl phthalate has antiandrogenic effects in fish. Environ Toxicol Chem 30(6):1338–1345
Arman S (2021) Effects of acute triclosan exposure on gill and liver tissues of zebrafish (Danio rerio). In: Annales de Limnologie-international journal of limnology, vol 57. EDP Sciences, p 6
Arukwe A, Förlin L, Goksøyr A (1997a) Xenobiotic and steroid biotransformation enzymes in Atlantic salmon (Salmo salar) liver treated with an estrogenic compound, 4-nonylphenol. Environmental Toxicology and Chemistry: An International Journal 16(12):2576–2583
Arukwe A, Knudsen FR, Goksøyr A (1997b) Fish zona radiata (eggshell) protein: a sensitive biomarker for environmental estrogens. Environ Health Perspect 105(4):418–422
Baumann L, Knörr S, Keiter S, Nagel T, Rehberger K, Volz S, Oberrauch S, Schiller V, Fenske M, Holbech H, Segner H (2014a) Persistence of endocrine disruption in zebrafish (Danio rerio) after discontinued exposure to the androgen 17β-trenbolone. Environ Toxicol Chem 33(11): 2488–2496
Baumann L, Knörr S, Keiter S, Rehberger K, Volz S, Schiller V, Fenske M, Holbech H, Segner H, Braunbeck T (2014b) Reversibility of endocrine disruption in zebrafish (Danio rerio) after discontinued exposure to the estrogen 17α-ethinylestradiol. Toxicol Appl Pharmacol 278(3): 230–237
Berg C, Halldin K, Brunström B (2001) Effects of bisphenol a and tetrabromobisphenol a on sex organ development in quail and chicken embryos. Environmental Toxicology and Chemistry: An International Journal 20(12):2836–2840
Bernhoft A, Skaare JU, Wiig O, Derocher AE, Larsen HJS (2000) Possible immunotoxic effects of organochlorines in polar bears (Ursus maritimus) at Svalbard. J Toxicol Environ Health A 59(7): 561–574

Birceanu O, Mai T, Vijayan MM (2015) Maternal transfer of bisphenol a impacts the ontogeny of cortisol stress response in rainbow trout. Aquat Toxicol 168:11–18

Blair RM, Fang H, Branham WS, Hass BS, Dial SL, Moland CL, Tong W, Shi L, Perkins R, Sheehan DM (2000) The estrogen receptor relative binding affinities of 188 natural and xenochemicals: structural diversity of ligands. Toxicol Sci 54(1):138–153

Cabaton NJ, Wadia PR, Rubin BS, Zalko D, Schaeberle CM, Askenase MH, Gadbois JL, Tharp AP, Whitt GS, Sonnenschein C, Soto AM (2011) Perinatal exposure to environmentally relevant levels of bisphenol a decreases fertility and fecundity in CD-1 mice. Environ Health Perspect 119(4):547–552

Cameron BE, Craig PM, Trudeau VL (2016) Implication of microRNA deregulation in the response of vertebrates to endocrine disrupting chemicals. Environ Toxicol Chem 35(4):788–793

Cardinali M, Maradonna F, Olivotto I, Bortoluzzi G, Mosconi G, Polzonetti-Magni AM, Carnevali O (2004) Temporary impairment of reproduction in freshwater teleost exposed to nonylphenol. Reprod Toxicol 18(4):597–604

Carnevali O, Santangeli S, Forner-Piquer I, Basili D, Maradonna F (2018) Endocrine-disrupting chemicals in aquatic environment: what are the risks for fish gametes? Fish Physiol Biochem 44: 1561–1576

Chakrabarty S, Rajakumar A, Raghuveer K, Sridevi P, Mohanachary A, Prathibha Y, Bashyam L, Dutta-Gupta A, Senthilkumaran B (2012) Endosulfan and flutamide, alone and in combination, target ovarian growth in juvenile catfish, Clarias batrachus. Comparative Biochemistry and Physiology Part C: Toxicology & Pharmacology 155(3):491–497

Chen J, Xiao Y, Gai Z, Li R, Zhu Z, Bai C, Tanguay RL, Xu X, Huang C, Dong Q (2015) Reproductive toxicity of low level bisphenol a exposures in a two-generation zebrafish assay: evidence of male-specific effects. Aquat Toxicol 169:204–214

Chen L, Au DW, Hu C, Zhang W, Zhou B, Cai L, Giesy JP, Qian PY (2017) Linking genomic responses of gonads with reproductive impairment in marine medaka (Oryzias melastigma) exposed chronically to the chemopreventive and antifouling agent, 3, 3′-diindolylmethane (DIM). Aquat Toxicol 183:135–143

Colborn T, Clement C (1992) Chemically-induced alterations in sexual and functional development: the wildlife/human connection. Princeton Scientific Pub, Co

Corradetti B, Stronati A, Tosti L, Manicardi G, Carnevali O, Bizzaro D (2013) Bis-(2-ethylexhyl) phthalate impairs spermatogenesis in zebrafish (Danio rerio). Reprod Biol 13(3):195–202

Corrales J, Kristofco LA, Steele WB, Yates BS, Breed CS, Williams ES, Brooks BW (2015) Global assessment of bisphenol a in the environment: review and analysis of its occurrence and bioaccumulation. Dose-Response 13(3):1559325815598308

Crain DA, Janssen SJ, Edwards TM, Heindel J, Ho SM, Hunt P, Iguchi T, Juul A, McLachlan JA, Schwartz J, Skakkebaek N (2008) Female reproductive disorders: the roles of endocrine-disrupting compounds and developmental timing. Fertil Steril 90(4):911–940

Dar SA, Chatterjee A, Rather MA, Chetia D, Srivastava PP, Gupta S (2020) Identification, functional characterization and expression profiling of cytochrome p450 1A (CYP1A) gene in Labeo rohita against emamectin benzoate. Int J Biol Macromol 158:1268–1278

Dar SA, Najar AM, Balkhi MH, Rather MA (2012) Histological observations of ovarian development in endemic snow trout, Schizopyge Esocinus (Heckel, 1838). Indian Veterinary Journal 89(8):110

Darbre PD (2006) Environmental oestrogens, cosmetics and breast cancer. Best practice & Research Clinical Endocrinology and Metabolism 20(1):121–143

Diamanti-Kandarakis E, Bourguignon JP, Giudice LC, Hauser R, Prins GS, Soto AM, Zoeller RT, Gore AC (2009) Endocrine-disrupting chemicals: an Endocrine Society scientific statement. Endocr Rev 30(4):293–342

Dietrich GJ, Dietrich M, Kowalski RK, Dobosz S, Karol H, Demianowicz W, Glogowski J (2010) Exposure of rainbow trout milt to mercury and cadmium alters sperm motility parameters and reproductive success. Aquat Toxicol 97(4):277–284

Ding N, Harlow SD, Randolph JF Jr, Loch-Caruso R, Park SK (2020) Perfluoroalkyl and polyfluoroalkyl substances (PFAS) and their effects on the ovary. Hum Reprod Update 26(5): 724–752

Donohoe RM, Curtis LR (1996) Estrogenic activity of chlordecone, o, p′-DDT and o, p′-DDE in juvenile rainbow trout: induction of vitellogenesis and interaction with hepatic estrogen binding sites. Aquat Toxicol 36(1–2):31–52

Durando M, Kass L, Piva J, Sonnenschein C, Soto AM, Luque EH, Muñoz-de-Toro M (2007) Prenatal bisphenol a exposure induces preneoplastic lesions in the mammary gland in Wistar rats. Environ Health Perspect 115(1):80–86

Ekelund R, Bergman Å, Granmo Å, Berggren M (1990) Bioaccumulation of 4-nonylphenol in marine animals—a re-evaluation. Environ Pollut 64(2):107–120

El-Sayed AFM, Abdel-Aziz ESH, Abdel-Ghani HM (2012) Effects of phytoestrogens on sex reversal of Nile tilapia (Oreochromis niloticus) larvae fed diets treated with 17-α-methyltestosterone. Aquaculture 360:58–63

Eo J, Lim H (2008) Effects of bisphenol a on gene expression and apoptosis of Leydig cells in the mouse testis. Korean Journal of Reproductive Medicine 35(3):181–191

Esteller M (2007) Epigenetic gene silencing in cancer: the DNA hypermethylome. Hum Mol Genet 16(R1):R50–R59

Fenske M, Segner H (2004) Aromatase modulation alters gonadal differentiation in developing zebrafish (Danio rerio). Aquat Toxicol 67(2):105–126

Fernández M, Bourguignon N, Lux-Lantos V, Libertun C (2010) Neonatal exposure to bisphenol a and reproductive and endocrine alterations resembling the polycystic ovarian syndrome in adult rats. Environ Health Perspect 118(9):1217–1222

Fitzgerald AC, Peyton C, Dong J, Thomas P (2015) Bisphenol a and related alkylphenols exert nongenomic estrogenic actions through a G protein-coupled estrogen receptor 1 (Gper)/epidermal growth factor receptor (Egfr) pathway to inhibit meiotic maturation of zebrafish oocytes. Biol Reprod 93(6):135–131

Forner-Piquer I, Maradonna F, Gioacchini G, Santangeli S, Allarà M, Piscitelli F, Habibi HR, Di Marzo V, Carnevali O (2017) Dose-specific effects of di-isononyl phthalate on the endocannabinoid system and on liver of female zebrafish. Endocrinology 158(10):3462–3476

Fossi MC, Marsili L (2003) Effects of endocrine disruptors in aquatic mammals. Pure Appl Chem 75(11–12):2235–2247

Gatidou G, Thomaidis NS, Stasinakis AS, Lekkas TD (2007) Simultaneous determination of the endocrine disrupting compounds nonylphenol, nonylphenol ethoxylates, triclosan and bisphenol a in wastewater and sewage sludge by gas chromatography–mass spectrometry. J Chromatogr A 1138(1–2):32–41

Giari L, Vincenzi F, Badini S, Guerranti C, Dezfuli BS, Fano EA, Castaldelli G (2016) Common carp Cyprinus carpio responses to sub-chronic exposure to perfluorooctanoic acid. Environ Sci Pollut Res 23(15):15321–15330

Gillesby BE, Zacharewski TR (1998) Exoestrogens: mechanisms of action and strategies for identification and assessment. Environmental Toxicology and Chemistry: An International Journal 17(1):3–14

Gimeno S, Komen H, Gerritsen AG, Bowmer T (1998) Feminisation of young males of the common carp, Cyprinus carpio, exposed to 4-tert-pentylphenol during sexual differentiation. Aquat Toxicol 43(2–3):77–92

Golshan M, Hatef A, Socha M, Milla S, Butts IA, Carnevali O, Rodina M, Sokołowska-Mikołajczyk M, Fontaine P, Linhart O, Alavi SMH (2015) Di-(2-ethylhexyl)-phthalate disrupts pituitary and testicular hormonal functions to reduce sperm quality in mature goldfish. Aquat Toxicol 163:16–26

Grun F, Watanabe H, Zamanian Z, Maeda L, Arima K, Cubacha R, Gardiner DM, Kanno J, Iguchi T, Blumberg B (2006) Endocrine-disrupting organotin compounds are potent inducers of adipogenesis in vertebrates. Mol Endocrinol 20(9):2141–2155
Guiguen Y, Fostier A, Piferrer F, Chang CF (2010) Ovarian aromatase and estrogens: a pivotal role for gonadal sex differentiation and sex change in fish. Gen Comp Endocrinol 165(3):352–366
Guillette LJ Jr, Crain DA, Rooney AA, Pickford DB (1995) Organization versus activation: the role of endocrine-disrupting contaminants (EDCs) during embryonic development in wildlife. Environ Health Perspect 103(suppl 7):157–164
Hajam YA, Rani R, Sharma P, Kumar R, Verma SK (2021) Zebrafish (Danio rerio): a versatile model for reproductive biology. In: Recent Updates in Molecular Endocrinology and Reproductive Physiology of Fish. Springer, Berlin, pp 105–120
Hajam YA, Kumar R, Rani R, Sharma P (2022) Efficacy of different treatments available against bacterial pathogens in fish. In: Bacterial fish diseases. Academic Press, pp 379–398
Hansen LG (1998) Stepping backward to improve assessment of PCB congener toxicities. Environ Health Perspect 106(suppl 1):171–189
Hansen PD, Von Westernhagen H, Rosenthal H (1985) Chlorinated hydrocarbons and hatching success in Baltic herring spring spawners. Mar Environ Res 15(1):59–76
Hara Y, Strüssmann CA, Hashimoto S (2007) Assessment of short-term exposure to nonylphenol in Japanese medaka using sperm velocity and frequency of motile sperm. Arch Environ Contam Toxicol 53(3):406–410
Harding LE, Harris ML, Stephen CR, Elliott JE (1999) Reproductive and morphological condition of wild mink (Mustela vison) and river otters (Lutra canadensis) in relation to chlorinated hydrocarbon contamination. Environ Health Perspect 107(2):141–147
Hatef A, Alavi SMH, Abdulfatah A, Fontaine P, Rodina M, Linhart O (2012a) Adverse effects of bisphenol a on reproductive physiology in male goldfish at environmentally relevant concentrations. Ecotoxicol Environ Saf 76:56–62
Hatef A, Alavi SMH, Butts IA, Policar T, Linhart O (2011) Mechanism of action of mercury on sperm morphology, adenosine triphosphate content, and motility in Perca fluviatilis (Percidae; Teleostei). Environ Toxicol Chem 30(4):905–914
Hatef A, Alavi SMH, Golshan M, Linhart O (2013) Toxicity of environmental contaminants to fish spermatozoa function in vitro—a review. Aquat Toxicol 140:134–144
Hatef A, Alavi SMH, Linhartova Z, Rodina M, Policar T, Linhart O (2010) In vitro effects of bisphenol a on sperm motility characteristics in Perca fluviatilis L.(Percidae; Teleostei). J Appl Ichthyol 26(5):696–701
Hatef A, Alavi SMH, Milla S, Křišťan J, Golshan M, Fontaine P, Linhart O (2012b) Anti-androgen vinclozolin impairs sperm quality and steroidogenesis in goldfish. Aquat Toxicol 122:181–187
Hatef A, Zare A, Alavi SMH, Habibi HR, Linhart O (2012c) Modulations in androgen and estrogen mediating genes and testicular response in male goldfish exposed to bisphenol a. Environ Toxicol Chem 31(9):2069–2077
Haynes D, Müller JF, McLachlan MS (1999) Polychlorinated dibenzo-p-dioxins and dibenzofurans in great barrier reef (Australia) dugongs (Dugong dugon). Chemosphere 38(2):255–262
Huang Q, Chen Y, Lin L, Liu Y, Chi Y, Lin Y, Ye G, Zhu H, Dong S (2017) Different effects of bisphenol a and its halogenated derivatives on the reproduction and development of Oryzias melastigma under environmentally relevant doses. Sci Total Environ 595:752–758
Huang YQ, Wong CKC, Zheng JS, Bouwman H, Barra R, Wahlström B, Neretin L, Wong MH (2012) Bisphenol a (BPA) in China: a review of sources, environmental levels, and potential human health impacts. Environ Int 42:91–99
ICES. ICES CM 1993/N:3 (1992)
Jarman WM, Norstrom RJ, Muir DCG, Rosenberg B, Simon M, Baird RW (1996) Levels of organochlorine compounds, including PCDDS and PCDFS, in the blubber of cetaceans from the west coast of North America. Mar Pollut Bull 32(5):426–436

Jobling S, Beresford N, Nolan M, Rodgers-Gray T, Brighty GC, Sumpter JP, Tyler CR (2002) Altered sexual maturation and gamete production in wild roach (Rutilus rutilus) living in rivers that receive treated sewage effluents. Biol Reprod 66(2):272–281

Jobling S, Nolan M, Tyler CR, Brighty G, Sumpter JP (1998) Ecotoxicology of human pharmaceuticals. Environ Sci Technol 32(17):2498–2506

Jobling S, Sumpter JP (1993) Detergent components in sewage effluent are weakly oestrogenic to fish: an in vitro study using rainbow trout (Oncorhynchus mykiss) hepatocytes. Aquat Toxicol 27(3–4):361–372

Jobling S, Sumpter JP, Sheahan D, Osborne JA, Matthiessen P (1996) Inhibition of testicular growth in rainbow trout (Oncorhynchus mykiss) exposed to estrogenic alkylphenolic chemicals. Environmental Toxicology and Chemistry: An International Journal 15(2):194–202

Karnam SS, Ghosh RC, Mondal S, Mondal M (2015) Evaluation of subacute bisphenol–a toxicity on male reproductive system. Veterinary world 8(6):738

Kavlock RJ, Daston GP, DeRosa C, Fenner-Crisp P, Gray LE, Kaattari S, Lucier G, Luster M, Mac MJ, Maczka C, Miller R (1996) Research needs for the risk assessment of health and environmental effects of endocrine disruptors: a report of the US EPA-sponsored workshop. Environ Health Perspect 104(suppl 4):715–740

Kime DE (1995) The effects of pollution on reproduction in fish. Rev Fish Biol Fish 5(1):52–95

Kime DE (1999) A strategy for assessing the effects of xenobiotics on fish reproduction. Sci Total Environ 225(1–2):3–11

Kinch CD, Ibhazehiebo K, Jeong JH, Habibi HR, Kurrasch DM (2015) Low-dose exposure to bisphenol a and replacement bisphenol S induces precocious hypothalamic neurogenesis in embryonic zebrafish. Proc Natl Acad Sci 112(5):1475–1480

King-Heiden TC, Mehta V, Xiong KM, Lanham KA, Antkiewicz DS, Ganser A, Heideman W, Peterson RE (2012) Reproductive and developmental toxicity of dioxin in fish. Mol Cell Endocrinol 354(1–2):121–138

Kollár T, Kása E, Ferincz Á, Urbányi B, Csenki-Bakos Z, Horváth Á (2018) Development of an in vitro toxicological test system based on zebrafish (Danio rerio) sperm analysis. Environ Sci Pollut Res 25(15):14426–14436

Kuriyama SN, Talsness CE, Grote K, Chahoud I (2005) Developmental exposure to low-dose PBDE-99: effects on male fertility and neurobehavior in rat offspring. Environ Health Perspect 113(2):149–154

Kusakabe M, Kobayashi T, Todo T, Mark Lokman P, Nagahama Y, Young G (2002) Molecular cloning and expression during spermatogenesis of a cDNA encoding testicular 11β-hydroxylase (P45011β) in rainbow trout (Oncorhynchus mykiss). Mol Reprod Dev 62(4):456–469

La Merrill MA, Vandenberg LN, Smith MT, Goodson W, Browne P, Patisaul HB, Guyton KZ, Kortenkamp A, Cogliano VJ, Woodruff TJ, Rieswijk L (2020) Consensus on the key characteristics of endocrine-disrupting chemicals as a basis for hazard identification. Nat Rev Endocrinol 16(1):45–57

Labbé C, Robles V, Herraez MP (2017) Epigenetics in fish gametes and early embryo. Aquaculture 472:93–106

Lahnsteiner F, Berger B, Kletzl M, Weismann T (2005) Effect of bisphenol a on maturation and quality of semen and eggs in the brown trout. Salmo trutta f fario Aquatic Toxicology 75(3): 213–224

Lahnsteiner F, Mansour N, Berger B (2004) The effect of inorganic and organic pollutants on sperm motility of some freshwater teleosts. J Fish Biol 65(5):1283–1297

Laing LV, Viana J, Dempster EL, Trznadel M, Trunkfield LA, Uren Webster TM, Van Aerle R, Paull GC, Wilson RJ, Mill J, Santos EM (2016) Bisphenol a causes reproductive toxicity, decreases dnmt1 transcription, and reduces global DNA methylation in breeding zebrafish (Danio rerio). Epigenetics 11(7):526–538

Li YJ, Song TB, Cai YY, Zhou JS, Song X, Zhao X, Wu XL (2009) Bisphenol a exposure induces apoptosis and upregulation of Fas/FasL and caspase-3 expression in the testes of mice. Toxicol Sci 108(2):427–436

Liao C, Kannan K (2011) Widespread occurrence of bisphenol a in paper and paper products: implications for human exposure. Environ Sci Technol 45(21):9372–9379
Linhart O, Alavi SMH, Rodina M, Gela D, Cosson J (2008) Comparison of sperm velocity, motility and fertilizing ability between firstly and secondly activated spermatozoa of common carp (Cyprinus carpio). J Appl Ichthyol 24(4):386–392
Liu S, Qin F, Wang H, Wu T, Zhang Y, Zheng Y, Li M, Wang Z (2012) Effects of 17-α-ethinylestradiol and bisphenol a on steroidogenic messenger ribonucleic acid levels in the rare minnow gonads. Aquat Toxicol 122:19–27
Liu Y, Chen S, Liu S, Zhang Y, Yuan C, Wang Z (2014a) DNA methylation in the 5′ flanking region of cytochrome P450 17 in adult rare minnow Gobiocypris rarus—tissue difference and effects of 17α-ethinylestradiol and 17α-methyltestoterone exposures. Comparative Biochemistry and Physiology Part C: Toxicology and Pharmacology 162:16–22
Liu Y, Yuan C, Chen S, Zheng Y, Zhang Y, Gao J, Wang Z (2014b) Global and cyp19a1a gene specific DNA methylation in gonads of adult rare minnow Gobiocypris rarus under bisphenol a exposure. Aquat Toxicol 156:10–16
Liu ZH, Kanjo Y, Mizutani S (2010) A review of phytoestrogens: their occurrence and fate in the environment. Water Res 44(2):567–577
Longnecker MP, Rogan WJ, Lucier G (1997) The human health effects of DDT (dichlorodiphenyl-trichloroethane) and PCBs (polychlorinated biphenyls) and an overview of organochlorines in public health. Annu Rev Public Health 18:211
Luzio A, Monteiro SM, Rocha E, Fontaínhas-Fernandes AA, Coimbra AM (2016) Development and recovery of histopathological alterations in the gonads of zebrafish (Danio rerio) after single and combined exposure to endocrine disruptors (17α-ethinylestradiol and fadrozole). Aquat Toxicol 175:90–105
Mahajan R, Blair A, Lynch CF, Schroeder P, Hoppin JA, Sandler DP, Alavanja MC (2006) Fonofos exposure and cancer incidence in the agricultural health study. Environ Health Perspect 114(12): 1838–1842
Maiti D, Ansari I, Rather MA, Deepa A (2019) Comprehensive review on wastewater discharged from the coal-related industries–characteristics and treatment strategies. Water Sci Technol 79(11):2023–2035
Mandich A, Bottero S, Benfenati E, Cevasco A, Erratico C, Maggioni S, Massari A, Pedemonte F, Viganò L (2007) In vivo exposure of carp to graded concentrations of bisphenol a. Gen Comp Endocrinol 153(1–3):15–24
Maradonna F, Polzonetti V, Bandiera SM, Migliarini B, Carnevali O (2004) Modulation of the hepatic CYP1A1 system in the marine fish Gobius Niger, exposed to xenobiotic compounds. Environ Sci Technol 38(23):6277–6282
Marchand O, Govoroun M, D'Cotta H, McMeel O, Lareyre JJ, Bernot A, Laudet V, Guiguen Y (2000) DMRT1 expression during gonadal differentiation and spermatogenesis in the rainbow trout, Oncorhynchus mykiss. Biochimica et Biophysica Acta (BBA)-Gene Structure and Expression 1493(1–2):180–187
Margiotta-Casaluci L, Hannah RE, Sumpter JP (2013) Mode of action of human pharmaceuticals in fish: the effects of the 5-alpha-reductase inhibitor, dutasteride, on reproduction as a case study. Aquat Toxicol 128:113–123
Martyniuk CJ, Denslow ND (2012) Exploring androgen-regulated pathways in teleost fish using transcriptomics and proteomics. Integr Comp Biol 52(5):695–704
Mason CF, Macdonald SM (1993) Impact of organochlorine pesticide residues and PCBs on otters (Lutra lutra): a study from western Britain. Sci Total Environ 138(1–3):127–145
Mazet JA, Gardner IA, Jessup DA, Lowenstine LJ (2001) Effects of petroleum on mink applied as a model for reproductive success in sea otters. J Wildl Dis 37(4):686–692
McLachlan JA (1985) Estrogens in the environment II. Influences in development–. Elsevier Science, New York.pp435

Meeker JD, Ehrlich S, Toth TL, Wright DL, Calafat AM, Trisini AT, Ye X, Hauser R (2010) Semen quality and sperm DNA damage in relation to urinary bisphenol a among men from an infertility clinic. Reprod Toxicol 30(4):532–539

Miao M, Yuan W, He Y, Zhou Z, Wang J, Gao E, Li G, Li DK (2011) In utero exposure to bisphenol-a and anogenital distance of male offspring. Birth Defects Res A Clin Mol Teratol 91(10):867–872

Migliarini B, Campisi AM, Maradonna F, Truzzi C, Annibaldi A, Scarponi G, Carnevali O (2005) Effects of cadmium exposure on testis apoptosis in the marine teleost Gobius Niger. Gen Comp Endocrinol 142(1–2):241–247

Minh TB, Nakata H, Watanabe M, Tanabe S, Miyazaki N, Jefferson TA, Prudente M, Subramanian A (2000) Isomer-specific accumulation and toxic assessment of polychlorinated biphenyls, including coplanar congeners, in cetaceans from the North Pacific and Asian coastal waters. Arch Environ Contam Toxicol 39(3):398–410

Molina A, Abril N, Morales-Prieto N, Monterde J, Ayala N, Lora A, Moyano R (2018) Hypothalamic-pituitary-ovarian axis perturbation in the basis of bisphenol a (BPA) reproductive toxicity in female zebrafish (Danio rerio). Ecotoxicol Environ Saf 156:116–124

Monod G (1985) Egg mortality of Lake Geneva charr (Salvelinus alpinus L.) contaminated by PCB and DDT derivatives. Bull Environ Contam Toxicol 35:531–536

Moyle, P.B. and Cech, J.J., 2000. Fishes: an introduction to ichthyology

Muir DCG, Ford CA, Rosenberg B, Norstrom RJ, Simon M, Beland P (1996) Persistent organochlorines in beluga whales (Delphinapterus leucas) from the St Lawrence river estuary-I. concentrations and patterns of specific PCBs, chlorinated pesticides and polychlorinated dibenzo-p-dioxins and dibenzofurans. Environ Pollut 93(2):219–234

Munshi AB, Karim N, Shaukat S, Hashmi D, Boardman GD, Flick GJ (2013) Toxicity of phthalate esters in fish and shellfish from Virginia Beach using matrix solid phase dispersion (MSPD) and GC-MS. J Chem Soc Pak 35(6):1463–1471

Nagato EG, Simpson AJ, Simpson MJ (2016) Metabolomics reveals energetic impairments in Daphnia magna exposed to diazinon, malathion and bisphenol-a. Aquat Toxicol 170:175–186

Nelson JS (1984) Fishes of the world, 2nd edn. Wiley-Inter Science, New York, p 521

Nimrod AC, Benson WH (1996) Environmental estrogenic effects of alkylphenol ethoxylates. Crit Rev Toxicol 26(3):335–364

Oberdörster E, Cheek AO (2001) Gender benders at the beach: endocrine disruption in marine and estuarine organisms. Environmental Toxicology and Chemistry: An International Journal 20(1): 23–36

Ortiz-Zarragoitia M, Bizarro C, Rojo-Bartolomé I, Diaz de Cerio O, Cajaraville MP, Cancio I (2014) Mugilid fish are sentinels of exposure to endocrine disrupting compounds in coastal and estuarine environments. Mar Drugs 12(9):4756–4782

Palmer BD, Palmer SK (1995) Vitellogenin induction by xenobiotic estrogens in the red-eared turtle and African clawed frog. Environ Health Perspect 103(suppl 4):19–25

Paul-Prasanth B, Shibata Y, Horiguchi R, Nagahama Y (2011) Exposure to diethylstilbestrol during embryonic and larval stages of medaka fish (Oryzias latipes) leads to sex reversal in genetic males and reduced gonad weight in genetic females. Endocrinology 152(2):707–717

Pelissero C, Bennetau B, Babin P, Le Menn F, Dunogues J (1991a) The estrogenic activity of certain phytoestrogens in the Siberian sturgeon Acipenser baeri. J Steroid Biochem Mol Biol 38(3):293–299

Pelissero C, Flouriot G, Foucher JL, Bennetau B, Dunogues JFLG, Le Gac F, Sumpter JP (1993) Vitellogenin synthesis in cultured hepatocytes; an in vitro test for the estrogenic potency of chemicals. J Steroid Biochem Mol Biol 44(3):263–272

Pelissero C, Le Menn F, Kaushick S (1991b) Estrogenic effect of dietary soya bean meal on vitellogenesis in cultured Siberian sturgeon Acipenser baeri. Gen Comp Endocrinol 83(3): 447–457

Pereira TSB, Boscolo CNP, Felício AA, Batlouni SR, Schlenk D, de Almeida EA (2016) Estrogenic activities of diuron metabolites in female Nile tilapia (Oreochromis niloticus). Chemosphere 146:497–502

Pierron F, Bureau du Colombier S, Moffett A, Caron A, Peluhet L, Daffe G, Lambert P, Elie P, Labadie P, Budzinski H, Dufour S (2014) Abnormal ovarian DNA methylation programming during gonad maturation in wild contaminated fish. Environ Sci Technol 48(19):11688–11695

Piferrer F, Guiguen Y (2008) Fish gonadogenesis. Part II: molecular biology and genomics of sex differentiation. Rev Fish Sci 16(sup1):35–55

Popek W, Dietrich G, Glogowski J, Demska-Zakeś K, Drag-Kozak E, Sionkowski J, Łuszczek-Trojan E, Epler P, Demianowicz W, Sarosiek B, Kowalski R (2006) Influence of heavy metals and 4-nonylphenol on reproductive function in fish. Reprod Biol 6:175–188

Power DM, Llewellyn L, Faustino M, Nowell MA, Björnsson BT, Einarsdóttir IE, Canario AV, Sweeney GE (2001) Thyroid hormones in growth and development of fish. Comparative Biochemistry and Physiology Part C: Toxicology & Pharmacology 130(4):447–459

Prisco M, Liguoro A, Comitato R, Cardone A, D'Onghia B, Ricchiari L, Angelini F, Andreuccetti P (2003) Apoptosis during spermatogenesis in the spotted ray Torpedo marmorata. Mol Reprod Dev 64(3):341–348

Rajakumar A, Singh R, Chakrabarty S, Murugananthkumar R, Laldinsangi C, Prathibha Y, Sudhakumari CC, Dutta-Gupta A, Senthilkumaran B (2012) Endosulfan and flutamide impair testicular development in the juvenile Asian catfish, Clarias batrachus. Aquat Toxicol 110:123–132

Rani R, Sharma P, Kumar R, Hajam YA (2022) Effects of heavy metals and pesticides on fish. In: Bacterial fish diseases. Academic Press, pp 59–86

Reijnders PJ (1980) Organochlorine and heavy metal residues in harbour seals from the Wadden Sea and their possible effects on reproduction. Neth J Sea Res 14(1):30–65

Reijnders PJH (1990) Progesterone and oestradiol-17β concentration profiles throughout the reproductive cycle in harbour seals (Phoca vitulina). Reproduction 90(2):403–409

Repossi A, Farabegoli F, Gazzotti T, Zironi E, Pagliuca G (2016) Bisphenol a in edible part of seafood. Ital J Food Saf 5(2):5666

Rime H, Nguyen T, Bobe J, Fostier A, Monod G (2010) Prochloraz-induced oocyte maturation in rainbow trout (Oncorhynchus mykiss), a molecular and functional analysis. Toxicol Sci 118(1):61–70

Roos A, Greyerz E, Olsson M, Sandegren F (2001) The otter (Lutra lutra) in Sweden—population trends in relation to ΣDDT and total PCB concentrations during 1968–99. Environ Pollut 111(3):457–469

Rurangwa E, Biegniewska A, Slominska E, Skorkowski EF, Ollevier F (2002) Effect of tributyltin on adenylate content and enzyme activities of teleost sperm: a biochemical approach to study the mechanisms of toxicant reduced spermatozoa motility. Comparative Biochemistry and Physiology Part C: Toxicology and Pharmacology 131(3):335–344

Safe S, Krishnan V (1995) Cellular and molecular biology of aryl hydrocarbon (Ah) receptor—mediated gene expression. Arch Toxicol Suppl 17:99–115

Safe SH (1995) Environmental and dietary estrogens and human health: is there a problem? Environ Health Perspect 103(4):346–351

Sandau CD, Meerts IA, Letcher RJ, McAlees AJ, Chittim B, Brouwer A, Norstrom RJ (2000) Identification of 4-hydroxyheptachlorostyrene in polar bear plasma and its binding affinity to transthyretin: a metabolite of octachlorostyrene? Environ Sci Technol 34(18):3871–3877

Santangeli S, Maradonna F, Gioacchini G, Cobellis G, Piccinetti CC, Dalla Valle L, Carnevali O (2016) BPA-induced deregulation of epigenetic patterns: effects on female zebrafish reproduction. Sci Rep 6(1):1–11

Santangeli S, Maradonna F, Zanardini M, Notarstefano V, Gioacchini G, Forner-Piquer I, Habibi H, Carnevali O (2017) Effects of diisononyl phthalate on Danio rerio reproduction. Environ Pollut 231:1051–1062

Schäfers C, Teigeler M, Wenzel A, Maack G, Fenske M, Segner H (2007) Concentration-and time-dependent effects of the synthetic estrogen, 17α-ethinylestradiol, on reproductive capabilities of the zebrafish, Danio rerio. J Toxic Environ Health A 70(9):768–779
Scholz S, Klüver N (2009) Effects of endocrine disrupters on sexual, gonadal development in fish. Sex Dev 3(2–3):136–151
Shaliutina O, Shaliutina-Kolešová A, Lebeda I, Rodina M, Gazo I (2017) The in vitro effect of nonylphenol, propranolol, and diethylstilbestrol on quality parameters and oxidative stress in sterlet (Acipenser ruthenus) spermatozoa. Toxicol In Vitro 43:9–15
Sharma P, Chadha P (2021) Bisphenol a induced toxicity in blood cells of freshwater fish Channa punctatus after acute exposure. Saudi Journal of Biological Sciences 28(8):4738–4750
Sharpe RM, Skakkebaek NE (1993) Are oestrogens involved in falling sperm counts and disorders of the male reproductive tract? Lancet 341(8857):1392–1396
Shukla KK, Mahdi AA, Rajender S (2012) Apoptosis, spermatogenesis and male infertility. Frontiers in Bioscience-Elite 4(2):746–754
Skaare JU, Bernhoft A, Wiig Ø, Norum KR, Haug E, Eide DM, Derocher AE (2001) Relationships between plasma levels of organochlorines, retinol and thyroid hormones from polar bears (Ursus maritimus) at Svalbard. J Toxicol Environ Health A 62(4):227–241
Skinner MK (2014) Endocrine disruptor induction of epigenetic transgenerational inheritance of disease. Mol Cell Endocrinol 398(1–2):4–12
Sohoni PCRT, Tyler CR, Hurd K, Caunter J, Hetheridge M, Williams T, Woods C, Evans M, Toy R, Gargas M, Sumpter JP (2001) Reproductive effects of long-term exposure to bisphenol a in the fathead minnow (Pimephales promelas). Environ Sci Technol 35(14):2917–2925
Sonnenschein C, Soto AM (1998) An updated review of environmental estrogen and androgen mimics and antagonists. J Steroid Biochem Mol Biol 65(1–6):143–150
Soto AM, Justicia H, Wray JW, Sonnenschein C (1991) P-nonyl-phenol: an estrogenic xenobiotic released from" modified" polystyrene. Environ Health Perspect 92:167–173
Spies RB, Rice DW (1988) Effects of organic contaminants on reproduction of the starry flounder Platichthys stellatus in San Francisco Bay. Mar Biol 98(2):191–200
Sridevi P, Chaitanya RK, Prathibha Y, Balakrishna SL, Dutta-Gupta A, Senthilkumaran B (2015) Early exposure of 17α-ethynylestradiol and diethylstilbestrol induces morphological changes and alters ovarian steroidogenic pathway enzyme gene expression in catfish. Clarias gariepinus Environmental Toxicology 30(4):439–451
Sruthisree C, Gowda G, Harish BD, Harsha N, Amin A (2015) Effect of heavy metal (lead) on oxygen consumption and ammonia excretion on iridescent shark (Pangusius hypohthalamus). J Exp Zool India 18(2):637–640
Sumpter JP, Jobling S (1995) Vitellogenesis as a biomarker for estrogenic contamination of the aquatic environment. Environ Health Perspect 103(suppl 7):173–178
Sun LN, Jiang XL, Xie QP, Yuan J, Huang BF, Tao WJ, Zhou LY, Nagahama Y, Wang DS (2014) Transdifferentiation of differentiated ovary into functional testis by long-term treatment of aromatase inhibitor in Nile tilapia. Endocrinology 155(4):1476–1488
Sundaray JK, Rather MA, Kumar S, Agarwal D (2021) Recent updates in molecular endocrinology and reproductive physiology of fish. Springer Singapore
Tanabe S, Watanabe S, Kan H, Tatsukawa R (1988) Capacity and mode of PCB metabolism in small cetaceans 1. Mar Mamm Sci 4(2):103–124
Thomas P, Doughty K (2004) Disruption of rapid, nongenomic steroid actions by environmental chemicals: interference with progestin stimulation of sperm motility in Atlantic croaker. Environ Sci Technol 38(23):6328–6332
Thorpe JE (1989) Developmental variation in salmonid populations. J Fish Biol 35:295–303
Tubbs CW, McDonough CE (2018) Reproductive impacts of endocrine-disrupting chemicals on wildlife species: implications for conservation of endangered species. Annual Review of Animal Biosciences 6(1):287–304
UNEP (2001) UNEP/pops/conf/4, 22, Stockholm, Sweden

Van Look KJW, Kime DE (2003) Automated sperm morphology analysis in fishes: the effect of mercury on goldfish sperm. J Fish Biol 63(4):1020–1033

Vandenberg LN, Maffini MV, Sonnenschein C, Rubin BS, Soto AM (2009) Bisphenol-a and the great divide: a review of controversies in the field of endocrine disruption. Endocr Rev 30(1): 75–95

Vonier PM, Crain DA, McLachlan JA, Guillette LJ Jr, Arnold SF (1996) Interaction of environmental chemicals with the estrogen and progesterone receptors from the oviduct of the American alligator. Environ Health Perspect 104(12):1318–1322

Wang H, Wu T, Qin F, Wang L, Wang Z (2012) Molecular cloning of Foxl2 gene and the effects of endocrine-disrupting chemicals on its mRNA level in rare minnow. Gobiocypris rarus Fish Physiology and Biochemistry 38(3):653–664

Wang P, Du Z, Gao S, Zhang X, Giesy JP (2016) Impairment of reproduction of adult zebrafish (Danio rerio) by binary mixtures of environmentally relevant concentrations of triclocarban and inorganic mercury. Ecotoxicol Environ Saf 134:124–132

Wang Q, Lam JC, Han J, Wang X, Guo Y, Lam PK, Zhou B (2015) Developmental exposure to the organophosphorus flame retardant tris (1, 3-dichloro-2-propyl) phosphate: estrogenic activity, endocrine disruption and reproductive effects on zebrafish. Aquat Toxicol 160:163–171

Wang Q, Liu Y, Peng C, Wang X, Xiao L, Wang D, Chen J, Zhang H, Zhao H, Li S, Zhang Y (2017) Molecular regulation of sex change induced by methyltestosterone-feeding and methyltestosterone-feeding withdrawal in the protogynous orange-spotted grouper. Biol Reprod 97(2):324–333

Wang X, Dong Q, Chen Y, Jiang H, Xiao Q, Wang Y, Li W, Bai C, Huang C, Yang D (2013) Bisphenol a affects axonal growth, musculature and motor behavior in developing zebrafish. Aquat Toxicol 142:104–113

Wester PW, Canton JH (1986) Histopathological study of Oryzias latipes (medaka) after long-term β-hexachlorocyclohexane exposure. Aquat Toxicol 9(1):21–45

Wetherill YB, Akingbemi BT, Kanno J, McLachlan JA, Nadal A, Sonnenschein C, Watson CS, Zoeller RT, Belcher SM (2007) In vitro molecular mechanisms of bisphenol a action. Reprod Toxicol 24(2):178–198

White R, Jobling S, Hoare SA, Sumpter JP, Parker MG (1994) Environmentally persistent alkylphenolic compounds are estrogenic. Endocrinology 135(1):175–182

Wiig Ø, Derocher AE, Cronin MM, Skaare JU (1998) Female pseudohermaphrodite polar bears at Svalbard. J Wildl Dis 34(4):792–796

Wintgens T, Gallenkemper M, Melin T (2003) Occurrence and removal of endocrine disrupters in landfill leachate treatment plants. Water Sci Technol 48(3):127–134

Wren CD (1991) Cause-effect linkages between chemicals and populations of mink (Mustela vison) and otter (Lutra canadensis) in the great lakes basin. Journal of Toxicology and Environmental Health, Part-A Current Issues 33(4):549–585

Wu S, Hu G, Zhao X, Wang Q, Jiang J (2018) Synergistic potential of fenvalerate and triadimefon on endocrine disruption and oxidative stress during rare minnow embryo development. Environ Toxicol 33(7):759–769

Xie M, Bu P, Li F, Lan S, Wu H, Yuan L, Wang Y (2016) Neonatal bisphenol a exposure induces meiotic arrest and apoptosis of spermatogenic cells. Oncotarget 7(9):10606–10615

Yin P, Li YW, Chen QL, Liu ZH (2017) Diethylstilbestrol, flutamide and their combination impaired the spermatogenesis of male adult zebrafish through disrupting HPG axis, meiosis and apoptosis. Aquat Toxicol 185:129–137

Yonkos LT, Friedel EA, Fisher DJ (2014) Intersex (testicular oocytes) in largemouth bass (Micropterus salmoides) on the Delmarva Peninsula, USA. Environ Toxicol Chem 33(5): 1163–1169

Yuan C, Zhang Y, Liu Y, Zhang T, Wang Z (2016) Enhanced GSH synthesis by bisphenol a exposure promoted DNA methylation process in the testes of adult rare minnow Gobiocypris rarus. Aquat Toxicol 178:99–105

Zhang T, Liu Y, Chen H, Gao J, Zhang Y, Yuan C, Wang Z (2017) The DNA methylation status alteration of two steroidogenic genes in gonads of rare minnow after bisphenol a exposure. Comparative Biochemistry and Physiology Part C: Toxicology & Pharmacology 198:9–18
Zhang Y, Cheng M, Wu L, Zhang G, Wang Z (2016) Bisphenol a induces spermatocyte apoptosis in rare minnow Gobiocypris rarus. Aquat Toxicol 179:18–26
Zhang Y, Gao J, Xu P, Yuan C, Qin F, Liu S, Zheng Y, Yang Y, Wang Z (2014a) Low-dose bisphenol a disrupts gonad development and steroidogenic genes expression in adult female rare minnow Gobiocypris rarus. Chemosphere 112:435–442
Zhang Y, Yuan C, Hu G, Li M, Zheng Y, Gao J, Yang Y, Zhou Y, Wang Z (2013) Characterization of four nr5a genes and gene expression profiling for testicular steroidogenesis-related genes and their regulatory factors in response to bisphenol a in rare minnow Gobiocypris rarus. Gen Comp Endocrinol 194:31–44
Zhang Y, Yuan C, Qin F, Hu G, Wang Z (2014b) Molecular characterization of gdf9 and bmp15 genes in rare minnow Gobiocypris rarus and their expression upon bisphenol a exposure in adult females. Gene 546(2):214–221
Zhu Y, Hua R, Zhou Y, Li H, Quan S, Yu Y (2016) Chronic exposure to mono-(2-ethylhexyl)-phthalate causes endocrine disruption and reproductive dysfunction in zebrafish. Environ Toxicol Chem 35(8):2117–2124
Zhuang W, Wu K, Wang Y, Zhu H, Deng Z, Peng L, Zhu G (2015) Association of serum bisphenol-a concentration and male reproductive function among exposed workers. Arch Environ Contam Toxicol 68(1):38–45

Water Contamination Through Xenobiotics and Their Toxic Effects on Aquatic Animals

Nafee Maqbool, Ishteyaq Majeed Shah, Shams M. Galib, and Fayaz Ahmad

1 Introduction

Industrialization, urbanization, and globalization are the great advancements and achievements of human race. These aspects with population growth are affecting our lives both positively and negatively (Gu 2019). Globalization has enormous positive and favorable benefits when seen through the economic as well as political perspective, but its effect on the environment is negative – a prerequisite for quality life. The twentieth century saw a rapid technological progress leading to development of many life-improving compounds such as antibiotics, dyes, PCPs, additives, etc. These compounds usually are not found naturally, or if found, their natural concentrations are significantly lower in the environment. The advances in technology, improved access to medicines, longer life, and use of daily grooming and personal care products lead to the introduction of new foreign substances into the environment. The use of crop-enhancing products like pesticides and fertilizers also leads to the addition of new foreign compounds into the environment (Ebele et al. 2017; Nikolaou et al. 2007).

The term "xenobiotic" is derived from two Greek words *xenos* and *bios*. *Xenos* refers to "foreign" and *bios* refers to "life." Xenobiotic thus itself means foreign thing in living form. Those chemicals which are foreign and are not naturally produced in organism or the environment are referred to as xenobiotic. It also includes some naturally found compounds when present in elevated concentrations.

N. Maqbool · I. M. Shah (✉) · F. Ahmad
Advanced Research Laboratory, Department of Zoology, University of Kashmir, Srinagar, Jammu and Kashmir, India
e-mail: ishteyaq.scholar@kashmiruniversity.net

S. M. Galib
Department of Biosciences, University of Durham, Durham, UK

Department of Fisheries, University of Rajshahi, Rajshahi, Bangladesh

M. A. Rather et al. (eds.), *Xenobiotics in Aquatic Animals*,
https://doi.org/10.1007/978-981-99-1214-8_3

These substances either as single or mixed with other substances can lead to serious problems including long-term and short-term effects on humans, animals, and the ecosystem (air, water, and soil). It is challenging to detect, measure, and eliminate xenobiotic due to their physicochemical properties like tiny molecular size, solubility in water, lipophilicity, volatile nature, polarity, and ionizability (de Oliveira et al. 2020). Wastewater systems and runoff from rainy weather are responsible for the presence of xenobiotic in freshwater (Roccaro et al. 2013; Pedersen et al. 2003). Manmade contaminants being nondegradable or taking a long time to degrade leads to pollution (Singh 2017; Dar et al. 2020). These substances which are foreign to the environment and are nondegradable or takes a very long time to degrade falls in the category of xenobiotic. The classification of xenobiotics according to Water Framework Directive is shown in Fig. 1. The sources and categories of xenobiotics is shown in Fig. 2.

2 Transport Route of Xenobiotic Compounds

The xenobiotics are introduced into the environment due to various human activities. Although their incorrect disposal is a contributing factor, the transit of these chemicals into the environment is also caused by manufacturing facilities, sewage treatment plants (STPs), wastewater treatment plants (WTPs), animal treatments, etc. (Fig. 1). In agriculture, insecticides and herbicides are applied directly on agricultural land and consequently end up in the soil. The rain further delivers these substances into groundwater, rivers, lakes, seas, etc. In addition, the majority of PPCPs are released into the environment indirectly. Animals and humans both consume PPCPs, but because they can't fully metabolize them, they either excrete these substances as such or their metabolites (Chopra and Kumar 2018). In this way, metabolites and parent molecules enter the food web. It needs to be mentioned that sometimes metabolites are more lethal/harmful than the original molecule and through excretion they enter into wastewater treatment plants, sewage treatment plants, etc. Finally, they reach the soil, groundwater, rivers, lakes, and seas. Unfortunately, plants and aquatic organisms take these pesticides and PPCPs – providing a route for them to enter the food chain. The different sources of xenobiotics is shown below in Fig. 3.

3 Classification of Xenobiotics

Xenobiotic can be classified on the basis of various characteristics like nature, uses, physical state, pathophysiological effects they cause, etc. Classification on the basis of some characteristics is given below:

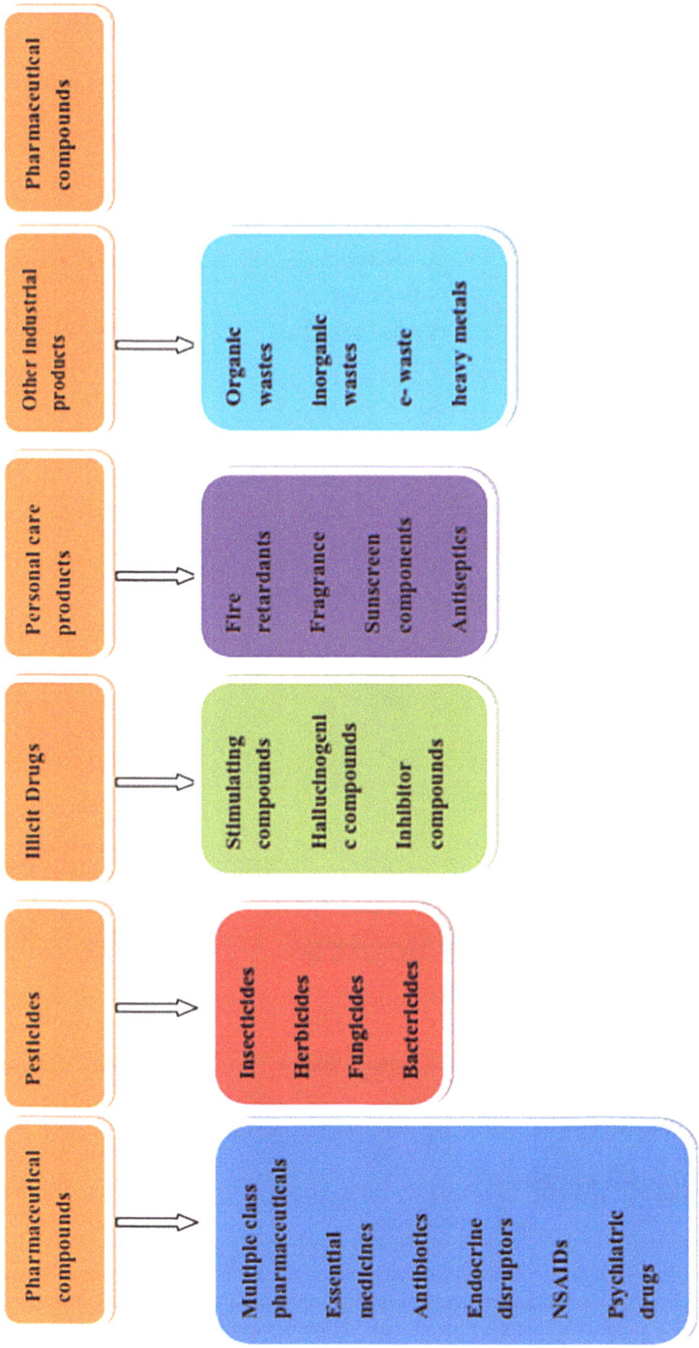

Fig. 1 Classification of xenobiotic according to Water Framework Directive

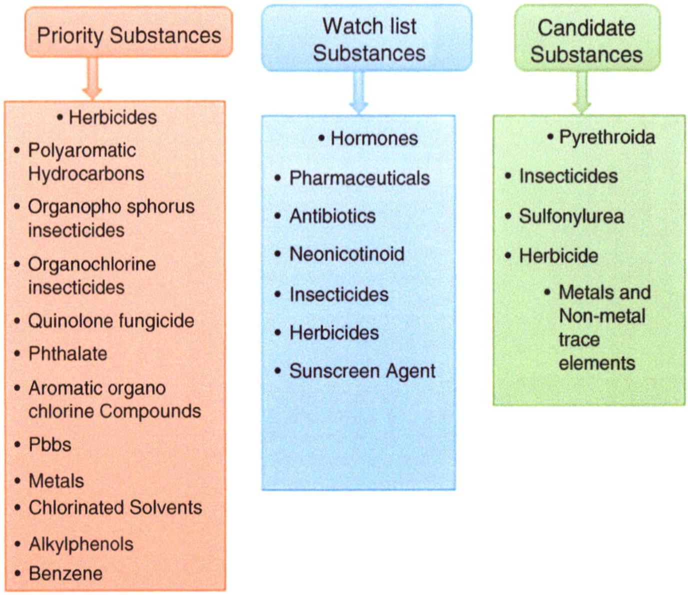

Fig. 2 Figure showing the categories and sources of xenobiotics

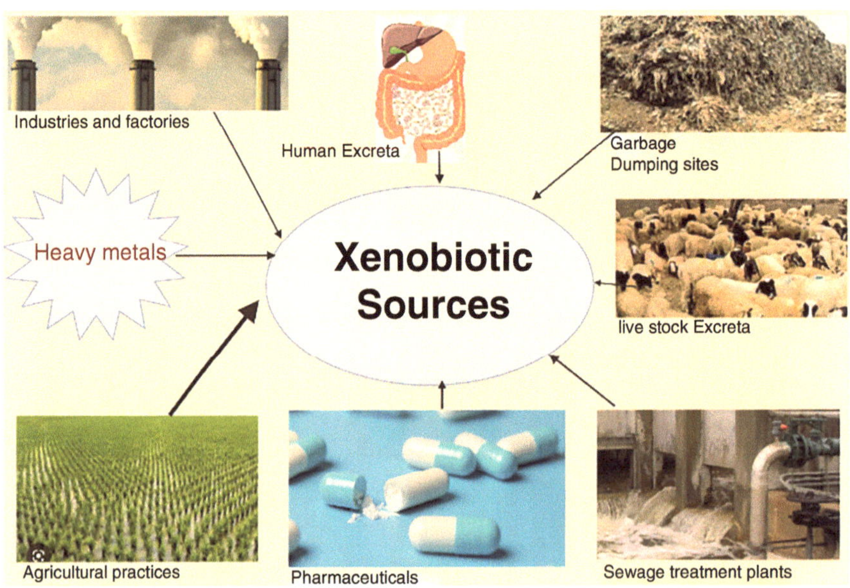

Fig. 3 Various sources of xenobiotic toxicants (industries and factories are the prime sources of xenobiotics. Garbage contained in dumping sites, heavy metals, sewage treatment plants also contribute to the xenobiotic sources. Pharmaceuticals, human excreta, livestock, and animal excreta all are powerful packs for xenobiotic formation

1. **Nature**: On the basis of nature, xenobiotics are classified as natural and synthetic.
   (a) Natural xenobiotics include the chemicals produced by living organisms mainly for defense purposes, such as bacteriotoxins, zootoxins, phytotoxins, serotonin, etc.
   (b) Synthetic xenobiotics include compounds generated by humans that are hazardous to living things or that become toxic as a result of transformation or accumulation (Gadzała-Kopciuch et al. 2004).
2. **Uses**: On the basis of use, xenobiotics are classified as active and passive.
   (a) Active xenobiotics include compounds such as paints, pesticides, and dyes.
   (b) Passive xenobiotics include additives and carrier molecules – used to facilitate the working of the active components (Donner et al. 2010).
3. **Physical state**: On the basis of physical state, xenobiotics are classified as gaseous, dust form, and liquid xenobiotic.
   (a) Gaseous xenobiotic includes benzene or aerosol form of pesticides. These xenobiotics can easily spread in the environment.
   (b) Dust form xenobiotics include solid particles that are very minute and lightweight, e.g., asbestos powder. These xenobiotics can be easily carried away with air or water.
   (c) Liquid xenobiotics include chemicals that get dissolved in water. These are generally effluents directly emitted into water bodies.
4. Pathophysiological effects they cause in an organism: Some xenobiotic affects any tissue or organ of an organisms like kidney toxins, while others affect the biochemical processes of an organism like methemoglobin-producing toxins.

4 Classification of Xenobiotic Sources

It is necessary to understand and study the sources from where they are released or emitted into the environment as xenobiotics are found everywhere. By a thorough understanding of the sources, new methods can be framed or developed to minimize their release. The various sources of xenobiotics are classified as follows.

4.1 Direct and Indirect Sources

(A) Direct sources include all those compounds which are released directly into the surrounding environment. Phenols from the pharma industries (Gayathri and Namasivayam 2010) and hydrocarbons from petroleum effluents (Whyte et al. 1997) are considered direct sources. Pesticides and insecticides like organophosphates, methyl parathion, morpholine, etc. degrade slowly and directly affect the environment (Hashmi et al. 2017). Other examples of direct pollutants that degrade slowly in the environment include plastics, paint containing emulsifiers and texturizers and dyes (Lazarevic et al. 2010; Abdelkader et al. 2011), and paper and pulp effluents (Hashmi et al. 2017).

(B) Indirect sources are substances which are directly released into the environment and do not act as a pollutant; however, its breakdown products or residues or combination products act as pollutants. Indirect sources include nonsteroidal or anti-inflammatory drugs, pharmaceutical compounds, and pesticide and herbicide residues. Their inclusion in the food chain is significantly influenced by bioaccumulation and biomagnification (Heberer 2002; Monadjem et al. 2004).

4.2 *Product and Processes as a Source*

(A) Product includes the end-product or outcome of any process or reaction. The end-products have enormous value as xenobiotics do not degrade easily, e.g., dyes, pesticides, etc.

(B) Process: Xenobiotics are released as a result of routine daily activities at home or at an industrial scale. It includes those xenobiotics that might be involved in the process from the beginning, such as pretreatment, through the end, such as packaging and transportation, or they might be involved in the manufacturing of finished items.

4.2.1 Deliberate and Accidental Causes

(A) Deliberate: Those xenobiotics which are intentionally or deliberately released into the environment such as those used in the pulp industries.

(B) Accidental: Some xenobiotics accidentally enter the environment as a result of an accident or a technical issue.

4.2.2 Moving and Stationary Sources

(A) Moving sources include those that discharge waste while they are moving, such as cars that generate substantial amounts of lead, carbon monoxide, and other pollutants.

(B) Stationary sources include those industries which are located in one location and generate and emit harmful pollutants.

4.2.3 Regulated and Unregulated Sources

(A) Regulated sources: By establishing some standards and enacting stringent laws and regulations, the effluent released can be controlled. Generally speaking, this legislation applies to big businesses and vehicles.

(B) Unregulated sources: It is exceedingly challenging to control residential waste generation in daily life.

5 Sources

5.1 *Personal Care Compounds*

Personal care products and cosmetics include surfactants, oils and waxes, fragrances, biocides, UV filters, and pigments. Most consumers prefer products with synthetic flavors and perfumes over those without. The majority of softeners, washing powders, cleaning detergents, etc. also include substantial amounts of fragrances like nitro and polycyclic musks. Cosmetics contain preservatives which give them longer shelf lives as they often contain high percentage of water. It has been demonstrated that biocides, perfumes, and UV blockers are released into water bodies (Bester 2007; Poiger et al. 2003; Balmer et al. 2005). The personal care compounds are being reported in the marine ecosystems (Andresen et al. 2007).

5.2 *Flame Retardants*

Flame retardants are utilized in various textiles, construction materials, and electronic devices. These compounds can be mobilized and released into surface or waste waters by a variety of activities, such as washing, cleaning, demolition, and construction work, as well as the disturbance of dusts (Marklund et al. 2003). Lipophilic polybrominated diphenyl ethers (PBDEs) have been used as flame retardant in the past particularly in textiles. Nowadays, the use of HBCD and decabromodiphenyl ethane has been reported as a replacement flame retardant for PBDE (Ricklund et al. 2009). This substance use, however, raises concerns because of their similar bioaccumulation and persistence qualities. The hydrophilic properties of organophosphate flame retardants, in contrast to PBDE flame retardants, make them more likely to be discovered in water than in the sludge portion (Meyer and Bester 2004). In addition to being detected in the wastewater (Bester 2007), they are additionally detected in the marine environment which has been identified as the final sink for these compounds (Andresen et al. 2007).

5.3 *Pharmaceuticals*

People who take medications release pharmaceuticals, with chemicals either exiting the body as the parent chemical or as a metabolite (Ternes and Joss 2007). Although excretion through feces is also significant, excretion through urine appears to be the primary method for the body to eliminate the majority of medicines (Lienert et al. 2007). The majority of these chemicals are not effectively eliminated quantitatively by current wastewater treatment methods, that is, they won't completely stop the entry of pharmaceuticals into wastewater. Pharmaceutical compounds are water loving in nature and are too hydrophilic to sorb to sludge in wastewater treatment (Joss et al. 2004, 2005; Ternes and Joss 2007; Reemtsma and Jekel 2006), and biotransformation appears to be insufficient for the complete breakdown of many compounds. According to Triebskorn et al. (2004)), surface waters in addition to estrogen ethinylestradiol contain many pharmaceuticals such as diclofenac. Diclofenac is shown to cause many effects in the laboratory tests, e.g., in gills of fish (Triebskorn et al. 2004) and swimming activity of *Gammarus* (De Lange et al. 2006). Reports indicate that most unneeded medicines are disposed via toilet or the sink. Therefore, strengthening disposal routes presents a potential to quickly ameliorate the problem.

5.4 Steroid Hormones

Due to typical human excretion, wastewater contains steroid hormones other than ethinylestradiol and mestranol. According to reports, effective STPs using activated sludge can lower these chemicals' concentrations to $\leq$10 ng/L (Joss et al. 2004; Schlüsener and Bester 2008). According to reports, male fish encounter estrogenic chemicals by generating egg yolk proteins in their testicles at level lower than this (Jobling et al. 1998; Desbrow et al. 1998). Synthetic steroid hormones like mestranol and ethinylestradiol are employed in hormone replacement therapy and birth control. Because of its most lethality and least biodegradability, ethinylestradiol's predestination and mass flow have been the subject of particular investigation (Bester et al. 2008a, b).

5.5 Biocides

Currently, biocides can be found in a diverse range of goods for a number of different reasons. These include improving consumer cleanliness, extending the shelf life of goods like paints and cosmetics (triclosan), preserving wood (thiocyanatomethylthiobenzothiazole), bolstering construction material (mold and algae growth inhibition), and inhibiting the growth of unwelcome vegetation on flat roofs. Wastewater and surface waters have been found to contain biocides like triclosan at amounts higher than those that are harmful.

5.6 *Polycyclic Aromatic Hydrocarbons (PAHs)*

PAHs, one of current and major environmental concerns, have sources from gas plants and blast furnaces of coke plants (belonging to coal processing units). Other sources like automobile industry (e.g., tires) also contribute to PAHs. According to Milukaite (2006), the level of PAHs is linked with industrial progression as they are found to be less with no industrial history (generally in low amounts and tend to sludge within STPs).

5.7 *Corrosion Inhibitors*

Benzotriazole and tolytriazole are the complexing agents that are routinely used as anticorrosives in aircrafts, de-icing, anti-icing fluids, and silver protection in dish-washer detergents. Benzotriazole is detected in surface water at unusually high median concentrations of 1 microgram per liter due to poor removal inside STPs (Giger et al. 2006; Weiss et al. 2006). These substances are identified as the pollutants with the overloads in groundwater that has been tainted by the intrusion of filthy water, together with medications like carbamazepine, amidotrizoic acid, etc. (Hollender et al. 2007). The chemical nature of different xenobiotics are represented in the Table 1.

6 Physicochemical Nature of Xenobiotics

Since majority of xenobiotic molecules share similitude/parallelism with parent compound, from where they arose, they are mainly joined by nonphysiological bonds which led to their complicated chemical framework. Although these compounds have highest degree of polymerization that leads to high molecular weight, at the same time, they are least soluble or partially soluble in water leading to their

Table 1 List of different xenobiotics and their chemical nature

Name	Chemical formula
Xylene	C_8H_{10}
PVC (polyvinyl chloride)	$(C_2H_3Cl)n$
Toluene	C_7H_8
Soda ash	Na_2CO_3
Paracetamol	$C_8H_9NO_2$
Acetaminophen	$C_8H_9NO_2$
Citric acid	$C_6H_8O_7$
Nerve gas (soman)	$C_7H_{16}F$
Caffeine	$C_8H_{10}N_4O_2$
Methyl cyclopropane	C_4H_{10}

condensation due to aromaticity and polycyclic rings. A vast array of compounds like aromatic hydrocarbons, halogen-substituted aliphatic compounds, azo chemicals, polycyclic aromatic hydrocarbons (PAHs), aromatic amines, and aromatic hydrocarbons belong to xenobiotics (Southorn and Powis 1988; Spain 1995). Given below are some of the xenobiotics and their chemical formulae.

Given the prevalence of these exogenous bodies all along the globe makes xenobiotics challenging to confine (Wu et al. 2017). There are generally four phases through which xenobiotics find their way in a living system, namely, absorption (first step), distribution and their metabolism, and finally elimination. Biotransformation process like hydrolysis, oxidation, and reduction is generally carried out by these xenobiotics in order to produce reactive functional groups like $-NH_2$, –COOH, and -OH preceded by complexation of hydrophilic molecules in order to boost the hydrophilicity of xenobiotics, which results in intestinal excretion (Mandlekar et al. 2006). Additionally, it has been discovered that xenobiotics can cause gene mutations that lead to cancer (Kamdem et al. 2006).

7 Water Contamination Through Xenobiotics

Graywater contains heterogeneous molecules called xenobiotic organic compounds (XOCs), which can be found in varying amounts due to detergents, skincare products, and hair colorants (Grčić et al. 2015). Various health effects may result from chronic exposure to these chemicals in human bodies due to their bioaccumulation in the environment and food chain (Kim and Zoh 2016). It is especially dangerous for people living in countries where surface water supplies most of their drinking water. Water quality is improved by treating surface water in certain nations before it is used as drinking water. For the treatment procedure to achieve primary objectives such as decreasing salinity, total suspended solids (TSS), chemical oxygen demand (COD), and biological oxygen demand (BOD), basic approaches are required. When it comes to getting rid of XOCs, these methods are ineffective.

Before municipal wastewater is released into aquatic systems, it must be treated with xenobiotic receptors found in traditional sewage treatment plants. It is possible to detect trace metals, xenobiotic substances, organic compounds, and pesticides in water bodies (Essumang and Ankrah 2010). Chemicals used in pharmaceuticals and xenobiotics may interfere with biological processes like nitrification in traditional sewage treatment facilities.

However, it is currently strictly forbidden and discouraged as the discharge of these toxicants could compromise the biological viability and structure of aquatic waterbodies. The leaching process may allow xenobiotic substances to reach groundwater. A number of aquatic creatures have been found to exhibit xenobiotic contamination (Fent et al. 2006). As metabolites or as their original forms, xenobiotic substances can enter the environment. Because of high resilience to biodegradation, leading to their buildup in greater quantities, these contaminants are highly toxic and lethal.

8 Impact of Xenobiotics on Aquatic Life

It is well known how xenobiotic contamination affects water bodies. Keeping in mind the boarder nature of xenobiotics, it is almost impossible to tackle down the implication caused by the constituents present in these xenobiotic compounds. Below listed are some of the major xenobiotics and their implication on aquatic life.

8.1 Microplastics

Microplastics can easily permeate marine environments like the ocean and sea. Credit goes to the high prevalence and extremely smaller size which is <5 mm. Implications of microplasticity on human health and other life forms are well documented. Several research indicate that marine creatures, specifically fish, absorb these microplastics (do Sul and Costa 2014). There are significant regional disparities in the distribution of microplastics on the surface of water bodies, near-shore beaches, and sea bottom sediments. Notable geographical differences are present in the distribution of microplastics on water bodies, sea bottom sediments, and beaches (near-shore). The impact of microplastic contamination on coastal waters is a serious concern, owing to increasing anthropogenic activities.

8.2 Mercury Accumulation

Mercury emissions from anthropogenic and natural sources, such as coal combustion and incineration, deplete aquatic ecosystems of invertebrates and vertebrates (Krabbenhoft 2004). The three categories of substances that mercury falls under are organic, inorganic, and elemental. The most mercury that is discharged into the environment is inorganic mercury (Li et al. 2019). The causes of the zonal variation of mercury have been determined by numerous models (Houssard et al. 2019). This is the major justification for locating and researching the distribution and pattern of mercury in aquatic environments. Both humans and animals are exposed to mercury through eating fish (Kim et al. 2016). Therefore, understanding the presence and abundance of mercury in aquatic environments is crucial. Humans are the primary consumers of mercury, so understanding the extent of the pollution is also useful. According to Zupo et al. (2019), around 13 species of marine fishes were tested for Hg, and many were found to be contaminated with the metal. The researchers also measured mercury concentrations in water, soil, feed items, and fish prey to demonstrate the bioaccumulation process. There were high levels of mercury deposition in marine fish compared to freshwater fish. Mercury concentrations rose as trophic level increased (Zupo et al. 2019). The total mercury concentration in market and coastal fish samples was higher than the legal limit.

8.3 Oxidative Stress

When the levels of both oxidants and antioxidants fall on the same balance, oxidative stress takes birth. Cells are damaged via oxidative stress when reactive oxygen species or oxidants are present at high levels. Free radicals are prevented from causing damage to the body by enzymes that are antioxidants. Oxidative stress is indicated by elevated enzyme activity levels. Healthy cells detoxify prooxidants and ROS with antioxidant defenses (Livingstone 2003). It is possible for redox-active metals, such as those that are redox-inactive, to generate ROS like hydrogen peroxide, superoxide radicals, and hydroxyl radicals. A cell's antioxidant defenses are overwhelmed by an increase in ROS production, resulting in "oxidative stress." ROS damage to lipids, proteins, and DNA causes various dysfunctions in cells under oxidative stress (Ercal et al. 2001).

8.4 Heavy Metals

A variety of heavy metals affect living things negatively, including arsenic, lead, cadmium, and mercury. Biological accumulation of heavy metals and their biomagnification are major concerns. It is critically significant to be aware of the high levels of toxicity and carcinogenicity associated with these metals. Further, they may cause morphological and functional defects, developmental delays, or even death in developing fish embryos, as well as adverse effects on several metabolic processes. Adnan Amin et al. (2015) found that developing fish (embryos and larvae) had altered body form, body anomalies, and delayed hatching.

8.5 Mercury

Freshwater is contaminated with mercury on a global scale. The presence of oxidative stress and damage after mercury exposure has been shown in numerous studies (Cappello et al. 2016; Larose et al. 2008). The expression of antioxidant enzymes and non-enzymatic scavenger molecules are mainly impacted by acute water exposure to Hg(II) (Kidd and Batchelar 2012). As a result of exposure to 150 mg/L of Hg (II) (MT) for 96 h, antioxidant enzymes such as GST, SOD, CAT, GSH-Px, GR, GSH, metallothionein, and glutathione are highly elevated.

8.6 Lead

Overdoses of lead can lead to anemia, which is characterized by microcytic and hypochromic red blood cells (Goyer and Clarkson 1996; Adnan Amin et al. 2017; Sruthisree et al. 2015). In numerous studies, waterborne Pb has been found to cause anemia in a variety of fish, including brown trout (*Salmo trutta*) (Mariussen et al. 2017), common carp (*Cyprinus carpio*), European catfish (*Silurus glanis*), and tench (*Tinca tinca*) (Shah 2006). Lead can cause oxidative damage via directly affecting the cell membrane (Sevcikova et al. 2011). A number of mechanisms can result in oxidative damage caused by lead, including interactions with SOD (superoxide dismutase), δ-aminolaevulinic acid, and methemoglobin (Costa et al. 1997; Adnan Amin et al. 2016).

8.7 Cadmium

Water organisms are toxic to Cd depending on their speciation, and the bioavailability of Cd depends on its free ion concentration. Ca^{2+} and Cd^{2+} compete for each other in complexing with Cd^{2+}, reducing its toxicity. Wood et al. (2011) found that cadmium poisoning adversely affected Ca, Na, and Mg ion homeostasis. There has been a growing body of evidence documenting cadmium accumulation in aquatic species, notably in the liver, spleen, gills, and muscle (Gomes et al. 2016; Olsvik et al. 2016). As a result of Cd exposure, they were able to verify that cytotoxicity and oxidative stress were caused. The dissociation of MT (metallothionein) by cadmium caused kidney injury in the cytoplasm of the proximal tubule cell where it induces oxidative stress. According to Verbost et al. (1988), Ca^{2+} and Cd^{2+} compete directly on ATPase binding sites, with Cd^{2+} having a high affinity for Ca^{2+}. A recent study found that exposure to Cd affects Na ions (Wood et al. 2011), decreases Na^+/K^+ ATPase function, and affects liver catalase action (Atli and Canli 2007). Effect of various heavy metal on fish health is shown in the following Fig. 4.

8.8 Pesticides

This is a concoction of chemicals meant to kill, subdue, or prevent the spread of pests (unwanted/undesirable organisms). Typically, nematodes, insects, microorganisms, and insects among others belong to the category of pests, meant to ruin crops and spread diseases, and compete in food chain (human food). Depending on the nature of origin, pesticides are often divided into biological and synthetic categories. While created through manufacturing techniques, synthetic pesticides differ in many ways from biological pesticides, isolated from natural sources (e.g., azadirachtin derived from neem). Another division of pesticides is based on working action/mechanism

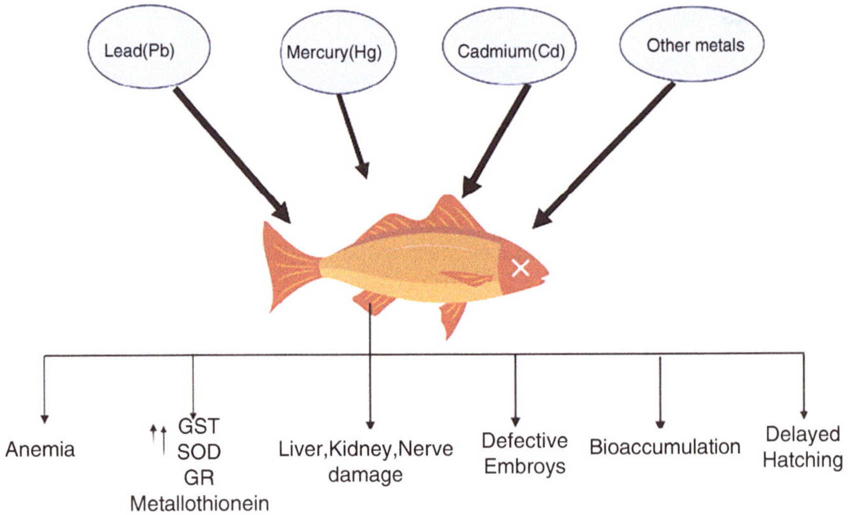

Fig. 4 Heavy metal effect on fish: heavy metals like Pb, Hg, Cd, etc. greatly impact the well-being of fishes, leading to anemia; liver, kidney, and nerve damage; delayed hatching; and elevation in many types of antioxidant enzymes like GST, SOD, GR, etc

with some as broad spectrum meant to control a wide range of pests/species and others as narrow spectrum (controls/kills or inhibits growth of few or limited species). Pesticide division is mainly based on the pests it controls or regulates. Pesticides are categorized as insecticides (meant to kill insects), Fungicides (control and growth of fungi) and herbicides kill unwanted plants known as weeds.

Meddling with nervous system at multiple areas, insecticides may alter many membrane transport channels like the ones of Na^+, K^+, chloride, and calcium, thus leading to inhibition of specific enzyme activities meant for nerve transmission (Correia et al. 2010).

8.8.1 Classification of Insecticides

(a) Organochlorides are chemically inert and have long-lasting effects on the nervous system. Examples include well-known dichlorodiphenyltrichloroethane (DDT), lindane, and endrin, all affecting the nervous system (DDT inhibits and targets release of neurotransmitters) (LeBlanc 2007).
(b) Organophosphates: After being banned by many developed countries, organochlorides were replaced by organophosphates (e.g., parathion and malathion). Symptoms caused by this typical class of insecticides are extended and consistent as they target enzyme acetylcholinesterase (neurotoxic in nature) (Ecobichon 1991; Sankhla et al. 2018; Alencar et al. 2020).

(c) Carbamates: By adhering to the enzyme's reactive site, this class of pesticide also inhibits AChE. AChE is quickly and permanently inhibited by this class of insecticide (Nie et al. 2020).
(d) Synthetic pyrethroids: A recent class of insecticides that have additional sites for enzyme binding and action, inhibiting calcium and magnesium ATPase. It also impedes with calcium removal from nerve endings, leading to release of neurotransmitters at postsynaptic gap.

8.8.2 Herbicides

Meant to eradicate noxious weeds, herbicides influence a number of systems that take part in growth, cell and nuclear division, photosynthesis, respiration, growth, and the production of proteins and lipids. Herbicides are categorized into different types, but two best studied herbicides are glyphosate and chlorophenoxy herbicides.

8.8.3 Fungicides

Fungicides are a class of insecticides that disrupt fungi's energy sources and prevent the development of their spores. Examples include dithiocarbamates, R-S-CCl3 compounds, etc. (Tsui and Chu 2003).

8.9 Dyes

The use of dyes is highly lethal and hazardous to both the environment and the well-being of humans. In the environment, dyes can last for many years since they are thermally and photostable to fend off biodegradation. When sunlight enters the water, it is absorbed and reflected by colors, which increases the level of environmental pollution. Absorption of light reduces algae's photosynthetic activity, affecting the food chain as well. Biological activity and photosynthesis of aquatic plants and algae can be disrupted by a high level of textile dyes in aquatic bodies. This is due to the dye blocking sunlight and preventing water from being reoxygenated (Zaharia et al. 2009).

Using textile dye wastewater (untreated and treated) on the *Gambusia affinis* freshwater fish, comparative toxicological studies found that mortality and cytotoxicity of RBCs decreased significantly, as well as their counts dropping, their appearance changing (poikilocytosis), and their size changing (Soni et al. 2006).

Algal growth characteristics, including protein, pigment, and other nutrients, are affected by an increasing concentration of dye in a body of water. The effects of different colors on algae may vary. The toxicity of pollutants in aquatic environments is 50% higher in algae than in typical test organisms (Hoffman et al. 2002)

9 Pesticides in the Aquatic Ecosystem

A wide variety of chemicals are used in agriculture and daily life as pesticides and herbicides, such as organophosphorus, endosulfan, nitrophenols, morpholine, synthetic pyrethroids (SPs), and carbamates. As a result of these substances, groundwater is contaminated, and runoff eventually reaches rivers and oceans. Pesticides, the most dangerous toxins posing substantial risks to biological organisms including humans, find their way into aquatic system through a number of channels including spills, industrial effluent, surface runoff, or soils that have been exposed to pesticides (Picó et al. 2020). The toxic consequences brought on by exposure to these hazardous substances can be divided into different categories according to exposure timeframe. Short-term exposure is defined as lasting $\leq$96 h, whereas long-term exposure is defined as lasting >96 h. Exposure type can be fatal or sublethal. Pesticides primarily reach aquatic fauna through three main ways, namely, through the skin, breathing, and orally. Aquatic organisms are typically exposed to pesticides through the consumption of pesticide exposed prey (secondary poisoning). For instance, if a fish consumes pesticide-exposed insects, the fish will become secondary poisoned.

Aquatic fauna can be negatively or positively impacted by pesticides; negative alterations include physiological changes in an organism (Acosta-Sánchez et al. 2020). As an illustration, the access of water flea to pesticides brings down their number, which may be viewed as a direct consequence of the pesticides, and also may elevate the biomass of algae because of release from grazing pressure, viewed as indirect effect. Globally, herbicide, mostly glyphosate, is mostly used in the management and regulation of aquatic and terrestrial weeds; however, its use skyrocketed recently. Originally, they were meant for plants (mainly unwanted weeds) but have been recognized to hit nontargets (aquatic fauna) by adversely impacting them. The impact on nontarget creatures (aquatic) is deadly for some and sublethal for others (Glusczak et al. 2007). It is believed that the insecticides cypermethrin and SP pose a significant risk to aquatic fish and invertebrates. According to Zhang et al. (2011), this substance can adversely affect reproductive, neurological, and developmental functions. Fertilizers with high nitrogen and phosphorus content cause algal blooms, which also cause hypoxia and lead to a loss of biodiversity in the ocean. DDT, for example, accumulates in fish's fatty tissues and moves up the food chain through bioconcentration and biomagnification. There is evidence that DDT can cause cancer and birth abnormalities. The route and effect of xenobiotics on aquatic organisms is shown below in Fig. 5.

10 Conclusion and Future Perspectives

The majority of research at the moment is concentrated on how xenobiotics affect aquatic life, since they have an adverse effect on humans, aquatic systems, and the environment as a whole, disrupting various food chains, leading to economic loss,

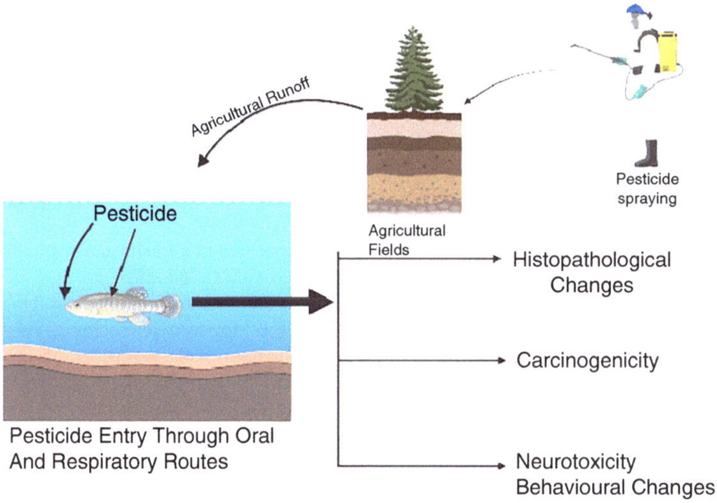

Fig. 5 Agricultural runoffs (mainly pesticides) finally reach the riverine system, where through oral, respiratory, and dermal routes they find their way in aquatic fauna mainly fish. The resultant impact of pesticide is histopathological changes, carcinogenicity, neurotoxicity, and behavioral changes in fishes among others

etc. Numerous reports have suggested that these toxicants may be poisonous, mutagenic, carcinogenic, and responsible for a number of other illnesses. The impact of xenobiotics on aquatic life finally ends up on impacting human health. Many researchers are currently working to find strategies to reduce the impact of xenobiotics on aquatic ecosystems and their fauna and also to use these xenobiotics sustainably for living in harmony with environment. Bioremediation is thought to be the most effective, environmentally beneficial, and economically viable method of reducing the impact of xenobiotics in the marine environment. It is the most well-known method for keeping aquaculture systems healthy and sustainable and, ultimately, regenerating aquatic ecosystems.

References

Abdelkader E, Nadjia L, Ahmed B (2011) Degradation study of phenazin neutral red from aqueous suspension by paper sludge. J Chem Eng Process Technol 2:109

Acosta-Sánchez A, Soto-Garita C, Masís-Mora M, Cambronero-Heinrichs JC, Rodríguez-Rodríguez CE (2020) Impaired pesticide removal and detoxification by biomixtures during the simulated pesticide application cycle of a tropical agricultural system. Ecotoxicol Environ Saf 195:110460

Adnan Amin AT, Naik R, Nayak H, Priyadarshini N (2016) Physiological responses of rohu (Labeo rohita) exposed to sub-lethal concentrations of Lead nitrate. Biochem Cell Arch 17(1):271–274

Adnan Amin AT, Naik R, Priyadarshini N (2015) Sub lethal effect of arsenic on oxygen consumption and behaviour of rohu, Labeo rohita. Asian J Biomed Pharm Sci 2(5):43–47

Adnan Amin AT, Naik R, Priyadarshini N, Nayak H, Sruthi Sree C (2017) Toxic effect of heavy metal Lead on oxygen consumption of rohu (Labeo rohita) fingerlings. Biochem Cell Arch 17(1):225–228

Alencar BTB, Ribeiro VHV, Cabral CM, dos Santos NMC, Ferreira EA, Francino DMT, dos Santos JB, Silva DV, de Freitas Souza M (2020) Use of macrophytes to reduce the contamination of water resources by pesticides. Ecol Indic 109:105785

Andresen JA, Muir D, Ueno D, Darling C, Theobald N, Bester K (2007) Emerging pollutants in the North Sea in comparison to Lake Ontario, Canada, data. Environ Toxicol Chem 26(6): 1081–1089

Atli G, Canli M (2007) Enzymatic responses to metal exposures in a freshwater fish Oreochromis niloticus. Comp Biochem Physiol C Toxicol Pharmaco 145(2):282–287

Balmer ME, Buser HR, Müller MD, Poiger T (2005) Occurrence of some organic UV filters in wastewater, in surface waters, and in fish from Swiss lakes. Environ Sci Technol 39(4):953–962

Bester K (2007) Personal care compounds in the environment: pathways, fate and methods for determination. John Wiley & Sons, New York

Bester K, Klasmeier J, Kupper T (2008a) Emissions of OTNE (iso-E-super)–mass flows in sewage treatment plants. Chemosphere 71(11):2003–2010

Bester K, Scholes L, Wahlberg C, McArdell CS (2008b) Sources and mass flows of xenobiotics in urban water cycles—an overview on current knowledge and data gaps. Water, Air, & Soil Pollution: Focus 8(5):407–423

Cappello T, Brandão F, Guilherme S, Santos MA, Maisano M, Mauceri A, Canário J, Pacheco M, Pereira P (2016) Insights into the mechanisms underlying mercury-induced oxidative stress in gills of wild fish (Liza aurata) combining ^{1}H NMR metabolomics and conventional biochemical assays. Sci Total Environ 548:13–24

Chopra S, Kumar D (2018) Pharmaceuticals and personal care products (PPCPs) as emerging environmental pollutants: toxicity and risk assessment. In: Advances in animal biotechnology and its applications. Springer, Singapore, pp 337–353

Correia TG, Narcizo ADM, Bianchini A, Moreira RG (2010) Aluminum as an endocrine disruptor in female Nile tilapia (Oreochromis niloticus). Comp Biochem Physiol C Toxicol Pharmacol 151(4):461–466

Costa CA, Trivelato GC, Pinto AM, Bechara EJ (1997) Correlation between plasma 5-aminolevulinic acid concentrations and indicators of oxidative stress in lead-exposed workers. Clin Chem 43(7):1196–1202

Dar SA, Chatterjee A, Rather MA, Chetia D, Srivastava PP, Gupta S (2020) Identification, functional characterization and expression profiling of cytochrome p450 1A (CYP1A) gene in Labeo rohita against emamectin benzoate. Int J Biol Macromol 158:1268–1278

De Lange HJ, Noordoven W, Murk AJ, Lürling MFLLW, Peeters ETHM (2006) Behavioural responses of Gammarus pulex (crustacea, Amphipoda) to low concentrations of pharmaceuticals. Aquat Toxicol 78(3):209–216

de Oliveira M, Frihling BEF, Velasques J, Magalhães Filho FJC, Cavalheri PS, Migliolo L (2020) Pharmaceuticals residues and xenobiotics contaminants: occurrence, analytical techniques and sustainable alternatives for wastewater treatment. Sci Total Environ 705:135568

Desbrow CEJR, Routledge EJ, Brighty GC, Sumpter JP, Waldock M (1998) Identification of estrogenic chemicals in STW effluent. 1. Chemical fractionation and in vitro biological screening. Environ Sci Technol 32(11):1549–1558

do Sul JAI, Costa MF (2014) The present and future of microplastic pollution in the marine environment. Environ Pollut 185:352–364

Donner E, Eriksson E, Holten-Lützhøft HC, Scholes L, Revitt M, Ledin A (2010) Identifying and classifying the sources and uses of xenobiotics in urban environments. In: Xenobiotics in the urban water cycle. Springer, Dordrecht, pp 27–50

Ebele AJ, Abdallah MAE, Harrad S (2017) Pharmaceuticals and personal care products (PPCPs) in the freshwater aquatic environment. Emerg Contam 3(1):1–16

Ecobichon DJ (1991) Toxic effects of pesticides in casarett and doull's toxicology: The Basic Science of Poisons. McGraw-Hill, New York

Ercal N, Gurer-Orhan H, Aykin-Burns N (2001) Toxic metals and oxidative stress part I: mechanisms involved in metal-induced oxidative damage. Curr Top Med Chem 1(6):529–539. https://doi.org/10.2174/1568026013394831

Essumang DK, Ankrah DA (2010) Distribution, levels, and risk assessment of polycyclic aromatic hydrocarbons (PAHs) in some water bodies along the coastal belt of Ghana. Sci World J 10: 972–985

Fent K, Weston AA, Caminada D (2006) Erratum to "ecotoxicology of human pharmaceuticals" [aquatic toxicology 76 (2006) 122–159]. Aquat Toxicol 2(78):207

Gadzała-Kopciuch R, Berecka B, Bartoszewicz J, Buszewski B (2004) Some considerations about bioindicators in environmental monitoring. Pol J Environ Stud 13(5):453–462

Gayathri KV, Namasivayam V (2010) Enrichment of phenol degrading moderately halophilic bacterial consortium from saline environment. J Bioremed Biodegr 1(1):1–6

Giger W, Schaffner C, Kohler HPE (2006) Benzotriazole and tolyltriazole as aquatic contaminants. 1. Input and occurrence in rivers and lakes. Environ Sci Technol 40(23):7186–7192

Glusczak L, dos Santos Miron D, Moraes BS, Simões RR, Schetinger MRC, Morsch VM, Loro VL (2007) Acute effects of glyphosate herbicide on metabolic and enzymatic parameters of silver catfish (Rhamdia quelen). Comp Biochem Physiol C Toxicol Pharmacol 146(4):519–524

Gomes JM, Donnici CL, Júnior JDC, da Silva JBB (2016) Validation of methods employing fast alkaline solubilization to determine cadmium in fish liver, spleen, gills and muscle by graphite furnace atomic absorption spectrometry. Microchem J 124:629–636

Goyer RA, Clarkson TW (1996) Toxic effects of metals. In: Casarett and Doull's toxicology: the basic science of poisons, vol 5. McGraw-Hill, New York, pp 691–736

Grčić I, Vrsaljko D, Katančić Z, Papić S (2015) Purification of household greywater loaded with hair colorants by solar photocatalysis using TiO_2-coated textile fibers coupled flocculation with chitosan. J Water Process Eng 5:15–27

Gu C (2019) Urbanization: processes and driving forces. Sci China Earth Sci 62(9):1351–1360

Hashmi MZ, Kumar V, Varma A (2017) Xenobiotics in the soil environment: monitoring, toxicity and management, vol 49. Springer, New York

Heberer T (2002) Occurrence, fate, and removal of pharmaceutical residues in the aquatic environment: a review of recent research data. Toxicol Lett 131(1–2):5–17

Hoffman DJ, Rattner BA, Burton GA Jr, Cairns J Jr (2002) Handbook of ecotoxicology. CRC Press, Boca Raton, Florida

Hollender J, McArdell CS, Escher B (2007) Mikroverunreinigungenaus der Siedlungsentwässerung in Gewässern der Schweiz: Vorkommen und Bewertung. GWA 11:843–852

Houssard P, Point D, Tremblay-Boyer L, Allain V, Pethybridge H, Masbou J, Ferriss BE, Baya PA, Lagane C, Menkes CE, Letourneur Y (2019) A model of mercury distribution in tuna from the western and Central Pacific Ocean: influence of physiology, ecology and environmental factors. Environ Sci Technol 53(3):1422–1431

Jobling S, Nolan M, Tyler CR, Brighty G, Sumpter JP (1998) Widespread sexual disruption in wild fish. Environ Sci Technol 32(17):2498–2506

Joss A, Andersen H, Ternes T, Richle PR, Siegrist H (2004) Removal of estrogens in municipal wastewater treatment under aerobic and anaerobic conditions: consequences for plant optimization. Environ Sci Technol 38(11):3047–3055

Joss A, Keller E, Alder AC, Göbel A, McArdell CS, Ternes T, Siegrist H (2005) Removal of pharmaceuticals and fragrances in biological wastewater treatment. Water Res 39(14): 3139–3152

Kamdem LK, Meineke I, Gödtel-Armbrust U, Brockmöller J, Wojnowski L (2006) Dominant contribution of P450 3A4 to the hepatic carcinogenic activation of aflatoxin B1. Chem Res Toxicol 19(4):577–586

Kidd K, Batchelar K (2012) Mercury. In: Wood CM, Farrell AP, Brauner CJ (eds) Fish Physiology-Homeostasis and Toxicology of Non-Essential Metals. Elsevier, Amsterdam, pp 238–284

Kim KH, Kabir E, Jahan SA (2016) A review on the distribution of hg in the environment and its human health impacts. J Hazard Mater 306:376–385

Kim MK, Zoh KD (2016) Occurrence and removals of micropollutants in water environment. Environ Eng Res 21(4):319–332

Krabbenhoft DP (2004) Methylmercury contamination of aquatic ecosystems: a widespread problem with many challenges for the chemical sciences. In water and sustainable development opportunities for the chemical sciences—a workshop report to the chemical sciences roundtable. National Academies Press, Washington, DC, pp 19–26

Larose C, Canuel R, Lucotte M, Di Giulio RT (2008) Toxicological effects of methylmercury on walleye (Sander vitreus) and perch (Perca flavescens) from lakes of the boreal forest. Comp Biochem Physiol C Toxicol Pharmacol 147(2):139–149

Lazarevic D, Aoustin E, Buclet N, Brandt N (2010) Plastic waste management in the context of a European recycling society: comparing results and uncertainties in a life cycle perspective. Resour Conserv Recycl 55(2):246–259

LeBlanc GA (2007) Crustacean endocrine toxicology: a review. Ecotoxicology 16(1):61–81

Li ZW, Huang M, Luo NL, Wen JJ, Deng CX, Yang R (2019) Spectroscopic study of the effects of dissolved organic matter compositional changes on availability of cadmium in paddy soil under different water management practices. Chemosphere 225:414–423

Lienert J, Güdel K, Escher BI (2007) Screening method for ecotoxicological hazard assessment of 42 pharmaceuticals considering human metabolism and excretory routes. Environ Sci Technol 41(12):4471–4478

Livingstone DR (2003) Oxidative stress in aquatic organisms in relation to pollution and aquaculture. Revue de MédecineVétérinaire (France) 154(6):427–430

Mandlekar S, Hong JL, Tony Kong AN (2006) Modulation of metabolic enzymes by dietary phytochemicals: a review of mechanisms underlying beneficial versus unfavorable effects. Curr Drug Metab 7(6):661–675

Mariussen E, Heier LS, Teien HC, Pettersen MN, Holth TF, Salbu B, Rosseland BO (2017) Accumulation of lead (pb) in brown trout (Salmo trutta) from a lake downstream a former shooting range. Ecotoxicol Environ Saf 135:327–336

Marklund A, Andersson B, Haglund P (2003) Screening of organophosphorus compounds and their distribution in various indoor environments. Chemosphere 53(9):1137–1146

Meyer J, Bester K (2004) Organophosphate flame retardants and plasticisers in wastewater treatment plants. J Environ Monit 6(7):599–605

Milukaite AR (2006) Long-term trends of benzo (a) pyrene concentration on the eastern coast of the Baltic Sea. Atmos Environ 40(11):2046–2057

Monadjem A, Anderson MD, Piper SE, Boshoff AF (2004) The vultures of southern Africa-quo vadis. In: Proceedings of a workshop on vulture research and conservation in southern Africa. Bird of Prey Working Group, Endangered Wildlife Trust, Johannesburg, South Africa

Nie J, Sun Y, Zhou Y, Kumar M, Usman M, Li J, Shao J, Wang L, Tsang DC (2020) Bioremediation of water containing pesticides by microalgae: mechanisms, methods, and prospects for future research. Sci Total Environ 707:136080

Nikolaou A, Meric S, Fatta D (2007) Occurrence patterns of pharmaceuticals in water and wastewater environments. Anal Bioanal Chem 387(4):1225–1234

Olsvik PA, Søfteland L, Hevrøy EM, Rasinger JD, Waagbø R (2016) Fish pre-acclimation temperature only modestly affects cadmium toxicity in Atlantic salmon hepatocytes. J Therm Biol 57:21–34

Pedersen JA, Yeager MA, Suffet IH (2003) Xenobiotic organic compounds in runoff from fields irrigated with treated wastewater. J Agric Food Chem 51(5):1360–1372

Picó Y, Alvarez-Ruiz R, Alfarhan AH, El-Sheikh MA, Alshahrani HO, Barceló D (2020) Pharmaceuticals, pesticides, personal care products and microplastics contamination assessment of Al-Hassa irrigation network (Saudi Arabia) and its shallow lakes. Sci Total Environ 701:135021

Poiger T, Buser HR, Müller MD, Balmer ME, Buerge IJ (2003) Occurrence and fate of organic micropollutants in the environment: regional mass balances and source apportioning in surface waters based on laboratory incubation studies in soil and water, monitoring, and computer modeling. CHIMIA Int J Chem 57(9):492–498

Reemtsma T, Jekel M (eds) (2006) Organic pollutants in the water cycle: properties, occurrence, analysis and environmental relevance of polar compounds. John Wiley & Sons, New York

Ricklund N, Kierkegaard A, McLachlan MS, Wahlberg C (2009) Mass balance of decabromodiphenyl ethane and decabromodiphenyl ether in a WWTP. Chemosphere 74(3): 389–394

Roccaro P, Sgroi M, Vagliasindi FG (2013) Removal of xenobiotic compounds from wastewater for environment protection: treatment processes and costs. Chem Eng Trans:32, 505–510

Sankhla MS, Kumari M, Sharma K, Kushwah RS, Kumar R (2018) Water contamination through pesticide & their toxic effect on human health. Int J Res Appl Sci Eng Technol 6(1):967–970

Schlüsener MP, Bester K (2008) Behavior of steroid hormones and conjugates during wastewater treatment–a comparison of three sewage treatment plants. CLEAN–Soil Air Water 36(1):25–33

Sevcikova M, Modra H, Slaninova A, Svobodova Z (2011) Metals as a cause of oxidative stress in fish: a review. Vet Med 56(11):537–546

Shah SL (2006) Hematological parameters in tench Tinca tinca after short term exposure to lead. J Appl Toxicol 26(3):223–228

Singh R (2017) Biodegradation of xenobiotics-a way for environmental detoxification. Int J Dev Res 7(1):14082–14087

Soni P, Sharma S, Sharma S, Kumar S, Sharma KP (2006) A comparative study on the toxic effects of textile dye wastewaters (untreated and treated) on mortality and RBC of a freshwater fish Gambusia affinis (Baird and Gerard). J Environ Biol 27(4):623–628

Southorn PA, Powis G (1988) Free radicals in medicine. I. Chemical nature and biologic reactions. In: Mayo Clinic Proceedings, vol 63. Elsevier, Amsterdam, pp 381–389

Spain JC (1995) Biodegradation of nitroaromatic compounds. Annu Rev Microbiol 49(1):523–555

Sruthisree C, Gowda G, Harish BD, Harsha N, Amin A (2015) Effect of heavy metal (lead) on oxygen consumption and ammonia excretion on iridescent shark (Pangasius hypothalamus). J Exp Zool India 18(2):637–640

Ternes T, Joss A (2007) Human pharmaceuticals, hormones and fragrances. IWA Publishing, London

Triebskorn R, Casper H, Heyd A, Eikemper R, Kohler HR, Schwaiger J (2004) Toxic effects of the non-steroidal anti-inflammatory drug diclofenac part II. Cytological effects in liver, kidney, gills and intestine of rainbow trout (Oncorhynchus mykiss). Aquat Toxicol 68:151–166

Tsui MT, Chu LM (2003) Aquatic toxicity of glyphosate-based formulations: comparison between different organisms and the effects of environmental factors. Chemosphere 52(7):1189–1197

Verbost PM, Flik G, Lock RAC, Wendelaar Bonga SE (1988) Cadmium inhibits plasma membrane calcium transport. J Membr Biol 102(2):97–104

Weiss S, Jakobs J, Reemtsma T (2006) Discharge of three benzotriazole corrosion inhibitors with municipal wastewater and improvements by membrane bioreactor treatment and ozonation. Environ Sci Technol 40(23):7193–7199

Whyte LG, Bourbonniere L, Greer CW (1997) Biodegradation of petroleum hydrocarbons by psychrotrophic pseudomonas strains possessing both alkane (alk) and naphthalene (nah) catabolic pathways. Appl Environ Microbiol 63(9):3719–3723

Wood CM, Farrell AP, Brauner CJ (2011) Fish physiology-homeostasis and toxicology of non-metals, vol 31. Elsevier, Amsterdam, pp 1–531

Wu JC, Lai CS, Tsai ML, Ho CT, Wang YJ, Pan MH (2017) Chemopreventive effect of natural dietary compounds on xenobiotic-induced toxicity. J Food Drug Anal 25(1):176–186
Zaharia C, Suteu D, Muresan A, Muresan R, Popescu A (2009) Textile waste water treatment by homogenous oxidation with hydrogen peroxide. Environ Eng Manag J 8:1359–1369
Zhang C, Wang S, Yan Y (2011) Isomerization and biodegradation of beta-cypermethrin by Pseudomonas aeruginosa CH7 with biosurfactant production. Bioresour Technol 102(14): 7139–7146
Zupo V, Graber G, Kamel S, Plichta V, Granitzer S, Gundacker C, Wittmann KJ (2019) Mercury accumulation in freshwater and marine fish from the wild and from aquaculture ponds. Environ Pollut 255:112975

Cypermethrin-Induced Reproductive Toxicity in Zebrafish: Biochemical and Molecular Perspective

Sana Aziz, Mumaiza Mumraiz, Fariha Latif, and Muhammad Sarfraz Ali

1 Introduction

The water bodies are persistently contaminated by millions of toxins due to the release of industrial, agricultural, and household wastes, which are already threatening aquatic life. Among these toxins, pesticides used for agricultural purposes are now a significant contributor to the decline in biodiversity in aquatic habitats. Generally, aquatic environments come into contact with several different types of pesticides. Previous literature revealed the occurrence of pesticides in all types of water systems, aquatic biota, and bottom residues in greater concentrations (Zheng et al. 2016). Cypermethrin, (R,S)-alpha-cyano-3-phenoxybenzyl(1RS)-cis,trans-3-(2,2-dichlorovinyl)- 2,2-dimethylcyclopropane-carboxylate, a widely used insecticide, is a type II synthetic halogenated pyrethroid pesticide of fourth generation (Kaviraj and Gupta 2014). The physical and chemical characteristics have been presented in Table 1. Cypermethrin has a 416.30 molecular weight, a 60–80 °C melting point, and a 6.6020 partition coefficient and shows colourless crystals in pure form (Kidd and James 1991; Ray 1991; and Wauchope et al. 1992a, b). These pyrethroid types are widely utilized in tropical regions to eradicate various insect

The original version of the chapter has been revised. A correction to this chapter can be found at https://doi.org/10.1007/978-981-99-1214-8_22

S. Aziz (✉) · M. Mumraiz
Department of Zoology, Wildlife and Fisheries, University of Agriculture, Faisalabad, Pakistan

F. Latif
Institute of Pure and Applied Biology, Bahauddin Zakariya University, Multan, Pakistan
e-mail: farihalatif@bzu.edu.pk

M. Sarfraz Ali
Islamabad, Department of Biosciences, COMSATS University, Islamabad, Pakistan
e-mail: sp20-pbs-002@isbstudent.comsats.edu.pk

M. A. Rather et al. (eds.), *Xenobiotics in Aquatic Animals*,
https://doi.org/10.1007/978-981-99-1214-8_4

Table 1 Physical and chemical characteristics of cypermethrin

Variables	Relevant Information
Appearance	Colorless crystals in pure form, viscous semisolid or viscous yellow liquid when mixed isomers are present
IUPAC name	Cyano-(3-phenoxyphenyl)methyl]3-(2,2-dichloroethenyl)-2,2-dimethylcyclopropane-1-carboxylate
Chemical abstract (CA) name	(R,S)-alpha-cyano-3-phenoxybenzyl(1RS)-cis,trans-3-(2,2-dichlorovinyl)- 2,2-dimethylcyclopropane-carboxylate
Chemical formula	$C_{23}H_{19}ClF_3NO_3$
Molecular weight	416.30
CAS number	52315-07-8
Solubility in water	Insoluble in water; 0.01 mg/L at 20 °C
Melting point (M.P)	60–80 °C of pure isomers
Partition coefficient	6.6020
Vapour pressure	5.1 9 10–7 nPa at 70 °C

pests from different crops such as wheat, okra, sunflower, sugarcane, cotton, brinjal, cabbage, and many more (Majumder and Kaviraj 2017). Ectoparasites that invade poultry, cattle, sheep, and some associated animals are also controlled with them (Velisek et al. 2006). This substance has been employed as a chemotherapeutic agent in the marine cage culture of Atlantic salmon (*Salmo salar*) to manage its ectoparasitic infections (*Caligus elongatus* and *Lepeophtheirus salmonis*) (Treasurer and Wadsworth 2004).

It has been abundantly found in all types of environments and is also found in fruits, vegetables, dust, and air in lower or higher concentrations (Li et al. 2014, 2018; Hung et al. 2018). Moreover, cypermethrin from fishery medications, rain erosion, and spray drift has been exposed in different water sources, i.e., surface water, contaminated water, and even drinking water (Feng et al. 2016). For instance, industrial effluents had a maximum cypermethrin concentration of 0.969 μg/L in Beijing, China, and in Roseville, USA, sewerage water during the arid conditions had a maximum cypermethrin content of 25.9 ng/L (Weston et al. 2009; Ge et al. 2010).

The amount of cypermethrin in aquatic environments from Pampa Ondulada, Argentina, can reach 194 g/L after rain (Marino and Ronco 2005). The Environmental Quality Standard (EQS) states that the extreme permissible concentration and annual average levels of cypermethrin in the water bodies are 10 ng/L and 1 ng/L, respectively. These levels are referred to as operating values and are employed by the organization to provide help in control of cypermethrin in various water bodies (Anonymous 1995). This suggests that nontarget animals might be in danger from the presence of such high levels of cypermethrin in aquatic ecosystems. Its growing use has drawn the attention of several toxicologists dealing with fish vulnerability.

As compared to fish, birds and mammals are less harmful to cypermethrin because of their slower metabolic rate and easier elimination. The pyrethroids take 48 h to eliminate from rainbow trout, while these substances are eliminated from mammals and birds only within 6–12 h (Bradbury and Coats 1989a, b). It is reported that cypermethrin is very harmful to fish, and due to the stereochemical arrangement, mixtures, and isomer formation, its toxic effects differ from species to species

(Kumar et al. 2007; Saha and Kaviraj 2008; Polat et al. 2002; Yılmaz et al. 2004). It has been established that common marketable formulations are more harmful than their active constituents, which have increased efficacy in agriculture (Demetrio et al. 2014; Majumder and Kaviraj 2015; Puglis and Boone 2011).

Zebrafish are now rapidly emerging as an essential model species in ecotoxicological studies due to their intrinsic benefits, e.g., their low cost, short reproductive period, abundant egg synthesis with a high level of fertilization, hatching, and production of many developing embryos, etc. They can absorb tiny compounds from their surroundings via skin, gills, and chorion (Parng 2005). Pesticides are frequently used in field works, and as fish are sensitive to low doses of hazardous substances, there is a need for a method to assess the effects of these pesticides on fish species as a biomarker for aquatic toxicity. So, it is essential to conduct studies on the harmful effects of pesticides on fish populations.

2 Applications and Pathways of Cypermethrin in Aquatic Systems

Cypermethrin is considered one of the most abundantly found toxicants in freshwater bodies, and its concentration is as high as 3 g/L in surface water (Carriquiriborde et al. 2007; Jaensson et al. 2007). It is not naturally occurring in the environment; instead, its presence in the ecosystem is a result of anthropogenic activities. It is extensively utilized in forestry, veterinary treatments, biocides, plant protection products, and chemical industry (Fig. 1) and discharged in the aquatic environments by various channels. Due to its wide range of applications, it may enter water systems through runoff from agriculture or from rain on treated outdoor surfaces

Fig. 1 Applications of cypermethrin

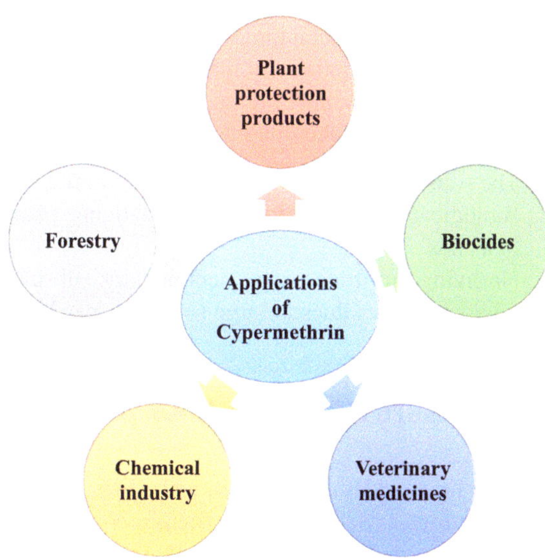

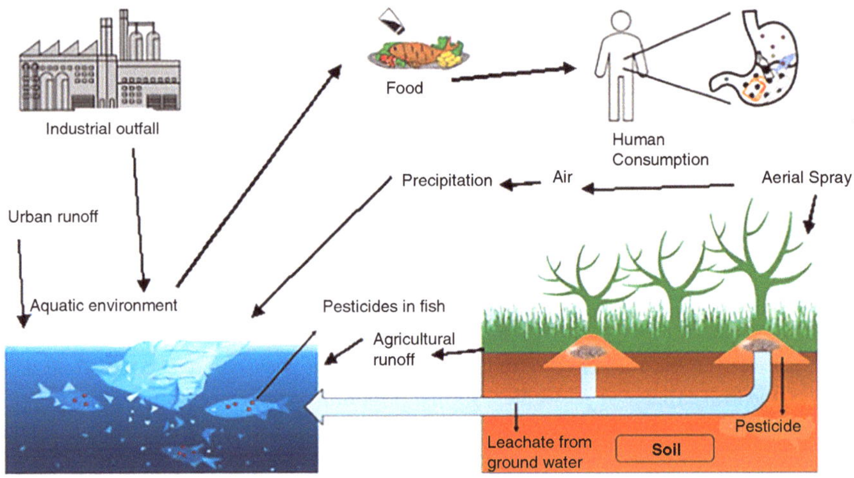

Fig. 2 Entrance pathway of cypermethrin into water bodies

(Defra 2009; ADAS 2004). Pyrethroids can start accumulation in water bodies that destroy zooplankton, the key source of natural food for fish, at an early stage (Pesticide Action Network North America 1999). Pyrethroids can also kill the different types of insects on which some fish species feed, triggering the fish to travel distantly for food that enhances the risk of predators attacking. Wet cleaning is regarded as a route of exposure because the EU recognized it as a potential route for cypermethrin to enter the environment (EU 2015). Due to the disposal of hazardous substances, leachate from landfills is a possible source of cypermethrin in the surrounding environment. The discharge of unhealthy materials and the release of pyrethroids from agricultural runoff may contaminate aquatic habitats, which will then cause adverse toxic effects in exposed aquatic organisms, including fish and other aquatic nontarget animals (Farag et al. 2021) (Fig. 2).

In order to recognize the major pathways by which pesticides are exposed to aquatic ecosystems and organisms, the following components of a worldwide biocycle should be taken into account:

1. The water systems are the first that are initially exposed to pesticides.
2. Biological substrates such as water-living plants, algae, fungi, leaf litter, and branches.
3. No-living substrates, e.g., sedimentary substances, which range in size from microscopic residues to large soil particles (Murthy et al. 2013).

3 Exposure Routes for Absorption of Cypermethrin

Various insecticides can be taken up by fish through integumentary, respiratory, and digestive routes (Schlenk 2005; Banaee et al. 2011) (Fig. 3). They can be absorbed by the epidermis and collaborate with the blood-carrying proteins. Afterward, these

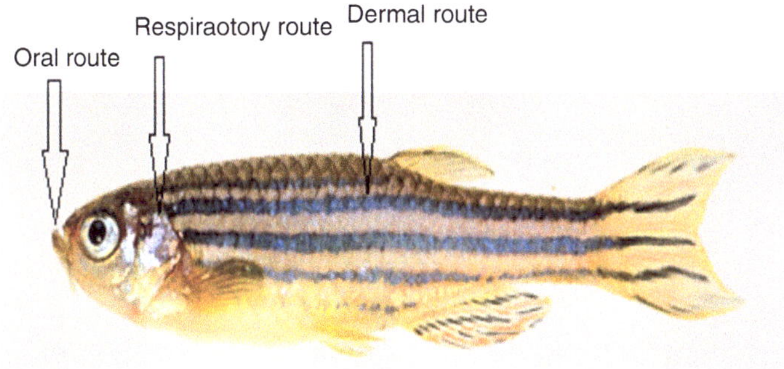

Fig. 3 Different exposure routes of cypermethrin

dispersed substances with skin cells disturb the CNS (central nervous system) directly by the connection with peripheral nervous system (PNS) (Neal et al. 2010). Moreover, they may enter the organism's body in low concentrations by inhalation of smoke during breathing mechanisms. Furthermore, they can enter the circulatory system directly via the alimentary canal during the digestive process (Prusty et al. 2015; Farag et al. 2021).

Fish organs, especially gills, have direct contact with water; therefore, they are considered the target organ which would be estimated to have a greater proportion of cypermethrin (Wendelaar Bonga 1997). Cypermethrin rapidly invades the cell membrane due to its lipophilic nature, which increases the sensitivity of fish against waterborne exposure to pyrethroid (Mishra et al. 2005). Therefore, gill tissues are thought to be well suited to evaluate the genotoxicity caused by water pollutants (Sharma et al. 2007). Gill tissues have a high proportion of pyrethroids even at lower levels because they are not capable of properly breaking the pyrethroids which adversely affect the performance of enzymes and rate of transport (Viran et al. 2003). Srivastava and Kaushik (2001) described that pest-killing agents accumulate in important organs (liver and muscles) causing dysfunction of organs and resulting in fish mortality.

4 Toxicological Mechanisms of Cypermethrin

There are many fish species which are vulnerable to various kinds of pesticides at different levels. All aquatic animals show a wide range of toxic effects caused by pesticides which depends on species, particular class of pesticides, exposure time, and water quality (Coppage and Matthews 1974). Cypermethrin, a specific type of pyrethroid, causes adverse effects by various toxicity processes (Fig. 4), and some of them have described below to elaborate on it further.

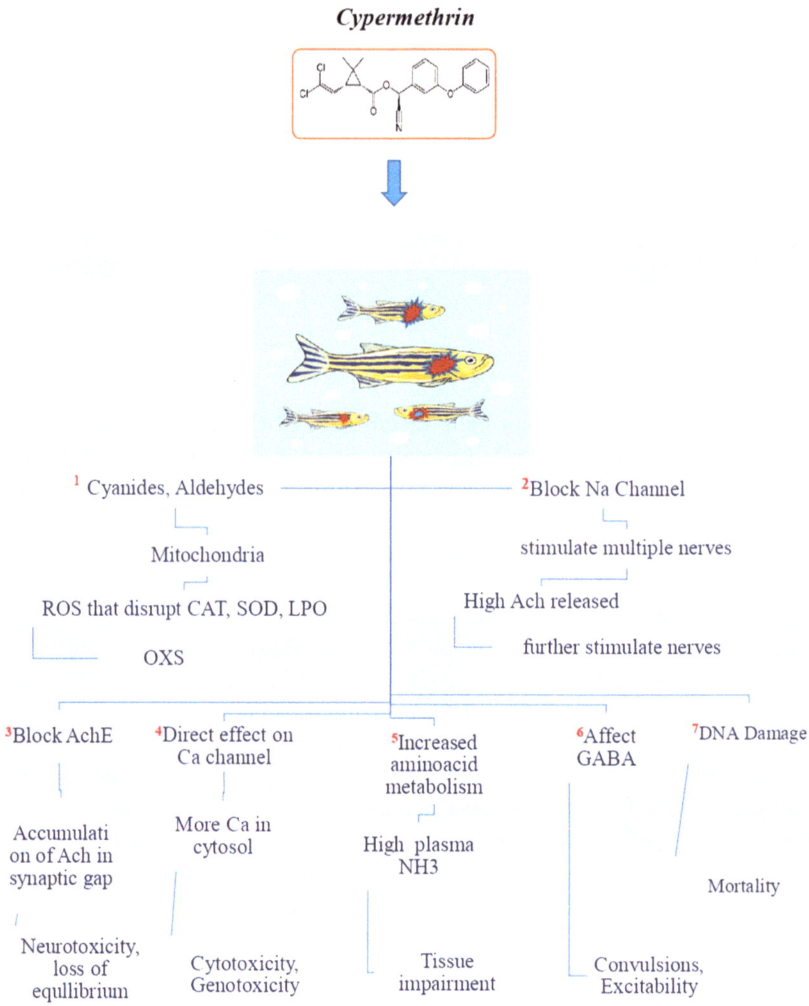

Fig. 4 The toxicological mechanism of cypermethrin. *4*. Cypermethrin resulted in aldehydes and cyanides that induce ROS that disrupt antioxidant enzymes and cause oxidative stress. *2*. Cypermethrin blocks the Na^{+2} channel gate closing leading to multiple nerve impulses, which in turn leads to the release of the acetylcholine neurotransmitters and stimulation of other nerves. *3*. Cypermethrin inhibits AChE, which results in ACh pathological retention in synaptic gaps. *4*. Cypermethrin has a direct effect on calcium channels leading to an increased concentration of cytosolic calcium that leads to genotoxicity and cytotoxicity. *5*. Cypermethrin increased amino acid catabolism that raised plasma NH_3 and caused tissue impairment. *6*. Cypermethrin inhibits GABA receptors, thus causing convulsions and excitability. *7*. Cypermethrin caused DNA damage that induced mortality

4.1 Induction of Oxidative Stress

Oxidative stress is one of the important processes that have a concern for nervous damage caused by cypermethrin toxicity. Oxidative stress is triggered by an imbalance in oxygen-free radicals that can cause destruction in micro- and macromolecules. Generally, the unusual increase in reactive oxygen species (ROS) is an important factor that results in oxidative stress, and can also destroy the cell structure (Barzilai and Yamamoto 2004; Amin and Indulkar 2017).

When cypermethrin is metabolized, it converts into cyanohydrines and then further breaks down into aldehydes and cyanides; all these compounds contribute in the formation of ROS (Wielgomas and Krechniak 2007). Neurotoxicity caused by cypermethrin also takes part in the production of oxygen free radicals (Kale et al. 1999; Giray et al. 2001). Cypermethrin exposure leads to the disruption in mitochondria, and causes cell death as well as oxidative damage that subsequently facilitates nigrostriatal dopaminergic neurodegeneration (Agrawal et al. 2015).

Waterborne exposure to certain chemicals might be responsible for the stimulation of oxidative damage in aquatic biota, if these chemicals cause an imbalance in ROS production (Jin et al. 2010; Amin et al. 2016). In controlled biological conditions, free radicals are rapidly removed from fish and other organisms with the help of antioxidant enzymes such as catalase (CAT), superoxide dismutase (SOD), glutathione peroxidase (GPx), etc. (Valavanidis et al. 2006; Zhang et al. 2009). In organisms that are exposed to harmful toxicants, the antioxidant defence system may reduce its activity, resulting in oxidative stress (Valavanidis et al. 2006).

However, according to previous reports, cypermethrin has the capacity to cause oxidative damage in fish. Zebrafish exposed to different levels of cypermethrin for 96 h showed a dose-dependent increase in the activity of CAT, SOD, and lipid peroxidation levels in relation to the control group (Shi et al. 2011). The increased level of lipid peroxidation after exposure to cypermethrin may contribute in the production of reactive oxygen species, which may lead to the oxidation of polyunsaturated fats and ultimately cause lipid peroxidation (Valavanidis et al. 2006). In fish, oxidative damage caused by ROS has been reported to facilitate developmental abnormalities in the embryo (Yabu et al. 2001; Yamashita 2003). According to Uner et al. (2001), exposure to 3 l g/L CYP for 10 days dramatically increases the GPx, SOD, and CAT activity as well as MDA levels in the liver and kidney of fish species such as *Oreochromis niloticus* and *Cyprinus carpio*.

Numerous studies have demonstrated the adverse effects of pest-killing agents in many biochemical processes (Ullah et al. 2014). The disturbances in antioxidant enzymes in fish are frequently tissue specific and found in the liver, brain, gills, muscles, viscera, and kidneys of various fish species. The antioxidant activity has presented variable results in various organs of each fish, for example, in tilapia, peroxidase activities were found to be higher in the brain, gills, viscera, and muscles, but the gills have experienced maximum change in peroxidase levels (Ahmad et al. 2000). In the same way, changes in lipid peroxidase levels have been noted in response to several pesticides and environmental contaminants. Cypermethrin caused significant alterations in peroxidase, lipid peroxidase, catalase, and glutathione reductase levels in the liver, brain, gills, and muscles of mahseer (*Tor putitora*) (Ullah et al. 2014).

4.2 Disturbances in Ion Channels

Pyrethroids are characterized as neurotoxins, targeting the axons of the central as well as peripheral nervous systems (CNS and PNS) by interfering with ionic (Na^+, K^+) pathways. Many studies have documented the harmful effects of pyrethroids on various shell and fin fishes, as well as their interference with ion exchange mechanisms in nervous and mitochondrial cell membranes (Lutnicka and Kozińska 2009; Lidova et al. 2016; Cárcamo et al. 2017; Wang et al. 2017). Na^+ channels are categorized as important proteins of the nervous system that regulate coordination as well as control the vital processes, e.g., heart rate, salt and water balance, and the functioning of the brain. Like mammals, zebrafish also exhibit a voltage-gated Na^+ pathway gene expression process (Novak et al. 2006). All the pyrethroid chemicals result in continued Na^+ discharge and delay the sodium activation channel closure, causing elongated and reduced Na^+ tail outflow (Wang et al. 2006). The association between cypermethrin and Na^+ channels creates a hyperexcitable condition, and it is the important part of its neurotoxic activity; changes in Na^+ pathways also affect other nerve cells.

The important mechanism of pesticide-mediated toxicity is due to its blockage of AChE (acetylcholinesterase) which results in an abundance of acetylcholine, a neurotransmitter, at the end of neuronal and neuromuscular interchanges. Pesticides also exhibit disturbances in animal's behaviour that are normally controlled by acetylcholinesterase (AChE) at the synaptic region. The results showed that cypermethrin ceases the functioning of AChE, an enzyme located at synapsis, and is involved in the regulation of nerve impulses by breaking down the acetylcholine into choline and acetic acid. The accumulation of acetylcholine at various sites in neurons results in hyper-excitability which leads to behavioural alterations and may cause fish mortality (Rao 1990).

Das and Mukherjee (2003) also revealed that cypermethrin hinders the working of AChE in the brain of *Labeo rohita* fingerlings. Cypermethrin works as a neurotoxin that disturbs the conduction of nerve impulses and affects the permeability of nerve cell membranes (Rao 1990). Cypermethrin also causes alterations in swimming performance, equilibrium loss, hyperexcitability, and sinking toward the bottom in *Tor putitora* and *Labeo rohita* (Marigoudar et al. 2009; Ullah et al. 2014). Pesticides, especially cypermethrin, change the migratory patterns of fish, which may cause disturbance in their whole life cycle, for example, in salmonids, movement from fresh to marine water can be changed as a result of AChE inhibition.

Some pyrethroids restricted the calmodulin protein, which increased the discharge of neurotransmitters in the postsynaptic region. This protein is also responsible for the interaction of Ca^+ ions and intracellular membranes and controls the Ca^+

elimination from the nerve endpoints, which decreases the immediate neurotransmitter discharge (Wang 2008). The nervous system and outer cell membrane become excited after chronic exposure to certain pest-killing agents. Moreover, GABA (γ-aminobutyric acid) neurotransmitters in the brain are adversely affected by pyrethroids (Coats et al. 1989; Bradbury and Coats 1989a, b; Richterova and Svobodová 2012). Furthermore, they change the functioning of voltage-gated Ca^{+2} pathways and restrict the transfer of Cl^{-} ions into the nervous tissues (Soderlund 2010). The beneficial effects of pyrethroids were less with regard to Ca-Mg ATPase receptors, peripheral benzodiazepine neurotransmitters, and calcium ATPase (Breckenridge et al. 2009).

4.3 Induction of DNA Damage

The exposure of zebrafish to different sublethal concentrations of cypermethrin increases the functioning of antioxidant enzymes towards reactive oxygen species and can result in DNA damage in gill cells. Paravani et al. (2018) observed significant alterations in expression of genes responsible for DNA repair, retinal structural layers, programmed cell death (γ-H2AX and caspase-3), DNA fragmentation and also in gene expression related to antioxidant enzymes. These outcomes are helpful in showing the genotoxicity and oxidative damage in retinal cells of zebrafish after exposure to environmentally relevant concentrations of cypermethrin.

Most importantly, it has been proven by earlier studies that oxidative stress causes DNA damage and cell death. Cypermethrin has a close interaction with nucleic acids like DNA, causing damage to them. According to Mitchelmore and Chipman (1998), DNA fragmentation, specifically assessed by a comet assay, can be used as an important biomarker of genetic toxicity in fish as well as other aquatic organisms. In this regard, cypermethrin bioassays displayed very high values of DNA damage index in the gill cells of zebrafish, followed by dose- and time-dependent patterns. One possible explanation for the genotoxicity of cypermethrin is that, due to its tiny size and hydrophobic nature, cypermethrin can easily penetrate cellular membranes, contacting and reacting with DNA through its acidic components (Saxena et al. 2005). Therefore, this interaction with DNA could result in unwinding of double helix and structural deformation, which results in chromosomal disruption.

5 Reproductive and Developmental Toxicity in Zebrafish

The toxicity of cypermethrin in the reproductive system is an important harmful effect (Al-Hamdani and Yajurvedi 2010; Yuan et al. 2010). Pesticides are endocrine-disturbing agents (Chatterjee et al. 2001; Rajakumar et al. 2012) and have the potency to disturb the pattern of fish development and reproduction that initiates

Table 2 Cypermethrin-mediated reproductive toxicity in zebrafish

Toxic effects	Exposure time	Exposure concentration	Reference
Increase in VTG content	21 days	0.1, 1, and 4 μg/L	Guo et al. (2021)
Large yolk sac in embryo	96 h	25, 50, 100, 200, and 400 μg/L	Shi et al. (2011)
No effects on cumulative fecundity and hatchability rate (%), lethargic movements, and immediate cessation in spawning fishes	21 days	0.1, 1, and 10μg/L	Pitchika et al. (2019)
No significant changes in vitellogenin levels in both male and female fish	21 days	10 μg/L	Pitchika et al. (2019)
Gonadal-somatic index (GSI) changed	21 days	0.1, 1, and 4 μg/L	Guo et al. (2021)
Significantly	21 days	0.1, 1, and 10μg/L	Pitchika et al. (2019)
Induce spermatotoxic effects and no effect on structural integrity of ovary reduction in cumulative egg production, no change in GSI, the percentages of CAO increased, the percentages of EVO and LO decreased	21 days	0.1, 0.5, and 2.5 μg/L	Lu et al. (2021)
The Sz percentages exhibited significant decrease, the percentage of Sc was increased while the percentage of Sg was increased	21 days	0.1, 0.5, and 2.5 μg/L	Lu et al. (2021)

GSI gonadal somatic index, *Sz* spermatids, *Sc* spermatocytes, *Sg* spermatogonia, *VTG* vitellogenin, *CAO* cortical alveolar oocyte, *EVO* early vitellogenic oocyte, *LO* late/mature oocyte

reproductive endocrine disruptions (Singh and Canario 2004; Lal et al. 2013; Eni et al. 2019). These types of endocrine disruptions affect immunity in fish, which initiates diseases and ultimately can lead to death. Pesticides affected reproductive success mainly causing disturbance in hypothalamic hypophyseal-gonadal-axis secretions (Singh and Singh 1982). During the early life stages of development, fish are also susceptible to endocrine disturbing impacts. A high level of pesticides in the blood of vertebrates has been reported to cause damage in gonads of fish, such as inhibition of oocyte maturational processes, intersex, and disorganization of ovarian structure, including atretic follicles, breeding, hatching, and slow spermatogenesis (Chatterjee et al. 1997; Deka and Mahanta 2012; Agbohessi et al. 2013; Chukwuka et al. 2019; Sundaray et al. 2021).

Cypermethrin also exhibits reproductive toxicity in zebrafish (Table 2). Both in vitro fertilization and embryogenesis mark the zebrafish as a more attractive and simple animal model to investigate toxicity in the reproductive system. Zebrafish can reduce cost, shorten the testing period, and increase throughput to evaluate the reproductive toxicity of different toxicants. Pesticides also block sex hormones, which cause anomalous sexual growth, male feminization, irregular ratios in sex, and disturbed breeding. It can also change other fish's hormonal processes like bone development and the proper functioning of the thyroid (Murthy et al. 2013). Several studies have reported that pollutants disturb levels of thyroid hormone that lead to thyroid dysfunction in fish. Pesticides disrupted mating and reproductive behaviour because they blocked sex hormones (Hoeger et al. 2005).

To determine its molecular mechanism, the levels of key genes to the hypothalamic hypophyseal-gonadal-axis play a main controlling role in the production and discharge of sex hormones (Sanderson 2006). Hypothalamus synthesizes gonadotropin-releasing hormone (GnRH), and after binding with its specific receptor, it regulates the production and release of pituitary gonadotropins (GtHs), that is, luteinizing hormone (LH) and follicle-stimulating hormone (FSH). Any change in GnRH might influence the two GtHs and ultimately disturb the sex-steroid hormones balance in fish (Liu et al. 2013; Shi et al. 2015). Lu et al. (2021) observed the significant downregulation in expression of gonadotropin-releasing hormone-2, follicle-stimulating hormone receptor, and luteinizing hormone receptor in zebrafish, which contributed to the changes in sex hormone levels (Liu et al. 2013).

6 Impacts in Female Zebrafish

Cypermethrin can destroy the endometrium, decrease the number of follicles and the pregnancy rate, and disrupt the secretions of estrogen and follicle-stimulating hormone in females (Liu et al. 2006; Sangha et al. 2013). Studies also reported that zebrafish exposed to cypermethrin showed disruption of their gills, ovary, kidney, and liver architecture (Rajini et al. 2015). Pitchika et al. (2019) observed lethargic activities and an abrupt cessation in spawning fish at a concentration of 10 μg L^{-1}, and cumulative fecundity and rate of hatchability (%) did not affect female zebrafish after exposure to 0.1, 1 or 10 μg L^{-1} cypermethrin which suggested normal fertilization efficiency and spawning of eggs. Vitellogenin is a process that nourishes developing oocyte and acts as a specific marker in females. In female zebrafish, no significant variations were detected in the plasma vitellogenin levels and structural integrity of ovaries after 10 μg L^{-1} cypermethrin exposure. Histology of ovaries showed the toxicological impacts of xenoestrogens and was therefore supposed to be a valued result of reproduction.

In China, due to its quick action, effectiveness, and low price, beta-cypermethrin has half the market share (Chen et al. 2002). Even though beta-cypermethrin is measured as safe for mammals, large-scale and long-term use of it has caused harmful effects in both natural ecosystems and other nontargeted organisms. In adult zebrafish, beta-cypermethrin had an adverse effect on the gonadal development and quantity of oocytes in each period. Nevertheless, after the 0.5 and 2.5 μg/L exposure, the CAO (cortical alveolar oocyte) percentages were remarkably augmented by 9.2% and 28.3%, while the LO (late/mature oocyte) and EVO (early vitellogenic oocyte) percentage were notably decreased (LO, 49.1%, 53.2%; EVO, 40.8%, 31.0%) (Lu et al. 2021). A 2-week pre-exposure to beta-cypermethrin showed no clear change in collective egg production in the unexposed zebrafish group. However, after a 21-day exposure, it caused a significant reduction in egg production by 24.5% and 34.6% in 0.5 and 2.5 μg/L groups, respectively. Zhou et al. (2018) reported that beta-cypermethrin may lessen the fecundity of mice by disrupting the levels of reproductive hormones and blocking the endometrium. Recent studies have proved that beta-cypermethrin may induce adverse effects on the reproductive system of zebrafish (Zhang et al. 2018).

7 Impacts in Male Zebrafish

Cypermethrin exposure may reduce the capability of male fish to sense female pheromones and affect serum sex hormone levels (Moore and Waring 2001). Pitchika et al. (2019) reported no significant alterations in male plasma vitellogenin levels in zebrafish after exposure to cypermethrin at 10 μg l^{-1} compared to the control group. Cypermethrin showed no structural alterations in the testis of zebrafish, although spermatozoans relative frequency declined by up to 12% in the exposed group as compared to controls. In male zebrafish, it has the ability to start spermatotoxic special effects without disrupting the histological design of testis.

Shi et al. (2011) showed that cypermethrin at 10 μg L^{-1} might induce spermatotoxic effects as directed by lessening sperm number in male zebrafish, but the similar concentration was ineffective to disrupt fertility rates. Spermatogenesis is a process that can be controlled by different factors including estrogens (Schulz and Miura 2002; Rather et al. 2017). Cypermethrin interferes with estrogen-mediated receptors in male zebrafish that cause a reduction in sperm count. This could be a clue to explain cypermethrin-mediated toxicity in the male reproductive system of zebrafish (Pitchika et al. 2019). Likewise, it was stated that beta-cypermethrin might induce reproductive toxicity and may decrease production of sperms in rats through reduction of the expression of androgen receptors (ARs) (Liu et al. 2010). Previous studies have indicated that 0.02 ppb cypermethrin showed no histopathology in the testis architecture of freshwater fish (*Esomus danricus*). Lu et al. (2021) reported that the spermatids in the testes markedly decreased, while spermatocytes significantly increased after exposure to cypermethrin at 0.5 and 2.5 μg/L, while spermatogonia was significantly increased only in high-dose group.

8 Impacts in Zebrafish Embryo

The early life stages of organisms are mostly vulnerable to the harmful effects of drugs and chemicals (Makri et al. 2004). Additionally, in vivo tests of zebrafish embryo are considered to be pain-free, and development of embryo is delicate to ecological stresses. Teratological structures and embryo-relevant toxicity can be scored easily from zebrafish due to its development outside of the mother body.

Other benefits of using zebrafish is the genetic basis of development (Kimmel 1989), and its main features of embryogenesis (Sayim et al. 2005), cDNA clone collections, mutant strains availability, physical map (Phillips and Reed 2000), and completed sequenced genome have been well studied. Zebrafish embryos showed different morphological anomalies such as pericardial edema, body-axis curves, and bulky yolk sac after exposure to 25 μg L^{-1} cypermethrin (Shi et al. 2011). The teratogenic abrasions made by cypermethrin in embryo of zebrafish were according to the results described by DeMicco et al. (2010). During embryogenesis, oxidative stress induced by reactive oxygen species has been reflected to induce different morphological anomalies in fishes (Yabu et al. 2001; Yamashita 2003). Different doses of cypermethrin induced a clear increase in catalase and superoxide dismutase activities in zebrafish embryos until 96 h postfertilization. Suppressed DNA repair capability and augmented caspase and p53 pathways were observed in the zebrafish embryonic development after cypermethrin exposure. Hence, it is likely that the high activities of these antioxidant enzymes might be due to removal of reactive oxygen species in the organisms developed by cypermethrin. Similar consequences have also been found reported in studies. For instance, cypermethrin exposure for 60 days at 25 mg kg^{-1} increased level of catalase and superoxide dismutase in rats (Nasuti et al. 2003).

The adverse effects of different pesticides have been reported on embryonic hatchability. Lower and Moore (2003) observed earlier hatching of salmon embryos after exposure to 0.05 μg/L cypermethrin. It has been reported that pesticides could increase the hatchability of zebrafish (Yu et al. 2015). Furthermore, more concentration of pesticides caused an inhibitory influence on hatchability.

Sathya et al. (2014) confirmed 0.05 μg/L of cypermethrin as LC_{50} value (lethal concentration at which more than 50% mortality occur) and observed no significant effects at the lowest 0.001 μg/L tested concentration. Thus, the sublethal and lethal effects of the zebrafish embryos increased with an increase in exposure time and concentration. In laboratory testing, the typical range of acute toxicity of cypermethrin of fishes has been reported to be 1.8–8.2 μg/L by the USDA National Agricultural Pesticide Impact (USEPA 1998). The LC_{50} of cypermethrin for fresh water prawn and carp have stated to be 12.6 μg/L (Wang et al. 2006) and 0.0031 μg/L (Collins and Cappello 2006), respectively. It is confirmed from the studies that contamination with cypermethrin even at low levels in the aquatic environment would affect the earlier developmental stages of fishes at various levels. The observations recorded in this study were in accordance with the Zebrafish embryo toxicity test described earlier (OECD 2013; Carlsson et al. 2013). The research conducted with the beta cypermethrin displayed increased number of dead embryos of the common carp (*C. carpio*) even after exposure to very low concentration of cypermethrin (0.0001, 0.001, 0.01, 0.1, 1, 2, 4, and 8 μg/L) (Polat et al. 2002). Coagulation, lack of heart beat, disrupting hatching rate, and lack of somite formation were observed in zebrafish embryos for lethal endpoints (OECD 2013). At very low concentrations, cypermethrin crosses the placental barrier and induces hazardous effects. Cypermethrin exposure decreases the DNA content and may lead to transformations, particularly mutations in the germline leading to teratological malformations (Anwar 2003; Bhunya and Pati 1988).

Shi et al. (2011) have found evidence that cypermethrin induced developmental toxicity by oxidative stress in the embryo-larval stages of zebrafish as indicated by increased malformations and apoptosis signal in the nervous system and reported that oxidative stress, repressed DNA repair capacity, and increased p53 and caspase pathway are possible mechanisms in terms of standard toxicological parameter. However, the level of cypermethrin utilized to induce acute developmental toxicity was significantly higher than those that have been investigated in animals, in order to better elicit evidently distinguishable effects and determine possible toxicity mechanisms.

9 Conclusion

In this chapter, the toxicity of cypermethrin to zebrafish has been discussed. The overproduction of reactive oxygen species and disturbance in ion channels due to its neurotoxin capability caused the induction of apoptosis or DNA damage in the target organs of fish. However, the number of knowledge gaps required to understand its exact mechanisms of toxicity in complex exposure media is significant. Future ecotoxicological research of pesticides should focus on the effects of pesticides as well as to confirm its exact mechanism in freshwater environments to fill the knowledge gaps. Cooperation is required from biologists, environmentalists, and toxicologists for more extensive studies on pesticides. Furthermore, standard techniques for testing are required to make toxicological studies more comparable.

References

ADAS (2004) Development of a design manual for agricultural pesticide handling and washdown areas

Agbohessi PT, Imorou TI, Houndji A, Gillardin V, Mandiki SN, Kestemont P (2013) Acute toxicity of agricultural pesticides to embryo-larval and juvenile African catfish *Clarias gariepinus*. Arch Environ Contam Toxicol 64(4):692–700

Agrawal S, Singh A, Tripathi P, Mishra M, Singh PK, Singh MP (2015) Cypermethrin-induced nigrostriatal dopaminergic neurodegeneration alters the mitochondrial function: a proteomics study. Mol Neurobiol 51(2):448–465

Ahmad S, Scopes RK, Rees GN, Patel BKC (2000) *Saccharococcus caldoxylosilyticus* sp. nov., an obligately thermophilic, xylose-utilizing, endospore-forming bacterium. Int J Syst Evol Microbiol 50(2):517–523

Al-Hamdani NM, Yajurvedi HN (2010) Cypermethrin reversibly alters sperm count without altering fertility in mice. Ecotoxicol Environ Saf 73(5):1092–1097

Amin A, Indulkar ST (2017) Chronic effect of Imidacloprid pesticide, on carcass composition of common carp fingerlings. Int J Agric Sci Res 7(2):343–348

Amin A, Indulkar ST, Pai R (2016) Sublethal effect of Buprofezin pesticide on carcass composition of *Cyprinus carpio communis* fingerlings. J Exp Zool India 20(1):269–272

Anonymous (1995) Pesticides in the aquatic environment, Water Quality Series No. 26, Report of the National Rivers Authority. HMSO, London

Anwar K (2003) Cypermethrin, a pyrethroid insecticide induces teratological and biological changes in young chick embryos. Pak J Biol Sci 6(19):1698–1705

Banaee M, Mirvaghefi A, Amiri BM, Rafiee G, Nematdost B (2011) Hematological and histopathological effects of diazinon poisoning in common carp (*Cyprinus carpio*). J Fish 64(1):1–12

Barzilai A, Yamamoto K (2004) DNA damage responses to oxidative stress. DNA Repair 3(8–9): 1109–1115

Bhunya SP, Pati PC (1988) Genotoxic effects of a synthetic pyrethroid insecticide, cypermethrin, in mice in vivo. Toxicol Lett 41(3):223–230

Bonga SW (1997) The stress response in fish. Physiol Rev 77:591–625

Bradbury SP, Coats JR (1989a) Toxicokinetics and toxicodynamics of pyrethroid insecticides in fish. Environ Toxicol Chem 8(5):373–380

Bradbury SP, Coats JR (1989b) Comparative toxicology of the pyrethroid insecticides. Rev Environ Contam Toxicol 108:133–177

Breckenridge CB, Holden L, Sturgess N, Weiner M, Sheets L, Sargent D et al (2009) Evidence for a separate mechanism of toxicity for the type I and the type II pyrethroid insecticides. Neurotoxicology 30:S17–S31

Cárcamo JG, Aguilar MN, Carreño CF, Vera T, Arias-Darraz L, Figueroa JE et al (2017) Consecutive emamectin benzoate and deltamethrin treatments affect the expressions and activities of detoxification enzymes in the rainbow trout (*Oncorhynchus mykiss*). Comp Biochem Physiol C Toxicol Pharmacol 191:129–137

Carlsson G, Patring J, Kreuger J, Norrgren L, Oskarsson A (2013) Toxicity of 15 veterinary pharmaceuticals in zebrafish (*Danio rerio*) embryos. Aquat Toxicol 126:30–41

Carriquiriborde P, Díaz J, Mugni H, Bonetto C, Ronco AE (2007) Impact of cypermethrin on stream fish populations under field-use in biotech-soybean production. Chemosphere 68(4): 613–621

Chatterjee S, Dutta AB, Ghosh R (1997) Impact of carbofuran in the oocyte maturation of catfish, *Heteropenustes fossilis* (Bloch). Arch Environ Contam Toxicol 32(4):426–430

Chatterjee S, Dasmahapatra AK, Ghosh R (2001) Disruption of pituitary-ovarian axis by carbofuran in catfish, *Heteropneustes fossilis* (Bloch). Comp Biochem Physiol C Toxicol Pharmacol 129(3):265–273

Chen H, Xiao J, Hu G, Zhou J, Xiao H, Wang X (2002) Estrogenicity of organophosphorus and pyrethroid pesticides. J Toxicol Environ Health A 65(19):1419–1435

Chukwuka A, Ogbeide O, Uhunamure G (2019) Gonad pathology and intersex severity in pelagic (tilapia zilli) and benthic (*Neochanna diversus* and *Clarias gariepinus*) species from a pesticide-impacted agrarian catchment, souths Nigeria. Chemosphere 225:535–547

Coats JR, Symonik DM, Bradbury SP, Dyer SD, Timson LK, Atchison GJ (1989) Toxicology of synthetic pyrethroids in aquatic organisms: an overview. Environmental Toxicology and Chemistry: An International Journal 8(8):671–679

Collins P, Cappello S (2006) Cypermethrin toxicity to aquatic life: bioassays for the freshwater prawn *Palaemonetes argentinus*. Arch Environ Contam Toxicol 51(1):79–85

Coppage DL, Matthews E (1974) Short-term effects of organophosphate pesticides on cholinesterases of estuarine fishes and pink shrimp. Bull Environ Contam Toxicol 11(5):483–488

Das BK, Mukherjee SC (2003) Toxicity of cypermethrin in *Labeo rohita* fingerlings: biochemical, enzymatic and haematological consequences. Comp Biochem Physiol C Toxicol Pharmacol 134(1):109–121

Defra (2009) Importance of surface run-off as a route of aquatic exposure to pesticides in the UK

Deka S, Mahanta R (2012) A study on the effect of organophosphorus pesticide malathion on hepato-renal and reproductive organs of *Heteropneustes fossilis* (Bloch). Science 1(1):1–13

Demetrio PM, Bonetto C, Ronco AE (2014) The effect of cypermethrin, chlorpyrifos, and glyphosate active ingredients and formulations on *Daphnia magna* (Straus). Bull Environ Contam Toxicol 93(3):268–273

DeMicco A, Cooper KR, Richardson JR, White LA (2010) Developmental neurotoxicity of pyrethroid insecticides in zebrafish embryos. Toxicol Sci 113(1):177–186

Eni G, Ibor OR, Andem AB, Oku EE, Chukwuka AV, Adeogun AO, Arukwe A (2019) Biochemical and endocrine-disrupting effects in *Clarias gariepinus* exposed to the synthetic pyrethroids, cypermethrin and deltamethrin. Comp Biochem Physiol C Toxicol Pharmacol 225:108584

EU (2015) Commission Implementing Regulation (EU) 2015/405 of 11 March 2015 approving alpha-cypermethrin as an active substance for use in biocidal products for product- type 18

Farag MR, Alagawany M, Bilal RM, Gewida AG, Dhama K, Abdel-Latif HM, Amer MS, Rivero-Perez N, Zragoza-Bastida A, Binnaser YS, Batiha GE, Naiel MA (2021) An overview on the potential hazards of pyrethroid insecticides in fish, with special emphasis on cypermethrin toxicity. Animals 11(7):1880

Feng L, Yang G, Zhu L, Xu J, Xu X, Chen Y (2016) Distribution and risk assessment of endocrine-disrupting pesticides in drinking water sources from agricultural watershed. Water Air Soil Pollut 227(1):1–10

Ge J, Cong J, Sun Y, Li G, Zhou Z, Qian C, Liu F (2010) Determination of endocrine disrupting chemicals in surface water and industrial wastewater from Beijing, China. Bull Environ Contam Toxicol 84(4):401–405

Giray B, Gürbay A, Hincal F (2001) Cypermethrin-induced oxidative stress in rat brain and liver is prevented by vitamin E or allopurinol. Toxicol Lett 118(3):139–146

Guo D, Liu W, Yao T, Ma M, Wang Q, Qiu J, Qian Y (2021) Combined endocrine disruptive toxicity of malathion and cypermethrin to gene transcription and hormones of the HPG axis of male zebrafish (*Danio rerio*). Chemosphere 267:128864

Hoeger B, Hitzfeld B, Köllner B, Dietrich DR, Van-den Heuvel MR (2005) Sex and low-level sampling stress modify the impacts of sewage effluent on the rainbow trout (*Oncorhynchus mykiss*) immune system. Aquat Toxicol 73(1):79–90

Hung CC, Huang FJ, Yang YQ, Hsieh CJ, Tseng CC, Yiin LM (2018) Pesticides in indoor and outdoor residential dust: a pilot study in a rural county of Taiwan. Environ Sci Pollut Res 25(23):23349–23356

Jaensson A, Scott AP, Moore A, Kylin H, Olsén KH (2007) Effects of a pyrethroid pesticide on endocrine responses to female odours and reproductive behaviour in male parr of brown trout (*Salmo trutta* L.). Aquat Toxicol 81(1):1–9

Jin Y, Zhang X, Shu L, Chen L, Sun L, Qian H et al (2010) Oxidative stress response and gene expression with atrazine exposure in adult female zebrafish (*Danio rerio*). Chemosphere 78(7): 846–852

Kale M, Rathore N, John S, Bhatnagar D (1999) Lipid peroxidative damage on pyrethroid exposure and alterations in antioxidant status in rat erythrocytes: a possible involvement of reactive oxygen species. Toxicol Lett 105(3):197–205

Kaviraj A, Gupta A (2014) Biomarkers of type II synthetic pyrethroid pesticides in freshwater fish. Biomed Res Int 2014:928063

Kidd H, James DR (eds) (1991) The agrochemicals handbook, 3rd edn. Royal Society of Chemistry Information Services, Cambridge, pp 2–13

Kimmel CB (1989) Genetics and early development of zebrafish. Trends Genet 5:283–288

Kumar A, Sharma B, Pandey RS (2007) Preliminary evaluation of the acute toxicity of cypermethrin and λ-cyhalothrin to *Channa punctatus*. Bull Environ Contam Toxicol 79(6): 613–616

Lal B, Sarang MK, Kumar P (2013) Malathion exposure induces the endocrine disruption and growth retardation in the catfish, *Clarias batrachus* (Linn.). Gen Comp Endocrinol 15(181): 139–145

Li H, Ma H, Lydy MJ, You J (2014) Occurrence, seasonal variation and inhalation exposure of atmospheric organophosphate and pyrethroid pesticides in an urban community in South China. Chemosphere 95:363–369

Li Z, Nie J, Yan Z, Cheng Y, Lan F, Huang Y et al (2018) A monitoring survey and dietary risk assessment for pesticide residues on peaches in China. Regul Toxicol Pharmacol 97:152–162

Lidova J, Stara A, Kouba A, Velisek J (2016) The effects of cypermethrin on oxidative stress and antioxidant biomarkers in marbled crayfish (*Procambarus fallax* f. virginalis). Neuroendocrinol Lett 37:53–59

Liu P, Song X, Yuan W, Wen W, Wu X, Li J, Chen X (2006) Effects of cypermethrin and methyl parathion mixtures on hormone levels and immune functions in Wistar rats. Arch Toxicol 80(7): 449–457

Liu L, Hu JX, Wang H, Chen BJ, He Z, Xu LC (2010) Effects of beta-cypermethrin on male rat reproductive system. Environ Toxicol Pharmacol 30(3):251–256

Liu X, Ji K, Jo A, Moon HB, Choi K (2013) Effects of TDCPP or TPP on gene transcriptions and hormones of HPG axis, and their consequences on reproduction in adult zebrafish (Danio rerio). Aquat Toxicol 134:104–111

Lower N, Moore A (2003) Exposure to insecticides inhibits embryo development and emergence in Atlantic salmon (*Salmo salar* L.). Fish Physiol Biochem 28(1):431–432

Lu J, Wu Q, Yang Q, Li G, Wang R, Liu Y et al (2021) Molecular mechanism of reproductive toxicity induced by beta-cypermethrin in zebrafish. Comp Biochem Physiol C Toxicol Pharmacol 239:108894

Lutnicka H, Kozińska A (2009) Pyrethroids as a predisposing factor in fish diseases. Ochr Środ Zasobów Nat 41:285–292

Majumder R, Kaviraj A (2015) Variation in acute toxicity between technical grade and commercial formulation of cypermethrin to some non-target freshwater organisms. Int J Curr Res 7(6): 16755–16759

Majumder R, Kaviraj A (2017) Cypermethrin induced stress and changes in growth of freshwater fish *Oreochromis niloticus*. Int Aquat Res 9(2):117–128

Makri A, Goveia M, Balbus J, Parkin R (2004) Children's susceptibility to chemicals: a review by developmental stage. J Toxicol Environ Health B Crit Rev 7(6):417–435

Marino D, Ronco A (2005) Cypermethrin and chlorpyrifos concentration levels in surface water bodies of the Pampa Ondulada, Argentina. Bull Environ Contam Toxicol 75(4):820–826

Marigoudar SR, Ahmed RN, David M (2009) Cypermethrin induced: in vivo inhibition of the acetylcholinesterase activity in functionally different tissues of the freshwater teleost, Labeo rohita (Hamilton). Toxicol Environ Chem 91(6):1175–1182

Mishra D, Srivastav SK, Srivastav AK (2005) Effects of the insecticide cypermethrin on plasma calcium and ultimobranchial gland of a teleost, Heteropneustes fossilis. Ecotoxicol Environ Saf 60(2):193–197

Mitchelmore CL, Chipman JK (1998) DNA strand breakage in aquatic organisms and the potential value of the comet assay in environmental monitoring. Mutat Res 399(2):135–147

Moore A, Waring CP (2001) The effects of a synthetic pyrethroid pesticide on some aspects of reproduction in Atlantic salmon (*Salmo salar* L.). Aquat Toxicol 52(1):1–12

Murthy KS, Kiran BR, Venkateshwarlu M (2013) A review on toxicity of pesticides in fish. Int J Open Sci Res 1(1):15–36

Nasuti C, Cantalamessa F, Falcioni G, Gabbianelli R (2003) Different effects of type I and type II pyrethroids on erythrocyte plasma membrane properties and enzymatic activity in rats. Toxicology 191(2-3):233–244

Neal AP, Yuan Y, Atchison WD (2010) Allethrin differentially modulates voltage-gated calcium channel subtypes in rat PC12 cells. Toxicol Sci 116(2):604–613

Novak AE, Taylor AD, Pineda RH, Lasda EL, Wright MA, Ribera AB (2006) Embryonic and larval expression of zebrafish voltage-gated sodium channel α-subunit genes. Dev Dyn 235(7): 1962–1973

OECD (2013) OECD guidelines for the testing of chemicals. Section 2: effects on biotic systems test no. 236: Fish embryo acute toxicity (FET) test

Paravani EV, Simoniello MF, Poletta GL, Zolessi FR, Casco VH (2018) Cypermethrin: oxidative stress and genotoxicity in retinal cells of the adult zebrafish. Mutat Res 826:25–32

Parng C (2005) In vivo zebrafish assays for toxicity testing. Curr Opin Drug Discov Devel 8 (1):100–106

Pesticide Action Network North America (1999) Pesticides threaten birds and fish in California

Phillips RB, Reed KM (2000) Localization of repetitive DNAs to zebrafish (*Danio rerio*) chromosomes by fluorescence in situ hybridization (FISH). Chromosom Res 8(1):27–35

Pitchika GK, Swamy KV, Ranjani TS, Tyler CR, Sainath SB, Phillip GH (2019) Effect of cypermethrin on reproductive efficacy in zebrafish (*Danio rerio*): in-vivo and in-silico studies. J Environ Biol 40(5):985–994

Polat H, Erkoc FU, Viran R, Koçak O (2002) Investigation of acute toxicity of beta-cypermethrin on guppies *Poecilia reticulata*. Chemosphere 49(1):39–44

Prusty AK, Meena DK, Mohapatra S, Panikkar P, Das P, Gupta SK, Behera BK (2015) Synthetic pyrethroids (type II) and freshwater fish culture: perils and mitigations. Int Aquat Res 7(3): 163–191

Puglis HJ, Boone MD (2011) Effects of technical-grade active ingredient vs. commercial formulation of seven pesticides in the presence or absence of UV radiation on survival of green frog tadpoles. Arch Environ Contam Toxicol 60(1):145–155

Rajakumar A, Singh R, Chakrabarty S, Murugananthkumar R, Laldinsangi C, Prathibha Y, Senthilkumaran B (2012) Endosulfan and flutamide impair testicular development in the juvenile Asian catfish, *Clarias batrachus*. Aquat Toxicol 110-111:123–132

Rajini A, Revathy K, Selvam G (2015) Histopathological changes in tissues of *Danio rerio* exposed to sub lethal concentration of combination pesticide. Indian J Sci Technol 8:1–12

Rao KRS (1990) Pesticide impact on fish metabolism. Discovery Publishing House, 232 pp

Rather MA, Basha SH, Bhat IA, Sharma N, Nandanpawar P, Badhe M et al (2017) Characterization, molecular docking, dynamics simulation and metadynamics of kisspeptin receptor with kisspeptin. Int J Biol Macromol 101:241–253

Ray DE (1991) Pesticides derived from plants and other organisms. In: Hayes WJ Jr, Laws ER Jr (eds) Handbook of pesticide toxicology. Academic Press, New York, pp 2–3

Richterova Z, Svobodová Z (2012) Pyrethroids influence on fish. Slov Vet Res 49(2):63–72

Saha S, Kaviraj A (2008) Acute toxicity of synthetic pyrethroid cypermethrin to some freshwater organisms. Bull Environ Contam Toxicol 80(1):49–52

Sanderson JT (2006) The steroid hormone biosynthesis pathway as a target for endocrine-disrupting chemicals. Toxicol Sci 94(1):3–21

Sangha GK, Kaur K, Khera KS (2013) Cypermethrin induced pathological and biochemical changes in reproductive organs of female rats. J Environ Biol 34(1):99–105

Sathya K, Thiripurasundari M, Uma A, Srinivasan MR, Subas CBM (2014) The toxic effect of cypermethrin on zebrafish embryo-an alternative model for assessing the embryo toxicity of environmental chemicals. Indo Am J Pharm Res 4(4):2316–2322

Saxena PN, Chauhan LKS, Gupta SK (2005) Cytogenetic effects of commercial formulation of cypermethrin in root meristem cells of *Allium sativum*: spectroscopic basis of chromosome damage. Toxicology 216(2–3):244–252

Sayim F, Yavasoglu NÜK, Uyanikgil Y, Aktug H, Yavasoglu A, Turgut M (2005) Neurotoxic effects of cypermethrin in Wistar rats: a haematological, biochemical and histopathological study. J Health Sci 51(3):300–307

Schlenk D (2005) Pesticide biotransformation in fish. In: Mommsen TP, Moon TW (eds) Biochemistry and molecular biology of fishes, vol 6. Elsevier, San Diego, pp 171–190

Schulz RW, Miura T (2002) Spermatogenesis and its endocrine regulation. Fish Physiol Biochem 26(1):43–56

Sharma S, Nagpure NS, Kumar R, Pandey S, Srivastava SK, Singh PJ, Mathur PK (2007) Studies on the genotoxicity of endosulfan in different tissues of fresh water fish *Mystus vittatus* using the comet assay. Arch Environ Contam Toxicol 53(4):617–623

Shi X, Gu A, Ji G, Li Y, Di J, Jin J et al (2011) Developmental toxicity of cypermethrin in embryo-larval stages of zebrafish. Chemosphere 85(6):1010–1016

Shi J, Jiao Z, Zheng S, Li M, Zhang J, Feng Y, Yin J, Shao B (2015) Long-term effects of bisphenol AF (BPAF) on hormonal balance and genes of hypothalamus-pituitary-gonad axis and liver of zebrafish (Danio rerio), and the impact on offspring. Chemosphere 128:252–257

Singh PB, Canario AV (2004) Reproductive endocrine disruption in the freshwater catfish, *Heteropneustes fossilis*, in response to the pesticide γ-hexachlorocyclohexane. Ecotoxicol Environ Saf 58(1):77–83
Singh H, Singh TP (1982) Effect of some pesticides on hypothalamo-hypophyseal-ovarian axis in the freshwater catfish *Heteropneustes fossilis* (Bloch). Environ Pollut Ser A Ecol Biol 27(4): 283–288
Soderlund DM (2010) State-dependent modification of voltage-gated sodium channels by pyrethroids. Pestic Biochem Physiol 97(2):78–86. [CrossRef]
Srivastava N, Kaushik N (2001) Use of fish as bioindicator of aquatic pollution. In: Proceedings of the international congress of chemistry and environment, pp 227–229
Sundaray JK, Rather MA, Kumar S, Agarwal D (2021) Recent updates in molecular endocrinology and reproductive physiology of fish. Springer Singapore
Treasurer JW, Wadsworth SL (2004) Interspecific comparison of experimental and natural routes of *Lepeophtheirus salmonis* and *Caligus elongatus* challenge and consequences for distribution of chalimus on salmonids and therapeutant screening. Aquac Res 35(8):773–783
Ullah R, Zuberi A, Ullah S, Ullah I, Dawar FU (2014) Cypermethrin induced behavioral and biochemical changes in mahseer, *Tor putitora*. J Toxicol Sci 39(6):829–836
Uner N, Oruc EO, Canli M, Sevgiler Y (2001) Effects of cypermethrin on antioxidant enzyme activities and lipid peroxidation in liver and kidney of the freshwater fish, *Oreochromis niloticus* and *Cyprinus carpio* (L.). Bull Environ Contam Toxicol 67(5):657–664
USEPA (1998) Guidelines for ecological risk assessment, United States Environmental Protection Agency (USEPA). Washington, D.C.: Government Printing Office. http://cfpub.epa.gov/ncea/cfm/recordisplay.cfm?deid=12460. Accessed 19 June 2009
Valavanidis A, Vlahogianni T, Dassenakis M, Scoullos M (2006) Molecular biomarkers of oxidative stress in aquatic organisms in relation to toxic environmental pollutants. Ecotoxicol Environ Saf 64(2):178–189
Velisek J, Wlasow T, Gomulka P, Svobodova Z, Dobsikova R, Novotny L, Dudzik M (2006) Effects of cypermethrin on rainbow trout (*Oncorhynchus mykiss*). Veterinarni Medicina 51(10): 469–476
Viran R, Erkoç FÜ, Polat H, Koçak O (2003) Investigation of acute toxicity of deltamethrin on guppies (*Poecilia reticulata*). Ecotoxicol Environ Saf 55(1):82–85
Wang ZW (2008) Regulation of synaptic transmission by presynaptic CaMKII and BK channels. Mol Neurobiol 38(2):153–166
Wang Y, Xiong L, Liu X (2006) Subacute toxicity of cypermethrin to carp. J Agro Environ Sci 25(1):200–203
Wang Y, Lv L, Yu Y, Yang G, Xu Z, Wang Q, Cai L (2017) Single and joint toxic effects of five selected pesticides on the early life stages of zebrafish (*Denio rerio*). Chemosphere 170:61–67
Wauchope RD, Buttler TM, Hornsby AG, Augustijn-Beckers PWM, Burt JP (1992a) SCS/ARS/CES pesticides properties database for environmental decision making. Rev Environ Contam Toxicol 123:7–22
Wauchope RD, Buttler TM, Hornsby AG, Augustijn-Beckers PWM, Burt JP (1992b) The SCS/ARS/CES pesticide properties database for environmental decision-making. Rev Environ Contam Toxicol 123:1–155
Weston DP, Holmes RW, Lydy MJ (2009) Residential runoff as a source of pyrethroid pesticides to urban creeks. Environ Pollut 157(1):287–294
Wielgomas B, Krechniak J (2007) Effect of α-Cypermethrin and Chlorpyrifos in a 28-day study on free radical parameters and cholinesterase activity in Wistar rats. Pol J Environ Stud 16(1): 91–95
Yabu T, Kishi S, Okazaki T, Yamashita M (2001) Characterization of zebrafish caspase-3 and induction of apoptosis through ceramide generation in fish fathead minnow tailbud cells and zebrafish embryo. Biochem J 360(1):39–47
Yamashita M (2003) Apoptosis in zebrafish development. Comp Biochem Physiol B: Biochem Mol Biol 136(4):731–742

Yılmaz M, Gül A, Erbaşlı K (2004) Acute toxicity of alpha-cypermethrin to guppy (*Poecilia reticulata*, Pallas, 1859). Chemosphere 56(4):381–385

Yu K, Li G, Feng W, Liu L, Zhang J, Wu W et al (2015) Chlorpyrifos is estrogenic and alters embryonic hatching, cell proliferation and apoptosis in zebrafish. Chem Biol Interact 239:26–33

Yuan C, Wang C, Gao SQ, Kong TT, Chen L, Li XF et al (2010) Effects of permethrin, cypermethrin and 3-phenoxybenzoic acid on rat sperm motility in vitro evaluated with computer-assisted sperm analysis. Toxicol In Vitro 24(2):382–386

Zhang X, Xie P, Li D, Tang R, Lei H, Zhao Y (2009) Time-dependent oxidative stress responses of crucian carp (*Carassius auratus*) to intraperitoneal injection of extracted microcystins. Bull Environ Contam Toxicol 82(5):574–578

Zhang Y, Zhou Y, Tang Q, Hu F, Feng L, Shen J, Huang B (2018) The protective effects of selenium-enriched spirulina on the reproductive system of male zebrafish (*Danio rerio*) exposed to beta-cypermethrin. Food Funct 9(11):5791–5804

Zheng S, Chen B, Qiu X, Chen M, Ma Z, Yu X (2016) Distribution and risk assessment of 82 pesticides in Jiulong River and estuary in South China. Chemosphere 144:1177–1192

Zhou YJ, Wang XD, Xiao S, Yu DE, Wang LQ, Wang JH, Zhu HQ (2018) Exposure to beta-cypermethrin impairs the reproductive function of female mice. Regul Toxicol Pharmacol 95: 385–394

Impact of Pesticide Application on Aquatic Environments and Biodiversity

Fariha Latif, Sana Aziz, Rehana Iqbal, Saman Iram, Maria Nazir, and Muhammad Shakeel

1 Introduction

Water is a necessary component for all living organisms because it is one of the basic compounds without which no life can exist. With reference to other hydrological features, water quality is concerned with the chemical, physical, and biological aspects. Diverse water sources are used in aquaculture, and it is normal for some of these sources to become contaminated with different kinds of xenobiotics (Çok et al. 2011). Heavy metals, pesticides, suspended solids, organic matter, and dyes are examples of water contaminants (Yohannes et al. 2013). A decline in water quality would have a substantial impact on aquatic biodiversity, at both individual and population levels, as well as on water quality.

Some persistent organic pollutants (POPs) and heavy metallic ions found in various fish and shellfish have exceeded the guideline limits for human consumption. It has been described that more than 90% of POPs in organs, such as chlorinated hydrocarbons, are derived from food; however, seafood is a big source of toxicity in the human body (Easton et al. 2002). The current agricultural system has been strained by the recent rise in global population, and the main objective of major nations is to raise food production for the expanding population, which is predicted to reach over ten billion by 2050 (Saravi and Shokrzadeh 2011; Yadav et al. 2020). Herbicides, fungicides, nematicides, and fertilizers are just a few of the agrochemicals that are used throughout crop production. Through the food chain, exposure of aquatic organisms to genotoxic substances may endanger human health. It may also have an adverse effect on the environment by causing transmissible mutations that

F. Latif (✉) · R. Iqbal · S. Iram · M. Nazir · M. Shakeel
Institute of Zoology, Bahauddin Zakariya University, Multan, Pakistan
e-mail: farihalatif@bzu.edu.pk

S. Aziz
Department of Zoology, Wildlife, and Fisheries, University of Agriculture, Faisalabad, Pakistan

M. A. Rather et al. (eds.), *Xenobiotics in Aquatic Animals*,
https://doi.org/10.1007/978-981-99-1214-8_5

result in the loss of biodiversity. Fish health is a suitable marker for environmental pollution-level monitoring because it concentrates toxicants in the fish tissues directly from surrounding water and food, allowing the assessment of routes of exposure of toxicants that have been transferred through the food web. Fish are relatively sensitive to even minute changes in their surroundings, and their health is an appropriate marker for environmental pollution-level monitoring (Jabeen et al. 2012; Laxmi et al. 2019).

Since the use of pesticides for crop protection has increased along with agricultural expansion, they represent a substantial source of toxicants in freshwater environments. There are over 1000 active components in this broad collection of synthetic organic chemicals, which are sold as fungicides, insecticides, and herbicides (Mostafalou and Abdollahi 2013). Pesticides pose a major hazard to the higher trophic levels of aquatic habitats due to their comparative higher bioaccumulation capacity in the food web (Dar et al. 2015). Due to their lengthy persistence and highly toxic potential, pesticide residues may be present at levels that are hazardous to the public's health in the soil, water, sediments, and fish. Depending on their chemical makeup, pesticides can be categorized as pyrethroids (natural or synthetic), organophosphates (phosphoric acid esters), organochlorines (chlorinated hydrocarbons), and carbamates.

2 Pesticides

A class of chemicals known as pesticides is employed to suppress, eradicate, or stunt the growth of pests: insecticides (viz., organophosphates, organochlorines, carbamates), rodenticides (viz., anticoagulants), herbicides (viz., paraquat, diquat, chlorophenoxyacetic acid), fungicides (viz., dithiocarbonates, captan), and fumigants (viz., ethylene). The term "pesticides" refers to a broad category of chemicals, the majority of which are insecticides (organophosphates (OPs), organochlorines, carbamates, and pythids) and herbicides (bipyridyl compounds).

Chemicals called pesticides are employed to get rid of pests. These pests include common bacteria, plant diseases, nematodes, and other insects that compete with human food sources, spread disease, and ruin crops. Pesticides are further divided into broad-spectrum and narrow-spectrum categories depending on how many species they are used to control and how many species they are used to control in total. In general, pesticides can be categorized as biological or synthetic. While synthetic pesticides are created in factories, biological pesticides come from naturally occurring sources like plant extracts (such as pyrethrin from chrysanthemums and azadirachtin from neem plants). The following are a few pesticide classifications (Table 1):

Table 1 Major classes of pesticides

Pesticide category	Principal classes	Purpose	Mode of action	Examples
Insecticides	Organophosphates, carbamates, pyrethroids, organochlorines, neonicotinoids	Kill or repel insects	Neurotoxic, bioaccumulates, and biomagnifies	Malathion, methyl parathion, aldicarb, carbaryl, methomyl
Fungicides	Thiocarbamates, triazoles, strobilurins	Kills, molds, and other fungi	Stop the development of fungus spores and plant illnesses	Metarn sodium, fluconazole, myclobutanil, triadimefon
Herbicides	Phosphonates Chlorophenoxy herbicides Dipyridyl herbicides	Kill weeds or unwanted plants	Neurotoxin to certain insect developmental stages	Glyphosate, 2,4-D, mecoprop, diquat, paraquat

2.1 Insecticides

The majority of insecticides have a variety of effects on the nervous system; they impede the transport of sodium (Na+), potassium (K+), calcium (Ca+), or chloride ions (Cl-) across membranes, which in turn prevents certain enzymes from carrying out their specific functions for chemical transmission at nerve endings (Correia et al. 2010; Amin et al. 2016).

2.1.1 Carbamate

Carbamate inhibits acetylcholinesterase (AChE) by attaching to the enzyme's reactive site. It inhibits AChE in a short and reversible manner (Jeon et al. 2013).

2.1.2 Organophosphate

Prior to the bans on DDT and other organochlorine pesticides, organophosphorus insecticides like malathion and parathion were used instead. Additionally hazardous, this class of insecticides prevents the activity of the enzyme acetylcholinesterase (AChE), extending and aggravating the effects of intoxication (Sankhla et al. 2018; Laxmi et al. 2020).

2.1.3 Pyrethroids

The two distinct acidic sections of chrysanthemic or pyrethric acids that cause type I and type II disease are the most modern insecticide group (Correia et al. 2010). The modes of action of pyrethroid insecticides are varied. One of these is Ca2+-Mg2+-

ATPase inhibition, which prevents calcium from leaving nerve endings and triggers the release of neurotransmitters in the postsynaptic gap.

2.1.4 Organochlorine

They are a particular class of pesticide that has long-lasting effects on the nervous system because they are chemically inert and stable. Among all insecticides that prevent neurotransmitter release, DDT is the chemical that has received the most research. In addition to the deadly DDT, additional organochlorine insecticides include endrin and lindane (LeBlanc 2007).

2.2 *Fungicides*

The electron transport chain in the respiration process is constrained by fungicides (Leroux 1996). These are a type of pesticide that stops the growth of spores by interfering with the fungi's energy source. A different class of fungicides known as phenylpyrroles (fenpiclonil and iprodione) inhibits spore germination and causes a number of morphological changes to limit the germ tube elongation, while dithiocarbamates (maneb and thiram), captan, and dichlofluanid inhibit respiratory process enzyme activities at various sites.

2.3 *Herbicides*

Herbicides are created in order to kill harmful plants (weeds). They are linked to a variety of mechanisms involved in the various cellular processes such as respiration, photosynthesis, development, growth, nuclear division, and protein or lipid synthesis (Mondal and Subramaniam 2020).

2.3.1 Chlorophenoxy-Based Herbicides

These herbicides, which primarily consist of dichlorophenoxyacetic acid, trichlorophenoxyacetic acid, and 4-chloro-o-toloxyacetic acid, mimic the function of growth hormones like auxin in plants (Kanan et al. 2020), disrupting normal growth processes and causing curling of leaves or stems, limiting the growth of roots and shoots, necrosis, and plant death (Cole 1985).

2.3.2 Glyphosate

This herbicide is most commonly used to control weeds by inhibiting the enzyme responsible for catalyzing amino acid synthetic pathways, thus disrupting the protein synthesis mechanism (Wafford et al. 1989). Due to glyphosate's widespread use in shallow water environments and high-water solubility, ecotoxicologists' primary worry is the exposure of aquatic creatures that are not intended for it (Sikorski and Gruys 1997).

3 Sources of Pesticides into Aquatic Ecosystems

The nature and qualities of the active ingredient and the current agroclimatic conditions have all been taken into account in the research over the past few years to identify a variety of potential entrance pathways. Several academics have studied the processes that control the fate of pesticides after application to agricultural land in order to monitor and comprehend these processes. It is now commonly accepted that, in addition to diffuse losses, there are a number of other entry pathways for pesticides into the water that result from improper usage, carelessness, unlawful activity, or misuse. Pesticides have been found to enter the water in several conditions (Table 2).

Table 2 Sources of pesticides into the aquatic environments

Entry source	Entry route/cause	Main water source type affected
Point	Washings and disposal of garbage	Surface water/groundwater
	Spillages	Surface water/groundwater
	Sumps, soakaways, and drainage	Groundwater/surface water
	Tank filling	Surface water/groundwater
	Faulty equipment	Groundwater/surface water
	Consented discharges	Streams, rivers
	Direct entry including overspray	Ditches, ponds, streams, rivers
Nonpoint sources	Spray drift	Ditches, streams, ponds, rivers
	Surface runoff or overland flow	Ditches, ponds, streams, rivers
	Volatilization and precipitation	Ditches, ponds, rivers, streams
	Leaching	Groundwater
	Drain flow	Streams, ditches, ponds and rivers
	Throughflow/interflow	Streams, ditches, ponds and rivers
	Base flow seepage	Groundwater and surface water

3.1 Point Sources

Water pollution reaches a water body from a single location or a small number of locations. In contrast to semi-point sources, which can happen when pesticides are administered to isolated or restricted areas like roadsides or railroads, approved point source contamination takes the form of authorized discharges, such as those from vegetable washing plants. Spills or discharges of product, tank mix, trash, or washings directly to surfaces or drainage systems, which can reach surface water or groundwater via soakaways, result in non-approved contamination occurrences.

3.2 Nonpoint Sources

There is a microbiologically active soil layer present, where degradation and dissipation processes can occur. Before entering water via artificial drainage systems or as surface or subsurface flow, leaching, or bypass flow, the active substance and/or its metabolites usually have the opportunity to move through the soil layers in solution or sorb to soil particles. Both groundwater and surface water bodies may be harmed. Spray drift and pesticides in precipitation are also examples of diffuse sources of surface water contamination.

The pesticide damages the health of aquatic life when it enters the aquatic ecosystem by runoff, vaporization to the atmosphere, agricultural runoff, groundwater intrusions, or adsorption or plant uptake. Outside applications of pesticides or fertilizer may cause some of the substance to shift. Streets are typically where you'll find. storm drains. Rain and runoff from street gutters into storm drains from irrigating gardens and lawns Sewers originate from drains inside the home and transport wastewater from sinks, toilets, and showers to treatment facilities, where they somehow contaminate river water (Fig. 1). Through crop field application, seepage of tainted surface water, unintentional spills and leaks, incorrect disposal, and even injection of waste material into wells, pesticides can enter underground water-bearing aquifers. Pesticides are more likely to get into groundwater and surface water when there is irrigation. Runoff that can carry pesticides is produced when moist soils are irrigated or when runoff rates are increased. Pesticides and other contaminants are more likely to seep into groundwater during irrigation that encourages the regular downward movement of water past the zone of plants. This is especially concerning in regions with rocky soils where irrigation is more important. Proper irrigation management is crucial to lowering the possibility of pesticides contaminating groundwater.

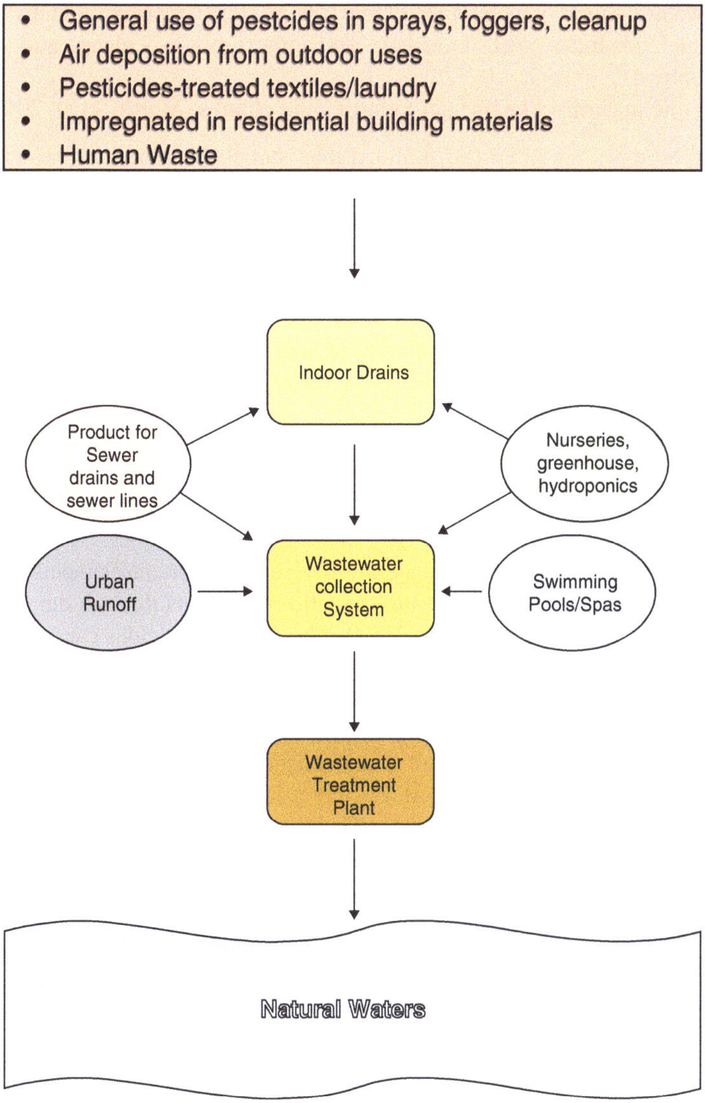

Fig. 1 Entry of pesticides in the rivers

4 Pesticides in Aquatic Environments

The following elements of a global bicycle should be taken into account when establishing the main paths of pesticide exposure to aquatic systems and biota:

1. The water column, which typically receives pesticide exposure initially
2. Organic substrates such branches, leaf litter, mosses, algae, and vascular hydrophytes
3. Inorganic materials like fine silt to large sand chunks (Murthy et al. 2013)

Lithic biotopes are often less polluted than standing waters, and pesticide concentrations in interstitial water and sediments are frequently lower than in the water column. Fish in particular are hazardous to pesticides at both sublethal and deadly concentrations (Khafaga et al. 2020). Aquatic animals' biological availability (bioavailability), bioconcentration, biomagnification, and persistence in the environment all play a role in how long they survive when exposed to pesticides. Bioavailability is the term used to describe the amount of pesticide in the environment that is accessible to fish, plants, and animals. After use, some pesticides quickly deteriorate.

5 Behavior and Fate of Pesticides in Aquatic Ecosystems

Pesticide persistence and mobility are the two most important elements that affect how a pesticide behaves when it enters aquatic ecosystems through dumping, crop application, or spillage (Kerle et al. 1994). The fate of pesticides can be somewhat predicted based on the compound's qualities. Pesticides' key characteristics, such as water solubility, degradation half-life, vapor pressure, soil sorption coefficient, and Henry's law constant, can be used to forecast how they will behave in the environment. Pesticide characteristics, soil characteristics, crop management techniques, and the loading of water on the soil are only a few of the variables that determine pesticide destiny (Kerle et al. 1994). The following variables affect how long pesticides remain in natural water:

There are following factors that decide the fate of pesticides in the aquatic environments:

1. Release of pesticides in the water
2. Mobility
3. Persistence
4. Patterns of application of pesticides
5. Site condition

5.1 Release of Pesticides in the Water

Determining the amount of contamination depends on how much of a pesticide is discharged into the environment. The kind and quantity of ground cover and vegetation, formulation properties, terrain, rate and methods of application, and weather conditions will all affect how much pesticide is released into the environment.

5.2 Mobility

Pesticide mobility refers to the possibility of a pesticide moving away from its original location. Pesticide mobility is influenced by the compound's volatilization, adsorption behavior in soil, and water solubility. The Henry's law constant describes volatilization (Kh) from moist soil. Kh is described as the ratio of pesticide's concentration in the water to the concentration in air at equilibrium. The pesticide vapor pressure and solubility can be used to calculate this value. The higher Kh causes greater rate of volatilization of pesticides from the moist soil (Kerle et al. 1994).

Adsorption is a critical process that influences pesticide fate. Many environmental factors influence the adsorption of pesticides, such as temperature, soil pH, and water content, as well as the type and amount of organic matter present. Pesticide adsorption is generally inversely related to pesticide solubility in water. The solubility of pesticides that are weak bases or acids is affected by pH. Furthermore, the water solubility of the pesticide influences plant uptake. According to researchers, highly soluble pesticides are more likely to move within the site or offsite via leaching or runoff.

5.3 Persistence

Persistence of the pesticides is frequently expressed in terms of half-life. The half-life of a pesticide is the amount of time that is required for one-half of the original time required to degrade. This is a constant that occurs under specific conditions for a given environmental degradation process and a given compound (Connell et al. 2009). The half-life can be used to categorize substances based on their overall persistence properties.

The longer a pesticide remains active before breaking down, the greater the risk. Pesticides have the potential to contaminate both surface water and groundwater. The rate of pesticide degradation is slower in sediments and deep. Pesticide degradation takes place primarily in the biologically active zone of soils, where the plant roots are plentiful. Pesticides must be kept from leaching out of the rooting zone (Mulla et al. 1996).

5.4 Patterns of Application

Regardless of the frequency or severity of a pest infestation, pesticides are generally sprayed in a set order. In addition, pesticide exposure in the environment typically happens in pulses and involves rain-induced runoff or spray drift. The length of a pulse sequential pesticide application can be between a few hours and 1–2 days, and the pulse concentration of the pesticide depends on both the kind of pesticide and the features of the recipient (Rosenkrantz et al. 2013).

5.5 Site Condition

Pesticides that are dissolved in water and those that have been attached to crumbling soil can both be transported by runoff, which is water flowing across a sloped surface (Taylor et al. 1991). In regions with high rates of rainfall or irrigation, a lot of water may be percolating through the soil, raising the possibility that pesticides will contaminate both groundwater and surface water.

6 The Impacts of Pesticides on Aquatic Ecosystems

Usually, the effectiveness or cost of pesticides is considered, not their impact on aquatic ecosystems (Kovach et al. 1992). Numerous of these pesticides are challenging to break down. These pesticides consequently penetrate aquatic habitats through the numerous routes mentioned above. These pesticides can enter aquatic organisms directly through absorption or ingestion of contaminated water or indirectly through feeding on previously contaminated organisms, depending on their chemical qualities that allow them to survive in the aquatic environment. Pesticides can modify metabolic pathways, affect membrane permeability, and limit enzyme function in addition to preventing photosynthesis, cellular development, and cell division.

7 Impact on Aquatic Flora

Aquatic animals may become exposed to pesticides after they enter water ecosystems in a number of ways, including direct chemical absorption into their habitats or movement of species into previously polluted areas as a result of pesticide retention. As a result, pesticides may put aquatic life at danger (Wilson and Koch 2013). Freshwater environments support a wide variety of plants and animals all around the world. Freshwater environments sustain a variety of species populations and give birds and mammals access to water and food. *Lemna gibba* and *Lemna minor*, two species of the aquatic vascular plant genus *Lemna* (duckweed), have frequently been employed as test subjects for phytotoxicity. *Lemna* has several benefits for ecotoxicity, including its tiny size, simplicity in handling, laboratory culture, quick growth rate, and sensitivity to a variety of contaminants (Zezulka et al. 2013). The European Union (EU) calculates toxicity exposure ratios (TERs) between toxicity endpoints (EC50 values) determined from normal laboratory work with algae or

Lemna species and the projected ambient concentration in order to assess the danger of herbicides on aquatic plants and algae (PECs). The resultant TER is evaluated against a 10-trigger. A TER score larger than 10 indicates that the chemical poses an acceptable danger to aquatic plants, whereas one less than 10 indicates a potential unacceptable risk and the necessity for a higher-tier risk assessment (Maltby et al. 2009).

8 Impact on Aquatic Fauna

Exposure to pesticides hurts a variety of nontarget organisms as well as the intended targets, with fish being the most notable. In some cases, fish mortality was caused by acute exposure to a number of pesticides, whereas deadly changes were caused by reduced exposure to the same chemicals. In many species of fish exposed to different pesticides, changes in hematological parameters such as red blood cells, white blood cells, plasma, and serum-level variations result in histological abnormalities affecting the liver, kidneys, gills, muscles, brain, and gut (Tahir et al. 2021).

Fish are at the bottom of the aquatic food chain and are a good indicator of how clean or contaminated the water is. They can gather and store substances like heavy metals and pesticides thanks to submissive phenomenon, which helps them detect pollutants in their environment. Fish eat more algae, phytoplankton, and other aquatic plants contaminated with pesticides, which causes deadly poisons to build up in the fish's tissues and organs. The fish's gills, skin, and alimentary canal absorb pollutants, which subsequently spread to different organs and tissues and change physiological and natural phenomena (Banaee et al. 2011). Because they are fully submerged in water, the gills are the most contaminated organs. Through the gills, toxins enter the body, raising the need for oxygen. Monitoring any potentially dangerous stress in the aquatic environment is therefore an important metric (Panigrahi et al. 2014).

9 Routes of Exposure to Fish

Fish and other aquatic animals can take in pesticides in three different ways: dermally (directly through the skin while swimming in pesticide-contaminated waters), bronchially (directly through the gills while breathing), and orally (through drinking pesticide-contaminated water or eating pesticide-contaminated prey). "Secondary poisoning" is the term used to describe eating an animal that has been poisoned by a chemical. If the insects they eat contain significant concentrations of pesticides or harmful metabolites, for instance, fish that feed on dying insects poisoned by insecticides may be killed (Fig. 2). There are several different chemicals that harm fish populations in various ways (Table 3).

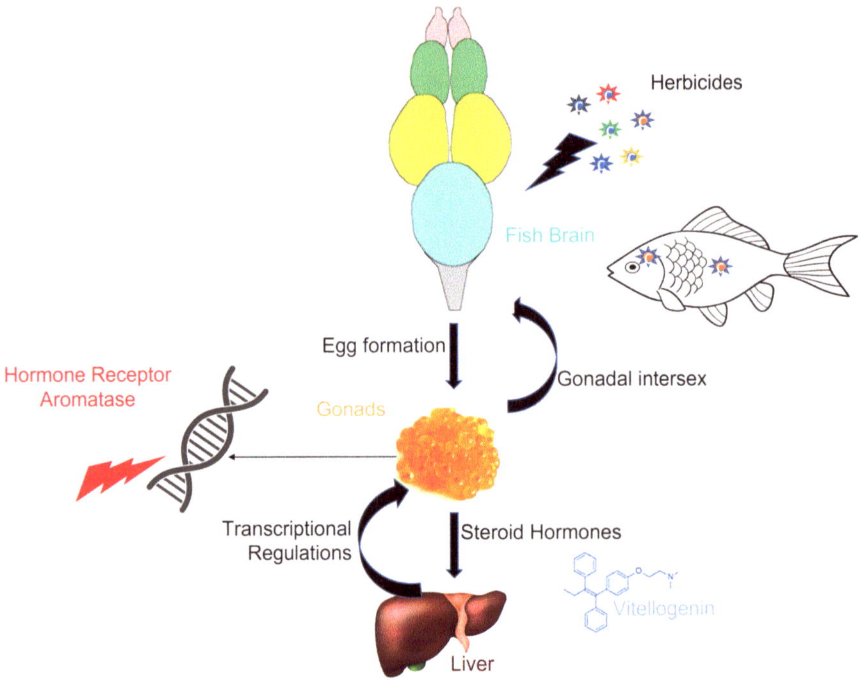

Fig. 2 Reproductive indicators that change in fish by exposure to herbicides

Table 3 Impact of different pesticides on fish

Pesticide	Fish	Effects	Reference
Diazinon	*Poecilia reticulata*	Disrupts activity of acetylcholinesterase (AChE) and neurological system, triggering numerous neurotoxic effects	Sharbidre et al. (2011)
Monocrotophos	*T. putitora*, *Lepomis macrochirus*, *Hoplias malabaricus*, *Oreochromis niloticus*, *Clarias gariepinus*, and *L. rohita*	Affects the activity of antioxidant defense enzymes and lowers the levels of glutathione, glutathione-S-transferase, glutathione reductase, and the lipid peroxidation marker malondialdehyde	Anusiya et al. (2015)
Carbamate and organophosphates	*Rhamdia quelen*, *C. carpio*, *Colisa fasciatus*, *Oreochromis mossambicus*, *L. rohita*	Alteration in the AChE activity	Joseph and Raj (2011)
Dimethoate and lambda-cyhalothrin	*Oncorhynchus mykiss* and *L. rohita*	Disrupt the normal activity of thyroid hormones	Dey and Saha (2014)
Atrazine	*Rhinella arenarum*	Disturb fish normal endocrine systems and metamorphosis hormones	Brodeur et al. (2013)
Cypermethrin	*Catla catla*	Neurotoxicity and apoptosis in the brain cells	Jindal and Sharma (2019)

10 Effect on the Hematological Parameters of Fish

According to Rios et al. (2002), a number of genetic and environmental factors changed the fish's blood parameters. Pesticides have an effect on a number of fish traits, with blood parameters receiving special attention. Pesticides quickly changed the hematological characteristics of fish (Rezania et al. 2018). The hematologic index can therefore be used to effectively track the health and response of fish and aquatic life to different toxicants, showing the ecological context of the environment and offering a standard technique to quantify the contaminant's sublethal effects (Pimpao et al. 2007). Fish hematological research has become more significant as a sensitive and accurate indicator for evaluating biological and pathological changes brought on by anthropogenic or natural variables, such as microbial infection or levels of contamination in aquatic sources. Hematological indicators are therefore thought to be a crucial tool for assessing how the body is responding to diverse stresses (Ali and Rani 2009). *Barbonymus gonionotus* subjected to quinalphos (Mostakim et al. 2015), *Cyprinus carpio* exposed to monocrotophos (Vaiyanan et al. 2015), *Ctenopharyngodon idella* exposed to dichlorvos (Kumari et al. 2018), and *Oreochromis mossambicus* exposed to quinalphos have all been found to have anemia caused by organophosphates (Ghayyur et al. 2019) (Table 4).

11 Pesticides as Genotoxic Agents

In addition to hematological and serological indicators, pesticides had an impact on fish genetic systems. Genetic harm was mentioned in a few instances. In *Channa punctata*, naphthalene-2-sulfonate was discovered to be genotoxic. The fish had LC50 values of 2.38 g/15 g BW and 4.77 g/15 g BW. Using 1/10th (0.238 g/L) and 1/20th (0.119 g/L) of the safe application rate, subchronic exposure was evaluated (SAR). A 60-day exposure demonstrated increased DNA damage in a time- and dose-dependent manner using the comet assay and the micronucleus assay.

When exposure was stopped after 30 days, the species appeared to recover during this time. Using attenuated total reflection-Fourier transform infrared (ATR-FTIR), the genotoxicity was further evaluated (Mehra and Chadha 2021). Using the comet test, the pesticide cocktail (endosulfan+chlorpyrifos) caused DNA damage in *Oreochromis niloticus*, a freshwater fish (Ambreen and Javed 2018). In fish erythrocytes, they found a dose-dependent response, with the highest concentration of the pesticide mixture (1/3rd of the LC50) causing the most DNA damage. When compared to the control group, treated fish had statistically significant effects for both concentrations and period of exposure in terms of DNA damage.

Table 4 Effects of pesticides on the hematological parameters of fish

Pesticides used	Fish species	Hematological findings	References
Atrazine	*Cyprinus carpio*	Decline of WBC, lymphocytes, Hb, and HCT. Rise in monocyte	Blahova et al. (2014)
2,4-Dichlorophenoxyacetate	*Carassius auratus*	Reduction in lymphocytes	Kubrak et al. (2013)
Butachlor	*Labeo rohita*	Decreased RBC, Hb, HCT, and lymphocyte. Increased TLC. Morphological and nuclear changes like pear shape erythrocyte, microcyte, tear shape erythrocyte, micronuclei	Ghaffar et al. (2015)
Carbaryl	*Channa punctatus*	Decrease in RBC, HB, and HCT	Johal and Grewal (2004)
Diazinon	*Oncorhynchus mykiss*	Reduction of WBC, RBC, PCV, Hb. Fluctuation in lymphocyte and neutrophil level	Far et al. (2012)
Chlorpyrifos	*Oreochromis mossambicus*	Decline in RBC, Hb, and HCT. Rise in WBC and platelets	Ghayyur et al. (2019)
Fipronil	*Rhamdia quelen*	Decline of HCT and platelets	Fredianelli et al. (2019)
Methyl parathion	*Mystus keletius*	Reduction of MCV, RBC, Hb, and thrombocyte. Increased TLC, ESR. Anemia, inhibition of erythropoiesis, and hemodilution	Sampath et al. (2003)
Chlorinated pesticides	*Heteropneustes fossilis*	Decline level of RBC, Hb, PCV. Rise in WBC. Altered level of MCH, MCV, and MCHC	Maurya et al. (2019)
Glyphosate	*Anabas testudineus*	Increase in MCV, MCH, WBC, and platelets and reduction of RBC, Hb, PCV, and lymphocytes	Samanta et al. (2019)

12 Effect of Pesticides on the Reproductive System of Fish

In fish, the endocrine system—which is managed by the brain—develops the gonads. The gonads' production of steroid hormones triggers the liver's creation of vitellogenin. Vitellogenin production is very high when there are reproductive issues brought on by pollution. Herbicides lead to aberrant oogenesis and gonadal intersex in fish.

The vulnerability of fish to the harmful effects of pesticides during their reproductive and developmental stages is well documented among numerous freshwater species. Generally speaking, reproductive and embryonic developments are impacted by pesticides that interfere with the endocrine system. Fish reproductive toxicity is characterized by elevated vitellogenin protein levels, aberrant gonadosomatic indices, and alterations in steroid hormone levels (Nichols et al. 2001).

Pyrethroids are detrimental to fish reproduction and early embryonic development, according to numerous studies. Pyrethroids like bifenthrin and permethrin can impede the development of the egg proteins choriogenin and vitellogenin in young fish (Brander et al. 2012). Deltamethrin second-generation (type II) pyrethroid neurotoxic insecticide was discovered by Wu et al. (2020) to have negative effects on the growth of the swim bladder in zebrafish embryos at doses of 20 and 40 g/L. Pesticides may prevent brown trout (*Salmo trutta*) and Atlantic salmon (*Salmo salar*) from reproducing (Jaensson et al. 2007). A range of developmental anomalies were also found, according to supplementary studies, in fish exposed to the pesticide (Dawar et al. 2016).

In fish, steroid hormones control the reproductive process. In addition, a rare fish reproductive condition called gonadal intersex exists. The most prevalent form of gonadal intersex, testicular oocytes, or oocytes produced in the testes has been connected to endocrine disruptors dispersed in aquatic environments (Blazer et al. 2007). Herbicide levels in aquatic environments are most likely connected to fish intersex cases among the various types of EDC (Abdel-Moneim et al. 2017). Fish reproduction could be hampered by contaminants that regulate sex hormone levels or vitellogenin gene expression (Hou et al. 2016). Therefore, it is believed that the expression of vitellogenin in juvenile and male fish serves as a marker of exposure to estrogenic compounds (Jobling and Tyler 2003). Developmental toxicity in children is also a risk when parents are exposed to contaminants in aquatic environments (Li et al. 2019). Intuitive signs of reproductive issues brought on by low levels of toxins include impaired egg production and quantity. The co-formulation of several ingredients in commercially available pesticides exposes aquatic organisms to multiple pollutants at once (Muschal and Warne 2003).

13 Effect of Pesticides on Various Biochemical Processes in Fish

Various pesticides affect different biochemical processes in fish in terms of enzyme actions and hormone protocols. Table 5 shows how different pesticides affect fish differently:

Table 5 Effect of pesticides on the biochemical processes in different fish species

Pesticides used	Fish species	Biochemical findings	References
Diazinon	*Clarias gariepinus*	Rise in level of glucose, AST, ALT	Al-Otaibi et al. (2018)
Deltamethrin	*Oreochromis niloticus*	Boosted blood urea, bilirubin, ALP, AST, and ALT. Decline in blood total protein, globulin, albumin, cortisol, and glucose	Dawood et al. (2020)
Boscalid	*Danio rerio*	Decline in level of glucose	Qian et al. (2019)
Chlorpyrifos	*Oreochromis mossambicus*	Blood glucose, cortisol, and cholesterol increased. Reduced total plasma proteins and triglyceride level	Ghayyur et al. (2019)
Propanil	*Oreochromis niloticus*	Rise in level of total proteins, phosphoglycerate kinase, and triglycerides and decrease of cholesterol	Abubakar et al. (2018)
Cypermethrin	*Brycon amazonicus*	Rise in level of sodium, glucose, and chloride	De Moraes et al. (2018)
Dimethoate	*Clarias batrachus*	Rise in serum level of glucose, creatinine, peroxidase, AST and decline globulin and albumin levels	Narra (2017)
Malathion	*Cirrhinus mrigala*	Rise in sodium and potassium levels in serum and decline of chloride and calcium	Rani et al. (2017)
Dioxin	*Oncorhynchus mykiss*	LDH, AST, and total protein plasma decreased. Interaction with DNA in a complex pathway alters how genes control the formation of protein such as vitellogenin protein for egg development	Zorriehzahra (2008)
Lead nitrate	*Mystus cavasius*	Decreased protein content in the liver and kidney due to proteolysis. Reduced liver glycogen	Jain and Batham (2016)

14 Impact on the Behavioral Changes of Fish

Several fish species, including *Tor putitora* and *Cyprinus carpio*, can exhibit schooling behavior, mucus formation through the skin's goblet cells (sliminess), motionlessness, changes in migration activities, tumbling toward base, jumping, nonresponsiveness with hyperexcitability, irregular activities, increased opercular rate (increased respiration), and body color changes as a result of pesticides.

Lufenuron-induced behavioral alterations in *Colossoma macropomum* included loss of stability, irregular swimming, motionlessness, and lying down, according to Soares et al. (2016). Additionally, they have the power to interfere with the reproduction and growth rates of aquatic vertebrates like fish and amphibians, as well as change and disrupt their swimming behavior (Stehle and Schulz 2015). Khalil et al. (2017)'s investigation into the effects of chlorpyrifos on social behavior in *Oryzias latipes* revealed dramatically decreased schooling and shoaling behavior that varied with time.

The behavior's duration alters. Pyrethroid exposure reduced the dopamine active transporter's efficiency, which led to unpredictable behavior (Pitzer et al. 2021).

15 Effect of Pesticides on Animal Diversity

A variety of creatures, including fish, amphibians, invertebrates, plants, and bacteria, make up the freshwater community. Both direct and indirect effects of pesticides on these organisms are possible. A pesticide's physiological activity within an organism has an immediate impact. For instance, the direct exposure of pesticides leads in the death of water fleas (direct effect), which may eventually cause algal blooms to be intensified as a result of reduced grazing pressure for fleas (indirect effect).

Pesticides may lessen the number of aquatic plants and insects that serve as a habitat and a source of food for fish and other aquatic animals. Fish that consume insects may experience a reduction in their food supply when pesticides are used. Fish can become more vulnerable to predators if there is an abrupt, insufficient supply of insects, which forces them to move farther in search of food.

Up to 80% of the dissolved oxygen needed by aquatic life in ponds and lakes can be supplied by aquatic plants. The use of herbicides to eradicate all aquatic plants can result in fish suffocation and severe oxygen shortage. When a pond is thoroughly "cleaned up" with herbicides, it dramatically affects the habitat, food source, dissolved oxygen, and fish production.

Spraying herbicides can also affect how successfully fish and other aquatic creatures reproduce. Young fish can find enough of food and shelter in the shallow, weedy nursery zones of many different fish species. Spraying herbicides close to weedy nurseries can lessen the amount of cover and protection young fish need to remain hidden from predators and forage. The majority of newborn fish in their nursery habitats rely on aquatic vegetation for safety.

Due to their widespread use in farmlands to kill off the intended animals, plants, and fungi—which may also damage unintended organisms—pesticides can have a considerable impact on biodiversity. Experimental investigations show that freshwater biodiversity is decreased by pesticide contamination (Relyea 2005). Although population structure may stay affected as the toxicant demonstrates age-dependent mortality or developmental delays, toxicant-induced impacts on individuals in high-density populations might limit intraspecific interaction and therefore be compensated (Berenzen et al. 2005). Although it is well known that pesticide contamination in the environment causes communities to drastically change and become dominated by pesticide-tolerant species, it has only recently come to light that pesticide concentrations have a significant negative impact on biodiversity in freshwater invertebrates. Because many freshwater bodies of water experience continuous inflows and outflows of water and organisms, it is challenging to gather accurate measurements of the conditions of communities and environmental variables at each sampling point given the spatiotemporal scale of pesticide application and residual effects.

Certain pesticides have been shown to reduce the species diversity of aquatic organisms and predatory insects when introduced into aquatic environments. Pesticide exposure causes a 42% loss in species richness in Europe, even at concentrations deemed environmentally safe by current regulations. Beneficial arthropod species richness, such as spiders, bees, and beetles, is found to be more in population on untreated or organic fields than on insecticide-treated fields, which is common in chemical-dependent agriculture. The diversity of these arthropods not only aids agricultural production through pollination but also creates a balance that prevents crop-damaging animals from overpopulating.

References

Abdel-Moneim A, Deegan D, Gao J, De Perre C, Doucette JS, Jenkinson B et al (2017) Gonadal intersex in smallmouth bass Micropterus dolomieu from northern Indiana with correlations to molecular biomarkers and anthropogenic chemicals. Environ Pollut 230:1099–1107

Abubakar YAJI, Iheanacho S, Ogueji E (2018) Sublethal exposure and toxicity effect of propanil on hematology and serum biochemistry in Oreochromis niloticus in a static bioassay. Gazi Univ J Sci 31(4):1048–1062

Ali H, Rani V (2009) Effect of phosalone on haematological indices in the tilapia, Oreochromis mossambicus. Turk J Vet Anim Sci 33(5):407–411

Al-Otaibi AM, Al-Balawi HFA, Ahmad Z, Suliman EM (2018) Toxicity bioassay and sub-lethal effects of diazinon on blood profile and histology of liver, gills and kidney of catfish, Clarias gariepinus. Braz J Biol 79:326–336

Ambreen F, Javed M (2018) Pesticide mixture induced DNA damage in peripheral blood erythrocytes of freshwater fish, Oreochromis niloticus. Pak J Zool 50(1)

Amin A, Indulkar ST, Pai R (2016) Sublethal effect of Buprofezin pesticide on carcass composition of *Cyprinus carpio communis* fingerlings. J Exp Zool India 20(1):269–272

Anusiya Devi K, Binukumari S, Vasanthi J (2015) Effects of monocrotophos on haematological analysis of blood in an Indian freshwater fish, Labeo rohita. Int J Recent Adv Multidiscip Res 2:870–874

Banaee M, Mirvaghefi AR, Amiri BM, Rafiee GR, Nematdost B (2011) Hematological and histopathological effects of diazinon poisoning in common carp (Cyprinus carpio). J Fish 64(1)

Berenzen N, Kumke T, Schulz HK, Schulz R (2005) Macroinvertebrate community structure in agricultural streams: impact of runoff-related pesticide contamination. Ecotoxicol Environ Saf 60(1):37–46

Blahova J, Modra H, Sevcikova M, Marsalek P, Zelnickova L, Skoric M, Svobodova Z (2014) Evaluation of biochemical, haematological, and histopathological responses and recovery ability of common carp (Cyprinus carpio L.) after acute exposure to atrazine herbicide. Biomed Res Int 2014:980948

Blazer VS, Iwanowicz LR, Iwanowicz DD, Smith DR, Young JA, Hedrick JD et al (2007) Intersex (testicular oocytes) in smallmouth bass from the Potomac River and selected nearby drainages. J Aquat Anim Health 19(4):242–253

Brander SM, He G, Smalling KL, Denison MS, Cherr GN (2012) The in vivo estrogenic and in vitro anti-estrogenic activity of permethrin and bifenthrin. Environ Toxicol Chem 31(12):2848–2855

Brodeur JC, Sassone A, Hermida GN, Codugnello N (2013) Environmentally-relevant concentrations of atrazine induce non-monotonic acceleration of developmental rate and increased size at metamorphosis in Rhinella arenarum tadpoles. Ecotoxicol Environ Saf 92:10–17

Çok İ, Ulutas OK, Okusluk Ö, Durmaz E, Demir N (2011) Evaluation of DNA damage in common carp (Cyprinus carpio L.) by comet assay for determination of possible pollution in Lake Mogan (Ankara). TheScientificWorldJOURNAL 11:1455–1461

Cole DJ (1985) Mode of action of glyphosate-a literature analysis. Herbicide Glyphosate 48:745

Connell DW, Lam P, Richardson B, Wu R (2009) Introduction to ecotoxicology. Wiley

Correia TG, Narcizo ADM, Bianchini A, Moreira RG (2010) Aluminum as an endocrine disruptor in female Nile tilapia (Oreochromis niloticus). Comp Biochem Physiol C Toxicol Pharmacol 151(4):461–466

Dar SA, Yousuf AR, Ganai FA, Bhat FA (2015) Assessment of endosulfan induced genotoxicity and mutagenicity manifested by oxidative stress pathways in freshwater cyprinid fish crucian carp (Carassius carassius L.). Chemosphere 120:273–283

Dawar FU, Zuberi A, Azizullah A, Khattak MNK (2016) Effects of cypermethrin on survival, morphological and biochemical aspects of rohu (Labeo rohita) during early development. Chemosphere 144:697–705

Dawood MA, AbdEl-Kader MF, Moustafa EM, Gewaily MS, Abdo SE (2020) Growth performance and hemato-immunological responses of Nile tilapia (Oreochromis niloticus) exposed to deltamethrin and fed immunobiotics. Environ Sci Pollut Res 27(11):11608–11617

De Moraes FD, Venturini FP, Rossi PA, Avilez IM, da Silva de Souza, N. E., & Moraes, G. (2018) Assessment of biomarkers in the neotropical fish Brycon amazonicus exposed to cypermethrin-based insecticide. Ecotoxicology 27(2):188–197

Dey C, Saha SK (2014) A comparative study on the acute toxicity bioassay of dimethoate and lambda-cyhalothrin and effects on thyroid hormones of freshwater teleost fish Labeo rohita (Hamilton). Int J Environ Res 8(4):1085–1092

Easton MDL, Luszniak D, Von der Geest E (2002) Preliminary examination of contaminant loadings in farmed salmon, wild salmon and commercial salmon feed. Chemosphere 46(7): 1053–1074

Far MS, Roodsari HV, Zamini A, Mirrasooli E, Kazemi R (2012) The effects of diazinon on behavior and some hematological parameters of fry rainbow trout (Oncorhynchus mykiss). World J Fish Mar Sci 4:369–375

Fredianelli AC, Pierin VH, Uhlig SC, Galeb L, Rocha DCC, Ribeiro DR et al (2019) Hematologic, biochemical, genetic, and histological biomarkers for the evaluation of the toxic effects of fipronil for Rhamdia quelen. Turk J Vet Anim Sci 43(1):54–59

Ghaffar A, Hussain R, Khan A, Abbas RZ, Asad M (2015) Butachlor induced clinico-hematological and cellular changes in fresh water fish Labeo rohita (Rohu). Pak Vet J 35(2): 201–206

Ghayyur S, Tabassum S, Ahmad MS, Akhtar N, Khan MF (2019) Effect of chlorpyrifos on hematological and seral biochemical components of fish Oreochromis mossambicus. Pak J Zool 51(3):1047–1052

Hou J, Li L, Wu N, Su Y, Lin W, Li G, Gu Z (2016) Reproduction impairment and endocrine disruption in female zebrafish after long-term exposure to MC-LR: a life cycle assessment. Environ Pollut 208:477–485

Jabeen G, Javed M, Azmat H (2012) Assessment of heavy metals in the fish collected from the river Ravi, Pakistan. Pak Vet J 32(1):107–111

Jaensson A, Scott AP, Moore A, Kylin H, Olsén KH (2007) Effects of a pyrethroid pesticide on endocrine responses to female odours and reproductive behaviour in male parr of brown trout (Salmo trutta L.). Aquat Toxicol 81(1):1–9

Jain S, Batham D (2016) Acute toxicity and biochemical studies of lead nitrate on the liver and kidney of fresh water fish Mystus cavasius. J Glob Biosci 5(9):4590–4597

Jeon J et al (2013) Characterization of acetylcholinesterase inhibition and energy allocation in Daphnia magna exposed to carbaryl. Ecotoxicol Environ Saf 98:28–35

Jindal R, Sharma R (2019) Neurotoxic responses in brain of Catla catla exposed to cypermethrin: a semiquantitative multibiomarker evaluation. Ecol Indic 106:105485

Jobling S, Tyler CR (2003) Endocrine disruption, parasites and pollutants in wild freshwater fish. Parasitology 126(7):S103–S107
Johal MS, Grewal H (2004) Toxicological study on the blood of Channa punctatus (Bloch) upon exposure to carbaryl. Pollut Res 23(1):601–606
Joseph B, Raj SJ (2011) Impact of pesticide toxicity on selected biomarkers in fishes. Int J Zool Res 7(2):212
Kanan S, Moyet MA, Arthur RB, Patterson HH (2020) Recent advances on TiO2-based photocatalysts toward the degradation of pesticides and major organic pollutants from water bodies. Catal Rev 62(1):1–65
Kerle EA, Jenkins JJ, Vogue PA (1994) Understanding pesticide persistence and mobility for groundwater and surface water protection
Khafaga AF, Naiel MA, Dawood MA, Abdel-Latif HM (2020) Dietary Origanum vulgare essential oil attenuates cypermethrin-induced biochemical changes, oxidative stress, histopathological alterations, apoptosis, and reduces DNA damage in common carp (Cyprinus carpio). Aquat Toxicol 228:105624
Khalil F, Qiu X, Kang IJ, Abo-Ghanema I, Shimasaki Y, Oshima Y (2017) Comparison of social behavior responses of Japanese medaka (Oryzias latipes) to lethal and sublethal chlorpyrifos concentrations at different exposure times. Ecotoxicol Environ Saf 145:78–82
Kovach J, Petzoldt C, Degni J, Tette J (1992) A method to measure the environmental impact of pesticides. NY Food Life Sci Bull 139:1–8
Kubrak OI, Atamaniuk TM, Storey KB, Lushchak VI (2013) Goldfish can recover after short-term exposure to 2, 4-dichlorophenoxyacetate: use of blood parameters as vital biomarkers. Comp Biochem Physiol C Toxicol Pharmacol 157(3):259–265
Kumari SR, Kumar MV, Rao GN (2018) Haematological changes induced in the fish Ctenopharyngodon idella (Valenciennes) exposed to organophosphate dichlorvos both technical and 76% EC (Nuvan). J Innov Pharm Biol Sci 5:58–63
Laxmi B, Madhavi K, Amin A, Chamundeswari Devi B, Dhanapal K, Ramana TV (2019) Sub lethal effects of Dichlorvos on physiological parameters in fingerlings of *Cyprinus carpio*. Int J Curr Microbiol Appl Sci 8(8):372–377
Laxmi B, Madhavi K, Dhanapal K, Amin A, Sudhakar O, Jesintha N (2020) Biochemical responses of fingerling *Cyprinus carpio* (Linnaeus, 1758) exposed to sub-lethal concentrations of Dichlorvos. Int J Curr Microbiol App Sci 9(10):3250–3257
LeBlanc GA (2007) Crustacean endocrine toxicology: a review. Ecotoxicology 16(1):61–81
Leroux P (1996) Recent developments in the mode of action of fungicides. Pestic Sci 47(2): 191–197
Li S, Wu Q, Sun Q, Coffin S, Gui W, Zhu G (2019) Parental exposure to tebuconazole causes thyroid endocrine disruption in zebrafish and developmental toxicity in offspring. Aquat Toxicol 211:116–123
Maltby L, Arnold D, Arts G, Davies J, Heimbach F, Pickl C, Poulsen V (2009) Aquatic macrophyte risk assessment for pesticides. CRC Press
Maurya PK, Malik DS, Yadav KK, Gupta N, Kumar S (2019) Haematological and histological changes in fish Heteropneustes fossilis exposed to pesticides from industrial waste water. Hum Ecol Risk Assess Int J 25:1251–1278
Mehra S, Chadha P (2021) Naphthalene-2-sulfonate induced toxicity in blood cells of freshwater fish Channa punctatus using comet assay, micronucleus assay and ATIR-FTIR approach. Chemosphere 265:129147
Mondal S, Subramaniam C (2020) Xenobiotic contamination of water by plastics and pesticides revealed through real-time, ultrasensitive, and reliable surface-enhanced raman scattering. ACS Sustain Chem Eng 8(20):7639–7648
Mostafalou S, Abdollahi M (2013) Pesticides and human chronic diseases: evidences, mechanisms, and perspectives. Toxicol Appl Pharmacol 268(2):157–177

Mostakim GM, Zahangir MM, Mishu MM, Rahman MK, Islam MS (2015) Alteration of blood parameters and histoarchitecture of liver and kidney of silver barb after chronic exposure to quinalphos. J Toxicol 2015:415984

Mulla DJ, Hermanson RERE, Maxwell RC (1996) Pesticide movement in soils-groundwater protection

Murthy KS, Kiran BR, Venkateshwarlu M (2013) A review on toxicity of pesticides in fish. Int J Open Sci Res 1(1):15–36

Muschal M, Warne MSJ (2003) Risk posed by pesticides to aquatic organisms in rivers of northern inland New South Wales, Australia. Hum Ecol Risk Assess Int J 9(7):1765–1787

Narra MR (2017) Haematological and immune upshots in Clarias batrachus exposed to dimethoate and defying response of dietary ascorbic acid. Chemosphere 168:988–995

Nichols KM, Snyder EM, Snyder SA, Pierens SL, Miles-Richardson SR, Giesy JP (2001) Effects of nonylphenol ethoxylate exposure on reproductive output and bioindicators of environmental estrogen exposure in fathead minnows, Pimephales promelas. Environ Toxicol Chem 20(3): 510–522

Panigrahi AK, Choudhury N, Tarafdar J (2014) Pollutional impact of some selective agricultural pesticides on fish Cyprinus carpio. IMPACT IJRANS 2(2):71–76

Pimpao CT, Zampronio AR, De Assis HS (2007) Effects of deltamethrin on hematological parameters and enzymatic activity in Ancistrus multispinis (Pisces, Teleostei). Pestic Biochem Physiol 88(2):122–127

Pitzer EM, Williams MT, Vorhees CV (2021) Effects of pyrethroids on brain development and behavior: deltamethrin. Neurotoxicol Teratol 87:106983

Qian L, Zhang J, Chen X, Qi S, Wu P, Wang C, Wang C (2019) Toxic effects of boscalid in adult zebrafish (Danio rerio) on carbohydrate and lipid metabolism. Environ Pollut 247:775–782

Rani M, Gupta RK, Kumar S, Yadav J, Rani S (2017) Pesticides induced alterations in blood serum ions of Indian major carps. Bioscan 12:847–850

Relyea RA (2005) The impact of insecticides and herbicides on the biodiversity and productivity of aquatic communities. Ecol Appl 15(2):618–627

Rezania S, Park J, Din MFM, Taib SM, Talaiekhozani A, Yadav KK, Kamyab H (2018) Microplastics pollution in different aquatic environments and biota: a review of recent studies. Mar Pollut Bull 133:191–208

Rios FS, Kalinin AL, Rantin FT (2002) The effects of long-term food deprivation on respiration and haematology of the neotropical fish Hoplias malabaricus. J Fish Biol 61(1):85–95

Rosenkrantz RT, Baun A, Kusk KO (2013) Growth inhibition and recovery of Lemna gibba after pulse exposure to sulfonylurea herbicides. Ecotoxicol Environ Saf 89:89–94

Samanta P, Pal S, Mukherjee AK, Senapati T, Jung J, Ghosh AR (2019) Assessment of adverse impacts of glyphosate-based herbicide, Excel Mera 71 by integrating multi-level biomarker responses in fishes. Int J Environ Sci Technol 16(10):6291–6300

Sampath K, James R, Velammal S (2003) Effect of methyl parathion on blood parameters and its recovery in a catfish, Mystus keletius. Indian J Fish 50(2):191–197

Sankhla MS, Kumari M, Sharma K, Kushwah RS, Kumar R (2018) Water contamination through pesticide & their toxic effect on human health. Int J Res Appl Sci Eng Technol 6(1):967–970

Saravi SSS, Shokrzadeh M (2011) Role of pesticides in human life in the modern age: a review. In: Pesticides in the modern world-risks and benefits, vol 5, pp 3–12

Sharbidre AA, Metkari V, ka Patode, P. (2011) Effect of Diazinon on acetylcholinesterase activity and lipid. Res J Environ Toxicol 5(2):152–161

Sikorski JA, Gruys KJ (1997) Understanding glyphosate's molecular mode of action with EPSP synthase: evidence favoring an allosteric inhibitor model. Acc Chem Res 30(1):2–8

Soares PRL, de Andrade ALC, Santos TP, da Silva SCBL, da Silva JF, Dos Santos AR et al (2016) Acute and chronic toxicity of the benzoylurea pesticide, lufenuron, in the fish, Colossoma macropomum. Chemosphere 161:412–421

Stehle S, Schulz R (2015) Agricultural insecticides threaten surface waters at the global scale. Proc Natl Acad Sci U S A 112(18):5750–5755

Tahir R, Ghaffar A, Abbas G, Turabi TH, Kausar S, Xiaoxia D et al (2021) Pesticide induced hematological, biochemical and genotoxic changes in fish: a review. Agrobiol Records 3:41–57
Taylor EJ, Maund SJ, Pascoe D (1991) Toxicity of four common pollutants to the freshwater macroinvertebratesChironomus riparius Meigen (Insecta: Diptera) and Gammarus pulex (L.) (Crustacea: Amphipoda). Arch Environ Contam Toxicol 21(3):371–376
Vaiyanan V, Sridharan G, Raveendran S, Chairman K (2015) Impact of pesticide on haematological parameters of Cyprinus carpio. World J Pharm Pharm Sci 4(8):1424–1430
Wafford KA, Sattelle DB, Gant DB, Eldefrawi AT, Eldefrawi ME (1989) Noncompetitive inhibition of GABA receptors in insect and vertebrate CNS by endrin and lindane. Pestic Biochem Physiol 33(3):213–219
Wilson PC, Koch R (2013) Influence of exposure concentration and duration on effects and recovery of Lemna minor exposed to the herbicide norflurazon. Arch Environ Contam Toxicol 64(2):228–234
Wu Y, Li W, Yuan M, Liu X (2020) The synthetic pyrethroid deltamethrin impairs zebrafish (Danio rerio) swim bladder development. Sci Total Environ 701:134870
Yadav H, Kumar R, Sankhla MS (2020) Residues of pesticides and heavy metals in crops resulting in toxic effects on living organism. J Seybold Rep 1533:9211
Yohannes YB, Ikenaka Y, Nakayama SM, Saengtienchai A, Watanabe K, Ishizuka M (2013) Organochlorine pesticides and heavy metals in fish from Lake Awassa, Ethiopia: insights from stable isotope analysis. Chemosphere 91(6):857–863
Zezulka Š, Kummerová M, Babula P, Váňová L (2013) Lemna minor exposed to fluoranthene: growth, biochemical, physiological and histochemical changes. Aquat Toxicol 140:37–47
Zorriehzahra MJ (2008) Aetiologic agents of fry mortality syndrome in the rainbow trout (*Oncorhynchus mykiss*) in Iran. PhD thesis, University of Putra Malaysia, p 270

Impact of Microplastics on Reproductive and Physiological Aspects of Aquatic Inhabitants

Menakshi Dhar, Raman Jasrotia, Seema Langer, and Nurul Suwartiningsih

1 Introduction

Food along with various dietary components, pollutants, chemicals, drugs, and medicines persistent in the environment are regarded as xenobiotics for aquatic organisms. Xenobiotics are chemical substances, adventitious for standard catabolism and anabolism in any biological entity. The term xenobiotics is generally used regarding the synthetic compounds that are generated in profuse amounts from agricultural, industrial, and domestic wastes (Atashgahi et al. 2018). These xenobiotics originating from the synthetic and natural sources infiltrate the ecosystem at very high or low concentrations (Schwarzenbach et al. 2006). Pollution due to the release of plastics has become a great environmental menace that has fascinated the considerable attention of the research worker analysts worldwide. Jambeck et al. (2015) estimated that in the year 2010, huge amounts of plastic trash has accumulated in the aquatic bodies and by the year 2025 there will be a significant rise in plastic waste if proper waste management is not implemented. A plastic substance is an organic polymer that is constructed from the polymerization of small molecular fragments, called monomers, which are extracted from gas or oil (Derraik 2002; Thompson et al. 2009).

Optimization of the numerous techniques associated with the mass production of durable, lightweight, and inert plastics has been done since the development of Bakelite in the year 1907, and this has accelerated the ample plastic utilization in innumerable practices. There has been a considerable increase in plastic production since the 1940s, and approximately 230 million tons of plastics were generated

M. Dhar · R. Jasrotia (✉) · S. Langer
Department of Zoology, University of Jammu, Jammu,, Jammu and Kashmir, India

N. Suwartiningsih
Laboratory of Ecology and Systematics, Study Program of Biology, Ahmad Dahlan University, Yogyakarta, Indonesia

M. A. Rather et al. (eds.), *Xenobiotics in Aquatic Animals*,
https://doi.org/10.1007/978-981-99-1214-8_6

worldwide in the year 2009 (Plastics Europe 2010). This commodity has far-reaching societal benefits, but it has become a subject of concern due to environment-related problems. Being highly durable, it is resistant to degradation, and so it remains persistent in the environment for a long period (Andrady and Neal 2009; Barnes et al. 2009). Plastics that enter the aquatic ecosystem have raised much concern among the aquatic ecologists. The large plastic debris has a detrimental effect on the aquatic ecosystem, where its existence proffers an aesthetic problem, thus having serious repercussions on the aquatic organisms. Their existence in a water body leads to the injury and death of birds, fishes, reptiles, and mammals due to the ingestion and entanglement of discarded plastic products (Lozano and Mouat 2009; Gregory 2009). Besides this, the debris of plastic materials floating on the surface of water plastic debris that floats on the water surface transports non-native species to new habitats (Moore 2008). Plastic particles which are less than 5 mm or 1 μm in size are defined as microplastics, and those particles whose size range varies between 1 μm and 100 nm are classified as nanoparticles (Wright et al. 2013; Rochman et al. 2016). Rise in contamination of freshwater ecosystems with numerous natural and synthetic chemical compounds across the globe is the major environment-related problem that humanity is facing at present. In the last few years, the contamination of the environment by xenobiotics and their uptake by living organisms is increasing at a rapid rate. The entry of such xenobiotic compounds into the ecosystem causes alterations in genetic makeup, lowers immunity, increases allergic reactions, disturbs metabolic activities, and may even lead to the mortality of an organism (Kovaleva et al. 2019). Transmission of microplastics from aquatic ecosystem to the aquatic inhabitants occurs by the ingestion of various food items, thus enkindling the numerous impacts on the organisms inhabiting the aquatic ecosystem. At present, plastic debris is among the most tenacious pollutant in the environment, and consequently, it acts as a great stressor for the environment by acting as a menace for aquatic inhabitants (Gallo et al. 2018; Oliveira et al. 2019). Low density and resistance to corrosion make them apt for utilization of various consumer products (Kannan and Vimalkumar 2021).

2 Occurrence of Microplastics in Aquatic Systems

The degradation of plastics is based on various factors such as the type of polymer and ecological conditions like temperature, pH, weathering, and irradiation. Microplastics are known to enter the aquatic systems as fragments, microbeads, and thread-like strands through numerous pathways which appear in all the environmental matrices such as sediments, coastlines, water columns, sea floor, and surface water. Inadequate management of plastic waste disposal leads to its entry into the aquatic system through wastewater flow, inland waterways, or through winds or tides. Hidalgo-Ruz et al. (2012) proposed that some microplastics such as polyethylene and polystyrene keep on floating on the surface of water bodies, whereas few of them such as acrylic and polyamides settle down in the bottom of the

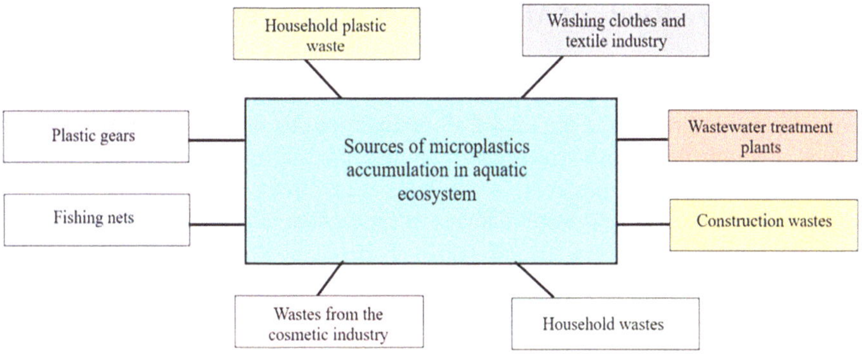

Fig. 1 Inception of microplastics accumulation in aquatic ecosystem

water body. Most of microplastics in freshwater emanate from terrestrial systems, with maximum loadings from local urban areas.

Breakdown of various plastic items occurs due to the physical, chemical, and biological processes, and this leads to the generation of secondary microplastics in freshwater ecosystems (Fadare et al. 2020; Horton et al. 2017). The degradation of fishing ropes, nets, boats, and various other fishing gears also contributes to the generation of microplastics in the lakes and rivers. Some of the smaller particles of plastics get carried away by the wind to remote areas from where they are washed into the water bodies after ample rainfall (Chen et al. 2020). Another source of microplastic pollution is textile industry. The fibers released from the washing of the clothes reach the sewage system from where they enter the aquatic systems (Guerranti et al. 2019) (Fig. 1).

Because of inappropriate ways of disposal, plastic fibers have become a grave problem as they finally reach aquatic environments, where they remain undegraded for years (Rios Mendoza et al. 2018). According to Geyer et al. (2017), roughly 8300 million metric tons of plastics were produced from 1950 to 2015, and this rate of plastic manufacturing is increasing and has reached about 368 million metric tons per year in 2019. According to the survey conducted by Plastics Europe (2015), the annual production of plastic materials across the world was 1.7 million tons in the year 1950, and it was gone up by 300 million metric tons in the year 2014. A profuse amount of this plastic material enters the aquatic system from land, and every year, 4.8–12.7 million tons of plastics reach the oceans per year (Jambeck et al. 2015). By the year 2030, the emission of plastics to aquatic ecosystems will be approximately 50 million metric tonnes every year (Borrelle et al. 2020). The use of plastic products has increased by 25-fold in the last 40 years due to their durability, lightweight, elasticity, and minimal cost (Sutherland et al. 2016).

3 Exposure Route of Microplastics in Aquatic Inhabitants

Primary way of being exposed to microplastics occurs via the ingestion of contaminated prey or directly via the intake of the microplastics from the natural environment although fishes and crustaceans can also take up these compounds through respiration via gills (Nelms et al. 2018; Watts et al. 2016). Upon consumption, they get transferred to the next trophic levels in food chain and accumulate in predator species (Farrell and Nelson 2013; Setälä et al. 2014). 90% of microplastics ingested by fishes, crustaceans, and mollusks include synthetic fibers and microfibers, and a small proportion includes foams, films, and fragments (Mizraji et al. 2017; Jabeen et al. 2017). Manufacturing of plastics involves the use of bisphenol A, parabens and phthalates are utilized in plastic manufacturing (Teuten et al. 2009), and when microplastics enter the body of an organism, these additives and monomers get released and absorbed by the predators (Browne et al. 2013). Reports have been made on the translocation of smaller microplastics to the liver and hepatopancreas in crabs and fish. In vivo studies have demonstrated that after the ingestion of nanoplastics by any organism, they get transported to all the organs of that organism (Lusher et al. 2017; Akbay and Özdemir 2016). Various spectroscopy techniques like Fourier-transform infrared and Raman spectroscopy have determined presence of foams, fragments, beads, fibers of polyethylene, terephthalate, nylon, polyurethane, polystyrene, and polypropylene in the alimentary canal of aquatic organisms (Wagner et al. 2017; Pinto da Costa et al. 2019). Microplastics are similar in size to that of particulate organic matter, and so they are consumed by prawns, crabs, crayfishes, and planktonic organisms, and this way they influence the numerous activities in the body of these organisms (Huang et al. 2022).

3.1 Effect of Microplastics on Reproductive Aspects

Polyethylene microplastics influence growth and reproductive behavior in *Hyalella azteca*, a large freshwater amphipod (Au et al. 2015). Chen et al. (2017) reported that the association of macroplastics with 17 α-ethinyl estradiol affects the reproductivity of *Danio rerio*. Development at the early stages is a critical point during the life cycle of any organism (Sundaray et al. 2021). Li et al. (2020) revealed that occurrence of microplastics hinders egg production, reduces the hatching time, and represses the larval growth in *Oryzias melastigma*. Exposure to microplastics hinders the process of sex cell formation in zebrafish, due to the increased expression of genes in the gastrointestinal tract and liver (Mak et al. 2019). Japanese medaka on exposure to polystyrene microplastics depicted the reduction in hatching process (Zhu et al. 2020). Batel et al. (2018) and Duan et al. (2020) revealed that microplastics bind to the eggs of a fish which alter the exchange of gases and slow down their time of hatching. Microplastics induce reproductive stress in the fish, thereby disrupting the process of reproduction. They also obstruct the steroidogenesis pathway and hypothalamus-pituitary-gonadal axis, thus impairing the endocrine system that is accountable for regulation of reproduction process (Rochman et al. 2014; Rather et al. 2017; Chen et al. 2019). Recent research has suggested the delay in the process of maturation of gonads and decreased fecundity in some aquatic inhabitants due to the existence of microplastics. Besides this, microplastics surge the numerous reactive oxygen species in the gonads of both male and female zebrafish, thus responsible for causing reproductive stress by elevating the levels of apoptosis and altering the histology of the testes. Levels of reactive oxygen species increase on exposure to waterborne microplastics (Wan et al. 2019; Qiang and Cheng 2021). Microplastics also lessen the levels of different hormones such as testosterone and 17beta-estradiol in female individuals of *Oryzias melastigma*. Studies done by Ismail et al. (2021) suggested that microplastic exposure lowers reproductive ability of male Nile tilapia (*Oreochromis niloticus*) due to oxidative stress. Pitt et al. (2018) documented the transmission of microplastics from mother to offspring along with its effect on the development of early fry stages (Wang et al. 2019). Results of the study conducted by Aryani et al. (2021) revealed that microplastics penetrate the gills, liver, muscles, stomach, intestines, and gonads of Nile tilapia. Incessant exposure to polystyrene microplastics hampers the functioning of reproductive organs in *Danio rerio*.

3.2 Effect of Microplastics on Various Physiological Aspects

After ingesting the microplastics, any aquatic organism can get affected in different ways (Strungaru et al. 2019). The microplastics may block their digestive tract, and some life-threatening additives may percolate from these plastics which may lead to the mortality of an organism. Assas et al. (2020) reported that Japanese medaka and Java medaka when exposed to 2 μ of fluorescent polystyrene microplastics for 3 weeks had a significant impact on the gene expression involved in brain development, cell adhesion, and various other metabolic processes. Cole et al. (2015) concluded in their study that microscopic plastic debris similar in size to an algal prey impedes the process of feeding in *Calanus helgolandicus* and thus they suffer from energy depletion over a period of time. Lönnstedt and Eklöv (2016) suggested that pollutants released from the microplastics impair the olfactory sense in the larval stages of fish, as a result of which they are not able to perceive the inhabitance of predators in the surrounding environment, so the presence of microplastics affects their survival rate. The physiology and behavior of aquatic organisms are affected by the existence of microplastics in the ecosystem (Espinosa et al. 2016). Yin et al. (2018) analyzed the effect of polystyrene toxicity on the behavior and functioning of *Sebastes schlegelii*. After getting accumulated in the digestive system and gills, these microplastics affect their feeding and swimming activity. This has a remarkable effect on the liver and gall bladder of fish. In addition to all of this, the presence of such contaminant particles in an ecosystem decreases the lipid and protein content in a consumable aquatic organism. Koongolla et al. (2020) detected the presence of various types of microplastics in the different organ systems of the fishes in Beibu Gulf in the South China. The microplastics were found in gills, dorsal muscles, and the gastrointestinal tract of fish species inhabiting the North Atlantic Ocean by Barboza et al. (2020). Increased lipid peroxidation levels and acetylcholine esterase activity were recorded in the brain, muscles, and gills of fish contaminated with microplastics. The results of their study suggested that microplastics cause neurotoxicity and lipid oxidative damage in the muscles of the fish. In the last few decades, numerous studies on toxicity caused by the ingestion of microplastics have increased rapidly. Exposure to microplastics in mollusks, crustaceans, and fish indicated that these microfibers incite oxidative stress and mortality and decrease the population growth rate in these species (Avio et al. 2015a, b; Fonte et al. 2016; Gambardella et al. 2017; Yin et al. 2018 and Zhu et al. 2019).

Barboza et al. (2018a, b) reported that ingestion of microplastics by European sea bass, *Dicentrarchus labrax*, causes severe effects on their behavior. It causes neurotoxicity and alterations in their alimentary tract. The reduction in swimming behavior and the alteration in an action of energy-coupled enzymes were reported in the juveniles of *D. labrax* on exposure to microplastics. Large amount of microplastics were collected from the fish sold in Shanghai fish market, China (Jabeen et al. 2017). Fishes inhabiting the coastal regions of western and central Guangdong Province in China depicted the presence of microplastics in their alimentary tract and gills (Pan et al. 2021). Polystyrene particles detected in *Danio rerio* had toxic effects on the liver of the fish. Histopathological analysis revealed that microplastics led to the accumulation of lipids and inflammation of the liver of fish. It also increased the activities of catalase and superoxide dismutase which caused oxidative stress by disturbing the energy and lipid metabolism (Lu et al. 2016). Microplastics act as the carriers of pollutants and disease-causing organisms from surroundings to the aquatic inhabitants (Alimba and Faggio 2019). Magni et al. (2019) reported that protein modification occurs in the gills of a mussel due to the presence of microplastics and this leads to oxidative stress in them. Handy et al. (2008) found that the plastic particles have detrimental impacts such as oxidative stress, disturbance in an ion regulation mechanism, vascular injuries, and the formation of tumors in fish. Studies conducted by Jabeen et al. (2018) on the goldfish have depicted that microplastics led to scrapping and swelling of abdominal epithelium and distended liver. Exposure to microplastics markedly affects the homeostasis in the brain, muscles, intestines, and liver tissues (Rainieri et al. 2018). Polystyrene microplastics caused alterations in the tissues of the gut, spleen, pharynx, and nephrogenesis in Japanese medaka (Zhu et al. 2020). Exposure to polystyrene induced neurotoxicity in the encephalon of freshwater *Oreochromis niloticus* as it inhibits the activity of acetylcholinesterase enzyme in them (Ding et al. 2018). In order to determine the microplastics level in the brain, muscles, and gills of *Alosa immaculata* and *Mullus barbatus*, Atamanalp et al. (2021) utilized various techniques such as attenuated total reflectance (ATR)-Fourier-transform infrared (FTIR) spectroscopy. Additives and plasticizers bound to the microplastics get liberated within the body of a fish where they alter the activity of the immune system and blood biochemistry (Parker et al. 2021). The processing of microplastics causes alteration in the morphology and anatomy of the body of an organism (Lu et al. 2016). Alterations in the morphology of the alimentary tract change activities of gut microbes, thus leading to dysbiosis (Jabeen et al. 2018; Zhao et al. 2020).

Microplastics have a great influence on blood biochemistry which leads to anemia and alteration in the functioning of the immune system of the body. Microplastics impede working of the central nervous system and endocrine system which affects the behavior, survival, and growth of an individual organism (LaPlaca and van den Hurk 2020; Lei et al. 2018; Karami et al. 2016). Studies done by Mak et al. (2019) revealed that behavior of fish gets altered on exposure to microplastics as they affect the brain cells which have a negative impact on their swimming activity. Polyethylene and polyvinyl chloride microplastics which are approximately 40–150 μm in size cause oxidative stress in the leucocytes of gilt-head bream, *Sparus aurata*, leading to immunotoxicity in fish. The existence of these microplastics in the sand and silt deposits hampers the growth and development of *Arenicola marina*, and the degree of growth inhibition is directly linked to the microplastic concentration (Li et al. 2021). The effect of microplastics (<3 μm size) on the adult fish *Serranus scriba* was assessed by Zitouni et al. (2020). They observed that the presence of small microplastics induced neurotoxicity and oxidative stress in fishes. The uptake of microplastics by the fish causes blockage and histopathological alterations in the gastrointestinal tract and behavioral changes and gets translocated to the hepatic region (Avio et al. 2015a, b; Jovanović 2017). Transcriptome findings have suggested that such plastic particles alter the gene expression of the body's defense mechanism in adult zebrafish. Besides this, these particles that are less than 5 mm in size also downregulate the genes associated with the integrity of epithelial tissue and metabolism of lipids in the fish (Limonta et al. 2019). Various investigations with regard to the impact of such plastic particles on aquatic inhabitants have been done in the laboratories. Lusher et al. (2013) and Wright et al. (2013) reported that after consumption, microplastics get accumulated in the digestive system of the fish which may lead to blockage across the alimentary tract and thus limit their feeding. Functional and anatomical changes in the alimentary tract of the fish due to the intake of microplastics may cause developmental and dietary problems in them (Huang et al. 2021; Jabeen et al. 2018; Borrelle et al. 2017). Detailed representation of the toxicological impact of microplastics on various fish organ systems has been given in Fig. 2.

4 Conclusion

Presence of microplastics has become part of the ecosystem in the past few decades, and it has been predicted that they will continue to surge in the coming generations. In the last few decades, studies on the toxicity of microplastics have expeditiously increased (Jeong and Choi 2019). Laboratory studies done on various organisms such as mollusks, fish, and crustaceans were suggestive of the fact that microplastics

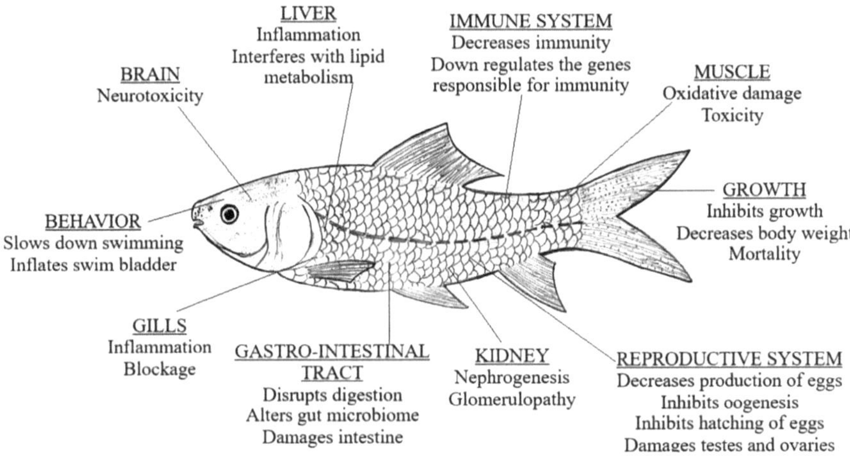

Fig. 2 Toxicological impacts of microplastics

incite the chemical and physical toxicity that includes oxidative stress, behavioral changes, obstruction in reproductive activity, and even mortality (Avio et al. 2015a, b; Fonte et al. 2016; Gambardella et al. 2017; Barboza et al. 2018a, b; Yin et al. 2018; Zhu et al. 2019). These plastics affect both the environment and the organisms inhabiting that environment. The pervasiveness of microplastics causes serious health problems in aquatic inhabitants after exposure of them. By examining the literature, it can be inferred that contamination due to microplastics occurs virtually in all kinds of aquatic ecosystems across the world. In aquatic organisms, microplastics get accumulated in the body and affect their various physiological and reproductive processes (Table 1).

Various remedial measures can be set down to lessen the impact of microplastics on water-living creatures:

- Assessment of the effect of micro-pollutants on the health of an aquatic ecosystem.
- Exploring the technologies for the cost-effective treatment of contaminated water.
- Minimizing the entry of pollutants into an aquatic ecosystem by designing various strategies for the disposal of wastes.
- Policies should be strengthened at local, national, and international levels.
- Awareness-related programs should be launched to curb the pollution caused due to the presence of microplastics.
- Development of the easy, fast, and reliable methods for the microplastic detection and quantification.

Table 1 Overview of the effect of microplastics on various physiological and reproductive aspects of fish

Fish	Family	Effect of microplastics	References
Danio rerio	Cyprinidae	Damages the gastrointestinal tract (GIT), alters the gut microbiome in larval stages, alters swimming behavior, increases the rate of apoptosis in the testis	Limonta et al. (2019), Jin et al. (2018), Qiang and Cheng (2019)
Clarias gariepinus	Clariidae	Alters the level of proteins and blood biochemistry; liver and GIT damage in juveniles	Karami et al. (2016), Iheanacho and Odo (2020)
Barbodes gonionotus	Cyprinidae	Thickens epithelial lining and increases the levels of proteins in juveniles	Romano et al. (2018)
Oreochromis niloticus	Cichlidae	Induces anemia by changing blood biochemistry and alters brain activity and metabolism in adults	Hamed et al. (2019), Ding et al. (2020)
Oryzias latipes	Adrianichthyidae	Changes in the morphology in the embryo and larval forms, decreases the egg production and hatching rate	Pannetier et al. (2020), Chisada et al. (2019)
Lates calcarifer	Latidae	Impairs feeding and swimming behavior in juveniles	Guven et al. (2018)
Dicentrarchus labrax	Moronidae	Alters the levels of protein and leads to larval mortality	Mazurais et al. (2015)
Cyprinus carpio	Cyprinidae	Induces oxidative stress and alters biochemistry levels, reduces the larval growth	Xia et al. (2020), Hatami et al. (2019)

References

Akbay İK, Özdemir T (2016) Monomer migration and degradation of polycarbonate via UV-C irradiation within aquatic and atmospheric environments. J Macromol Sci A 53(6):340–345

Alimba CG, Faggio C (2019) Microplastics in the marine environment: current trends in environmental pollution and mechanisms of toxicological profile. Environ Toxicol Pharmacol 68:61–74

Andrady AL, Neal MA (2009) Applications and societal benefits of plastics. Philos Trans R Soc Lond B Biol Sci 364(1526):1977–1984

Aryani D, Khalifa MA, Herjayanto M, Solahudin EA, Rizki EM, Halwatiyah W, Istiqomah H, Maharani SH, Wahyudin H, Pratama G (2021) Penetration of microplastics (polyethylene) to several organs of Nile tilapia (Oreochromis niloticus). In: IOP conference series: earth and environmental science, vol 715, no 1. IOP Publishing, p 012061

Assas M, Qiu X, Chen K, Ogawa H, Xu H, Shimasaki Y, Oshima Y (2020) Bioaccumulation and reproductive effects of fluorescent microplastics in medaka fish. Mar Pollut Bull 158:111446

Atamanalp M, Köktürk M, Uçar A, Duyar HA, Özdemir S, Parlak V, Esenbuğa N, Alak G (2021) Microplastics in tissues (brain, gill, muscle and gastrointestinal) of *Mullus barbatus* and *Alosa immaculata*. Arch Environ Contam Toxicol 81(3):460–469

Atashgahi S, Sánchez-Andrea I, Heipieper HJ, van der Meer JR, Stams AJM, Smidt H (2018) Prospects for harnessing biocide resistance for bioremediation and detoxification. Science 360(6390):743–746. https://doi.org/10.1126/science.aar3778

Au SY, Bruce TF, Bridges WC, Klaine SJ (2015) Responses of *Hyalella azteca* to acute and chronic microplastic exposure. Environ Toxicol Chem 34:2564–2572

Avio CG, Gorbi S, Milan M, Benedetti M, Fattorini D, D'Errico G, Pauletto G, Bargelloni L, Regoli F (2015a) Pollutants bioavailability and toxicological risk from microplastics to marine mussels. Environ Pollut 198:211–222

Avio CG, Gorbi S, Regoli F (2015b) Experimental development of a new protocol for extraction and characterization of microplastics in fish tissues: first observations in commercial species from Adriatic Sea. Mar Environ Res 111:18–26

Barboza LG, Vieira LR, Branco V, Carvalho C, Guilhermino L (2018a) Microplastics increase mercury bioconcentration in gills and bioaccumulation in the liver, and cause oxidative stress and damage in *Dicentrarchus labrax* juveniles. Sci Rep 8(1):1–9

Barboza LGA, Vieira LR, Branco V, Figueiredo N, Carvalho F, Carvalho C, Guilhermino L (2018b) Microplastics cause neurotoxicity, oxidative damage and energy-related changes and interact with the bioaccumulation of mercury in the European seabass, *Dicentrarchus labrax* (Linnaeus, 1758). Aquat Toxicol 195:49–57

Barboza LGA, Lopes C, Oliveira P, Bessa F, Otero V, Henriques B, Guilhermino L (2020) Microplastics in wild fish from North East Atlantic Ocean and its potential for causing neurotoxic effects, lipid oxidative damage, and human health risks associated with ingestion exposure. Sci Total Environ 717:134625. https://doi.org/10.1016/j.scitotenv.2019.134625

Barnes DKA, Galgani F, Thompson RC, Barlaz M (2009) Accumulation and fragmentation of plastic debris in global environments. Philos Trans R Soc Lond B Biol Sci 364:1985–1998

Batel A, Borchert F, Reinwald H, Erdinger L, Braunbeck T (2018) Microplastic accumulation patterns and transfer of benzo[a]pyrene to adult zebrafish (*Danio rerio*) gills and zebrafish embryos. Environ Pollut 235:918–930

Borrelle SB, Rochman CM, Liboiron M, Bond AL, Lusher A, Bradshaw H et al (2017) Opinion: why we need an international agreement on marine plastic pollution. Proc Natl Acad Sci U S A 114(38):9994–9997. https://doi.org/10.1073/pnas.1714450114

Borrelle SB, Ringma J, Law KL, Monnahan CC, Lebreton L, McGivern A, Murphy E, Jambeck J, Leonard GH, Hilleary MA, Eriksen M, Possingham HP, De Frond H, Gerber LR, Polidoro B, Tahir A, Bernard M, Mallos N, Barnes M, Rochman CM (2020) Predicted growth in plastic waste exceeds efforts to mitigate plastic pollution. Science 369(6510):1515–1518. https://doi.org/10.1126/science.aba3656

Browne MA, Niven SJ, Galloway TS, Rowland SJ, Thompson RC (2013) Microplastic moves pollutants and additives to worms, reducing functions linked to health and biodiversity. Curr Biol 23:2388–2392

Chen Q, Gundlach M, Yang S, Jiang J, Velki M, Yin D, Hollert H (2017) Quantitative investigation of the mechanisms of microplastics and nanoplastics toward zebrafish larvae locomotor activity. Sci Total Environ 584-585:1022–1031

Chen Q, Allgeier A, Yin D, Hollert H (2019) Leaching of endocrine disrupting chemicals from marine microplastics and mesoplastics under common life stress conditions. Environ Int 130: 104938

Chen G, Feng Q, Wang J (2020) Mini-review of microplastics in the atmosphere and their risks to humans. Sci Total Environ 703:135504

Chisada S, Yoshida M, Karita K (2019) Ingestion of polyethylene microbeads affects the growth and reproduction of medaka, *Oryzias latipes*. Environ Pollut 254(Pt B):113094

Cole M, Lindeque P, Fileman E, Halsband C, Galloway TS (2015) The impact of polystyrene microplastics on feeding, function and fecundity in the marine copepod *Calanus helgolandicus*. Environ Sci Technol 49(2):1130–1137

Derraik JGB (2002) The pollution of the marine environment by plastic debris: a review. Mar Pollut Bull 44:842–852

Ding J, Zhang S, Razanajatovo RM, Zou H, Zhu W (2018) Accumulation, tissue distribution, and biochemical effects of polystyrene microplastics in the freshwater fish red tilapia (*Oreochromis niloticus*). Environ Pollut 238:1–9

Ding J, Huang Y, Liu S, Zhang S, Zou H, Wang Z, Zhu W, Geng J (2020) Toxicological effects of nano- and micro-polystyrene plastics on red tilapia: are larger plastic particles more harmless? J Hazard Mater 396:122693

Duan Z, Duan X, Zhao S, Wang X, Wang J, Liu Y, Peng Y, Gong Z, Wang L (2020) Barrier function of zebrafish embryonic chorions against microplastics and nanoplastics and its impact on embryo development. J Hazard Mater 395:122621

Espinosa C, Esteban MÁ, Cuesta A (2016) Microplastics in aquatic environments and their toxicological implications for fish. Toxicology–new aspects to this scientific conundrum. InTech, Rijeka, pp 113–145

Fadare OO, Wan B, Zhao L, Guo LH (2020) Microplastics from consumer plastic food containers: are we consuming it? Chemosphere 253:126787

Farrell P, Nelson K (2013) Trophic level transfer of microplastic: *Mytilus edulis*(L.) to *Carcinus maenas* (L.). Environ Pollut 177:1–3

Fonte E, Ferreira P, Guilhermino L (2016) Temperature rise and microplastics interact with the toxicity of the antibiotic cefalexin to juveniles of the common goby (*Pomatoschistus microps*): post-exposure predatory behaviour, acetylcholinesterase activity and lipid peroxidation. Aquat Toxicol 180:173–185

Gallo F, Fossi C, Weber R, Santillo D, Sousa J, Ingram I, Nadal A, Romano D (2018) Marine litter plastics and microplastics and their toxic chemicals components: the need for urgent preventive measures. Environ Sci Eur 30:13. https://doi.org/10.1186/s12302-018-0139-z

Gambardella C, Morgana S, Ferrando S, Bramini M, Piazza V, Costa E, Garaventa F, Faimali M (2017) Effects of polystyrene microbeads in marine planktonic crustaceans. Ecotoxicol Environ Saf 145:250–257

Geyer R, Jambeck JR, Law KL (2017) Production, use, and fate of all plastics ever made. Sci Adv 3(7):e1700782. https://doi.org/10.1126/sciadv.1700782

Gregory MR (2009) Environmental implications of plastic debris in marine settings: entanglement, ingestion, smothering, hangers-on, hitch-hiking and alien invasions. Philos Trans R Soc Lond B Biol Sci 364:2013–2025

Guerranti C, Martellini T, Perra G, Scopetani C, Cincinelli A (2019) Microplastics in cosmetics: environmental issues and needs for global bans. Environ Toxicol Pharmacol 68:75–79

Guven O, Bach L, Munk P, Dinh KV, Mariani P, Nielsen TG (2018) Microplastic does not magnify the acute effect of PAH pyrene on predatory performance of a tropical fish (*Lates calcarifer*). Aquat Toxicol 198:287–293

Hamed M, Soliman HAM, Osman AGM, Sayed AEDH (2019) Assessment the effect of exposure to microplastics in Nile Tilapia (*Oreochromis niloticus*) early juvenile: I. blood biomarkers. Chemosphere 228:345–350

Handy RD, Henry TB, Scown TM, Johnston BD, Tyler CR (2008) Manufactured nanoparticles: their uptake and effects on fish - a mechanistic analysis. Ecotoxicology 17:396–409

Hatami M, Banaee M, Nematdoost Haghi B (2019) Sub-lethal toxicity of chlorpyrifos alone and in combination with polyethylene glycol to common carp (*Cyprinus carpio*). Chemosphere 219: 981–988

Hidalgo-Ruz V, Gutow L, Thompson RC, Thiel M (2012) Microplastics in the marine environment: a review of the methods used for identification and quantification. Environ Sci Technol 46(6): 3060–3075. https://doi.org/10.1021/es2031505

Horton AA, Svendsen C, Williams RJ, Spurgeon DJ, Lahive E (2017) Large microplastic particles in sediments of tributaries of the river Thames, UK – abundance, sources and methods for effective quantification. Mar Pollut Bull 114:218–226

Huang W, Song B, Liang J, Niu Q, Zeng G, Shen M et al (2021) Microplastics and associated contaminants in the aquatic environment: a review on their ecotoxicological effects, trophic transfer, and potential impacts to human health. J Hazard Mater 405:124187. https://doi.org/10.1016/j.jhazmat.2020.124187

Huang CH, Chu TW, Kuo CH, Hong MC, Chen YY, Chen B (2022) Effects of microplastics on reproduction and growth of freshwater live feeds *Daphnia magna*. Aust Fish 7:181. https://doi.org/10.3390/fishes7040181

Iheanacho SC, Odo GE (2020) Neurotoxicity, oxidative stress biomarkers and haematological responses in African catfish (*Clarias gariepinus*) exposed to polyvinyl chloride microparticles. Comp Biochem Physiol C Toxicol Pharmacol 232:108741

Ismail RF, Saleh NE, Sayed AE (2021) Impacts of microplastics on reproductive performance of male tilapia (*Oreochromis niloticus*) pre-fed on *Amphora coffeaeformis*. Environ Sci Pollut Res 28(48):68732–68744

Jabeen K, Su L, Li J, Yang D, Tong C, Mu J, Shi H (2017) Microplastics and mesoplastics in fish from coastal and fresh waters of China. Environ Pollut 221:141–149

Jabeen K, Li B, Chen Q, Su L, Wu C, Hollert H, Shi H (2018) Effects of virgin microplastics on goldfish (*Carassius auratus*). Chemosphere 213:323–332. https://doi.org/10.1016/j.chemosphere.2018.09.031

Jambeck JR, Geyer R, Wilcox C, Siegler TR, Perryman M, Andrady A, Narayan R, Law KL (2015) Plastic waste inputs from land into the ocean. Science 347(6223):768–771

Jeong J, Choi J (2019) Adverse outcome pathways potentially related to hazard identification of microplastics based on toxicity mechanisms. Chemosphere 231:249–255

Jin Y, Xia J, Pan Z, Yang J, Wang W, Fu Z (2018) Polystyrene microplastics induce microbiota dysbiosis and inflammation in the gut of adult zebrafish. Environ Pollut 235:322–329

Jovanović B (2017) Ingestion of microplastics by fish and its potential consequences from a physical perspective. Integr Environ Assess Manag 13(3):510–515

Kannan K, Vimalkumar K (2021) A review of human exposure to microplastics and insights into microplastics as obesogens. Front Endocrinol 12:724989. https://doi.org/10.3389/fendo.2021.724989

Karami A, Romano N, Galloway T, Hamzah H (2016) Virgin microplastics cause toxicity and modulate the impacts of phenanthrene on biomarker responses in African catfish (*Clarias gariepinus*). Environ Res 151:58–70

Koongolla JB, Lin L, Pan YF, Yang CP, Sun DR, Liu S, Xu XR, Maharana D, Huang JS, Li HX (2020) Occurrence of microplastics in gastrointestinal tracts and gills of fish from Beibu Gulf, South China Sea. Environ Pollut 258:113734

Kovaleva TN, Shtofer LL, Stukalov SY, Spektor LA, Chernov IV (2019) Ecological safety and hazards to public health: the search for balance within the "human-biosphere" system. Ekoloji 28(107):5039–5043

LaPlaca SB, van den Hurk P (2020) Toxicological effects of micronized tire crumb rubber on mummichog (*Fundulus heteroclitus*) and fathead minnow (*Pimephales promelas*). Ecotoxicology 29:524–534

Lei L, Wu S, Lu S, Liu M, Song Y, Fu Z, Shi H, Raley-Susman KM, He D (2018) Microplastic particles cause intestinal damage and other adverse effects in zebrafish *Danio rerio* and nematode *Caenorhabditis elegans*. Sci Total Environ 619–620:1–8

Li Y, Wang J, Yang G, Lu L, Zheng Y, Zhang Q, Zhang X, Tian H, Wang W, Ru S (2020) Low level of polystyrene microplastics decreases early developmental toxicity of phenanthrene on marine medaka (*Oryzias melastigma*). J Hazard Mater 385:121586

Li Y, Sun Y, Li J, Tang R, Miu Y, Ma X (2021) Research on the influence of microplastics on marine life. 3rd international conference on air pollution and environmental engineering IOP conference series: earth and environmental science, vol 631. p 012006

Limonta G, Mancia A, Benkhalqui A, Bertolucci C, Abelli L, Fossi MC, Panti C (2019) Microplastics induce transcriptional changes, immune response and behavioral alterations in adult zebrafish. Sci Rep 9(1):1–1. https://doi.org/10.1038/s41598-019-52292

Lönnstedt OM, Eklöv P (2016) Environmentally relevant concentrations of microplastic particles influence larval fish ecology. Science 352:1213–1316. https://doi.org/10.1126/science.aad8828

Lozano RL, Mouat J (2009) Marine litter in the North-East Atlantic region: assessment and priorities for response. KIMO International

Lu Y, Zhang Y, Deng Y, Jiang W, Zhao Y, Geng J, Ren H (2016) Uptake and accumulation of polystyrene microplastics in zebrafish (*Danio rerio*) and toxic effects in liver. Environmental Science and Technology 50:4054–4060. https://doi.org/10.1021/acs.est.6b00183

Lusher AL, McHugh M, Thompson RC (2013) Occurrence of microplastics in the gastrointestinal tract of pelagic and demersal fish from the English Channel. Mar Pollut Bull 67:94–99

Lusher A, Hollman P, Mendoza-Hill J (2017) Microplastics in fisheries and aquaculture: status of knowledge on their occurrence and implications for aquatic organisms and food safety. FAO Fisheries and Aquaculture Technical Paper 615

Magni S, Della Torre C, Garrone G, D'Amato A, Parenti CC, Binelli A (2019) First evidence of protein modulation by polystyrene microplastics in a freshwater biological model. Environ Pollut 250:407–415

Mak CW, Yeung KCF, Chan KM (2019) Acute toxic effects of polyethylene microplastic on adult zebrafish. Ecotoxicol Environ Saf 182:1–10

Mazurais D, Ernande B, Quazuguel P, Severe A, Huelvan C, Madec L, Mouchel O, Soudant P, Robbens J, Huvet A, Zambonino-Infante J (2015) Evaluation of the impact of polyethylene mi-crobeads ingestion in European sea bass (*Dicentrarchus labrax*) larvae. Mar Environ Res 112:78–85

Mizraji R, Ahrendt C, Perez-Venegas D, Vargas J, Pulgar J, Aldana M, Patricio Ojeda F, Duarte C, Galbán-Malagón C (2017) Is the feeding type related with the content of microplastics in intertidal fish gut? Mar Pollut Bull 116(1–2):498–500. https://doi.org/10.1016/j.marpolbul.2017.01.008

Moore CJ (2008) Synthetic polymers in the marine environment: a rapidly increasing, long-term threat. Environ Res 108:131–139

Nelms SE, Galloway TS, Godley BJ, Jarvis DS, Lindque PK (2018) Investigating microplastic trophic transfer in marine top predators. Environ Pollut 238:999–1007

Oliveira M, Almeida M, Miguel I (2019) A micro(nano)plastic boomerang tale: a never ending story? Trends Anal Chem 112:196. https://doi.org/10.1016/j.trac.2019.01.005

Pan Z, Zhang C, Wang S, Sun D, Zhou A, Xie S, Xu G, Zou J (2021) Occurrence of microplastics in the gastrointestinal tract and gills of fish from Guangdong, South China. J Mar Sci Eng 9:981. https://doi.org/10.3390/jmse9090981

Pannetier P, Morin B, Le Bihanic F, Dubreil L, Clérandeau C, Chouvellon F, Van Arkel K, Danion M, Cachot J (2020) Environmental samples of microplastics induce significant toxic effects in fish larvae. Environ Int 134:105047

Parker B, Andreou D, Green ID, Britton JR (2021) Microplastics in freshwater fishes: occurrence, impacts and future perspectives. Fish Fish 22(3):467–488

Pinto da Costa J, Reis V, Paço A, Costa M, Duarte AC, Rocha-Santos T (2019) Micro(nano)-plastics–analytical challenges towards risk evaluation. Trends Anal Chem 111:173–184. https://doi.org/10.1016/j.trac.2018.12.013

Pitt JA, Trevisan R, Massarsky A, Kozal JS, Levin ED, Di Giulio RT (2018) Maternal transfer of nanoplastics to offspring in zebrafish (*Danio rerio*): a case study with nanopolystyrene. Sci Total Environ 643(1):324–334. https://doi.org/10.1016/j.scitotenv.2018.06.186

Plastics Europe (2010) Plastics – the facts 2010

Plastics Europe (2015) Plastics-the facts 2014/2015, an analysis of European plastics production, demand and waste data. Plastics Europe, Brussels, Belgium

Qiang L, Cheng J (2019) Exposure to microplastics decreases swimming competence in larval zebrafish (*Danio rerio*). Ecotoxicol Environ Saf 176:226–233

Qiang L, Cheng J (2021) Exposure to polystyrene microplastics impairs gonads of zebrafish (*Danio rerio*). Chemosphere 263:128161

Rainieri S, Conlledo N, Larsen BK, Granby K, Barranco A (2018) Combined effects of microplastics and chemical contaminants on the organ toxicity of zebrafish (*Danio rerio*). Environ Res 162:135–143

Rather MA, Basha SH, Bhat IA, Sharma N, Nandanpawar P, Badhe M et al (2017) Characterization, molecular docking, dynamics simulation and metadynamics of kisspeptin receptor with kisspeptin. Int J Biol Macromol 101:241–253

Rios Mendoza LM, Karapanagioti H, Álvarez NR (2018) Micro(nanoplastics) in the marine environment: current knowledge and gaps. Curr Opin Environ Sci Health 1:47–51. https://doi.org/10.1016/J.COESH.2017.11.004

Rochman C, Kurobe T, Flores I, Teh SJ (2014) Early warning signs of endocrine disruption in adult fish from the ingestion of polyethylene with and without sorbed chemical pollutants from the marine environment. Sci Total Environ 493:656–661

Rochman CM, Anna-Marie Cook AM, Koelmans AA (2016) Plastic debris and policy: using current scientific understanding to invoke positive change. Environ Toxicol Chem 35:1617–1626

Romano N, Ashikin M, Teh JC, Syukri F, Karami A (2018) Effects of pristine polyvinyl chloride fragments on whole body histology and protease activity in silver barb *Barbodes gonionotus* fry. Environ Pollut 237:1106–1111

Schwarzenbach RP, Escher BI, Fenner K, Hofstetter TB, Johnson CA, von Gunten U, Wehrli B (2006) The challenge of micropollutants in aquatic systems. Science 313(5790):1072–1077

Setälä O, Fleming-Lehtinen V, Lehtiniemi M (2014) Ingestion and transfer of microplastics in the planktonic food web. Environ Pollut 185:77–83

Strungaru SA, Jijie R, Nicoara M, Plavan G, Faggio C (2019) Micro-(nano) plastics in freshwater ecosystems: abundance, toxicological impact and quantification methodology. Trends Anal Chem 110:116–128

Sundaray JK, Rather MA, Kumar S, Agarwal D (2021) Recent updates in molecular endocrinology and reproductive physiology of fish. Springer, Singapore

Sutherland WJ, Broad S, Caine J, Clout M, Dicks LV, Doran H, Entwistle AC, Fleishman E, Gibbons DW, Keim B, LeAnstey B, Lickorish FA, Markillie P, Monk KA, Mortimer D, Ockendon N, Pearce-Higgins JW, Peck LS, Pretty J, Rockström J, Spalding MD, Tonneijck FH, Wintle BC, Wright KE (2016) A horizon scan of global conservation issues for 2016. Trends Ecol Evol 31(1):44–53. https://doi.org/10.1016/j.tree.2015.11.007

Teuten EL, Saquing JM, Knappe DR, Barlaz MA, Jonsson S, Björn A, Rowland SJ, Thompson RC, Galloway TS, Yamashita R, Ochi D, Watanuki Y, Moore C, Viet PH, Tana TS, Prudente M, Boonyatumanond R, Zakaria MP, Akkhavong K, Ogata Y, Hirai H, Iwasa S, Mizukawa K, Hagino Y, Imamura A, Saha M, Takada H (2009) Transport and release of chemicals from plastics to the environment and to wildlife. Philos Trans R Soc Lond Ser B Biol Sci 364(1526): 2027–2045. https://doi.org/10.1098/rstb.2008.0284

Thompson RC, Moore CJ, vom Saal FS, Swan SH (2009) Plastics, the environment and human health: current consensus and future trends. Philos Trans R Soc Lond B Biol Sci 364:2153–2166

Wagner J, Wang ZM, Ghosal S, Rochman C, Gassel M, Wall S (2017) Novel method for the extraction and identification of microplastics in ocean trawl and fish gut matrices. Anal Methods 9:1479–1490

Wan Z, Wang C, Zhou J, Shen M, Wang X, Fu Z, Jin Y (2019) Effects of polystyrene microplastics on the composition of the microbiome and metabolism in larval zebrafish. Chemosphere 217: 646–658

Wang J, Li Y, Lu L, Zheng M, Zhang X, Tian H, Wang W, Ru S (2019) Polystyrene microplastics cause tissue damages, sex-specific reproductive disruption and transgenerational effects in marine medaka (*Oryzias melastigma*). Environ Pollut 254:113024. https://doi.org/10.1016/j.envpol.2019.113024

Watts AJR, Urbina MA, Goodhead R, Moger J, Lewis C, Galloway TS (2016) Effect of microplastic on the gills of the shore crab *Carcinus maenas*. Environ Sci Technol 50:5364–5369

Wright SL, Thompson RC, Galloway TS (2013) The physical impacts of microplastics on marine organisms: a review. Environ Pollut 178:483–492

Xia X, Sun M, Zhou M, Chang Z, Li L (2020) Polyvinyl chloride microplastics induce growth inhibition and oxidative stress in *Cyprinus carpio* var. larvae. Sci Total Environ 716:136479

Yin L, Chen B, Xia B, Shi X, Qu K (2018) Polystyrene microplastics alter the behavior, energy reserve and nutritional composition of marine jacopever (*Sebastes schlegelii*). J Hazard Mater 15(360):97–105

Zhao Y, Bao Z, Wan Z, Fu Z, Jin Y (2020) Polystyrene microplastic exposure disturbs hepatic glycolipid metabolism at the physiological, biochemical, and transcriptomic levels in adult zebrafish. Sci Total Environ 710:136279

Zhu ZL, Wang SC, Zhao FF, Wang SG, Liu FF, Liu GZ (2019) Joint toxicity of microplastics with triclosan to marine microalgae *Skeletonema costatum*. Environ Pollut 246:509–517. https://doi.org/10.1016/j.envpol.2018.12.044

Zhu M, Chernick M, Rittschof D, Hinton DE (2020) Chronic dietary exposure to polystyrene microplastics in maturing Japanese medaka (*Oryzias latipes*). Aquat Toxicol 220:105396

Zitouni N, Bousserrhine N, Belbekhouche S, Missawi O, Alphonse V, Boughatass I, Banni M (2020) First report on the presence of small microplastics ($\leq$ 3 μm) in tissue of the commercial fish *Serranus scriba* (Linnaeus, 1758) from Tunisian coasts and associated cellular alterations. Environ Pollut 263:114576

The Impact of Xenobiotics in Development and Reproduction of Freshwater Fishes

Partigya Sharma, Shriya Purohit, Sachin Kothiyal, and Indrashis Bhattacharya

1 Introduction

Xenobiotics are exogenous compounds not produced naturally within living bodies and are considered as detrimental particles to the ecosystem (Ravindra and Haq 2019). These chemicals are generated from pharmaceutical industries and get released as wastewater and solid residues. Paper/pulp mills and textile industries or agricultural products, e.g., pesticides and herbicides, discharge such waste products which are also potential sources of xenobiotics. Furthermore, chemicals, including heavy metals used and emitted by coal refineries, phenol manufacturing companies, nuclear power plant, etc., get released to the aquatic environments as industrial effluents (Ravindra and Haq 2019; Yousuf et al. 2012). All such xenobiotics are harmful to fish health and show adverse impact on freshwater ecosystems. For example, water contaminated with insecticide fipronil decreases the dissolved oxygen level and induces hypoxia in fishes resulting in elevated ammonia secretion in the aquatic environment (Dhamgaye et al. 2020). Organophosphate compounds are broadly utilized as pesticides in the aquaculture and/or agriculture industries (Kunwar et al. 2021; Dar et al. 2020). Chlorpyrifos, which belongs to the organophosphate group, is a major popular pesticide found in aquatic systems, and its deposits have been observed in both captured (wild) and cultured fishes (Kunwar et al. 2021). One of the most toxic and profoundly used pesticides in developing countries is the organochlorine which has an abundant capability of accumulation in biomembranes and subsequently gets biomagnified across the food chains. Another class of pesticides includes triazine, which predominantly consists of herbicides and

P. Sharma · S. Purohit · S. Kothiyal
Department of Zoology, HNB Garhwal University, Pauri Garhwal, Uttarakhand, India

I. Bhattacharya (✉)
Department of Zoology, HNB Garhwal University, Pauri Garhwal, Uttarakhand, India

Department of Zoology, The Central University of Kerala, Kasaragod, Kerala, India

M. A. Rather et al. (eds.), *Xenobiotics in Aquatic Animals*,
https://doi.org/10.1007/978-981-99-1214-8_7

selectively inhibits electron transport during photophosphorylation and thus contributes as a critical environmental concern regarding nontarget photosynthetic organisms (Gonçalves et al. 2021). Apart from the critical role in respiration (absorbing oxygen), gills also perform essential physiological functions like regulating acid-base equilibrium, exchange of ions, and removal of harmful substances from the body. Upon exposure to insecticides, the initial target tissues are fish gills (Akhtar et al. 2021a). Metallic exposure to aquatic organisms can cause long-term irreversible effects (Yousuf et al. 2012). Other organs, viz., the intestine, kidney, liver, and muscles, further get affected with the exposure of such xenobiotics (Akhtar et al. 2021a).

Xeno-estrogens have the ability to mimic the natural estrogens (E_2) (Arukwe and Goksøyr 1998). These mediate their effects via binding to estrogen receptor (ER) with high affinity and generate E_2 like responses in fishes such as the induction of vitellogenic and zona-radiata proteins, etc. Estrogenic xenobiotics have been shown to sex reversal in medaka (*Oryzias latipes*) and roach (*Rutilus rutilus*) (Arukwe and Goksøyr 1998).

As per food and agriculture survey of the United Nations 2022, the total aquatic production in 2020 was around 178 million tons globally, of which capture and culture fisheries constitute around 51% and 49%, respectively (FAO 2022). Of this total yield, 63% (112 million tons) was harvested in marine waters (70% from capture and 30% from cultured fisheries), and the rest 37% (66 million tons) was produced in inland waters (83% from cultured and 17% from capture fisheries) (FAO 2022). During the past two decades, the rise in uncontrolled anthropogenic activities (e.g., urbanization, industrialization, water pollution, construction of roads/hydroelectric projects, etc.) has potentially damaged the natural habitat of fishes with a dramatic change in the feeding and breeding pattern leading to an excruciating decline in the fish population (Agarwal and Singh 2009; Nautiyal et al. 2019).We briefly discuss the adverse effects of xenobiotics in freshwater fishes found in the Indo-Asian region reported during the recent past.

1.1 Types or Chemical Nature and the Applications of Xenobiotics

Wastewater and solid residues emitted through pharmaceutical and other chemical industries primarily contain phenols, hydrocarbons, and dyes, which are considered to be the primary source of xenobiotics. Other sources of xenobiotics include plastics; wastes released from different mills, viz., paper, pulp, and textiles; and agricultural products, e.g., pesticides and herbicides (Ravindra and Haq 2019). Chemicals such as petroleum hydrocarbon, phenol derivatives, heavy metals, radioactive substances, stains and paints, and oil spills are effluent discharge by coal refineries, phenol manufacturing, nuclear power plant, pharmaceuticals, etc., which reach the aquatic environments by land runoffs and as industrial discharges

(Ravindra and Haq 2019). Petroleum industries emit polycyclic aromatic hydrocarbons such as naphthalene, phenanthrene, biphenyl, phenol, anthracene, and pyrene are responsible for polluting the aquatic habitats (Ravindra and Haq 2019). Commercial fertilizers, urban wastes, sewage sludge, liming/sticky material, agrichemicals, and additional wastes used as soil amendments are the source of metals (Yousuf et al. 2012). Pesticides are classified considering the substance target like bactericides, fungicides, insecticides, herbicides, etc. or of chemical class. Most commonly and widely used chemical class pesticides include organophosphorus pesticides, carbamates, and pyrethroid. The examples of organophosphorus pesticides include chlorpyrifos, diazinon, triazophos, parathion, malathion, dichlorvos, methyl parathion, phosmet, tetrachlorvinphos, oxydemeton, and azinphos methyl. Organophosphate compounds are most extensively utilized as pesticides in agriculture and aquaculture (Gonçalves et al. 2021). Dichlorvos and chlorpyrifos are widely used as pest management products in the larger section of the developing nations. Chlorpyrifos [O,O-diethyl-O-(3,5,6-trichloro-2-pyridyl) phosphorothioate] is a synthetic, nonsystemic, wide-spectrum pesticide and is also one of the commonly used pesticides in aquatic systems, and its remnants were found in water/sediments and in the bodies of both captured (wild) and/or cultured fishes (Kunwar et al. 2021). Organochlorine pesticides (OCPs) are extremely toxic and persistent pesticides in the environment. OCPs include chlordane, hexachlorobenzene, chlordecone, lindane, mirex, aldrin, endrin, alpha- and β-hexachlorocyclohexane (HCH), dieldrin, pentachlorobenzene, heptachlor, toxaphene, and dichlorodiphenyltrichloroethane (DDT). These get accumulated in cell/organelle membranes, biomagnifying along with food chains (Gonçalves et al. 2021).

Table 1 describes the type(s) and chemical nature of potential xenobiotics predominantly found in freshwater ecosystems.

1.2 Impact on Different Freshwater Fish Species

Teleosts are supra-sensitive toward such environmental contaminants of water (Dhamgaye et al. 2020). Here, we have summarized some of the major toxicants/pollutants/contaminants to aquatic ecosystems having a direct severe impact on fish physiology.

1.2.1 Cypermethrin

Gills of the fishes when exposed to cypermethrin show shortening of lamellae, blood congestion, lamellar curling, erosion of arch, necrosis, hypertrophy of epithelial cells, degeneration of epithelial cells, shortening of secondary gill lamellae, degenerative changes, and lamellar destruction (Akhtar et al. 2021a). On the other hand, the liver is another major target organ due to its critical role in metabolism. Bioaccumulations of insecticides in hepatocytes result in significant structural

Table 1 Types/chemical nature and applications of xenobiotics

S. no.	Name of xenobiotic	Chemical nature and applications	Key references
1	Formaldehyde	Formaldehyde (HCHO), commonly known as formalin, usually commercialized as a 37% (w/w) aqueous solution of HCHO gas; it is applied as prophylactic anti-parasitic agent	Jerbi et al. (2011)
2	Oxytetracycline (OTC)	OTC is an antibiotic (usually administrated via diet or immersion) extensively utilized in aquaculture	Jerbi et al. (2011)
3	Peracetic acid (PAA)-based compounds	Peracetic acid (PAA)-based disinfectants are taken as sustainable substitutions in aquaculture because of their nontoxic remains from spontaneous decay	Liu et al. (2020)
4	Alkylphenolpolyethoxylates (APEs)	Alkylphenolpolyethoxylates (APEs) represent an important class of nonionic surfactants that are widely used as detergents, emulsifiers, and wetting and dispersing agents and also in plastic products for industrial, agricultural, and domestic use. Alkylphenolpolyethoxylates (APEs) is non-ionic surfactants used as wetting and dispersing agents, detergents, and emulsifiers and also in plastic products for industrial agricultural/or domestic purpose. The microbial degradation of APEs leads to the formation of alkylphenols (APs). APs are most critical biodegradable/nondegradable metabolites of exhibiting estrogenic effects	Zaki et al. (2014)
5	Non-phenol	Non-phenol functions by binding with the E_2 response element	Carnevali and Maradonna (2003)
6	β-Naphtho-flavone	β-Naphtho-flavone, a dioxin-like compound that exerts its toxic action through the aryl hydrocarbon receptor	Carnevali and Maradonna (2003)
7	Chlorpyrifos [O,O-diethyl-O-(3,5,6-trichloro-2-pyridyl) phosphorothioate]	A synthetic, nonsystemic, wide-spectrum pesticide. In the aquatic system, its residues were found in the water, sediments, and both wild and cultured fishes	Kunwar et al. (2021)

(continued)

Table 1 (continued)

S. no.	Name of xenobiotic	Chemical nature and applications	Key references
8	Fipronil 5-amino1-[2,6-dichloro-4-(trifluoromethyl) phenyl-4 (trifluoromethylsulfonyl) pyrazole-3-carbonitrile	The stock solution is made having strength of 1000 ppm with the addition of 20-mL fipronil (5% active component) in 1-L distilled water	Dhamgaye et al. (2020)

alterations leading to malfunctions. Exposure of the liver to different concentrations of cypermethrin results in impairment of hepatic cell membrane, compactly arranged hepatocytes, bile pigment, necrosis, infiltration of leukocyte, and pyknosis (Akhtar et al. 2021a). Kidney treated with different concentrations of cypermethrin shows vacuolation, interstitial hemorrhage, and multifocal granulomas. The intestine too is the pivotal fish organ, having principal contribution to ingestion and adjustment of food. It is extensively being utilized as a target organ for eco-toxicology because of its sensitivity toward all types of toxic substances. Under different concentrations of cypermethrin, fish intestines show destructive changes like blood congestion, necrosis, loss of structural integrity, pyknosis and shrinkage of the mucosa, and necrosis of the tip (Akhtar et al. 2021a). Exposure to cypermethrin leads to muscle necrosis and fragmentation of sarcoplasm (Akhtar et al. 2021a). Rohu (*Labeo rohita*) when introduced to different concentrations of cypermethrin results in impairment of intestines including pyknosis, fusion of villi, construction of cup cells, hemorrhage, and necrosis (Akhtar et al. 2021a). Furthermore, there are dramatic changes in liver tissues with cypermethrin treatment such as damage in cell membrane, congestion, pyknosis, necrosis, hyperplasia, and vacuolization in rohu (Akhtar et al. 2021a).

1.2.2 Chlorpyrifos

Nile tilapia (*Oreochromis niloticus*) when exposed to chlorpyrifos shows abnormality and merging of secondary lamellae, severe epithelial hyperplasia, lamellar necrosis, degradation in respiratory epithelial cells, and accumulation of mucus in gill tissue cells (Kunwar et al. 2021). Notably, several studies have suggested chlorpyrifos mediated behavioral changes in tilapia (*Oreochromis mossambicus*), African catfish (*Clarias gariepinus*), and common carp (*Cyprinus carpio*) (Kunwar et al. 2021). Chlorpyrifos also has been reported to disturb the hematological and biochemical parameters of common carp, African catfish, Asian catfish, and mrigal carp (*Cirrhinus mrigala*) (Kunwar et al. 2021).

1.2.3 Dichlorvos

Dichlorvos (2,2-dichlorovinyl dimethyl phosphate) is another widely exploited organophosphate pesticide (Kunwar et al. 2021). Harmful results of dichlorvos on behavioral and hematological changes in rohu, behavioral changes of guppy (*Poecilia reticulata*), and changes in energy metabolism of zebrafish (*Danio rerio*) were also documented. Similarly, recent evidence indicates that the treatment with dichlorvos can change oxygen consumption in grass carp (*Ctenopharyngodon idella*) and instigate/induce histopathological changes in mrigal and rohu (Kunwar et al. 2021). Several studies performed on common carp described the harmful effect of dichlorvos on behavior, ammonia excretion rate, metabolism, food consumption, and immune response (Kunwar et al. 2021).

1.2.4 Fipronil

Oxygen consumption gets reduced progressively, and the rate of ammonia secretion is also found to be elevated on fipronil exposure (Dhamgaye et al. 2020). Fipronil has been found as a highly toxic compound for rainbow trout (*Oncorhynchus mykiss)*, common carp, sheephead minnow (*Cyprinodon variegatus*), bluegill sunfish (*Lepomis macrochirus*), and Nile tilapia (*Oreochromis niloticus*).

1.2.5 Atrazine

Adult or juvenile common carp, Nile tilapia, and Golden Nile catfish (*Chrysichthys auratus*) exposed to atrazine shows significant reduction in plasma parameters, viz., erythrocyte count, serum glucose, and total protein levels after exposure to atrazine (Akhtar et al. 2021b). Atrazine-induced damage of gill lamellae leading to decreased respiratory volume in *Tilapia mossambica* has been reported. Appearance of micronuclei and nuclear aberrations has been reported in Nile tilapia after treatment with various doses of atrazine. Common carp introduced to different concentrations of atrazine results into atrophy in gills with extreme lamellar telangiectasia, dilated lamellar capillaries filled with erythrocytes, rupture of dilated lamellar capillary, and pooling of the blood with formation of thrombi (Akhtar et al. 2021b). Common carp treated with atrazine showed abrasions in the liver including hepatocyte vacuolar degeneration (Akhtar et al. 2021b). Neotropical fish curimbata/curimba (*Prochilodus lineatus*) exposed to different concentrations of atrazine shows increased DNA breakdown/degradation.

1.2.6 Others

Common carp treated to quinalphos results in structural damage, mild degenerative changes in neural cells, necrotic changes in neural cells and intracellular edema, increased necrotic condition of neural cells and cytoplasmic vacuolization observed, slight degenerative changes, and vacuolization (Akhtar et al. 2021a). DNA damage is noted due to significant oxidative stress, mediated by xenobiotic metabolites or xenobiotics. Common carp exposed to fipronil and buprofezin collectively exhibit

Table 2 Xenobiotics and impact on freshwater fishes

S. no.	Xenobiotics	Potential ill-impact	Key references
1.	Fipronil	Oxygen consumption rate of fish reduces progressively, and ammonia secretion gets elevated	Dhamgaye et al. (2020)
2.	Atrazine	It is not biodegradable by microorganisms in aquatic ecosystem	Akhtar et al. (2021b)
3	Cypermethrin	The gills, kidney, liver, muscle, and intestine get affected	Akhtar et al. (2021a)
4	Chlorpyrifos	Behavioral changes in common carp (*Cyprinus carpio*), African catfish (*Clarias gariepinus*), and tilapia (*Oreochromis mossambicus*)	Kunwar et al. (2021)
5	Dichlorvos (2,2-dichlorovinyl dimethyl phosphate)	Behavior and hematological changes in rohu (*Labeo rohita*), behavioral changes in guppy (*Poecilia reticulate*), and changes in energy metabolism of zebrafish (*Danio rerio*)	Kunwar et al. (2021)

increased genotoxicity in whole blood DNA content, and exposure to oxadiazon shows DNA damage (Akhtar et al. 2021a).

Table 2 describes the impact of some of the xenobiotics on freshwater fishes.

1.3 Impact in Cold Water Hill Stream Fishes

The term cold water refers to the high-altitude aquatic ecosystem which maintains high transparency, typical low thermal (5 °C–20 °C) conditions with high dissolved oxygen levels and generally known for the natural habitat of trouts, minnows, mahseers, etc. (Dhamgaye et al. 2020). In the recent past, there have been multiple reports coming regarding the adverse impact of environmental contaminants/xenobiotics, etc. in cold water fisheries. For example, when the juveniles of golden mahseer (*Tor putitora*) have been treated with sublethal doses of dichlorvos and chlorpyrifos either alone or in combination, the blood glucose got elevated in dichlorvos-treated fishes, alanine aminotransferase and alkaline phosphatase were elevated in chlorpyrifos and dichlorvos treated fishes, and finally the aspartate aminotransferase and urea were elevated in combined pesticide treatment. Blood albumin and triglycerides were found to be reduced in the combined group (Kunwar et al. 2022).With increasing concentrations of fipronil and exposure period, the oxygen consumption rate has been shown to get reduced in mahseer (*Tor* sp.) fries (Dhamgaye et al. 2020). Chlorpyrifos has been classified as extremely toxic and dichlorvos as a moderately toxic pesticide for golden mahseer (Kunwar et al. 2021). Kupffer cell hyperplasia has been noticed in the liver of snow trout (*Schizothorax niger)* (Yousuf et al. 2012). Exposure of atrazine to snow trout (*Schizothorax plagiostomus*) caused major changes in all hematological parameters, e.g., remarkable decline in hemoglobin, leukocytes, erythrocytes, monocytes, and lymphocytes was detected in treated fishes (Akhtar et al. 2021b). Snow trout (*Schizothorax esocinus*) treated with different concentrations of cypermethrin resulted in DNA damage of peripheral blood erythrocytes (Akhtar et al. 2021a).

1.4 Impact on Fish Reproduction

The fish reproduction, including gamete maturation, spawning, and fertilization, can easily get disturbed and impaired by small concentrations of xenobiotic contaminations. Xenobiotics either operate directly, if having hormone live activity (like estrogenic response), or indirectly via altering the gonadal hormonal milieu critical for gametogenic development (Arukwe and Goksøyr 1998; Sundaray et al. 2021). Zinc is a crucial element for the development of functional gametes, and its natural balance is sustained by metallothioneins (Kime 1999). Heavy metals interfering metallothionein synthesis might disrupt development of gamete through impaired zinc homeostasis.

Xeno-estrogens are categorized as endocrine-disrupting compounds (EDCs) having the ability to mimic the natural E_2. These includes synthetic steroids such as those used in contraceptive pills; surfactants; detergents; organochlorine pesticides like DDT, hexachlorocyclohexanes, etc.; plasticizers; polychlorinated biphenyls (PCB); and other natural chemicals, e.g., phytoestrogens, myco-estrogens, etc. (Arukwe and Goksøyr 1998). EDCs mediate their effects by binding with high affinity to the ER showing estrogenic responses in fishes like the induction of synthesis of zona radiata and vitellogenin (Vtg) proteins (Arukwe and Goksøyr 1998). In teleosts, hatching eggs or developing larvae exposed to such environmental endocrine disruptors shows sex reversal. Estrogenic xenobiotics like organochlorine γ-BHC and the detergent derivatives of nonylphenol can also produce intersex in medaka and roach (Kime 1999). The recent concerns about a possible decline in sperm counts and increased incidence of males exhibiting female characteristics have been reported with contamination of xeno-estrogens, viz., organochlorine pesticides, such as DDT, PCBs, and breakdown products of the alkylphenoxy detergents as E_2 of human and animal origin (Agarwal and Singh 2009). In medaka, there is development of testis-ova in males and induction of vitellogenesis in either sex, when exposed to β-hexachlorocyclohexane and 4-nonylphenol indicating the estrogenic effects of this compound (Arukwe and Goksøyr 1998). In Atlantic salmon (*Salmo salar*), multiple doses of 4-nonylphenol substantially elevated the plasma levels of zona radiata and Vtg proteins (Arukwe and Goksøyr 1998). Several in vivo studies have also demonstrated Vtg induction by xenobiotic E_2, like alkyl phenolic chemicals, DDT, and 1,1-dichloro-2,2-bis(p-chlorophenyl) ethylene (DDE) in rainbow trout (Arukwe and Goksøyr 1998). Polycyclic aromatic hydrocarbons (PAHs) and halogenated aromatic hydrocarbons are inducers of cytochrome P4501A. They have been shown to modulate E_2-induced Vtg synthesis in rainbow trout (Arukwe

Table 3 Xenobiotics on fish reproduction

S. no.	Name of xenobiotics	Effect on fish reproduction	Key references
1	Xeno-estrogens	Xeno-estrogen imitates the natural hormone estrogen	Zaki et al. (2014)
2	Polychlorinated biphenyls (PCB)	In *Platichthys flesus* and *Clupea harengus*, there is decrease in viable hatching of eggs	Zaki et al. (2014)
3	Dichlorodiphenyltrichloroethane (DDT)	High mortality of eggs	Zaki et al. (2014)
4	β-Hexachlorocyclohexanes and nonylphenol	In medaka and roach, shows sexual alternations	Kime (1999)
5	Alkyl phenolic chemicals and 1,1-dichloro-2,2-bis (p-chlorophenyl) ethylene (DDE)	Vitellogenin induction by xenobiotic estrogen in rainbow trout (*Oncorhynchus mykiss*)	Arukwe and Goksøyr (1998)
6	Polycyclic aromatic hydrocarbons (PAHs) and halogenated aromatic hydrocarbons	Cytochrome P4501a gets induced, in rainbow trout; these compounds regulate vitellogenic synthesis	Arukwe and Goksøyr (1998)

and Goksøyr 1998). By using Vtg as a marker of estrogenic potency, it has been shown that the following five substances/products viz., 1-nonyl-4-phenol, 4-octylphenol (OP), DDT, Aroclor 1221, and bisphenol A, collectively induced Vtg synthesis in cultured hepatocytes in the same fish species (Arukwe and Goksøyr 1998). Since xeno-estrogens are known to stimulate the synthesis of yolk protein Vtg, detection of such protein in male or immature fish provides a rapid and easy way for evaluating the presence of environmental pollutants having estrogenic activity in aquatic ecosystems (Kime 1999; Zafar et al. 2021).

Table 3 summarizes the adverse role of xenobiotics on fish reproduction.

1.5 Concluding Remarks

In summary, xenobiotics are foreign exogenous compounds originating from pharmaceutical industries, paper/pulp/textile mills, or agricultural products like pesticides and other anthropogenic wastes. An inappropriate waste management and uncontrolled release of such xenobiotics into the aquatic ecosystem result into impaired ecological balance and poor fish health. Bioaccumulations of such toxicants/pollutants/contaminants in the freshwater bodies affect fish organs like the liver, gills, kidney, brain, muscle, etc. EDCs like xeno-estrogens act via ER leading to dysregulation in metabolism and development.

New environmental policies should be given immediate attention on an urgent basis by government and nongovernment organizations (NGOs) in order to regulate the waste management for protecting the freshwater ecosystem and fishery industries for long-term benefit. The use of plastic and other biodegradable/nondegradable compounds should be prohibited/banned, and eco-friendly natural compounds (jute/bamboo/other organic materials) should be promoted. Finally, sincere attention should be given toward mass awareness as it will be the major driving force for the successful and adequate implementation of such policies for protecting the aquatic ecosystem.

Funding and Acknowledgments This work was fully funded by the Department of Science and Technology (DST-SERB, file number ECR/2018/000868), Government of India, to IB. The authors are grateful to the Honorable Vice Chancellor, HNB Garhwal University, Uttarakhand (HNBGU), and past head(s), Department of Zoology, HNBGU, for the valuable support. IB is also thankful to the Honorable Vice Chancellor and the Department of Zoology, Central University of Kerala, Periya Campus, Kasaragod, Kerala, for the necessary administrative and infrastructural support.

References

Agarwal NK, Singh H (2009) Snow trout fishery in Uttaranchal: causes of depletion & strategy for propagation. Environ Conserv J 10(1 & 2):135–140

Akhtar N, Khan MF, Tabassum S, Ahmad MS, Badshah KD (2021a) Effects of cypermethrin on the hematological parameters, biochemical components of blood and histopathological changes in different organs of chirruh snow trout (Schizothoraxesocinus). Pak J Zool 53(3):943–953

Akhtar N, Khan MF, Tabassum S, Zahran E (2021b) Adverse effects of atrazine on blood parameters, biochemical profile and genotoxicity of snow trout (Schizothoraxplagiostomus). Saudi J Biol Sci 28(3):1999–2003

Arukwe A, Goksøyr A (1998) Xenobiotics, xenoestrogens and reproduction disturbances in fish. Sarsia 83(3):225–241

Carnevali O, Maradonna F (2003) Exposure to xenobiotic compounds: looking for new biomarkers. Gen Comp Endocrinol 131(3):203–208

Dar SA, Chatterjee A, Rather MA, Chetia D, Srivastava PP, Gupta S (2020) Identification, functional characterization and expression profiling of cytochrome p450 1A (CYP1A) gene in Labeorohita against emamectin benzoate. Int J Biol Macromol 158:1268–1278

Dhamgaye H, Jadhao S, Meshram S, Chavan B, Sadawarte R (2020) Effect of fipronil on oxygen consumption and ammonia excretion of mahseer (tor KHUDREE) fry. Indian J Anim Health 59(1):91–96

FAO (2022) The state of world fisheries and aquaculture 2022. Towards Blue transformation. FAO, Rome. https://doi.org/10.4060/cc0461en

Gonçalves AM, Rocha CP, Marques JC, Gonçalves FJ (2021) Fatty acids as suitable biomarkers to assess pesticide impacts in freshwater biological scales–a review. Ecol Indic 122:107299

Jerbi MA, Ouanes Z, Besbes R, Achour L, Kacem A (2011) Single and combined genotoxic and cytotoxic effects of two xenobiotics widely used in intensive aquaculture. Mutat Res Genet Toxicol Environ Mutagen 724(1-2):22–27

Kime DE (1999) A strategy for assessing the effects of xenobiotics on fish reproduction. Sci Total Environ 225(1-2):3–11

Kunwar PS, Parajuli K, Badu S, Sapkota B, Sinha AK, De Boeck G, Sapkota K (2021) Mixed toxicity of chlorpyrifos and dichlorvos show antagonistic effects in the endangered fish species golden mahseer (Tor putitora). Comp Biochem Physiol Pt C Toxicol Pharmacol 240:108923

Kunwar PS, Sinha AK, De Boeck G, Sapkota K (2022) Modulations of blood biochemical parameters of golden mahseer, tor putitora following exposures to single and mixed organophosphate. Comp Biochem Physiol Pt C Toxicol Pharmacol 251:109207

Liu D, Lazado CC, Pedersen LF, Straus DL, Meinelt T (2020) Antioxidative, histological and immunological responses of rainbow trout after periodic and continuous exposures to a peracetic acid-based disinfectant. Aquaculture 520:734956

Nautiyal P, Kumar S, Sharma N, Chauhan K, Kala K (2019) Coldwater capture fishery in the ganga headwaters having operational hydropower projects: a failing co-existence. J Inland Fisheries Soc India 51:06–23

Ravindra PA, Haq SA (2019) Effects of xenobiotics and their biodegradation in marine life. In: Bhatt P (ed) Smart bioremediation technologies: microbial enzymes. Academic Press, New York

Sundaray JK, Rather MA, Kumar S, Agarwal D (2021) Recent updates in molecular endocrinology and reproductive physiology of fish. Springer, Singapore

Yousuf R, Mir SH, Tanveer S, Darzi MM, Mir MS (2012) Metals and histopathological alterations in the liver of Schizothoraxniger, Heckel from the Dal Lake of Kashmir Valley. J Environ Sci Nat Resour 5(2):231–237

Zafar I, Saba M, Raza MA, Rather MA, Rayan RA, Fatima M, Ishneiwra RM (2021) Roles of bioinformatics in reproductive science. In: Recent updates in molecular endocrinology and reproductive physiology of fish. Springer, Singapore, pp 203–228

Zaki MS, Hammam AM, Nagwa S (2014) Ata. Xenobiotics as stressful factors in aquatic system (in fish). Life Sci J 11(4):188–197

Contaminant Mixtures and Reproduction in Aquatic Organisms

Melissa K. Driessnack

1 Introduction

Before large-scale migration, increasing human population, and industrialization, many of the world's ecosystems were what we would consider pristine. However, over the last 100 years, we have polluted most of our freshwater and marine systems with pesticides, urban waste, industrial effluents, automotive chemicals, and more (Covert et al. 2020; de Zwart et al. 2018; Müller et al. 2020; Sumpter 2009), highlighting that one of the most adverse consequences of industrial practices is the release of contaminants into the environment and surrounding watersheds (Cirillo et al. 2012; Knapen et al. 2004). Whether these exposures are the result of regulated effluent discharges, drainage from mining practices, spill events, or runoff from urban and agricultural areas, the implications of both intentional and unintentional release in aquatic environments can severely impact the organisms inhabiting those waters (Cirillo et al. 2012; Knapen et al. 2004; Meyers et al. 2005; Weir et al. 2016).

To highlight the challenges facing our aquatic systems, imagine a flowing river. At the source, it may be fed by annual glacial melt or an underwater spring. That water will begin to flow across a complex landscape on its journey to the mouth. During the river's journey, it will likely flow by agricultural fields, receiving inputs of herbicides, fungicides, and insecticides along with fertilizers and animal waste (Burkholder et al. 2007; Covert et al. 2020). When this hypothetical river flows past our urban centers, it will be inundated by our municipal sewage systems, which contain pharmaceuticals, personal care products, household cleaners, and industrial effluents (Arnold et al. 2014; de Zwart et al. 2018; Wilkinson et al. 2022). Each time it rains, our roads serve as chemical expressways for vehicle and tire debris, oil,

M. K. Driessnack (✉)
Puyallup Research and Extension Center, Washington State University, Puyallup, WA, USA
e-mail: melissa.driessnack@wsu.edu

M. A. Rather et al. (eds.), *Xenobiotics in Aquatic Animals*,
https://doi.org/10.1007/978-981-99-1214-8_8

gasoline, and metals (McIntyre et al. 2021; Müller et al. 2020). All of these "ingredients" added to this imaginary river result in a complex mixture of pollutants. However, this river is not imaginary—it is the reality of our waterways.

Advancing societies tend to exploit more natural resources to meet their demands for expansion and innovation. For example, humans learned agriculture and mining to meet their requirements; however, in this process, the environment was damaged. To curb the wanton destruction of nature through the release of persistent complex mixtures into ecosystems, effluent discharge is often regulated and held to standards of release in places like the European Union, Canada, and the United States with many other countries also employing such practices (Environment Canada 2000; US EPA 2000). Although there is value in this approach, it is unrealistic because organisms are exposed to a "pollutant soup," not singular chemicals. Furthermore, chemicals can interact with each other and alter each other's toxic effects. This interactive effect can also be applied to environmental stressors (e.g., temperature and hypoxia) (Hooper et al. 2013). Chemicals in mixtures are dynamic and interact not only with each other in water but also with the sediments of the water body and also with the living organisms inhabiting the water (de Zwart and Posthuma 2005; Walker et al. 2005). Furthermore, there is doubt in the applicability of acute single chemical and single organism laboratory exposure data to effluents that are chronic, low-dose mixtures in real-world receiving environments, where the endpoint of lethality is often not always the most relevant (de Zwart et al. 2018; Kamunde and MacPhail 2011; Meyers et al. 2005).

Though single chemical tests have provided valuable information regarding the mechanistic underpinnings of the detrimental effects of toxic chemicals and cause-and-effect relationships, these simplified tests do not appropriately reflect actual low-dose and long-term mixture exposure conditions in the environment (Gauthier et al. 2006; Weir et al. 2016). The potential for mixtures to have long-term interactive effects on organisms, especially their reproductive potential, is recognized and may ultimately result in additive, synergistic, or antagonistic toxicity on the health, physiology, and reproductive success of exposed organisms (Moreau et al. 1999; Norwood et al. 2003; Weir et al. 2016).

Since the aquatic contaminants often coexist in the environment at low concentrations, toxicity tests of mixtures at chronic low doses is more applicable to real-world scenarios. From the literature on contaminant mixture tests, it becomes evident the need to investigate the effects of complex mixtures on biological endpoints in fish with a greater emphasis on chronic endpoints, such as reproduction—which is critical for long-term population stability. Increased consideration should be placed on understanding how chemicals in mixtures can modulate the uptake, distribution, and metabolism of others in a mixture. Even as societal concern for healthy environments rises, industrial and agricultural production continues to expand as does the risk of impacting our freshwater systems (Dethloff et al. 1999; de Zwart and Posthuma 2005; Connon et al. 2012).

2 Ecotoxicology Approaches to Mixtures

Contaminants in mixtures, whether natural, synthetic, or a combination thereof, possess the potential to interact in organisms (Connon et al. 2012; Hooper et al. 2013). However, at present, mixtures remain a challenge both in assessing for toxic impacts and in regulating from the perspectives of human, environmental, and ecological health (Mumtaz 2010). The number of studies assessing the impacts of mixtures is growing using primarily aquatic crustaceans (e.g., daphnia), invertebrates (e.g., chironomids), and to a lesser extent fishes (e.g., fathead minnows, zebrafish). However, most of these studies are conducted under acute exposure regimes and focused on how different combinations and dose ratios affect lethality. A very limited number of studies have assessed the effects of low-dose chemical mixtures on chronic endpoints, including reproductive performance and adverse effects such as induction or inhibition of endpoints (e.g., plasma estradiol levels).

While there is no doubt that chemicals in combination can experience interactions when the presence of one modulates the toxicity of another, the consensus, especially in risk assessment, is to assign safety factors or assume additivity (Jonker et al. 2005; Newman 2010). To understand additivity, think of this simplified hypothetical effluent that is found to contain Chemicals A, B, and C, at concentrations that are equal to its single chemical LC10 (lethal concentration to 10% of the population) for a local fish species; assuming additivity we should expect to see 30% lethality in that population based on the following: 10% lethality (Chemical A) + 10% (Chemical B) + 10% (Chemical C). A primary challenge regarding mixtures arises if we observed 10% lethality or 70% lethality in this simplified scenario; those responses do not match the predicted 30% and does not match the assumption of additivity. Therefore, applying the concept of simple additivity to all mixtures is likely to miss interactions that are antagonistic, synergistic, and even potentiation in some cases (Walker et al. 2005).

Before assessing the models available for investigating mixtures, an understanding of the four primary potential interactions is required—antagonism, addition (described above), synergism, and potentiation (Walker et al. 2005; Newman 2010). Antagonism occurs when the observed effect of a mixture is lower than what is predicted by additivity. Thinking about the above example again, an observed lethality of 10% would suggest that the presence of one chemical modulated the toxicity of another resulting in a lower than predicted lethality of 30% (Newman 2010; Landis et al. 2010). The reverse would be synergism, where the observed lethality or toxicity is much greater than what is predicted, for example, a 70% observed lethality, whereby the presence of one chemical seems to increase the toxicity of the mixture based on assumed additivity (Newman 2010).

The fourth potential interaction is that of potentiation. This can occur when a chemical is not known to be toxic or lethal at the exposure concentration or dose, but its presence enhances the toxicity of another chemical in the mixture. An environmentally relevant example of potentiation involves the use of pesticides, where the compound piperonyl butoxide is added to insecticide formulations to inhibit the breakdown of the active chemical by detoxifying pathways, allowing for less pesticide to be applied to the environment while still achieving pest management goals (Newman 2010). It should also be noted that some authors choose to use the terms potentiation, synergism, and greater than additive interchangeably, while others define them separately (Walker et al. 2005; Newman 2010; Landis et al. 2010).

With the potential interactions defined, we can begin investigating the tools we have to assess mixtures. The field of toxicology has its roots in medicine and pharmacology, and much of the principles used to assess toxicity are based on a dose-response relationship. While these valuable tools in pharmacology to assess for joint action of drugs, dosage, site of action, and physiological changes have served as the basis for mixture assessments in environmental toxicology (Connon et al. 2012; Jonker et al. 2005; Landis et al. 2010), the challenge occurs when we use a mathematical model based on individuals and binary mixtures and apply them to multispecies ecosystems with complex mixtures and environmental stressors.

The additivity approach has served as the basis for evaluating mixtures, using either a concentration-based approach or a response-based approach, as both are seeking to relate a chemical exposure to an outcome (Norwood et al. 2003; de Zwart and Posthuma 2005; de Zwart et al. 2018). Both models function by assuming that interactions do not occur in the mixture and predict an outcome based on that assumption. That predicted value can then be compared to an observed outcome, so if the two values agree, we are assuming one just added to the other, but when a deviation is either greater (synergistic) or less (antagonistic) than expected, there is support that an interaction occurred (de Zwart et al. 2018).

In a concentration-based approach, also called toxic units (TU), the model assumes a similar site of action between the chemicals, and the mixture's toxicity is assessed based on converting individual chemical concentrations to an equitoxic dose (Sprague 1970). This is achieved by using the concentration of a chemical of interest in the mixture and dividing it by the reference toxic dose (e.g., LC50 or IC50) of an individual chemical for any organism. This allows for chemicals with varying potencies to be compared, an equitoxic concentration (Norwood et al. 2003). A commonly used equation to determine a chemical's TU using the concentration (c) and the defined toxicity threshold of interest here is listed as an effect concentration though lethality can also be employed (Jonker et al. 2005; Maloney et al. 2018):

$$\mathrm{TU} = \frac{C}{\mathrm{EC}_{50}} \tag{1}$$

A similar approach is used when assessing the organic pollutants dioxins, polychlorinated biphenyls (PCBs), polychlorinated dibenzo-*p*-dioxins (PCDDs), and polychlorinated dibenzofuran (PCDFs), where toxic equivalency factors (TEF) are used to assess complex mixtures (Bhavsar et al. 2008; Di Giulio and Hinton 2008). The use of TEF and in turn TEQs (toxic equivalents) is employed for a group of 75 structurally related chemicals that are known to induce toxicity through the aryl hydrocarbon receptor (AhR) pathway. The toxicity of these chemicals (dioxins, PCBs, PCDDs, PCDFs) are compared against the most toxic—2,3,7,8-tetrachlorodibenzo-*p*-dioxin (2,3,7,8-TCDD). The individual toxicity values are determined by multiplying the concentrations of the chemicals with the chemical-specific TEF. The resulting values are then summed to calculate a TEQ for the mixture (Bhavsar et al. 2008; van den Berg et al. 1998, 2006; Walker et al. 2005). This approach has become the standard practice for monitoring exposure to humans and levels in the environment for these ubiquitous chemicals (Bhavsar et al. 2008; Di Giulio and Hinton 2008, Gandhi et al. 2019; Liem et al. 2000); while there are several advantages to this method, it also assumes additivity is the only outcome and interactions occur only at a singular receptor.

The other commonly used approach is a response or effect-based model. This model is employed when the mode of action(s) is different or dissimilar between chemicals in the mixture, which is often more representative of environmental conditions (Balistrieri and Mebane 2014; de Zwart et al. 2018; de Zwart and Posthuma 2005). The response-based approach, also called the independent action model, seeks to relate the probability of a response to a model, which differs from the relative toxicity approach of the TU approach (Jonker et al. 2005). A response-based approach was employed by Driessnack et al. (2016, 2017a, b) to assess the impacts of binary metal mixtures on fathead minnow reproductive output. The authors employed a simple additivity model to compare the predicted (via the following equation) reproductive output compared to the observed for three binary metal mixtures (cadmium and copper, copper and nickel, and cadmium and zinc) (Driessnack et al. 2016, 2017a, b). This design was appropriate as a singular fixed dose was used to assess the interactive effects of metals on an important chronic endpoint—fish fecundity. The simple additivity equation, based on a response-addition model, can yield a predicted response:

$$\text{Predicted Simple additivity} = [(x + y) - f] \times 100 \quad (2)$$

where x and y are the percentage fraction of a measured response induced by Chemical A and B (metals in this case), respectively, and f is the interaction factor (estimated by multiplying x and y). The results of those studies found that the observed impacts on fecundity in fathead minnows tended to be impacted to a greater degree than what was predicted, especially regarding the mixture of cadmium and copper, where a total cessation of reproduction was noted. The work of these authors also sought to deduce underlying mechanisms that drove the altered fecundity rates and suggested molecular sites (e.g., vitellogenesis) critical to successful reproduction where the metals could be interacting (Driessnack et al. 2016, 2017a, b). The studies by Driessnack et al. (2011, 2016, 2017a, b) all used a standardized reproductive assay for fish, and that method and others are explained further in the next section.

3 Reproduction in Mixture Toxicology

The success of any species, whether it be a fish or an aquatic invertebrate, hinges upon the ability of its individuals to successfully reproduce in changing environments and exposure to stressors (e.g., fluctuating temperatures, pollution) to maintain a viable population (Moyle and Cech 2005; Orr et al. 2020). The US EPA considers an ecosystem as appropriately protected when species that belong in that system can survive and successfully reproduce (de Zwart et al. 2018; US EPA 1992, 1998).

When considering the reproduction of fish, each species' success can be greatly influenced by its fecundity and behavior (Moyle and Cech 2005). The life histories of fishes, and other aquatic organisms, are very diverse, for example, if you compare the reproductive behavior of a Cyprinidae, like the fathead minnow (*Pimephales promelas*), to a salmonid such as the Chinook salmon (*Oncorhynchus tshawytscha*). The first species can reach sexually maturing in under a year and begin spawning broods of 50–150 eggs every 3–4 days over its lifespan (Ankley et al. 2001; Parrott 2005). When compared to the Chinook salmon, their life history involves emerging from a redd in a freshwater system, migrating through an estuary before undergoing smolt transformation and progressing through a marine phase that can be years in duration, before returning to freshwater for a singular spawning event (Quinn 2018). Understanding these different life histories becomes important when considering exposure to pollution as a population's unique rates of development and reproduction determine its ability to recover from a toxic event. For example, if the minnow and salmon species from above are reported to have similar LC_{50} values (lethal concentration to 50% of a population) for a toxicant but different reproduction strategies, then each population's ability and time to recover may vary substantially after the same toxicity events (Stark et al. 2004; Feist et al. 2017). This brief comparison highlights that aquatic organisms have a range of reproductive strategies; however, some aspects of reproduction remain consistent. First, reproduction is vital in maintaining healthy populations, and second, it is energetically costly (Moyle and Cech 2005).

When considering reproduction, Suter et al. (1987) evaluated various chronic response endpoints in several fish species (e.g., FHM, rainbow trout, white sucker) and found that fecundity is on average the most sensitive effect for chronic exposures. This study also weighed the sensitivity of fecundity against early life stage assessments (e.g., fathead minnow, zebrafish), which are often an early step in assessing chemicals of interest or mixtures (Rand 1995; US EPA 1996a). While valuable information may be elucidated from these studies, as the viability of eggs and embryos is important to a population, it is an early life stage impact and not a chronic lifetime exposure (Suter et al. 1987).

A range of endpoints have been utilized to evaluate reproduction, including the number of eggs produced per female, the number of viable eggs produced per surviving female, the total number of spawns, spawns per female, eggs per female per day, and more (Driessnack et al. 2011, 2016, 2017a; Ouellet et al. 2013).

However, analyzing each of these endpoints separately potentially minimizes the total effect that a chemical or mixture may have on spawning (Suter et al. 1987). Using instead a measure that is intended to provide insight into future cohort size, as done by fishery managers, likely has the greatest strength when considering population and community-level effects. Assessing the number of eggs per surviving female is a strong endpoint that achieves that goal. Using surviving versus spawning females accounts for the potential that a chemical or a mixture may result in a 100% decline in fecundity, a result previously mentioned in a binary mixture of cadmium and copper using a 21-day fathead minnow reproductive bioassay (Driessnack et al. 2016; Suter et al. 1987).

Including measures of reproductive performance in single and chemical mixture assessments can be viewed as having two primary benefits: First, as noted by Suter et al. (1987), fecundity is a sensitive endpoint though often undervalued. Second, stable and successful reproduction is critical in maintaining healthy populations (Moyle and Cech 2005; Suter et al. 1987). However, the use of reproduction-based endpoints comes with the challenges of natural variability in healthy populations and different life history strategies (Suter et al. 1987). Below are examples of standardized laboratory tests used to assess for potential impacts on reproduction in aquatic organisms.

3.1 Standardized Toxicology Tests That Assess Reproduction

Cladocerans *Daphnia magna*, *Ceriodaphnia dubia*, and *Daphnia pulex* have been laboratory standards for decades for establishing both acute exposure dose-response curves and chronic exposures involving reproductive output (Conners et al. 2022; Versteeg et al. 1997). These standardized species can reach maturity and produce three broods in under 21 days in most cases. This allows for a controlled laboratory-based assessment of individual chemicals and chemical mixtures to determine alterations to the reproductive output of the daphniids in chronic exposure (Conners et al. 2022; US EPA 1996b). Work by Pérez and Hoang (2017) assessed how the chronic toxicity of cadmium (Cd) to *D. magna* might be modulated in the presence of zinc (Zn) over a range of concentrations. The results noted that Zn could either be protective of cadmium toxicity or also contribute to toxicity depending on the concentration. A range of 40–120 μg/L of Zn in the presence of 1.5 μg/L Cd yielded less than additive effects (antagonistic), while Zn concentrations at 160 and 200 μg/L contributed to the observed toxicity. This trend was noted for the reproductive endpoints of the total number of neonates and neonates/adult/day endpoints (Pérez and Hoang 2017). Results such as these highlight the importance of understanding how mixtures and the concentrations of the mixture's constituents are important.

While daphnia are commonly used to represent aquatic invertebrates, the midge larvae *Chironomus dilutus* are commonly employed to represent benthic macroinvertebrates in aquatic environments (Rubach et al. 2011; Cavallaro et al. 2016). Similarly, to daphniids, standardized tests are utilized to assess the chronic toxicity of chemicals alone and in combination. Maloney et al. (2018) used *C. dilutus* to determine whether three neonicotinoids of interest elicit interactive impacts (e.g., additive) employing a toxic unit approach. While fecundity was not an endpoint of interest in this study, interactions were noted for the mixture of imidacloprid and thiamethoxam with a greater than additive (synergistic) reduction in adult emergence and a shift in the sex ratio. While these results do not directly imply a change in a population, such as altered fecundity, they do still warrant consideration as the environmental impacts of altered sex ratios may have impacts on the environment (Maloney et al. 2018).

When considering the impacts of chemicals and mixtures on fishes, the fathead minnow reproductive bioassay is frequently employed (Ankley et al. 2001; Parrott 2005). In this assay, sexually mature fathead minnows are exposed to chemicals, controlled mixtures, and even whole industrial effluents of interest for 21 days. During which time, the survival, health status, and reproductive output of the fish can be assessed. The resulting eggs and larvae can also be assessed for the prevalence of a range of endpoints in more expansive studies (Driessnack et al. 2011; Ouellet et al. 2013; Yan et al. 2016). Several recent studies have begun looking at mixtures of metals, pesticides, and pharmaceuticals, due to their increasing prevalence in aquatic systems. Additionally, the tropical fish zebrafish (*Danio rerio*) has been increasingly employed in reproductive assays (Driessnack et al. 2016, 2017a, b; Hua et al. 2016; Thrupp et al. 2018).

The impacts of five synthetic steroidal pharmaceuticals on fathead minnow reproduction were assessed by Thrupp et al. (2018). Employing a range of concentrations both alone and in combination, the impacts of the steroids on egg production were assessed and compared against predicted values using both mixture methods (concentration and response). Interestingly, this study was designed in a way to assess the prediction of both response models. The authors found two very notable results: First, the impact on reproduction went from "nothing" to "a lot from a little" response in the mixture. Second, the response-based method (i.e., independent action) better predicted the additivity responses than the concentration model. The result, by Thrupp et al. (2018), that the independent action model best predicted the response in the five chemical mixtures is different than the work by Runnalls et al. (2015). Runnalls et al. (2015) also employed the fathead minnow to assess the impact of a binary mixture of two synthetic steroids. Again, the greatest reduction in fecundity was noted in the binary steroid mixture, but the concentration addition model was better at predicting the response, despite the assumed mode of altered fecundity being different between the two chemicals.

A different study using steroid hormones in a binary mixture, but with zebrafish, also noted an additive effect on fecundity when the chemicals megestrol acetate and 17α-ethinylestradiol were combined (Hua et al. 2016). The study by Hua et al. (2016) also began delving into potential underlying mechanisms, sites of interaction, that could be contributing to altered fecundity similar to Driessnack et al. (2016, 2017a, b). Hua et al. (2016) noted that plasma levels of key hormones used for signaling development in the gonads were altered by the single chemical exposures, with the greatest reduction found in the binary exposure, though this study did not compare the predictive outcome of the two standard mixture models as the above noted studies (Runnalls et al. 2015).

3.2 Expanding Beyond Fecundity in Reproduction

While egg production has been supported as a sensitive endpoint, if we employ fecundity as the only measure of reproductive impairment, we may miss more subtle changes with significant impacts. One example is the "silent" reproductive toxicity of selenium (Se) from a coal ash plant in Belews Lake, North Carolina, USA (Lemly 1993, 2002). A combination of complex waste from nearby coal-powered plants was discharged into the lake, and over a period of years, almost all species of fish were lost from the lake. Interestingly in this example, while selenium alone did not seem to impair fecundity, it did result in the larvae developing severe abnormalities. This led not to a loss of reproducing individuals but a loss in the ability of those adults to replace themselves in the population (Lemly 1993, 1999, 2002).

Another example involves pre-spawn mortality in coho salmon in the Pacific Northwest of the United States (Chow et al. 2019; Scholz et al. 2011). While not labeled as a reproductive toxicant, urban and stormwater runoff events that overlap with coho salmon spawning in the region can result in upward of 90% of female coho failing to spawn in the region, which could result in the eventual loss of the species (Spromberg and Scholz 2011). A series of investigations into the causal chemical in the complex mixtures eventually identified a quinone ozonation product of *N*-(1,3-dimethylbutyl)-*N*′-phenyl-*p*-phenylenediamine (6PPD)-6PPD-Q (McIntyre et al. 2021; Tian et al. 2021).

I highlight these two examples because the best approach for understanding how mixtures impair reproduction will require the use of several approaches such as traditional toxicology testing methods, mechanistic models (e.g., adverse outcome pathways), population-based models (e.g., life tables), using appropriate test species, and assessing for mixture interactions appropriately (Ankley et al. 2010b; Connon et al. 2012; Stark et al. 2004, 2020). Arcand-Hoy and Benson (1998) noted that while reproductive toxicity occurs in the adult stage, it can also impact larval and juvenile stages via altered development (e.g., Se in Belews Lake). Changes in males of a species may also drive reproductive failure, though linkages are still being established (e.g., intersex organisms, induced vitellogenesis). Thus, the inclusion of critical endpoints (e.g., hormone levels) in endocrine pathways such as the hypothalamus-pituitary-adrenal/interrenal axis (HPA/HPI), the hypothalamus-pituitary-thyroid (HPT) axis, and, especially, the hypothalamus-pituitary-gonad (HPG) axis, which directly supports reproduction, is necessary (Arcand-Hoy and Benson 1998; Carr and Patiño 2011; Norris and Lopez 2011; Ottinger et al. 2002; Vandenberg et al. 2013).

The inclusion of mechanistic endpoints, in conjunction with fecundity, is supported by the adverse outcome pathway (AOP) approach. An AOP seeks to link molecular initiating events, such as sites of chemicals interacting with receptors, to adverse outcomes (e.g., vitellogenesis in males, altered fecundity). The AOP framework has been invaluable in improving our mechanistic understanding and has been increasingly deployed to better assess mixtures and predict population- and community-level effects (Ankley et al. 2010a, b; Artigas et al. 2012; Kramer et al. 2011; Villeneuve and Garcia-Reyero 2011). Utilizing more than a singular endpoint such as fecundity, even if known to be highly sensitive, is of extreme value, but as noted by Suter et al. (1987) assessing for too many overlapping endpoints or asking unclear hypotheses may dampen their strength.

4 Mechanisms of Reproduction and Mixture Toxicology

Optimal fish reproduction is achieved mainly through actions of the HPG axis, though the HPA/HPI and HPT axes have important supporting roles (Carr and Patiño 2011; Norris and Lopez 2011; Ottinger et al. 2002; Vandenberg et al. 2013). Each of these axes is subject to alteration due to chemical exposure, which in aquatic environments can occur through the water column, contaminated sediments, and dietary sources (Jamwal and Shekh 2021; Walker et al. 2005). Alterations to the axes can occur via the action of chemicals mimicking natural hormones; blocking hormone action; interfering with hormone synthesis; altering the secretion, availability, and metabolism of endogenous hormones; and ultimately altering mechanisms of hormone action. These alterations, which can be upregulated or downregulated changes, can be transient and reversible or irreversible disruptions that impair reproduction, induce epigenetic change, and potentially culminate in localized population extinction (Norris and Lopez 2011; Scholz et al. 2013; Kerdivel et al. 2013).

4.1 Hypothalamus-Pituitary-Gonad Axis in Fish

Fish detect and respond to social and environmental cues via various chemical messengers and hormones in the brain, with those signals culminating in the release of the decapeptide gonadotropin-releasing hormone (GnRH) by the hypothalamus (Norris and Lopez 2011). The released GnRH acts upon the pituitary, which is directly innervated in teleosts, stimulating the release of two main gonadotropins (GtH) – follicle-stimulating hormone (FSH) and luteinizing hormone (LH) (Arcand-Hoy and Benson 1998). The GtHs are then circulated throughout the body and modulate gametogenesis and steroidogenesis in the gonads (ovaries for females and testes in males). Sex steroids produced by the gonads are also released into the circulation, where positive or negative feedback loops are initiated by binding to the hypothalamus and pituitary, depending on the reproductive stage (Weltzien et al. 2004; Zohar et al. 2010). This simplified description of the HPG pathway highlights the already vast number of sites where exogenous chemicals can interfere with the production, release, transport, metabolism, binding, action, or elimination of endogenous hormones responsible for maintaining homeostasis, regulating developmental processes, and facilitating reproduction (Kavlock and Daston 1996; Scholz et al. 2013).

4.2 Fish Gonads

In fish, the ovaries are normally paired organs that are linked to the body cavity wall, and its form and function may differ depending on the species' spawning technique (e.g., synchronous versus asynchronous). Histologically, ovaries of all fish are generally similar, comprising of germ cells, oogonia, and oocytes at various stages (Urbatzka et al. 2012). The fish oocyte development progresses along five stages. In stage I, the FSH stimulates the germ cells to undergo primary oocyte growth wherein the follicle cells surround the primary oocyte and begins forming the granulosa and theca cell layers. Stage II, also called the cortical alveoli growth stage, is marked by the formation of the vitelline envelope and membrane-limited glycoprotein vesicles. In the stage II, the oocyte growth continues with the development of theca and granulosa layer, which are critical for steroidogenesis. The GtH stimulates uptake of cholesterol via StAR (steroidogenic acute regulatory protein) into the inner mitochondrial membrane of theca cells, where cholesterol is converted first into pregnenolone and further into testosterone (T) (Nagahama 1994). The T is then released by the theca cells to the granulosa cells, which use cytochrome P450 aromatase to convert T into estrogen (Nagahama 1994; Norris and Lopez 2011).

The estrogen, as 17β-estradiol (E2), is released by the granulosa cells into the plasma, which carries it to the liver to stimulate vitellogenin (Vtg) production. Estrogen, in the plasma, also regulates the hypothalamic-pituitary-gonadal axis via modulating hypothalamic and pituitary feedback loops. Hepatocytes eject Vtg into the plasma for incorporation into the oocytes via endocytosis, initiating stage III of oocyte development. The accumulated Vtg is then processed into yolk protein for use by developing embryos upon successful fertilization. The increased LH and decreasing FSH secretion from the pituitary stimulates further growth and complete maturation of oocyte in stage IV. Finally, the mature oocyte detaches and ovulates as an egg in stage V (Arcand-Hoy and Benson 1998; Lubzens et al. 2010; Norris and Lopez 2011).

The testes in male fish are much like females in that they are a paired organ. They tend to be a smooth, white structure that when in a spawning condition can account for up to about 12% of the fish's weight (Moyle and Cech 2005). The development of the sperm also undergoes various stages of spermatogenesis, much like with the ovaries, and also begins with germ cells being stimulated by FSH and LH to progress through different stages of maturation (Norris and Lopez 2011). In males, the dominant circulating sex steroid hormones are testosterone (T), 11-keto-testosterone (11-KT), 11β-hydroxytestosterone (OHT), and 11β-hydroxyandrostenedione (Norris and Lopez 2011).

4.3 Mixture Studies Assessing Endpoints Beyond Fecundity

Referring back to the work by Driessnack et al. (2016, 2017a, b), they utilized several of these mechanistic tools to link changes in the gonad and livers of females to altered reproduction in single and binary metal mixture exposures. To summarize those three studies, the author reported that chronic waterborne binary metal exposures significantly impaired the fecundity in fathead minnows both additively (Exposure to Cd-Zn and Cu-NI; Driessnack et al. 2017a, b) and synergistically (exposure to Cd-Cu; Driessnack et al. 2016). All three studies demonstrated that metals in mixture (e.g., Cd and Zn) impaired reproduction in the fish, to a degree greater than any of the metals assessed during their respective individual exposures. The impaired reproductive output was postulated to be the result of two primary factors. First, metal exposure was found to alter reproduction by influencing the molecular and physiological parameters linked with fish reproductive functions. One notable change that was recorded was decreased serum estradiol induced by metal mixtures. In addition, associated effects such as increased follicular atresia, altered estrogen receptor (α and β) expression, and reduced Vtg gene expression in the liver supported the assumption that altered reproduction is likely linked to disruption of estrogen-mediated functions in fathead minnows.

The second factor harkens back to an aforementioned factor of reproduction – the energetic cost (Moyle and Cech 2005). Driessnack et al. (2016, 2017a, b) noted an induction of hepatic metallothionein expression, a key protein in metal detoxification, and differences in liver metal accumulation. The authors noted that the assumed increased energetic cost of producing metallothionein in response to metal exposure and combined with metal-induced alterations to estrogen signaling diminished the reproductive capacity of fathead minnow females. These findings support that assessments of fish fecundity when combined with sub-organismal endpoints, such as plasma estradiol, hepatic Vtg and MT expressions, and ovarian histopathology, are promising biomarkers for assessing the reproductive toxicity of metal mixtures, and other more complex mixtures, in fishes and aquatic organisms.

Work by Werner et al. (2010) assessed the impact of different pulp and paper mill effluents on fathead minnow reproduction and the expression of androgen- and estrogen-linked pathways in both females and males. While limited impacts were noted in the females, the authors did note changes in male hepatic expression of androgen and estrogen receptors, Vtg, and cytochrome p4501A between the different treated and untreated effluents. Their decision to also include male endpoints is of value as reproductive research often focuses on responses in females. Such approaches have been used in both standard lab species such as daphnia and zebrafish (Hua et al. 2016; Nkoom et al. 2022) and nonstandard fish species (Paschoalini et al. 2021).

5 The Challenges the Field of Ecotoxicology from Contaminant Mixtures

As noted frequently in this chapter, successful reproduction is necessary for populations to remain viable in aquatic environments. The reproductive strategies and success rates for aquatic species, whether it be fishes or invertebrates, are variable. This presents a challenge in terms of how we determine exposure to a singular chemical, nonetheless how a complex mixture is impacting the individuals to such a degree that failure of that population is a possibility. Additionally, laboratory-based tests are not natural systems, and despite best efforts to extrapolate data or use site-specific guidelines, until we can better define how and where chemical interactions occur in aquatic organisms, the current guidelines are likely lacking in their goals to effectively protect natural systems (Connon et al. 2012; de Zwart et al. 2018; Stark et al. 2004).

One of the emerging approaches to begin linking and strengthening the mathematical models with biological approaches (e.g., daphnia chronic tests) is the use of AOPs to gather and employ mechanistic data to link molecular/cellular changes (initiating events) to higher-level endpoints such as reproduction (Ankley et al. 2010a, b; Connon et al. 2012; Villeneuve and Garcia-Reyero 2011). This is essentially a bottom-up approach improved upon by using the most recent advances in high-throughput, high-output techniques. However, the results are a single snapshot of one moment in an artificial scenario. A return to or greater attention paid to top-down approaches, as used in ecology, still very much has a place in toxicology (Connon et al. 2012. This sentiment also applies to the models we use to assess mixtures, which are currently based on mathematics and pharmacology principles and less on biology and ecology (de Zwart and Posthuma 2005).

The expression of reproductive genes and circulating sex steroid hormone levels recommended in AOPs, and increasingly requested by regulatory agencies, are sources of invaluable data. However, as researchers seek to increasingly understand the physiology of aquatic organisms, we must not minimize the bigger picture—healthy populations, communities, and ecosystems. Integrated top-down and bottom-up approaches should utilize histological assessment of gonadal tissue to link molecular change to a realized outcome. For example, does a downregulation of Vtg hepatic expression in female fish result in altered oocyte development or reduce egg size to such a degree that fecundity or larval survival is impaired (Beketov and Liess 2012; Connon et al. 2012; Driessnack et al. 2011; Gessner and Tlili 2016)? The production of healthy embryos, larvae, fry, and juveniles also plays a role in successful long-term populations (Arcand-Hoy and Benson 1998). Successful mixture assessment will depend on linking quantitative results in the lab with increasingly environmental relevant exposures using tools such as mesocosms and field surveys and improving the mixture models (Connon et al. 2012; Driessnack et al. 2011; Miles et al. 2017).

The two dominant mixture model approaches are the concentration-based and effects or response-based approaches. Both models assess stressor interactions, where the stressors of interest are chemicals in a mixture. Stressor interactions are determined in references to a null model that predicts the joint effect assuming the absence of interactions, in which the models assume the stressors are operating independently. The default assumption is additivity, and any deviation could support an interaction of the mixture, but the challenge arises when those deviations are noted as more complex models accounting for stressor interactions are required for reliable predictions (Schäfer and Piggott 2018). The additive null model assumes that the combined effect of stressors equals the sum of their individual proportional effects; this is the concentration addition model of ecotoxicology or the toxic unit approach. The multiplicative null model assumes an independent mode of action, which is the effects or response-based approach noted in the prior mixture section (Thompson et al. 2018a). These models, especially the concentration or additive approach, have started to become the standard in toxicology. However, both models have limitations that have been minimized by researchers in attempts to better understand and predict mixture toxicity under chronic conditions (de Zwart and Posthuma 2005; Thompson et al. 2018a). One issue that faces both models is that they assess impacts on individuals and those results are being applied to populations. More specifically in the concentration approach, the model assigns equal potency of the chemicals or stressors, which can lead to a null expectation greater than 100%. The data may also be forced into assuming additive as the only possible response of the null model and ultimately miss the interactions the model seeks to identify (Schmidt et al. 2022; Thompson et al. 2018a). This often leads to the effect-based or multiplicative null models being used, as they are based on proportional responses, but again it is also biased in assuming additivity similar to the concentration-based approach (Thompson et al. 2018a).

For example, a study by Shadid et al. (2019), assessed the impacts of pesticide mixtures on *D. magna* using the two common toxicology mixture models—concentration and effects. The authors also assessed a more recently proposed stress addition model (SAM) to investigate the success of the different approaches by combining a binary pesticide mixture with the environmental stressor of food limitation. The authors reported synergistic interactions on daphnid survival in the mixture combined with food limitation and that all three models underestimated the observed response. This included the SAM model, which is being developed to predict synergism between toxicants and environmental stressors, and the only model that considers sensitivity distributions (Liess et al. 2016; Schäfer and Piggott 2018).

While these approaches have value in the establishment of mechanisms of action and laboratory-based responses, they are rooted in medicine, pharmacology, and environmental chemistry, rather than in ecology (Gessner and Tlili 2016). Models in the domain of ecology are intended to assess for patterns, processes, and relationships such as indirect effects and interspecies interactions (e.g., competition, predation) (Gessner and Tlili 2016; Thompson et al. 2018b). Interactions among stressors, both chemical and environmental (e.g., food scarcity), can produce "ecological surprises," whereby their combined effect is not predictable based on their known individual effects. These missed predictions can have damaging impacts to ecosystems, which makes the selection of our null models crucial, and the available toxicology standard models provide debatable assessments (Thompson et al. 2018a). So, if our current models are missing the transition from individuals to the realistic world of populations and communities, what approaches are available?

Thompson et al. (2018a) proposed a model that seeks to understand the toxicological importance of the response but employs ecology-based measures of stress. This model is termed the compositional null model. This model is proposed to predict the effects of combined stressors based on the species-specific additive effects of individual stressors that can be applied to community-based indicators, such as species richness (Thompson et al. 2018a, b). The promise of this model and others, such as SAM and species-sensitivity distributions, likely lies in integrating ecology into ecotoxicology (Gessner and Tlili 2016). Ultimately, our models are only as strong as the data supporting them, which requires us to use environmentally relevant endpoints such as reproductive output obtained from longer studies such as life table matrixes and mesocosms to achieve realistic and protective predictions (Connon et al. 2012; Schmidt et al. 2022; Stark et al. 2004). It is this author's belief that one of the great strengths of ecotoxicology is its interdisciplinary nature, and access to the vast skills and analytical tools utilized in the field of toxicology makes its researchers equipped to assess these complex questions of mixtures and reproduction.

References

Ankley GT, Jensen KM, Kahl MD, Korte JJ, Makynen EA (2001) Description and evaluation of a short-term reproduction test with the fathead minnow (*Pimephales promelas*). Environ Toxicol Chem 20:1276–1290

Ankley GT, Bennett RS, Erickson RJ, Hoff DJ, Hornung MW, Johnson RD, Mount DR, Nichols JW, Russom CL, Schmieder PK, Serrrano JA, Tietge JE, Villeneuve DL (2010a) Adverse outcome pathways: a conceptual framework to support ecotoxicology research and risk assessment. Environ Toxicol Chem 29:730–741. https://doi.org/10.1002/etc.34

Ankley GT, Jensen KM, Kahl MD, Durhan EJ, Makynen EA, Cavallin JE, Martinović D, Wehmas LC, Mueller ND, Villeneuve DL (2010b) Use of chemical mixtures to differentiate mechanisms of endocrine action in a small fish model. Aquat Toxicol 99:389–396. https://doi.org/10.1016/j.aquatox.2010.05.020

Arcand-Hoy LD, Benson WH (1998) Fish reproduction: an ecologically relevant indicator of endocrine disruption. Environ Toxicol Chem 17:49–57

Arnold KE, Brown AR, Brown AR, Ankley GT, Sumpter JP (2014) Medicating the environment: assessing risks of pharmaceuticals to wildlife and ecosystems. Philos Trans R Soc B Biol Sci 369:20130569. https://doi.org/10.1098/rstb.2013.0569

Artigas J, Arts G, Babut M, Caracciolo AB, Charles S, Chaumot A et al (2012) Towards a renewed research agenda in ecotoxicology. Environ Pollut 160:201–206

Balistrieri LS, Mebane CA (2014) Predicting the toxicity of metal mixtures. Sci Total Environ 466–467:788–799. https://doi.org/10.1016/j.scitotenv.2013.07.034

Beketov MA, Liess M (2012) Ecotoxicology and macroecology—time for integration. Environ Pollut 162:247–254. https://doi.org/10.1016/j.envpol.2011.11.011

Bhavsar SP, Reiner EJ, Hayton A, Fletcher R, MacPherson K (2008) Converting toxic equivalents (TEQ) of dioxins and dioxin-like compounds in fish from one toxic equivalency factor (TEF) scheme to another. Environ Int 34:915–921. https://doi.org/10.1016/j.envint.2008.02.001

Burkholder J, Libra B, Weyer P, Heathcote S, Kolpin D, Thorne PS, Wichman M (2007) Impacts of waste from concentrated animal feeding operations on water quality. Environ Health Perspect 2: 308–312

Carr JA, Patiño R (2011) The hypothalamus-pituitary-thyroid axis in teleosts and amphibians: endocrine disruption and its consequences to natural populations. Gen Comp Endocrinol 170: 299–312

Cavallaro MC, Morrissey CA, Headley JV, Peru KM, Liber K (2016) Comparative chronic toxicity of imidacloprid, clothianidin, and thiamethoxam to *Chironomus dilutus* and estimation of toxic equivalency factors. Environ Toxicol Chem 36:372–382

Chow MI, Lundin JI, Mitchell CJ, Davis JW, Young G, Scholz NL, McIntyre JK (2019) An urban stormwater runoff mortality syndrome in juvenile coho salmon. Aquat Toxicol 214:105231

Cirillo T, Amodio Cocchieri R, Fasano E, Lucisano A, Tafuri S, Ferrante MC, Carpenè E, Andreani G, Isani G (2012) Cadmium accumulation and antioxidant responses in sparus aurata exposed to waterborne cadmium. Arch Environ Contam Toxicol 62:118–126. https://doi.org/10.1007/s00244-011-9676-9

Conners KA, Brill JL, Norberg-King T, Barron MG, Carr G, Belanger SE (2022) *Daphnia magna* and *Ceriodaphnia dubia* have similar sensitivity in standard acute and chronic toxicity tests. Environ Toxicol 41:134–147

Connon RE, Geist J, Werner I (2012) Effect-based tools for monitoring and predicting the ecotoxicological effects of chemicals in the aquatic environment. Sens Switz 12:12741–12771

Covert SA, Shoda ME, Stackpoole SM, Stone WW (2020) Pesticide mixtures show potential toxicity to aquatic life in U.S. streams, water years 2013–2017. Sci Total Environ 745:141285

de Zwart D, Posthuma L (2005) Complex mixture toxicity for single and multiple species: proposed methodologies. Environ Toxicol Chem 24:2665–2676

de Zwart D, Adams W, Galay Burgos M, Hollender J, Junghans M, Merrington G, Muir D, Parkerton T, De Schamphelaere KAC, Whale G, Williams R (2018) Aquatic exposures of chemical mixtures in urban environments: approaches to impact assessment. Environ Toxicol Chem 37:703–714. https://doi.org/10.1002/etc.3975

Dethloff GM, Schlenk D, Hamm JT, Bailey HC (1999) Alterations in physiological parameters of rainbow trout (*Oncorhynchus mykiss*) with exposure to copper and copper/zinc mixtures. Ecotoxicol Environ Saf 42:253–264

Di Giulio RT, Hinton DE (eds) (2008) The toxicology of fishes, 1st edn. CRC. https://doi.org/10.1201/9780203647295

Driessnack MK, Dubé MG, Rozon-Ramilo LG, Jones PD, Wiramanaden CIE, Pickering IJ (2011) The use of field-based mesocosm systems to assess the effects of uranium milling effluent on fathead minnow (*Pimephales promelas*). Ecotoxicology 20:1209–1224

Driessnack MK, Matthews AL, Raine JC, Niyogi S (2016) Interactive effects of chronic waterborne copper and cadmium exposure on tissue-specific metal accumulation and reproduction in fathead minnow (Pimephales promelas). Comp Biochem Physiol Part C Toxicol Pharmacol 179:165–173

Driessnack MK, Jamwal A, Niyogi S (2017a) Interactions of waterborne cadmium and zinc during chronic exposure on tissue-specific metal accumulation and reproduction in fathead minnow (*Pimephales promelas*). Ecotoxicol Environ Saf 140:65–75

Driessnack MK, Jamwal A, Niyogi S (2017b) Effects of chronic exposure to waterborne copper and nickel in binary mixture on tissue-specific metal accumulation and reproduction in fathead minnow (*Pimephales promelas*). Chemosphere 185:964–974

Environment Canada (2000) Biological test method: acute lethality of effluents to *Daphnia magna*. Environment Canada, Ottawa

Feist BE, Buhle ER, Baldwin DH, Spromberg JA, Damm SE, Davis JW et al (2017) Roads to ruin: conservation threats to a sentinel species across an urban gradient. Ecol Appl 27:2382–2396

Gandhi N, Gewurtz SB, Drouillard KG, Kolic T, MacPherson K, Reiner EJ, Bhavsar SP (2019) Dioxins in Great Lakes fish: past, present and implications for future monitoring. Chemosphere 222:479–488. https://doi.org/10.1016/j.chemosphere.2018.12.139

Gauthier C, Couture P, Pyle GG (2006) Metal effects on fathead minnow (*Pimephales promelas)* under field and laboratory conditions. Ecotoxicol Environ Saf 63:353–364

Gessner MO, Tlili A (2016) Fostering integration of freshwater ecology with ecotoxicology. Freshw Biol 61:1991–2001

Hooper MJ, Ankley GT, Cristol DA, Maryoung LA, Noyes PD, Pinkerton KE (2013) Interactions between chemical and climate stressors: a role for mechanistic toxicology in assessing climate change risks. Environ Toxicol Chem 32:32–48

Hua J, Han J, Wang X, Guo Y, Zhou B (2016) The binary mixtures of megestrol acetate and 17a-ethynylestradiol adversely affect zebrafish reproduction. Environ Pollut 213:776–784

Jamwal A, Shekh K (2021) Endocrine disruption in freshwater fish from contaminants of emerging concern. In: Sundaray JK, Rather MA, Kumar S, Agarwal D (eds) Recent updates in molecular endocrinology and reproductive physiology of fish. Springer, Singapore

Jonker MJ, Svendsen C, Bedaux JJM, Bongers M, Kammenga JE (2005) Significance testing of synergistic/antagonistic, dose level-dependent, or dose ratio-dependent effects in mixture dose-response analysis. Environ Toxicol Chem 24:2701–2713

Kamunde C, MacPhail R (2011) Metal-metal interactions of dietary cadmium, copper and zinc in rainbow trout, Oncorhynchus mykiss. Ecotoxicol Environ Saf 74:658–667

Kavlock R, Daston G (1996) Research needs for the risk assessment of health and environmental effects of endocrine disruptors: a report of the US EPA-sponsored workshop. Environ Health 104:715

Kerdivel G, Habauzit D, Pakdel F (2013) Assessment and molecular actions of endocrine-disrupting chemicals that interfere with estrogen receptor pathways. Int J Endocrinol 2013: 501851

Knapen D, Bervoets L, Verheyen E, Blust R (2004) Resistance to water pollution in natural gudgeon (*Gobio gobio*) populations may be due to genetic adaptation. Aquat Toxicol 67:155–165

Kramer VJ, Etterson MA, Hecker M, Murphy CA, Roesijadi G, Spade DJ et al (2011) Adverse outcome pathways and ecological risk assessment: bridging to population-level effects. Environ Toxicol Chem 30:64–76

Landis W, Sofield R, Yu MH (2010) Introduction to environmental toxicology: molecular sub-structures to ecological landscapes, 4th edn. CRC, Boca Raton, FL. https://doi.org/10.1201/9781439804117

Lemly AD (1993) Teratogenic effects of selenium in natural populations of freshwater fish. Ecotoxicol Environ Saf 26:181–204

Lemly AD (1999) Selenium transport and bioaccumulation in aquatic ecosystems: a proposal for water quality criteria based on hydrological units. Ecotoxicol Environ Saf 42:150–156

Lemly AD (2002) Symptoms and implications of selenium toxicity in fish: the Belews Lake case example. Aquat Toxicol 57:39–49

Liem AKD, Fürst P, Rappe C (2000) Exposure of populations to dioxins and related compounds. Food Addit Contam 17:241–259

Liess M, Foit K, Knillmann S, Schäfer RB, Liess HD (2016) Predicting the synergy of multiple stress effects. Sci Rep 6:32965

Lubzens E, Young G, Bobe J, Cerdà J (2010) Oogenesis in teleosts: how fish eggs are formed. Gen Comp Endocrinol 165:367–389

Maloney EM, Morrissey CA, Headley JV, Peru KM, Liber K (2018) Can chronic exposure to imidacloprid, clothianidin, and thiamethoxam mixtures exert greater than additive toxicity in *Chironomus dilutus?* Ecotoxicol Environ Saf 156:354–365

McIntyre JK, Prat J, Cameron J, Wetzel J, Mudrock E, Peter KT, Tian Z, Mackenzie C, Lundin J, Stark JD, King K, Davis JW, Kolodziej EP, Scholz NL (2021) Treading water: tire wear particle leachate recreates an urban runoff mortality syndrome in coho but not chum Salmon. Environ Sci Technol 55:11767–11774. https://doi.org/10.1021/acs.est.1c03569

Meyers JS, Adams WJ, Brix KV, Luoma SN, Mount DR (2005) Toxicity of dietborne metals to aquatic organisms. SETAC Press, Pensacola, FL

Miles JC, Hua J, Sepulveda MS, Krupke CH, Hoverman JT (2017) Effects of clothianidin on aquatic communities: evaluating the impacts of lethal and sublethal exposure to neonicotinoids. PloS One 12:e0174171

Moreau CJ, Klerks PL, Haas CN (1999) Interaction between phenanthrene and zinc in their toxicity to the sheepshead minnow (*Cyprinodon variegatus*). Arch Environ Contam Toxicol 37:251–257

Moyle PB, Cech JJ (eds) (2005) Fishes: an introduction to ichthyology. Fifth Edition, Pearson Prentice Hall

Müller A, Österlund H, Marsalek J, Viklander M (2020) The pollution conveyed by urban runoff: a review of sources. Sci Total Environ 709. https://doi.org/10.1016/j.scitotenv.2019.136125

Mumtaz M (ed) (2010) Principles and practice of mixtures toxicology. Wiley, Hoboken, NJ. https://doi.org/10.1002/9783527630196

Nagahama Y (1994) Endocrine regulation of gametogenesis in fish. Int J Dev Biol 38:217–229

Newman MC (ed) (2010) Fundamentals of ecotoxicolgy, 3rd edn. CRC Press LLC, Boca Raton, FL

Nkoom M, Lu G, Liu J (2022) Chronic toxicity of diclofenac, carbamazepine and their mixture to Daphnia magna: a comparative two-generational study. Environ Sci Pollut Res 29:58963–58979

Norris DO, Lopez KH (eds) (2011) Hormones and reproduction of vertebrates. Fishes, vol 1. Academic Press/Elsevier, Waltham, MA

Norwood WP, Borgmann U, Dixon DG, Wallace A (2003) Effects of metal mixtures on aquatic biota: a review of observations and methods. Hum Ecol Risk Assess Int J 9:795–811

Orr JA, Vinebrooke RD, Jackson MC, Kroeker KJ, Kordas RL, Mantyka-Pringle C, Van den Brink PJ, De Laender F, Stoks R, Holmstrup M, Matthaei CD, Monk WA, Penk MR, Leuzinger S, Schäfer RB, Piggott JJ (2020) Towards a unified study of multiple stressors: divisions and common goals across research disciplines. Proc R Soc B 287:20200421. https://doi.org/10.1098/rspb.2020.0421

Ottinger MA, Wu J, Pelican K (2002) Neuroendocrine regulation of reproduction in birds and clinical applications of GnRH analogues in birds and mammals. Semin Avian Exotic Pet Med 11:71–79

Ouellet JD, Dubé MG, Niyogi S (2013) A single metal, metal mixture, and whole-effluent approach to investigate causes of metal mine effluent effects on fathead minnows (Pimephales promelas). Water Air Soil Pollut 224(3):1462

Parrott JL (2005) Overview of methodology and endpoints in fathead minnow lifecycle tests assessing pulp and paper mill effluent. Water Qual Res J Canada 40:334–346

Paschoalini AL, Savassi LA, Weber AA, Moreira DP, Ribeiro YM, Rizzo E et al (2021) Evaluation of the oestrogenic potential of oestrone and bisphenol-a on the reproduction of *Astyanax bimaculatus* males after subacute exposure. Fish Physiol Biochem 47:797–810

Pérez E, Hoang TC (2017) Chronic toxicity of binary-metal mixtures of cadmium and zinc to Daphnia magna. Environ Toxicol Chem 36:2739–2749. https://doi.org/10.1002/etc.3830

Quinn TP (2018) The behavior and ecology of pacific salmon and trout, 2nd edn. University of Washington Press, Seattle, WA

Rand GM (ed) (1995) Fundamentals of aquatic toxicology: effects, environmental fate, and risk assessment, 2nd edn. CRC, Boca Raton, FL. https://doi.org/10.1201/9781003075363

Rubach MN, Ashauer R, Buchwalter DB, De Lange HJ, Hamer M, Preuss TG, Töpke K, Maund SJ (2011) Framework for traits-based assessment in ecotoxicology. Integr Environ Assess Manag 7:172–186. https://doi.org/10.1002/ieam.105

Runnalls T, Beresford N, Kugathas S, Margiotta-Casaluci L, Scholze M, Scott AP, Sumpter JP (2015) From single chemicals to mixtures—reproductive effects of levonorgestrel and ethinylestradiol on the fathead minnow. Aquat Toxicol 169:152–167

Schäfer RB, Piggott JJ (2018) Advancing understanding and prediction in multiple stressor research through a mechanistic basis for null models. Glob Chang Biol 24:1817–1826
Schmidt TS, Miller JL, Mahler BJ, Van Metre PC, Nowell LH, Sandstrom MW et al (2022) Ecological consequences of neonicotinoid mixtures in streams. Sci Adv 8:eabj8182
Scholz NL, Myers MS, McCarthy SG, Labenia JS, McIntyre JK, Ylitalo GM, Rhodes LD, Laetz CA, Stehr CM, French BL, McMillan B, Wilson D, Reed L, Lynch KD, Damm S, Davis JW, Collier TK (2011) Recurrent die-offs of adult coho salmon returning to spawn in Puget Sound lowland urban streams. PloS One 6:e28013. https://doi.org/10.1371/journal.pone.0028013
Scholz S, Renner P, Belanger SE, Busquet F, Davi R, Demeneix BA, Denny JS, Léonard M, McMaster ME, Villeneuve DL, Embry MR (2013) Alternatives to *in vivo* tests to detect endocrine disrupting chemicals (EDCs) in fish and amphibians--screening for estrogen, androgen and thyroid hormone disruption. Crit Rev Toxicol 43:45–72
Shadid N, Liess M, Knillmann S (2019) Environmental stress increases synergistic effects of pesticide mixtures on Daphnia manga. Environ Sci Technol 53:12586–12593. https://doi.org/10.1021/acs.est.9b04293
Sprague JB (1970) Measurement of pollutant toxicity to fish. II. Utilizing and applying bioassay results. Water Res 4:3–32
Spromberg JA, Scholz NL (2011) Estimating the future decline of wild coho salmon populations resulting from early spawner die-offs in urbanizing watershed of the Pacific northwest, USA. Integr Environ Assess Manag 7:648–656
Stark JD, Banks JE, Vargas R (2004) How risky is risk assessment: the role that life history strategies play in susceptibility of species to stress. Proc Natl Acad Sci 101:732–736
Stark JD, McIntyre JK, Banks JE (2020) Population viability in a host-parasitoid system is mediated by interactions between population stage structure and life stage differential susceptibility to toxicants. Sci Rep 10:1–7
Sumpter JP (2009) Protecting aquatic organisms from chemicals: the harsh realities. Phil Trans R Soc A 367:3987–3894
Suter GW, Rosen AE, Linder E, Parkhurst DF (1987) Endpoints for responses of fish to chronic toxic exposures. Environ Toxicol Chem 6:793–809
Thompson PL, MacLennan MM, Vinebrooke RD (2018a) An improved null model for assessing the net effects of multiple stressors on communities. Glob Chang Biol 24:517–525
Thompson PL, MacLennan MM, Vinebrooke RD (2018b) Species interactions cause non-additive effects of multiple environmental stressors on communities. Ecosphere 9:e02518
Thrupp TJ, Runnalls TJ, Scholze M, Kugathas S, Kortenkamp A, Sumpter JP (2018) The consequences of exposure to mixtures of chemicals: something from 'nothing' and 'a lot from a little' when fish are exposed to steroid hormones. Sci Total Environ 619–620:1482–1492
Tian Z, Zhao H, Peter KT, Gonzalez M, Wetzel J, Wu C et al (2021) A ubiquitous tire rubber–derived chemical induces acute mortality in coho salmon. Science 371:185–189
Urbatzka R, Rocha E, Reis B, Cruzeiro C, Monteiro RAF, Rocha MJ (2012) Effects of ethinylestradiol and of an environmentally relevant mixture of xenoestrogens on steroidogenic gene expression and specific transcription factors in zebra fi sh. Environ Pollut 164:28–35
US Environmental Protection Agency (1992) Framework for ecological risk assessment. No. EPA/630/R-92/001. Washington, DC
US Environmental Protection Agency (1996a) Fish early-life stage toxicity test. Washington, DC
US Environmental Protection Agency (1996b) Ecological effects test guideline: daphnid chronic toxicity test. Washington, DC
US Environmental Protection Agency (1998) Guidelines for ecological risk assessment. No. EPA/630/R-95/002F. Washington, DC
US Environmental Protection Agency (2000) Method guidance and recommendations for whole effluent toxicity (WET) testing. Washington, DC
van den Berg M, Birnbaum L, Bosveld A, BrunstroemB CP, Feeley M (1998) Toxic equivalency factors (TEFs) for PCBs, PCDDs, PCDFs for humans and wildlife. Environ Health Perspect 106:775–792

van den Berg M, Birnbaum LS, DenisonM DVM, Farland W, Feeley M et al (2006) The 2005 World Health Organization re-evaluation of human and mammalian toxic equivalency factors for dioxins and dioxin-like compounds. Toxicol Sci 93:223–241

Vandenberg LN, Colborn T, Hayes TB, Heindel JJ, Jacobs DR, Lee DH et al (2013) Regulatory decisions on endocrine disrupting chemicals should be based on the principles of endocrinology. Reprod Toxicol 38:1–15

Versteeg DJ, Stalmans M, Janssen C (1997) Ceriodaphnia and daphnia: a comparison of their sensitivity to xenobiotics and utility as a test species. Chemosphere 340:869–892

Villeneuve DL, Garcia-Reyero N (2011) Vision & strategy: predictive ecotoxicology in the 21st century. Environ Toxicol Chem SETAC 30:1–8

Walker CH, Sibly RM, Sibly RM, Peakall DB (2005) Principles of ecotoxicology, 3rd edn. CRC, Boca Raton, FL. https://doi.org/10.1201/9781439894224

Weir SM, Flynn RW, Scott DE, Yu S, Lance SL (2016) Environmental levels of Zn do not protect embryos from cu toxicity in three species of amphibians. Environ Pollut 214:161–168. https://doi.org/10.1016/j.envpol.2016.04.005

Weltzien FA, Andersson E, Andersen O, Shalchian-Tabrizi K, Norberg B (2004) The brain-pituitary-gonad axis in male teleosts, with special emphasis on flatfish (*Pleuronectiformes*). Comp Biochem Physiol A 137:447–477

Werner J, Ouellet JD, Cheng CS, Ju YJ, Law RD (2010) Pulp and paper mill effluents induce distinct gene expression changes linked to androgenic and estrogenic responses in the fathead minnow (Pimephales promelas). Environ Toxicol Chem 29:430–439

Wilkinson JL, Boxall ABA, Kolpin DW, Leung KMY, Lai RWS et al (2022) Pharmaceutical pollution of the world's rivers. Proc Natl Acad Sci U S A 119:e2113947119

Yan Z, Guanghua L, Jianchao L (2016) Long-term effects of antibiotics, norfloxacin, and sulfamethoxazole, in a partial life-cycle study with zebrafish (Danio rerio): effects on growth, development, and reproduction. Environ Sci Pollut Res 23:18222–18228

Zohar Y, Muñoz-Cueto JA, Elizur A, Kah O (2010) Neuroendocrinology of reproduction in teleost fish. Gen Comp Endocrinol 165:438–455

Phytoestrogens as Endocrine-Disrupting Agents in Aquaculture

Nuzaiba P. Muhammad, T. Nirmal, Arya Prabhakaran, and Tincy Varghese

1 Introduction

Many estrogen-mimicking chemicals are introduced as endocrine disruptors in the aquatic environment, which may be synthetic or natural; the synthetic ones are called xenoestrogens. The natural estrogen-mimicking compounds could be of plant or animal origin. Phytoestrogens, which resemble mammalian estrogens, particularly 17-estradiol, structurally or functionally, are sizable heterogenic group of chemical substances derived from plants. Legumes are the main source of phytoestrogens as they are involved in plant-microbe interactions and the reason to induce nodulation in the roots of legumes. Phytoestrogens are also involved in the defense mechanisms in the plant kingdom but do not involve in the endocrine system of plants. The majority of phytoestrogens are not found in seed oil, and the concentration of whole or ground seeds in a given area can vary depending on the climate, crop maturity, storage time, and region. Phytoestrogenic compounds are being introduced in the aquatic environment majorly through the feed. Some of the common plant-based ingredients used in aquaculture feed contain phytoestrogens including soybean meal and alfalfa meal.

The low molecular weight and stable structure of phytoestrogens enable them to easily pass through the lipid bilayer and interact with the enzyme and receptors of the cell and induce proestrogenic or antiestrogenic effects. Several reports suggest that phytoestrogens have beneficial effects in humans, including hepatoprotective, anti-allergic, anti-inflammatory, antitumor, antithrombotic, and antioxidant properties

N. P. Muhammad · T. Nirmal
Centurion University of Technology and Management, Paralakhemundi, Odisha, India

A. Prabhakaran · T. Varghese (✉)
ICAR-Central Institute of Fisheries Education, Versova, Andheri (w), Mumbai, Maharashtra, India
e-mail: tincy@cife.edu.in

M. A. Rather et al. (eds.), *Xenobiotics in Aquatic Animals*,
https://doi.org/10.1007/978-981-99-1214-8_9

Table 1 Sources of phytoestrogens

Phytoestrogens	Source
Genistein	Soybean, broad bean, white bean, chickpeas, red bean
Daidzein	Soybean, black bean, and green split peas
Coumestrol	Alfalfa, clover, split peas, chickpeas, lima beans, soy sprouts, and pinto beans
Lignan (enterolactone and enterodiol)	Flaxseed, chickpeas, unhulled soybean, whole legumes, and cereal brans
Formononetin	Green bean, lima bean, broad bean, pink bean, mung bean, clover sprout, and alfalfa
Glycitin	Soybean, chick peas, and other legumes
Biochanin A	Clover sprout, chick peas, Chinese peas, pinto beans, and kidney bean
Resveratrol (matairesinol, secoisolariciresinol, lariciresinol, etc.)	Grape cane waste, peanut roots

(Ahmad et al. 2013; Roca et al. 2014). Phytoestrogens or their active metabolites typically have an estrogenic effect on the central nervous system and reproductive systems of both males and females in mammals and are studied extensively in mammals. This chapter concentrates mostly on the common phytoestrogens found in aquaculture feed ingredients and their effects on reproductive endocrine function.

2 Structure and Classification of Phytoestrogens

A phenolic ring and two hydroxyl groups, which are essential for binding with estrogen receptors (ER), are characteristics of phytoestrogens. The phenolic group determines whether phytoestrogens have agonist or antagonistic properties with animal estrogen (Roca et al. 2014; Torrens-Mas and Roca 2020). Based on their structural characteristics, phytoestrogens are divided into three main classes: flavonoids, lignans, and stilbenes (Cos et al. 2003; Nikolić et al. 2017). The classification of phytoestrogens is given in Fig. 2, and various sources of phytoestrogens are given in Table 1.

2.1 Flavonoids

Flavonoids have two benzene rings (A and B) in their structure, joined together by a heterocyclic (C) ring (Wang et al. 2018) as depicted in Fig. 1. Based on the position of B and C rings, oxidation and hydroxylation of the C ring, and the degree of saturation, they are typically divided into isoflavonoids, flavones, flavonols, flavan-3-ols (or catechins), flavanones, chalcones, and anthocyanins (Fig. 2).

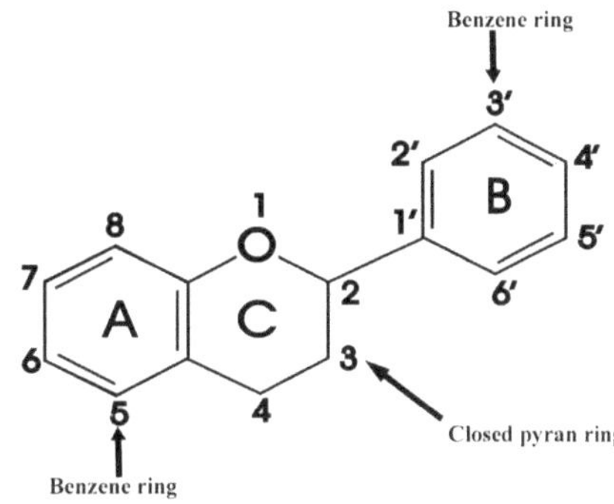

Fig. 1 General structure of flavonoids

Isoflavonoids comprise isoflavones and coumetrans. Isoflavones have a general backbone made of 3-phenylchromen-4-one. Genistein, daidzein (from soybean meal), formononetin, and biochanin A (from red clover) are included in the isoflavone category as they share the general backbone (Fig. 1). Coumestans have a 1-benzoxolo(3,2-c) chromen-6-one linked to the B ring at the C3 position instead of the C2 position.

2.1.1 Isoflavones

Isoflavones can be found in large quantities in soybean and its byproducts. Majorly three types of isoflavones have been found from soybean meal which exists in four chemical forms, namely:

1. Aglycones: genistein (4′,5,7-trihydroxyisoflavone), daidzein (4′,7-dihydroxyisoflavone), and glycitein (4′,7-dihydroxy-6-methoxy isoflavone).
2. Glucosides: genistein, daidzein, and glycitein.
3. Acetylglucosides: 6′-O-acetylgenistin, 6′-O-acetyldaidzin, and 6′-O-acetylglycitin.
4. Malonylglucosides: 6-O-malonylgenistin, 6-O-malonyldaidzin, and 6″-O-malonylglycitin.

According to Chakraborty et al. (2014), most dietary forms of soybean meal have a mixture of three bioavailable forms of aglycones, namely, genistein, daidzein, and glycitein. Glycoside forms are biologically inactive, which can become bioactive absorbable forms at the intestinal lumen, aglycones, when hydrolyzed by bacterial glucosidases (Setchell 1998). Genistein can be further metabolized to *p*-ethyl phenol, while daidzein can be processed to equol and O-demethyangolensin in animals (Soukup et al. 2016); however, the efficiency of metabolism varies in different organisms. Isoflavone content is known to vary based on the crop and products (Carrera et al. 2011; Medic et al. 2014). The total isoflavone concentrations in soybean meal were estimated to be 15.70–30.75 mg/g dry matter (Flachowsky et al. 2011).

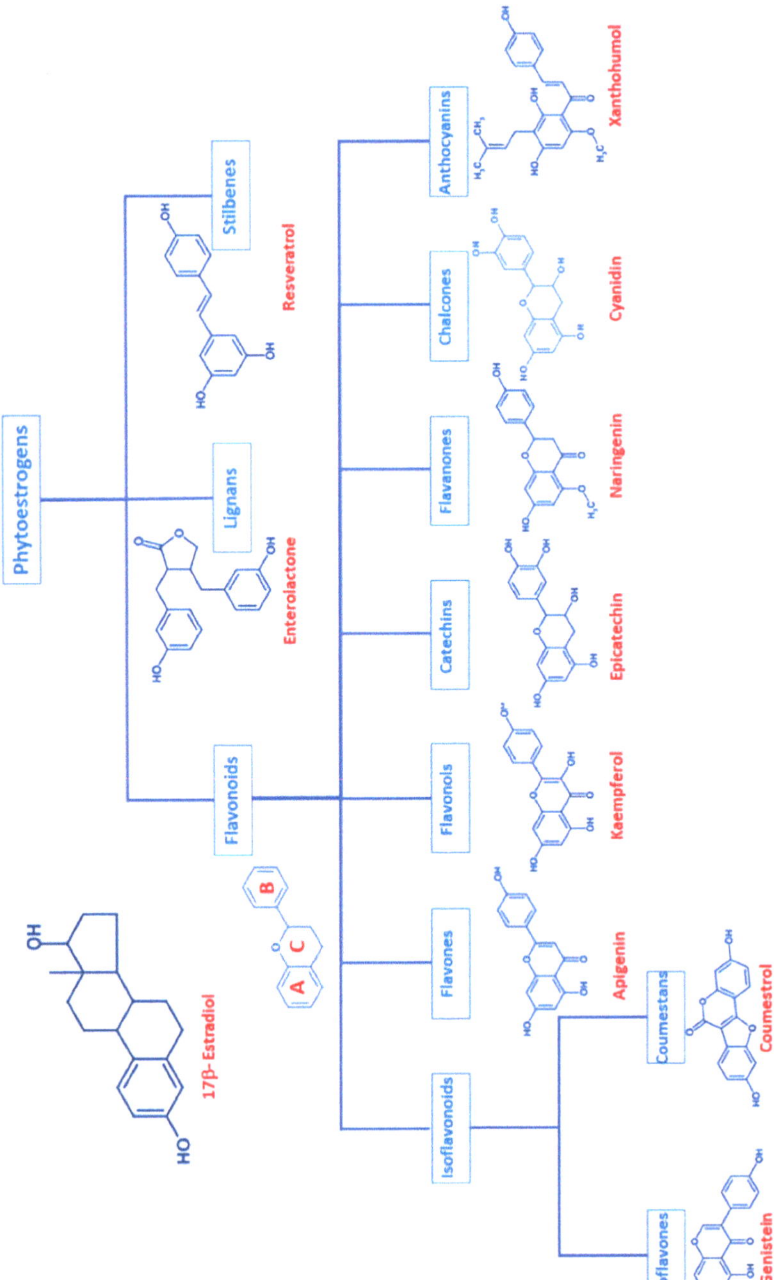

Fig. 2 Classification and structure of phytoestrogens (Torrens-Mas and Roca 2020)

Genistein

Among the total isoflavone content in soybean meal, genistein is recorded as the predominant one. According to reports, genistein content in soybean meal varies from 2.6 to 22.4 mg/100 g of soybean meal (Chen and Wei 2008) based on the crop, season, and type of processing. Since soybean meal is the widely utilized leguminous plant source of protein in feed, it is relevant to describe the genistein content reported in fish feed worldwide. The carp commercial diets were reported to have 6.76–23.71 mg, the trout diets were estimated to be 4.16–5.02 mg, and the diet of medaka was reported to have 0.93–5.85 mg of genistein/100 g feed (Xiao et al. 2018). Feeds manufactured in various countries, namely, Japan, the UK, and Korea, were determined to contain genistein 3.97–11.89 mg/100 g (Inudo et al. 2004; Xiao et al. 2018).

Daidzein

7,4′-dihydroxy isoflavone (daidzein) is found widely in plants belonging to legumes. Recent studies on humans have demonstrated the efficacy of daidzein in treating menopause, osteoporosis, and high blood cholesterol and reducing the risk of some hormone-related cancers and heart disease. Despite the known health advantages, it is also known for its potential to affect fertility and cause developmental toxicity to the reproductive tract in female rats. The effect of daidzein on fish reproductive endocrine system is still at the preliminary face.

Glycitein

In soy food products, glycitein, an O-methylated isoflavone, makes up 5–10% of all isoflavones. Glycitein comes under phytoestrogens similar to other soy isoflavones in that it has a weak estrogenic activity compared to other isoflavones.

Biochanin A and Formononetin

Biochanin A (5,7-dihydroxy-4-methoxy-isoflavone) is derived from edible and herbal plants such as peanuts, alfalfa sprouts, soy, and red clover. Formononetin, otherwise known as biochanin B, is a 7-hydroxyisoflavone substituted by a methoxy group at position 4′. The plant metabolite acts as an estrogenic or antiestrogenic compound in animals. It is a member of 7-hydroxyisoflavones and a member of 4′-methoxyisoflavones. It is functionally related to daidzein.

2.1.2 Coumestrol

Coumestrol was first discovered as an estrogenic compound in 1957 by E. M. Backoff in ladino clover and alfalfa. The compound is known to have high estrogenic activity and high binding affinity with estrogen receptors (binds with 94% affinity to ERα and 185% affinity to ERβ, when related to the binding affinity of estradiol to both). The compound, however, is having much less activity than estradiol while 30–100 times more activity than isoflavones. The compound is normally found in a limited range of food, and alfalfa sprouts are one among them.

2.2 Lignans

A large class of low molecular weight polyphenols called lignans is present in many types of plants, especially grain, nuts, coffee, tea, cocoa, flaxseed, and fruits. Lignans have two phenylpropane groups linked by a -β′-β′ bond formed between the central atoms of their side chains (eighth or position). Lignans function as anti-nutrients in the defense of seeds and plants against herbivores and are precursors to phytoestrogens. Lignans and lignin have different molecular weights; lignin is a high polymer that is indigestible, whereas lignans are small and soluble in water. High levels of lignans can be found in sesame and flax seeds. Cereals (rye, wheat, oat, and barley), soybeans, tofu, cruciferous vegetables (like broccoli and cabbage), and some fruits, particularly apricots and strawberries, are other foods that contain lignans. Lignan and its estrogenic effects are well-studied in humans and rats (Tou et al. 1998), especially in delaying puberty, lengthening diestrus, affecting the reproductive development of offspring, and so on. However, research on the effect of lignan in fish is null, so far (according to the Web of Science).

2.3 Stilbenes

The stilbenes (Fig. 2) have a 1,2-diphenylethylene nucleus and a polyphenolic structure. The stilbene resveratrol, which can be found in a variety of foods like nuts, grapes, berries, and wine, has received the most research. The most well-known stilbene, resveratrol, is used to prevent disease, particularly obesity. The compounds are well known for their antiaging activity in humans, and they act as an antioxidant and inhibit cyclooxygenases in zebrafish (Kowalska et al. 2011). Other stilbenes and their metabolites are also thought to be potential anti-obesity candidates. The main stilbene in grape plants is called piceid (5,4′-dihydroxystilbene-3-O-β-glucoside), and its concentration in red grape juice is almost six times higher than that of resveratrol monomers. Red wine also contains astringin (5,4′,3-trihydroxystilbene-3-O-β-glucoside) and isorhapontin (5,4′-dihydroxy-3′-methoxystilbene-3-β-Dglucoside), though they are rarely found in higher concentrations than trans-piceid or resveratrol monomeric forms. The studies on endocrine disruptive effects of stilbenes in fishes are limited to a few, which are conducted on medaka and zebrafish (Cavalcante et al. 2017; Rohmah et al. 2022).

3 Phytoestrogen on Reproduction Mechanism of Endocrine Disruption

A concern of successful reproduction and development of seeds for enhancing aquaculture production has been discussed in the recent past. The endocrine-disrupting compounds have the potential ability to perturb the sensitive hormone pathways that take part in reproduction. Phytoestrogen acts on the endocrine axis by (1) disrupting the enzymes involved in various hormone synthesis, action, and metabolism, (2) mimicking the endogenous estrogen due to the structural similarity of phytoestrogen with estradiol and competing for the receptor binding sites, and (3) inhibiting cell signaling activities at ER-deficient cells and consecutively inducing apoptosis (Sassi-Messai et al. 2009). These actions consequently result in hormone imbalance, reduced gonad size, decreased fertility and fecundity, lowered egg production, impaired sexual differentiation, and apoptosis of fish embryos (Tables 2, 3 and 4). However, it is known that the impact of these chemicals on fish reproductive function varies depending on several factors, including species, sex, age, type, and dose of phytoestrogens. The detailed account on mechanism of phytoestrogen on the reproductive endocrine system is given in the following subsections.

3.1 Effect on Gonadotropins and Thyroid Homeostasis

Gonadotropins are secreted from the adenohypophysis by the action of gonadotropin-releasing hormone (GnRH) from the hypothalamus to secret reproductive steroids from gonads. It is believed that phytoestrogens either by directly contributing or by disrupting the estrogen-degrading enzymes increase the circulating level of total estrogen, which in turn reduces the FSH secretion from the pituitary gland. Pelissero et al. (2001) noticed a slight decrease in βFSH levels in plasma at the end of spermatogenesis in male and ovulating female fishes fed with a diet containing 500-ppm genistein. Similarly, feeding Turkey berry leaf extract (which contains solasodine, a phytoestrogen) to *C. carpio* disrupted the balance of the gonadotropin hormone (Rahmadiah et al. 2019).

The enzyme 5′-iodothyronine deiodinase is responsible for the bioconversion of thyroxine (T4) in animals, resulting in most circulating triiodothyronine (T3). T3 is comparatively high in biological activity and induces vitellogenesis in female fishes

Table 2 Effect of phytoestrogens on female fish

Compound and dose	Effect	Species	Reference
100% mixed plant protein containing soybean meal	Reduced proportion of vitellogenic and mature oocytes	*Oreochromis niloticus*	Fontaınhas-Fernandes et al. (2000)
500 ppm of genistein	Increase in plasma VTG, decreased testosterone levels, βFSH, and βLH level	*Oncorhynchus mykiss*	Pelissero et al. (2001)
Genistein at 750 and 30,000 ng/fish	Increased testosterone and estradiol production by the ovaries	*Oryzias latipes*	Zhang et al. (2002)
1000 times higher phytoestrogen than E2 concentration	Binds with estrogen receptors and impairs its function	*Acipenser baeri*	Latonnelle et al. (2002)
1000 μg/L of genistein	Atretic oocytes, enlarged ovarian lumen, somatic stromal tissue proliferation, delayed oocyte maturation, and the presence of primordial germ cells in the ovary	*O. latipes*	Kiparissis et al. (2003)
Genistein at 415,800 ng/g	Increase in plasma Vtg	*Ictalurus punctatus*	Kelly and Green (2006)
Soybean meal at 160 and 340 g/kg of feed	Reduction in the average number of spawned eggs	*Carassius auratus*	Bagheri et al. (2013)
Genistein at 75.83 mg/g with daidzein at 67.82 mg/g	Increase in plasma E2	*C. auratus*	Bagheri et al. (2014)
Genistein 1.6 g per kg of diet	No significant changes in the DHP level	*Huso huso*	Jourdehi et al. (2014)
10 mg/L of genistein and daidzein	Decreased content of ovarian ERβ levels	*Cyprinus carpio*	Sarasquete et al. (2017)
17.5 and 35% of soybean meal	Reduced proportion of vitellogenic and mature oocytes	*C. carpio*	Banani (2019)
Genistein at 1,3, 6, 9 mg/100 g of feed	Increased serum estradiol and testosterone, reduced vitellogenesis and expression level of ovarian aromatase and hepatic estrogen receptors, and reduced vitellogenic oocytes in the ovary	*C. carpio*	Nuzaiba et al. (2020)
Genistein at 10 mg/L in water	Decreased levels of ERβ in the ovary	*Danio rerio*	Sarasquete et al. (2020)
Cacao bean meal at 10 g/kg diet	Increased oocyte granulation and follicle numbers	*C. auratus*	Al-Khalaifah et al. (2020)

(Nelson and Habibi 2016). Genistein disrupted thyroid homeostasis by disrupting the enzyme 5′-iodothyronine deiodinase (Schiller et al. 2013). Genistein also competes with thyroxine for binding to thyroxine-binding globulin (Schiller et al. 2013).

Table 3 Effect of phytoestrogens on male fish

Compounds and dose	Effects	Species	Reference
500 ppm of genistein	Increased vitellogenin synthesis, decrease in testosterone levels, plasma βFSH and βLH and 17a,20b(OH)2	*O. mykiss*	Pelissero et al. (2001)
1000 ppm of genistein	Decreased sperm motility and spermatocrit	*O. mykiss*	Pelissero et al. (2001)
Genistein at 750 and 30,000 ng/fish	E2 level was increased Testosterone level from the testis was decreased	*O. latipes*	Zhang et al. (2002)
1000 μg/L of genistein	Low densities of spermatozoa; 72% of male medaka showed feminized secondary sex characteristics	*O. latipes*	Kiparissis et al. (2003)
Genistein at 58.5 ± 0.6 μg/g with daidzein at 37.3 ± 0.2 μg/g	VTG production was induced, but there was no negative impact on the success of reproduction	*O. latipes*	Inudo et al. (2004)
1–10 mg/L genistein	Induced VTG gene expression and circulating vitellogenin	*O. latipes*	Scholz et al. (2004)
Genistein at 50–100 μg/mL	Increased circulating vitellogenin and a decrease in 11-KT level	*C. auratus*	Ishibashi et al. (2004)
Genistein at 415,800 ng/g	Increased in plasma Vtg	*I. punctatus*	Kelly and Green (2006)
Soybean meal at 46.4 g/100 g feed	Reduction in plasma Vg	*O. mossambicus*	Davis et al. (2009)
Soybean meal at 500 mg/kg	Lowered gonadosomatic index (GSI)	*C. carpio*	Turker and Bozcaarmutlu (2009)
Genistein, 75.833 μg/g with daidzein, 67.821 μg/g	Decrease in plasma testosterone level, while 17-estradiol levels were increased	*C. auratus*	Bagheri et al. (2013)
Soybean meal at 160 and 340 g/kg of feed	Reduction in sperm quality	*C. auratus*	Bagheri et al. (2013)
Genistein at 3 mg/L	Disrupt thyroid homeostasis	*C. auratus*	Nelson and Habibi (2016)
Daidzein at 10 mg/L in water	Increase in expression of the bromodomain testis-specific gene (BRDT) in gonad	*D. rerio*	Sarasquete et al. (2020)
10% inclusion of cacao bean meal	Increased testosterone levels	*C. auratus*	Al-Khalaifah et al. (2020)
1-mg genistein in 100-g feed	Increased serum estradiol and vitellogenin and expression of vtgb2 in the liver	*C. carpio*	Nuzaiba et al. (2022)

Table 4 Effect of phytoestrogens on early life stages

Compound and dose	Effect	Species	Reference
4 mg/g of genistein	Decreased circulating vitellogenin	Striped bass, *Morone saxatilis* juvenile	Pollack et al. (2003)
2 and 8 mg/g of genistein	Increase in circulating vitellogenin	Striped bass, *Morone saxatilis* juvenile	Pollack et al. (2003)
2 mg/kg of phytoestrogens	Increased proportion of females by 55%	*Anguilla anguilla* juvenile	Tzchori et al. (2004)
50% dilution of effluent water from pulp mill	Elevated concentrations of vitellogenin, male-biased sex ratios, and occurrence of inter-sex gonad	*Danio rerio* juvenile	Örn et al. (2006)
Genistein (1×10^{-4} M, 0.5×10^{-4} M, 0.25×10^{-4} M) for 60 h	Teratogenic effects in the embryo including dead cells in the brain, spinal kyphosis, yolk sac edema, and pericardial edema	*D. rerio* at 24 h postfertilization	Kim et al. (2009)
10 μM of genistein	Apoptosis in the embryo	*D. rerio* from 5 days post-hatch	Sassi-Messai et al. (2009)
Genistein at 2, 4, and 8 mg/g	Increased proportions of phenotypically male individuals	*Ictalurus punctatus* at 60 and 150 days post-hatch	Green and Kelly (2009)
17% of soybean meal with 0.05 g/100 g of genistein	Increased proportion of males	*Clarias gariepinus* fry	Ahmed et al. (2015)
3 and 10 mg/L genistein	Reduce thyroxine peroxidase and transthyretin (transfer protein)	*Senegalese sole* early life stages	Sarasquete et al. (2017)
Genistein at 500 ng/L	Increased 17β-estradiol and aromatase activity	*Rutilus kutum* Post-fertilized embryo	Mohammadrezaei and Nematollahi (2018)
Genistein and daidzein (from 1.25 mg/L to 20 mg/L)	Upregulation of estrogen receptor, *cyp1a* transcript levels, and death receptors (*fas*)	*D. rerio* at 2–3 h post-fertilization	Sarasquete et al. (2020)
Genistein @ 4.41 mg/L or daidzein 65.15 mg/L	The lethal concentration of genistein and daidzein	*D. rerio* at 2–3 h post-fertilization	Sarasquete et al. (2020)
Turkey berry (*Solanum torvum*) leaf extract at 300 mg/L	Increased the percentage of female common carp larvae	*Cyprinus carpio* post-hatching larvae	Rahmadiah et al. (2019)

3.2 Effect on Steroidogenesis

Sex steroid synthesis occurs in the gonads, the interrenal gland, and the brain on the action of gonadotropins. The enzymes including aromatases, dehydrogenases, and side chain cleavage enzymes involved in the biosynthesis of various steroids (including 17β-estradiol, progesterone, dehydroepiandrosterone, testosterone and 11-Keto testosterone) from the common precursor, cholesterol. Most of the steroids impart in gonad growth, maturation, ovulation, permeation, spawning, and fertilization. Phytoestrogens, especially isoflavones, are reported to block the transfer of cholesterol (Stevenson et al. 2011) and disrupt certain enzymes of the steroid synthesis pathway, including aromatases (cyp19a and cyp19b), 3-hydroxysteroid dehydrogenase (3βHSD), and 20β-hydroxysteroid dehydrogenase (20βHSD) enzymes and enzymes involved in estrogen-inactivating pathways. This disruption action of phytoestrogen induces an imbalance in steroid hormone levels in fish.

The gonad and bran aromatase enzymes are involved in converting androgens to estrogens, hence testosterone to estradiol during steroidogenesis. Disruption of aromatase enzymes causes an increase in total testosterone and a reduction in total estrogen levels, consequently disturbing the steroidogenesis pathway. An increase in testosterone was observed in *H. huso* (Jourdehi et al. 2014). According to Weber et al. (2002), genistein induced the level of endogenous testosterone as an indication of aromatase inhibition in rainbow trout. Serum level of testosterone was significantly increased in males *C. auratus* fed on the 10% inclusion of cacao bean meal (Al-Khalaifah et al. 2020). The enzyme 3βHSD catalyzes the biosynthesis of the steroids progesterone from pregnenolone, 17-hydroxyprogesterone from 17-hydroxypregnenolone, and androstenedione from dehydroepiandrosterone (DHEA) in the steroid-producing glands. Zhang et al. (2019) reviewed that flavones and isoflavone are interfere with testicular isoforms of 3βHSD and possibly with androgen synthesis. Synthesis of active progesterone in fish, 17α,20β-dihydroxy-4-pregnen-3-one (DHP), is catalyzed by the enzyme 20βHSD from 17-α-hydroxyprogesterone. DHP plays an important role in the final maturation of prophase I-arrested oocytes (Senthilkumaran 2011), testicular recrudescence (Sreenivasulu et al. 2012), and spermiation (Miura et al. 1992). A dose-dependent reduction in the expression of 20βHSD is observed in male *C. carpio* when genistein (1–9-mg/100-g feed) included in the diet (Nuzaiba et al. 2022). Supplementation of dietary genistein at 500 ppm reduced the DHP level in both males and females of *O. mykiss* (Pelissero et al. 2001).

Steroid catabolism occurs in two phases mainly in the liver with some contributions from the gills, intestine, and kidney. In phase I, catabolism occurs through monoxygenation, hydroxylation, and esterification, while phase II catabolic pathway majorly occurs through glucuronidation and sulfonation (James 2011). Enzymes involved in estrogen-inactivating pathways (phase II steroid estrogen-metabolizing enzymes) are glucuronosyltransferase (UGT) and sulfotransferase (SULT). The UGT catabolism of 17β-estradiol is reported in *C. carpio* and in red mullet (Solé et al. 2003; Daidoji et al. 2006; Martin-Skilton et al. 2006)), and the expression is higher in male and juvenile fishes. Similarly, 17β-estradiol catabolism by SULT has been found in zebrafish, Siberian sturgeon, channel catfish, red mullet, and four-spotted megrim (Perdu-Durand and Cravedi 1989; Martin-Skilton et al. 2006; Wang and James 2007; Yasuda et al. 2008). Expression of these enzymes is susceptible to xenobiotic exposure (James 2011), and genistein exposure to the salmonid fish disrupted estrogen-catabolizing enzymes (Ng et al. 2006). Degradation of estrogen-inactivating enzymes by phytoestrogens may significantly increase the total estrogen level in male and juvenile fish (Nezafatian et al. 2017; Mohammadrezaei and Nematollahi 2018).

3.3 Effect on Vitellogenesis in Fish

Estrogen secreted by the gonads is transported through the blood to the liver to synthesize vitellogenin in female fishes. ERβ1 and ERβ2 are two cytoplasmic estrogen receptors reported to exist in teleosts. Hepatic vitellogenesis is based on the type of vitellogenin, and they are more sensitive to the induction by E2. Vitellogenin is normally synthesized from the liver of sexually mature females and is released to the bloodstream and stored in the developing oocytes through receptor-mediated endocytosis (Wahli et al. 1981). Still, in the male, the vitellogenin-producing gene remains silent until they are exposed to any type of xenoestrogens. Hence, the measurement of plasma vitellogenin levels, mainly in males and immature females, is used as a biomarker of exposure to estrogenic compounds in aquatic environments (Cheek et al. 2001; Matozzo et al. 2008). Vitellogenin also aids in hemagglutinating and bacteriostatic functions in male fish if present (Reading et al. 2011).

Phytoestrogens may either have agonistic effects with endogenous estrogen (estrogenic or proestrogenic effect) or act antagonistically to E2 (antiestrogenic effect) (Pelissero et al. 2001; Green and Kelly 2009). The effect is reported to depend on the ratio of phytoestrogens to endogenous estrogens, aromatase activity, species, and reproductive status, length of exposure, and method of administration in fish (Tsai et al. 2000; Trant et al. 2001). Differential effects in both male and female fish conducted in recent studies indicated that the action of phytoestrogens, particularly genistein, depended on the endogenous estradiol level. The agonistic effect of phytoestrogen with animal estrogen is observed when the endogenous level of estradiol is low. Phytoestrogens disrupt the estrogen-metabolizing enzymes, and as a result, the total circulating estradiol level is increased, significant enough to induce the following cascade, including vitellogenesis in a low estradiol environment. The estrogenic potency of these compounds in fish has been documented during the last decades. Various types and levels of phytoestrogens are known to stimulate the vitellogenin concentration in fishes, viz., injection of genistein at 50–100 μg/mL increased the circulating vitellogenin in male *C. auratus* (Ishibashi et al. 2004; Nezafatian et al. 2017; Nezafatian and Zadmajid 2018), 1–10 mg/L of genistein-exposed juvenile and adult males and primary cultures of male liver cells of

O. latipes (Scholz et al. 2004), yearlings of *A. baeri*-fed soybean meal included diet (Pelissero et al. 1991), 1 mg genistein in 100-g feed in adult male *C. carpio* (Nuzaiba et al. 2022), and 2 and 8 mg/g of genistein in juvenile-striped bass, *Morone saxatilis* (Pollack et al. 2003).

When the endogenous estradiol level is high, as, in the case of female fish, antagonistic effects of phytoestrogens are observed, the increase in the estradiol level due to the disruption of estrogen-metabolizing enzymes becomes insignificant. Phytoestrogens in the circulation compete with endogenous estradiol for binding with its membrane and nuclear receptors (ERs), which in turn reduce the estradiol action (antagonistic activity of phytoestrogen). Low level of circulating vitellogenin on administering dietary phytoestrogens was observed in previtellogenic to vitellogenic phase in female *O. mykiss* (Bennetau-Pelissero et al. 2001; Pastore et al. 2018), female *C. carpio* (Turker and Bozcaarmutlu 2009; Nuzaiba et al. 2020), and female *O. latipes* (Zhang et al. 2002). Reduced vitellogenin protein and mRNA concentration in the liver were later described based on the activity and expression of estrogen receptors. Downregulation of *erβ* expression is observed on feeding phytoestrogens (Fritz et al. 2002; Sarasquete et al. 2017).

3.4 Effect on Gonadal Development and Sex Reversal

The effect of environmental estrogenic compounds, including phytoestrogens, is known for impairing gonadal maturation and inducing female characteristics in their male counterparts. The ovary of a 100% plant protein-based diet-fed female *O. niloticus* was observed to contain comparatively a lower number of vitellogenic oocytes (Fontaınhas-Fernandes et al. 2000). A similar result was obtained when soybean meal was fed to *C. carpio* at 17.5 and 35% by Banani (2019). At the same time, dietary cacao bean meal supplementation increased granulation and follicle numbers (Al-Khalaifah et al. 2020). Histological examination of the long-term effect of dietary soybean meal fed in *C. auratus* showed an impact on oocyte maturation progress and spermatogenesis process in female and male fish, respectively (Bagheri et al. 2013), at the inclusion of 35 and 65%. The mean number of eggs spawned by females and sperm quality in males was also reduced on feeding soybean meal at 35 and 65% (Bagheri et al. 2013). In his long-duration study, Bagheri et al. (2014) observed that isoflavone contents affect the GSI of males. Isoflavone extract from soybean-fed *C. carpio* lowered male gonadosomatic index (GSI) but female GSI at low-level extract up to 500 mg/kg supplementation (Turker and Bozcaarmutlu 2009). The fish gonadosomatic index (GSI) did not significantly differ among the three experimental groups; when *O. mykiss was* fed with 500- and 1500-ppm genistein (Pelissero et al. 2001), they also reported delay in spawning in long-term genistein-exposed fishes. Similarly, reproductive success and egg viability of soybean meal included diet in *C. auratus* (Bagheri et al. 2013) and flame angelfish, *Centropyge loriculus* (Callan et al. 2014.

Genistein feeding increased the proportion of females in juvenile eel, *Anguilla anguilla*; Southern flounder, *Paralichthys lethostigma*; and *C. gariepinus* (Tzchori et al. 2004; Ahmed et al. 2015). Kiparissis et al. (2003) observed oocyte development in the testes of adult male *O. latipes*. A greater number of male and intersex populations were observed by Örn et al. (2006) when *D. rerio* was fed with pulp mill effluent rich in xenoestrogens. Sayed et al. (2012) observed an abundance of 77% female population when *C. gariepinus* was fed with 4-nonylphenol.

4 Beneficial Effects of Phytoestrogens

Apart from the endocrine-disrupting activity, phytoestrogens are also reported to induce some beneficial effects in fish. Its effect on the endocrine system can be employed for modulating the endocrine system in super intensive aquaculture system. Phytoestrogens like genistein have induced aromatase inhibition and consecutively produced a monosex population of tilapia. The estrogenic effects of these compounds can be explored to produce a single-sex population in species that shows a higher growth rate in females than male counterparts, as observed in salmon and trout. The antioxidant property of these phytochemicals is focused on in recent research. The dietary genistein (100–500 mg/kg) was found to be instrumental in improving growth, antioxidant capacity, and lipid metabolism in common carp (Yang et al. 2022).

5 Conclusion

Phytoestrogens have been known to exert a major effect on the fish endocrine system and reproductive outcomes. Though the impact would not be visible in aquaculture grow-out systems, loss of awareness on the effect of individual compounds and transgenerational effects are yet to be evaluated. Studies indicated that the accumulation of phytoestrogens in the muscle of fish is insignificant to induce any effect in fish consumers. Further oil extraction does not remove the phytoestrogen contents from the feed ingredients, as these compounds are not readily soluble in lipids. Further, studies on their beneficial roles and use as sex reversal agents also need to be explored with dose and duration specificity for the oral route of administration.

References

Ahmad A, Biersack B, Li Y, Bao B, Kong D, Ali S, Banerjee S, Sarkar FH (2013) Perspectives on the role of isoflavones in prostate cancer. AAPS J 15(4):991–1000

Ahmed Y, Idowu T, Adedeji H, Sogbesan O (2015) Sex reversal and growth performances in *Clarias gariepinus* (Burchell, 1822) fry fed dietary genistein under laboratory conditions. J Agric Food Technol 5:1–4

Al-Khalaifah H, Amer SA, Al-Sadek DM, Khalil AA, Zaki EM, El-Araby DA (2020) Optimizing the growth, health, reproductive performance, and gonadal histology of Broodstock fantail goldfish (Carassius auratus, L.) by dietary cacao bean meal. Animals 10(10):1808

Bagheri T, Imanpoor MR, Jafari V, Bennetau-Pelissero C (2013) Reproductive impairment and endocrine disruption in goldfish by feeding diets containing soybean meal. Anim Reprod Sci 139(1–4):136–144. https://doi.org/10.1016/j.anireprosci.2013.02.003

Bagheri T, Imanpoor MR, Jafari V (2014) Effects of diets containing genistein and Diadzein in a long-term study on sex steroid dynamics of goldfish (Carassius auratus). Toxicol Ind Health 30(2):132–140

Banani M (2019) Comparative assessment of estrogenic activity in female *Cyprinus carpio* (Linnaeus, 1758) fed with soybean or synthetic Diadzein (M.f.Sc. Dissertation). ICAR-central Institute of Fisheries Education (deemed university), Versova, Mumbai

Bennetau-Pelissero C, Breton BB, Bennetau B, Corraze G, Le Menn F, Davail-Cuisset B, Helou C, Kaushik SJ (2001) Effect of genistein-enriched diets on the endocrine process of gametogenesis and on reproduction efficiency of the rainbow trout Oncorhynchus mykiss. Gen Comp Endocrinol 121(2):173–187. https://doi.org/10.1006/gcen.2000.7585

Callan CK, Laidley CW, Kling LJ, Breen NE, Rhyne AL (2014) The effects of dietary Hufa level on flame angelfish (Centropyge loriculus) spawning, egg quality and early larval characteristics. Aquacult Res 45(7):1176–1186

Carrera C, Martínez MJ, Dardanelli J, Balzarini M (2011) Environmental variation and correlation of seed components in nontransgenic soybeans: protein, oil, unsaturated fatty acids, tocopherols, and isoflavones. Crop Sci 51(2):800–809

Cavalcante AK, Lopes-Ferreira M, Rogero SO, Rogero JR (2017) Evaluation of resveratrol toxicity in the embryolarval stage of danio rerio fish. Ecotoxicol Environ Contam 12(1):133–139

Chakraborty SB, Horn P, Hancz C (2014) Application of phytochemicals as growth-promoters and endocrine modulators in fish culture. Rev Aquac 6(1):1–19. https://doi.org/10.1111/raq.12021

Cheek AO, Brouwer TH, Carroll S, Manning S, McLachlan JA, Brouwer M (2001) Experimental evaluation of vitellogenin as a predictive biomarker for reproductive disruption. Environ Health Perspect 109(7):681–690

Chen RT, Wei KQ (2008) Analysis of bioactive aglycone isoflavones in soybean and soybean products. Nutr Food Sci 38(6):540–547. https://doi.org/10.1108/00346650810920132

Cos P, De Bruyne T, Apers S, Berghe DV, Pieters L, Vlietinck AJ (2003) Phytoestrogens: recent developments. Planta Med 69(07):589–599

Daidoji T, Kaino T, Iwano H, Inoue H, Kurihara R, Hashimoto S, Yokota H (2006) Down regulation of bisphenol a glucuronidation in carp during the winter pre-breeding season. Aquat Toxicol 77(4):386–392

Davis LK, Fox BK, Lim C, Hiramatsu N, Sullivan CV, Hirano T, Grau EG (2009) Induction of vitellogenin production in male tilapia (Oreochromis mossambicus) by commercial fish diets. Comp Biochem Physiol A Mol Integr Physiol 154(2):249–254

Flachowsky G, Hünerberg M, Meyer U, Kammerer DR, Carle R, Goerke M, Eklund M (2011) Isoflavone concentration of soybean meal from various origins and transfer of isoflavones into milk of dairy cows. J Verbrauch Lebensm 6(4):449–456

Fontaınhas-Fernandes A, Monteiro M, Figueiredo A, Gomes E, Coimbra J, Reis-Henriques M (2000) Partial or total replacement of fish meal byplant protein affects gonadal development and plasma17 B-estradiol levels in female Nile tilapia. Aquac Int 8:299–313

Fritz WA, Wang J, Eltoum I-E, Lamartiniere CA (2002) Dietary genistein down-regulates androgen and estrogen receptor expression in the rat prostate. Mol Cell Endocrinol 186(1):89–99. https://doi.org/10.1016/S0303-7207(01)00663-3

Green C, Kelly A (2009) Effects of the Estrogen mimic genistein as a dietary component on sex differentiation and Ethoxyresorufin-O-Deethylase (Erod) activity in channel catfish (*Ictalurus punctatus*). Fish Physiol Biochem 35(3):377–384. https://doi.org/10.1007/s10695-008-9260-z

Inudo M, Ishibashi H, Matsumura N, Matsuoka M, Mori T, Taniyama S, Kadokami K, Koga M, Shinohara R, Hutchinson T (2004) Effect of estrogenic activity, and phytoestrogen and organochlorine pesticide contents in an experimental fish diet on reproduction and hepatic vitellogenin production in medaka (*Oryzias latipes*). Comp Med 54(6):673–680

Ishibashi H, Tachibana K, Tsuchimoto M, Soyano K, Tatarazako N, Matsumura N, Tomiyasu Y, Tominaga N, Arizono K (2004) Effects of nonylphenol and phytoestrogen-enriched diet on plasma vitellogenin, steroid hormone, hepatic cytochrome P450 1a, and glutathione-S-transferase values in goldfish (Carassius auratus). Comp Med 54(1):54–62

James MO (2011) Steroid catabolism in marine and freshwater fish. J Steroid Biochem Mol Biol 127(3–5):167–175

Jourdehi AY, Sudagar M, Bahmani M, Hosseini S, Dehghani A, Yazdani M (2014) Comparative study of dietary soy phytoestrogens genistein and Equol effects on growth parameters and ovarian development in farmed female beluga sturgeon, Huso huso. Fish Physiol Biochem 40(1):117–128. https://doi.org/10.1007/s10695-013-9829-z

Kelly AM, Green CC (2006) Binding affinities of hepatic nuclear estrogen receptors for genistein in channel catfish. N Am J Aquac 68(2):160–167. https://doi.org/10.1577/A05-081.1

Kim D-J, Seok S-H, Baek M-W, Lee H-Y, Na Y-R, Park S-H, Lee H-K, Dutta NK, Kawakami K, Park J-H (2009) Developmental toxicity and brain aromatase induction by high genistein concentrations in zebrafish embryos. Toxicol Mech Methods 19(3):251–256

Kiparissis Y, Balch GC, Metcalfe TL, Metcalfe CD (2003) Effects of the isoflavones genistein and Equol on the gonadal development of Japanese medaka, Oryzias latipes. Environ Health Perspect 111(9):1158–1163. https://doi.org/10.1289/ehp.5928

Kowalska A, Kowalski RK, Zakęś Z (2011) The effect of selective cyclooxygenase (cox) inhibitors on Japanese medaka (Oryzias latipes) reproduction parameters. Int J Anim Vet Sci 5(5): 261–265

Latonnelle K, Fostier A, Le Menn F, Bennetau-Pelissero C (2002) Binding affinities of hepatic nuclear estrogen receptors for phytoestrogens in rainbow trout (Oncorhynchus mykiss) and Siberian sturgeon (Acipenser baeri). Gen Comp Endocrinol 129(2):69–79. https://doi.org/10.1016/S0016-6480(02)00512-9

Martin-Skilton R, Lavado R, Thibaut R, Minier C, Porte C (2006) Evidence of endocrine alteration in the red mullet, Mullus barbatus from the Nw Mediterranean. Environ Pollut 141(1):60–68

Matozzo V, Gagné F, Marin MG, Ricciardi F, Blaise C (2008) Vitellogenin as a biomarker of exposure to estrogenic compounds in aquatic invertebrates: a review. Environ Int 34(4): 531–545

Medic J, Atkinson C, Hurburgh CR (2014) Current knowledge in soybean composition. J Am Oil Chem Soc 91(3):363–384

Miura T, Yamauchi K, Takahashi H, Nagahama Y (1992) The role of hormones in the acquisition of sperm motility in salmonid fish. J Exp Zool 261(3):359–363. https://doi.org/10.1002/jez.1402610316

Mohammadrezaei D, Nematollahi M (2018) Effects of genistein and B-sitosterol on early life of Caspian Kutum. ECOPERSIA 6(3):163–169

Nelson ER, Habibi HR (2016) Thyroid hormone regulates vitellogenin by inducing estrogen receptor alpha in the goldfish liver. Mol Cell Endocrinol 436:259–267. https://doi.org/10.1016/j.mce.2016.08.045

Nezafatian E, Zadmajid V (2018) The estrogenic effect of genistein on vitellogenin synthesis in the mature male goldfish (Carassius auratus). J Aquat Ecol 7(4):94–102

Nezafatian E, Zadmajid V, Cleveland BM (2017) Short-term effects of genistein on the reproductive characteristics of male Gibel carp, *Carassius auratus* Gibelio. J World Aquacult Soc 48(5): 810–820. https://doi.org/10.1111/jwas.12399

Ng Y, Hanson S, Malison JA, Wentworth B, Barry TP (2006) Genistein and other isoflavones found in soybeans inhibit estrogen metabolism in salmonid fish. Aquaculture 254(1-4):658–665. https://doi.org/10.1016/j.aquaculture.2005.10.039

Nikolić IL, Savić-Gajić IM, Tačić AD, Savić IM (2017) Classification and biological activity of phytoestrogens: a review. Adv Technol 6(2):96–106

Nuzaiba PM, Varghese T, Gupta S, Sahu NP, Banani M, Srivastava PP, Krishna G (2020) Dietary genistein disrupts sex steroid and vitellogenic response in female common carp, *Cyprinus carpio* L. Aquaculture 522:735062. https://doi.org/10.1016/j.aquaculture.2020.735062

Nuzaiba PM, Varghese T, Gupta S, Sahu NP, Srivastava PP (2022) Estrogenic and vitellogenic responses in genistein fed adult male *Cyprinus carpio*. Aquaculture 548:737559. https://doi.org/10.1016/j.aquaculture.2021.737559

Örn S, Svenson A, Viktor T, Holbech H, Norrgren L (2006) Male-biased sex ratios and vitellogenin induction in zebrafish exposed to effluent water from a Swedish pulp mill. Arch Environ Contam Toxicol 51(3):445–451

Pastore MR, Negrato E, Poltronieri C, Barion G, Messina M, Tulli F, Ballarin C, Maccatrozzo L, Radaelli G, Bertotto D (2018) Effects of dietary soy isoflavones on estrogenic activity, cortisol level, health and growth in rainbow trout, Oncorhynchus mykiss. Aquacult Res 49(4): 1469–1479. https://doi.org/10.1111/are.13602

Pelissero CB, Bennetau B, Babin P, Le Menn F, Dunogues J (1991) The estrogenic activity of certain phytoestrogens in the Siberian sturgeon, *Acipenser baeri*. J Steroid Biochem Mol Biol 38(3):293–299. https://doi.org/10.1016/0960-0760(91)90100-J

Pelissero CB, Breton B, Bennetau B, Corraze G, Le Menn F, Davail Cuisset B, Helou C, Kaushik S (2001) Effect of genistein-enriched diets on the endocrine process of gametogenesis and on reproduction efficiency of the rainbow trout, Oncorhynchus mykiss. Gen Comp Endocrinol 121(2):173–187. https://doi.org/10.1006/gcen.2000.7585

Perdu-Durand E, Cravedi J (1989) Characterization of xenobiotic metabolizing enzymes in sturgeon (*Acipenser baeri*). Comp Biochem Physiol B Biochem Mol Biol 93(4):921–928. https://doi.org/10.1016/0305-0491(89)90067-9

Pollack SJ, Ottinger MA, Sullivan CV, Woods LC III (2003) The effects of the soy isoflavone genistein on the reproductive development of striped bass. N Am J Aquac 65(3):226–234. https://doi.org/10.1577/C02-041

Rahmadiah T, Muslim M, Sasanti AD (2019) Effect of Turkey berry (Solanum torvum) leaf extract on feminization of common carp (Cyprinus carpio). Int J Fish Aquat Res 4(1):30–34

Reading B, Sullivan C, Schilling J (2011) Vitellogenesis in fishes. In: Encyclopedia of fish physiology: from genome to environment, vol 1, pp 635–646

Roca P, Sastre-Serra J, Nadal-Serrano M, Pons DG, del Mar Blanquer-Rossello M, Oliver J (2014) Phytoestrogens and mitochondrial biogenesis in breast cancer. Influence of estrogen receptors ratio. Curr Pharm Des 20(35):5594–5618

Rohmah MK, Salahdin OD, Gupta R, Muzammil K, Qasim MT, Al-Qaim ZH, Abbas NF, Jawad MA, Yasin G, Mustafa YF (2022) Modulatory role of dietary curcumin and resveratrol on growth performance, serum immunity responses, mucus enzymes activity, antioxidant capacity and serum and mucus biochemicals in the common carp, Cyprinus carpio exposed to abamectin. Fish Shellfish Immunol 129:221–230

Sarasquete C, Úbeda-Manzanaro M, Ortiz-Delgado JB (2017) Effects of the soya isoflavone genistein in early life stages of the Senegalese sole, *Solea senegalensis*: thyroid, estrogenic and metabolic biomarkers. Gen Comp Endocrinol 250:136–151. https://doi.org/10.1016/j.ygcen.2017.06.008

Sarasquete C, Úbeda-Manzanaro M, Ortiz-Delgado JB (2020) Soya isoflavones, genistein and daidzein, induce differential transcriptional modulation in the ovary and testis of zebrafish, *Danio rerio*. Aquat Biol 29:79–91. https://doi.org/10.3354/ab00726

Sassi-Messai S, Gibert Y, Bernard L, Nishio S, Ferri Lagneau K (2009) The phytoestrogen genistein affects zebrafish development through two different pathways. PloS One 4:e4935

Sayed AE-DH, Mahmoud UM, Mekkawy IA (2012) Reproductive biomarkers to identify endocrine disruption in Clarias gariepinus exposed to 4-nonylphenol. Ecotoxicol Environ Saf 78:310–319
Schiller V, Wichmann A, Kriehuber R, Muth-Köhne E, Giesy JP, Hecker M, Fenske M (2013) Studying the effects of genistein on gene expression of fish embryos as an alternative testing approach for endocrine disruption. Comp Biochem Physiol Pt C Toxicol Pharmacol 157(1): 41–53. https://doi.org/10.1016/j.cbpc.2012.09.005
Scholz S, Kordes C, Hamann J, Gutzeit HO (2004) Induction of vitellogenin in vivo and in vitro in the model teleost medaka (*Oryzias latipes*): comparison of gene expression and protein levels. Mar Environ Res 57(3):235–244. https://doi.org/10.1016/S0141-1136(03)00082-5
Senthilkumaran B (2011) Recent advances in meiotic maturation and ovulation: comparing mammals and pisces. Front Biosci 16(5):1898–1914
Setchell K (1998) Phytoestrogens: the biochemistry, physiology, and implications for human health of soy isoflavones. Am J Clin Nutr 68(6):1333S–1346S. https://doi.org/10.1093/ajcn/68.6.1333S
Solé M, Raldúa D, Piferrer F, Barceló D, Porte C (2003) Feminization of wild carp, *Cyprinus carpio*, in a polluted environment: plasma steroid hormones, gonadal morphology and xenobiotic metabolizing system. Comp Biochem Physiol Pt C Toxicol Pharmacol 136(2):145–156. https://doi.org/10.1016/S1532-0456(03)00192-3
Soukup ST, Helppi J, Müller DR, Zierau O, Watzl B, Vollmer G, Diel P, Bub A, Kulling SE (2016) Phase ii metabolism of the soy isoflavones genistein and daidzein in humans, rats and mice: a cross-species and sex comparison. Arch Toxicol 90(6):1335–1347
Sreenivasulu G, Senthilkumaran B, Sridevi P, Rajakumar A, Rasheeda M (2012) Expression and immunolocalization of 20β-hydroxysteroid dehydrogenase during testicular cycle and after Hcg induction, in vivo in the catfish, Clarias gariepinus. Gen Comp Endocrinol 175(1):48–54. https://doi.org/10.1016/j.ygcen.2011.09.002
Stevenson LM, Brown AC, Montgomery TM, Clotfelter ED (2011) Reproductive consequences of exposure to waterborne phytoestrogens in male fighting fish betta splendens. Arch Environ Contam Toxicol 60(3):501–510. https://doi.org/10.1007/s00244-010-9561-y
Torrens-Mas M, Roca P (2020) Phytoestrogens for cancer prevention and treatment. Biology 9(12): 427. https://doi.org/10.3390/biology9120427
Tou JC, Chen J, Thompson LU (1998) Flaxseed and its lignan precursor, Secoisolariciresinol Diglycoside, affect pregnancy outcome and reproductive development in rats. J Nutr 128(11): 1861–1868
Trant JM, Gavasso S, Ackers J, Chung BC, Place AR (2001) Developmental expression of cytochrome P450 aromatase genes (Cyp19a and Cyp19b) in zebrafish fry (Danio rerio). J Exp Zool 290(5):475–483
Tsai CL, Wang LH, Chang CF, Kao CC (2000) Effects of gonadal steroids on brain serotonergic and aromatase activity during the critical period of sexual differentiation in tilapia, Oreochromis mossambicus. J Neuroendocrinol 12(9):894–898
Turker H, Bozcaarmutlu A (2009) Effect of total isoflavones found in soybean on vitellogenin production in common carp. Kafkas Universitesi Veteriner Fakultesi Dergisi 15(4):561–568. https://doi.org/10.9775/kvfd.2009.062-A
Tzchori I, Degani G, Elisha R, Eliyahu R, Hurvitz A, Vaya J, Moav B (2004) The influence of phytoestrogens and oestradiol-17β on growth and sex determination in the European eel *(Anguilla anguilla*). Aquacult Res 35(13):1213–1219. https://doi.org/10.1111/j.1365-2109.2004.01129.x
Wahli W, Dawid IB, Ryffel GU, Weber R (1981) Vitellogenesis and the vitellogenin gene family. Science 212(4492):298–304
Wang L-Q, James MO (2007) Sulfonation of 17β-Estradiol and inhibition of sulfotransferase activity by polychlorobiphenylols and celecoxib in channel catfish, *Ictalurus punctatus*. Aquat Toxicol 81(3):286–292
Wang TY, Li Q, Bi KS (2018) Bioactive flavonoids in medicinal plants: structure, activity and biological fate. Asian J Pharmaceut Sci 13(1):12–23

Weber LP, Kiparissis Y, Hwang GS, Niimi AJ, Janz DM, Metcalfe CD (2002) Increased cellular apoptosis after chronic aqueous exposure to nonylphenol and quercetin in adult medaka (*Oryzias latipes*). Comp Biochem Physiol Pt C Toxicol Pharmacol 131(1):51–59. https://doi.org/10.1016/S1532-0456(01)00276-9

Xiao Y, Zhang S, Tong H, Shi S (2018) Comprehensive evaluation of the role of soy and isoflavone supplementation in humans and animals over the past two decades. Phytother Res 32(3): 384–394. https://doi.org/10.1002/ptr.5966

Yang L, Zhang W, Zhi S, Zhao M, Liu M, Qin C, Feng J, Yan X, Nie G (2022) Evaluation of dietary genistein on the antioxidant capacity, non-specific immune status, and fatty acid composition of common carp (Cyprinus carpio. L). Aquaculture 550:737822

Yasuda T, Yasuda S, Williams FE, Liu M-Y, Sakakibara Y, Bhuiyan S, Snow R, Carter G, Liu M-C (2008) Characterization and ontogenic study of novel steroid-Sulfating Sult3 sulfotransferases from zebrafish. Mol Cell Endocrinol 294(1-2):29–36

Zhang L, Khan IA, Foran CM (2002) Characterization of the estrogenic response to genistein in Japanese medaka (*Oryzias latipes*). Comp Biochem Physiol Pt C Toxicol Pharmacol 132(2): 203–211. https://doi.org/10.1016/S1532-0456(02)00067-4

Zhang S, Mo J, Wang Y, Ni C, Li X, Zhu Q, Ge R-S (2019) Endocrine disruptors of inhibiting testicular 3β-hydroxysteroid dehydrogenase. Chem Biol Interact 303:90–97

A Proteomic Approach to Studying the Effects of Xenobiotics on Aquatic Living Organisms

Nevil Pinto, B. S. Yashwanth, Gopika Radhakrishnan, and A. Sathiyanarayanan

1 Introduction

Fish play an essential role in food sectors and also emerged as a model organism for biomedical research. Fish habitats are mainly categorized as freshwater, marine, and brackish water. In recent intensive aquaculture production with various controlling measures against microbes such as using therapeutics and discharging of harmful chemicals to the aquatic system leads to deterioration in the quality of meat and wild fish both in capture and aquaculture species. In recent years, there have been an increasing awareness and a rising concern with regard to the disruption of the endogenous hormone system of wildlife exposed to environmental pollutants. The physiology and endocrinology of living organisms are disrupted by a variety of man-made chemicals. The reproductive process in fish is a continuous process throughout its life cycle. Thus, the reproductive system is susceptible to xenobiotics throughout its entire life cycle, i.e., fertilization, embryonic development, maturity, and breeding stage. There is ample evidence that aquatic organisms living in or around pollutants can be adversely affected by bioaccumulating xenobiotic chemicals. To understand the precise mechanism between the host reproductive development and environment, it is necessary to know the entire genome, proteome, and metabolome (Bhat et al. 2016). Before the entire genome, gene expression and microarray analysis have been a vital role in the understanding of molecular pathways underlying physiological responses. However, the recent "omics" (genomics, proteomics, and metabolomics) revolution greatly enhances the new integrative approach that has gained the right platform for understanding biomarker research.

N. Pinto (✉) · B. S. Yashwanth · A. Sathiyanarayanan
ICAR-CIFE, Versova, Andheri (W), Mumbai, Maharashtra, India

G. Radhakrishnan
Department of Mathematics and Natural Science, Institute of Marine Research, University of Bergen, Bergen, Norway

M. A. Rather et al. (eds.), *Xenobiotics in Aquatic Animals*,
https://doi.org/10.1007/978-981-99-1214-8_10

For environmental monitoring, the welfare condition of fish is crucial such as the consumption of food, hepatosomatic index (HSI) and gonadosomatic index (GSI) are the main physiological responses observed during xenobiotics. It is important to identify the relationship between physiological response and biochemical biomarkers (Solé et al. 2009; Sundaray et al. 2021). During exposure to xenobiotics, each biomarker shows stronger specificity in organisms; therefore, it has been highly considered to be included in monitoring programs (Cajaraville et al. 2000).

Proteomics is an upcoming research area, which appears as a strong branch for a global protein study, including their quantification and localization of protein in tissue, and possible post-modifications are being reflected the greatest set of technologies to prove protein changes that occurred in aquatic organisms. Proteomics is a complement to genomics, which is conceptually defined as "the use of quantitative protein-level measurements of gene expression to characterize biological processes (e.g., disease processes and drug effects) and decipher the mechanisms of gene expression control" (Anderson and Anderson 1998; Zafar et al. 2020). In toxicity lists, toxicology pathways are significant canonical pathways. A xenobiotic insult leads to adaptive, defensive, or reparative responses that are formed through functional groupings based on critical biological processes (Nuez-Ortín et al. 2018). Proteomics in aquaculture have also been entered recently to find antigenic proteins and to explore the physiology, detection of proteins which are differentially regulated, and impact of pollutants in aquatic organisms. The new zone of bioinformatics information through mass spectrometry (MS) in aquatic organisms is necessary for spontaneous data processing and validation to identify peptide/protein. After the invention of MS technique, proteome repositories came into existence publicly like PeptideAtlas under the Institute for Systems Biology for targeted proteomic workflow (Deutsch et al. 2008), Global Proteome Machine Database (GPMDB) (Craig et al. 2004), the University of Texas' Open Proteomics Database (opd), and PRIDE Archive http://www.ebi.ac.uk/pride (Martens et al. 2005) which gain value recently. Identification of proteins firstly emerged through Edman degradation as picomoles to advanced MS-based proteomic approaches in femtomole have evolved rapidly in clinical proteomics. However, proteomic concept debates the relationships of the DNA revolution and indicates some of the most fruitful directions for future assessment.

2 Proteomics

The word "proteomics" was first introduced in 1995 (Wilkins et al. 1996). It is an emergent technology utilized for the study of complete set of protein in the organism which is encoded by a genome. Technology has progressed in several studies such as to design vaccines for tick-borne diseases (Marcelino et al. 2012), meat science (Paredi et al. 2012, 2013), dairy science (E Hernandez-Castellano et al. 2014), establishing ideal stress markers and settling welfare issues in aquaculture (Marco-Ramell et al. 2016), diverse uses in expression work in aquatic organisms (Rodrigues et al. 2012), or setting up markers for pollution in aquatic organisms (Campos et al. 2012). A strong link of proteins in diseased conditions result in promising baseline applications of clinical proteomics for initial detection and target site for therapeutics (Petricoin et al. 2004). After successful entry of genomic field in the fishery DNA sequence database accumulated in enormous quantities, then researchers realize that merely having complete sequences of genomes will be not abundant to clarify the biological function of the organism. A cell which is usually for its survival depends upon an assembly of metabolic and regulatory pathways (Pandey and Mann 2000). Two major steps are necessary to understand the generation of manifold protein forms, the mRNA process, and the level of translation during process. Changes that occur in transcriptome level, i.e., the mRNA itself, can be exposed to frequent adjustments such as alternative splicing, editing of mRNA, and polyadenylation. Through this modification, single gene can provide numerous protein isoforms and later formed protein further subject to subject to posttranslational modifications (Ciereszko et al. 2012).

Genome programs for aquaculture species is still not well developed, but the growing number of recent projects in the world is based on RNA or DNA, thus allowing the considerable access to novel genomic facts from a significant number of aquatic organisms, especially in farmed species (Campos and De Almeida 2016). It is essential to obtain species-specific/family-specific genome sequence to produce an entire proteome map of a particular family. Fish as models (zebrafish) have been used as an effective way in biological aspects of many human diseases, in areas like toxicology, cancer, neurology, infectious diseases, and drug development (Amatruda and Patton 2008; Tognoli 2010). Proteomic studies that observed economically important species such as salmon, cod, and catfish were reviewed by Rodrigues et al. (2012)).

3 Evolution of Proteomic Research

The Human Genome Project succeeded in investigating the structure of the human genome provided for the study of genes, proteins, and metabolites at the same time. For the first time in the last 12 years, the proteomic community of India has overcome an extensive way in contributing to the first human draft map of 30 normal tissues to detect all the proteome encrypted by 17,294 genes in the human (Kim et al. 2014). The human proteome map publication is considered a crowning achievement for proteome study in the world. In non-model organisms such as fish, complete proteome data is not available, except the model organism such as zebrafish. The main areas of research in aquaculture are related to improving health status of fish, reproduction, and developmental biology (Schultz et al. 2013; Karlsen et al. 2011; Talakhun et al. 2014; Klinbunga et al. 2012; Marie et al. 2011) to understand the immunology and physiology for species that represent commercial and wild stock (Provan et al. 2013; Braceland et al. 2013). The first database of fish, FISHPROT (http://www.cifri.ernet.in/Fishprot/index.html), in India was developed by the

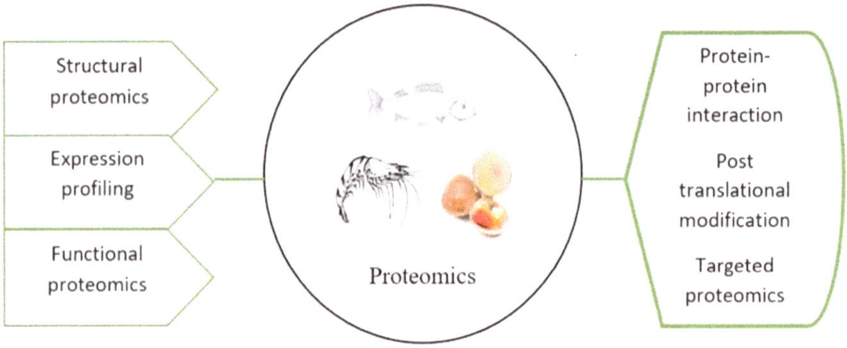

Fig. 1 Different types of proteomic approach

Central Inland Fisheries Research Institute (CIFRI), Barrackpore, especially for biomarker discovery and evaluation. FISHPROT is a web-based database created by open-source technologies Apache, PHP, and MySQL that contain liver proteome of the murrel *Channa striatus*, muscle proteome of catla, riverine catfishes *Sperata seenghala* and *Sperata aor*, plasma proteome of rohu, and muscle and lens proteome of riverine catfish *Rita rita*. The financial support of proteomic research in India was also focused by the government through its different departments such as the Department of Biotechnology (DBT), Department of Science and Technology (DST), Indian Council of Medical Research (ICMR), Council of Scientific and Industrial Research (CSIR), and Board of Research in Nuclear Sciences (BRNS). In India, a freshwater sector dominated with carp culture and marine area is recently improving. Recently, the whole proteome of *Labeo rohita* was developed for the aquaculture community using 19 tissue including gonads and the brain (Nissa et al. 2021). Figure 1 illustrates the different proteomic approaches for different applications depending on the target study.

4 Types of Proteomic

Structural proteomics is used to ascertain all the proteins that represented by a protein complex, define their location, and characterize all protein-protein interaction pathways. Through structural proteomics, isolation of subcellular-specific organelles by purification can be more comfortable (Jung et al. 2000). Functional proteomics provides information on the elucidation of the biological function of protein groups or proteins and classes on a proteome-wide level. This could include the use of ligand-specific protein isolation or the isolation of protein complexes (Graves and Haystead 2002) in differential expression quantitative study of expression between samples that differ by some variable. In this approach, protein expression of the entire proteome between samples can be matched and later recognize disease-specific markers for targeted proteomics (Graves and Haystead 2002).

Among all types, differential proteomics play a vital role in toxicology and disease management in aquaculture.

5 Biomarkers in Xenobiotics

Among all the tissue, the liver is known to be the most important eco-monitoring organ in fish. The liver was selected as a biomarker mainly as a metabolizing organ for foreign compounds (xenobiotics). During stress, antioxidants enzyme expressed and counteract the action of oxyradicals derived from any aerobic process, including xenobiotic exposure. In seminal plasma, saxitoxin- and tetrodotoxin-binding protein 1 identified in pikeperch suggests its role in protecting reproductive tract tissue spermatozoa by detoxification mechanisms (Dietrich et al. 2021). Exposure to bisphenol A (1 μg/L) activates the NRF2-mediated oxidative stress response in zebrafish ovaries (Molina et al. 2021). In Atlantic cod, candidate biomarkers such as alpha-1 antitrypsin, alpha-enolase, and apolipoproteins were expressed for crude oil, alkyl phenols, and polycyclic aromatic hydrocarbons (PAHs) (Bohne-Kjersem et al. 2009). In the case of caged flounder, proteins involved in the glycolysis pathway and in the Krebs cycle and meantime fatty acids and the metabolism of sterols were reduced. Apart from these pathways in flounder, stress proteins changed such as chaperones HSP70, DnaJ; oxidation with thioredoxins, peroxyredoxin, and recycling system like proteasome, 4 ubiquitin carboxy-terminal hydrolases was dysregulated after exposure of cages in polluted site (Borcier et al. 2019). In three species of South American freshwater fish, male specimens were treated with estradiol and observed phosphorylation ed. and glycosylation of vitellogenin (Urdaneta et al. 2018). In zebrafish larvae, 3,4-dichloroaniline was exposed for 96 h and modulated proteins belonging to lipids and hormone metabolism (Apoa1 and Apoa1b and vitellogenins), as well as proteins important for developmental processes and organogenesis (Myhc4, Acta2, Sncb, and Marcksb) (Vieira et al. 2020).

6 Methods Used in the Identification of Xenobiotic Markers

6.1 *Quantitative Proteomics*

Quantitative proteomics is used to reveal the species-specific responses of the molecule to bacterial infection. In this workflow, relative and absolute are the two forms of quantification used to express the protein. In relative quantification, it is associated with the protein fold change in abundance by estimating the levels of a specific protein in various samples, whereas in absolute quantitation the exact amount of protein has to be estimated. Traditional gel based techniques are preferred in quantitative proteomics, e.g., 1D and 2D gel electrophoresis and 2-D difference

Table 1 Available option for protein quantification

Peptide labeling	Protein labeling	Label-free LC-MS	DIGE using 2-DE
iTRAQ with four (or eight) samples labeled and then LC-MS	SILAC with two samples labeled, digested, and then LC-MS	Spectral counting or ion current measurements on MS	Two samples fluorescently labeled, 2-DE, and image analysis

gel electrophoresis (DIGE), and in gel-free methods, with the recent upgrading of databases and the relative decline in equipment price that can be separated further into label-free approaches (e.g., spectral counting and absolute quantitation) and labeling approaches such as stable isotope labeling (SILAC), isotope-coded affinity tags (ICAT), and isobaric tagging (iTRAQ®) (Martyniuk and Denslow 2009). In the gel-based approaches, the proteins isolated electrophoretically are later identified by MS to quantify the expression (Bagchi et al. 2015). The DIGE (chemical labeling) was familiarized to assess the special effects of xenobiotics in aquatic species (Apraiz et al. 2006). The different quantification platforms for proteomic analysis are shown in Table 1.

6.2 *Techniques Used to Quantify Proteome*

The greatest part in proteomics is handling and obtaining the protein sample. The contribution by the Indian proteomic sector has been securely growing with prominence on a biomarker for diverse queries related to clinical conditions using such as two-dimensional electrophoresis and MS (Sirdeshmukh 2006). In proteomic research, the steps are as follows: (1) isolation and separation of proteins, (2) determination of protein (the amino acid sequence), and (3) available database search for identified protein sequence (Graves and Haystead 2002).

6.3 *Gel-Based Quantification*

6.3.1 Two-Dimensional Gel Electrophoresis (2-DE)

Proteomics was originated in 1975 through the introduction of the 2-DE gel to identify proteins and initiated to build databases of proteins (O'Farrell 1975). Proteomic analysis requires protein isolation from a sample containing a complex mixture of protein, their separation by traditional 2D gel electrophoresis, and staining of the proteins in the gel. During 2-DE, proteins as per their respective net charge are resolved as isoelectric focusing (IEF) in the first dimension and according to their size/molecular mass (SDS-PAGE) for second-dimensional electrophoresis. As a result of utilizing these two different resolving techniques, much greater resolution as compared to normal SDS-PAGE is achieved by the detection of protein

Table 2 Software used in gel-based proteomics

Software	Company	Type
BioNumerics 2D	BioSystematica	Spot-based
Dymension	Syngene	Spot-based
ImageMaster	GE HealthCare	Spot-based
Melanie	GeneBio, GE HealthCare	Spot-based
PDQuest	Bio-Rad	Spot-based
Progenesis SameSpots	Nonlinear Dynamics	Warping
Progenesis and REDFIN	Nonlinear Dynamics	Warping
Delta 2D	DECODON	Warping
DeCyder	GE HealthCare	Spot-based

spots of 1500–3000 (Brewis and Brennan 2010). Protein spots are detected by classical protein staining methods; then, a software is used (Table 2) to pick the significant spot in the gel (Natale et al. 2011). There are some disadvantages, i.e., gel-to-gel dissimilarity is common, and limited throughput and linear dynamic range still pose challenges to gel-based proteomics. In aquaculture, the 2-DE method has increasingly become famous, and several results are observed (Link et al. 2006; Forne et al. 2011; Silvestre et al. 2010).

6.4 Gel-Free Quantification

The unceasing development and cost-effectiveness of MS-based methods were made recently in life science to mitigate the current apprehensions regarding reproducibility and restrictions on the ability to study certain classes of proteins (Abdallah et al. 2012). The gel-free is a promising strategy that employs a few stages in which proteins were firstly isolated, digested, eluted on chromatography, and detected by mass spectrometry (MS) for absolute or relative quantification. The gel-free method is also termed as "bottom-up" approach. Here, proteins are processed initially, and analyses are done at the peptide level for MS. Two main steps included in gel-free quantification are label and label-free strategies.

6.4.1 Label-Free-Based Method

Label-free methods are progressing in the proteomic field, but their innovation in Indian aquaculture is still underdeveloped. LFQ (label-free quantification) methods use in-solution shotgun applications, an alternative tool in making global proteome characterizations in a less time period. LFQ is most consistent, precise, and accurate that can allow a higher efficient range of quantification (Lundgren et al. 2010; Soares et al. 2012; Deracinois et al. 2013). The protein is digested enzymatically and exposed to chromatography for detaching peptides, and these peptides are transferred to mass spectrometry unit where *m/z* (mass to charge) ratios are analyzed. For higher resolution, these peptides are permitted to pass in triple-quadrupole mass analyzer, where corresponding collision energies are liable for collecting information for precursor referred to as tandem MS (MS/MS). The MS/MS data delivers the truthful information of proteins, and identification is done by comparing peptide masses obtained from analysis to that in the protein databases (Zargar et al. 2016). The positive benefit of this technique is that it can compare MS signal intensities and spectra after proper separation and detection as many as protein samples on LC-MS/MS.

LFQ is useful for relative quantification to ascertain global and targeted approaches. In relative quantification, data-dependent acquisition of MS and MS/MS-based label-free analysis has been broadly categorized into spectral intensity (no MS/MS) and spectral count (MS/MS) measurement (Chelius and Bondarenko 2002). LFQ can also be used for absolute quantification, which assesses the protein abundance based on the peptide numbers detected and hypothetically observed tryptic peptides for each protein. For this LFQ analysis, software such as MaxQuant, Progenesis QIP, Proteome Discoverer, and Scaffold Q+ (Al Shweiki et al. 2017) has been used to detect the spectral peak.

6.4.2 Label-Based Method

Label-based tactics rely on specific isotope tags having specific groups that can chemically/metabolically or enzymatically label the proteins and peptides (Bantscheff et al. 2007). The labeled peptides were separated on LC first and later analyzed by highly delicate MS technique that produces extremely useful data for retrieving accurate results (Zargar et al. 2016). The label-based technique promises the advanced well-programmed high-throughput quantitative proteomic analysis of anonymous proteins with reliable multiplexing and automated facilities. There are several labeling methods available such as chemical labeling (ICAT for proteins, iTRAQ, ICPL, TMT for peptide), tandem mass tag, enzymatic labeling, and metabolic labeling (SILAC). Protease, such as trypsin, is used in most of the research for digesting proteins due to some reasons. Trypsin cleaves at the carboxyl side of lysine and arginine and residues of proteins into its peptides with an average size ranging from 700 to 1500 Daltons. Trypsin is highly active and can tolerate some additives and can be altered by the methylation of lysines to prevent self-digestion at these sites.

6.5 Mass Spectrometry (MS)

The development of mass spectrometry, a highly reproducible technique, holds the key to fulfilling aspirations. MS deals with the accurate measurement of charged ions and mainly measures the mass-to-charge ratio of charged species under a vacuum.

MS prototype was used first to measure *m/z* of the electron, in which JJ Thomson was awarded Nobel Prize in 1906 (Thomson 1913), and to measure mass, in 1948–1952, time-of-flight (TOF) mass analyzers came into existence, and in 1955, quadrupole ion filters were introduced by W. Paul, who also invented the ion trap in 1983 (and won Nobel Prize in 1989) (Griffiths 2008). In the 1980s, two main approaches to MS protein identification, "peptide-mass mapping" approach, suggested by Henzel et al. (1993)) where matrix-assisted laser desorption/ionization (MALDI) is used that results in a time-of-flight (TOF) distribution of the peptides, compile the mixture. MALDI acquire hundreds of protein spots and can be excised, and digested enzymatically, obtained mass spectra were automatically searched against databases (Jensen et al. 1997; Berndt et al. 1999). Here, ion is formed through short laser pulses, and the sample is deposited in a sample plate and then embedded in a matrix that initiates ionization. Another method is by peptide sequencing known as electrospray ionization (ESI), achieved by spraying a solution through a charged needle at atmospheric pressure toward the inlet of the MS (Patterson and Aebersold 2003). ESI is coupled with ion trap or triple-quadrupole MS/MS spectrometers.

6.6 Mass Analyzer

Molecule that ionized produces molecular and fragment ions which are formed in the source region of a mass spectrometer and moved into a mass analyzer by an electric field. There are two broad categories of mass analyzers: ion beam and scanning mass spectrometers, such as time of flight (TOF) and quadrupoles (Q), and the trapping mass spectrometers, such as ion traps (IT), Orbitrap, Fourier transform (FT), and ion cyclotron resonance (ICR). There are advanced hybrid instruments which combine more than one mass analyzer (Yates et al. 2009). The first TOF analyzer isolates precursor ions through a velocity filter, and the second TOF separates the fragment ions and ion storage, ion isolation (the precursor ion/ions are trapped, while ions out of interest are destabilized and ejected out of the analyzer). The technique is used to fragment ions by collision-induced dissociation (CID), electron transfer dissociation (ETD), electron capture dissociation (ECD), and electron detachment dissociation (EDD) (Chalkley et al. 2005). Mass analyzers such as IT, O, and ICR separate ions based on their *m/z* resonance frequency, quadrupoles (Q) use *m/z* stability, and time-of-flight (TOF) analyzers use flight time. The mass analyzer can quantify the mass-to-charge ratio with high resolution (up to 150,000 $m/\Delta m$, where m denotes mass) and high mass accuracy (to <1 part per million) (Tyers and Mann 2003).

7 Database Searching and Software

The first software tool SEQUEST was introduced for peptide matching MS/MS spectra in 1994 (Eng et al. 1994). In 1999, Mascot was developed by Matrix Science, UK (http://www.matrixscience.com), and was described by Perkins et al. (1999)). To achieve this goal, a protein database was constructed in a FASTA format that consists of theoretical proteomes derived from all fully sequenced bacterial genomes. The large protein database Human Protein Reference Database (HPRD), a database initiated for human proteomics by the Institute of Bioinformatics (http://www.hprd.org/), was victorious in the country. For more substantial data set of the human proteome, the mass spectrometry-derived label-free approach was used in a human proteome map database (http://humanproteomemap.org/) (Gowda et al. 2015).

8 Proteomic Application in Fisheries

8.1 *In Culture Species*

Using proteomic tool, the quality of gametes, embryos, and dietary components, aspects directly related to fertility, developmental competence, and growth of alevins can be analyzed. In trout, the proteome of the coelomic fluid of females was studied to identify proteins that disappear or accumulate simultaneously with the decrease in viability of the egg during the postovulatory period (Rime et al. 2004). Sperm quality was analyzed in European sea bass (*Dicentrarchus labrax*) to verify whether the protein expression of sperm was affected by cryopreservation (Zilli et al. 2005). Maturation in the fishes is essential to avoid the current sexing procedure that requires an invasive surgical examination; in Persian sturgeon (*Acipenser persicus*), the identification of mature male and female gonads was determined to identify proteins involved in the gonad maturation and sex determination (Keyvanshokooh et al. 2009). For muscle growth, intact and proteolytic fragments of muscle-specific gene products are identified, and it has been investigated in yellow perch and sea bream (Reddish et al. 2008).

8.2 *Authentication and Food Safety in Aquaculture*

To ensure food quality and safety, authentication of aquaculture products represents a central issue for the assessment. Identification through visual inspection of processed products becomes quite a big task as anatomical and morphological characteristics. In 2-DE, isoelectric electric focusing has been applied to discriminate different fish and shrimp (Etienne et al. 2000; Piñeiro et al. 2000; Rehbein et al.

2000; Tepedino et al. 2001). However, the approach was available for raw fish and the baked products; the IEF method is only applicable to fish species which show a specific pattern with the heat-stable parvalbumins (PRVBs) (Rehbein 1992). The PRVBs, used as a biomarker, which showed high variability in their primary structure give differences in protein patterns. FDA have collected the profiles of different fish species IEF in the internet library Regulatory Fish Encyclopedia (https://www.fda.gov/Food/FoodScienceResearch/RFE/ucm071647.htm). Allergens are another concern in the seafood; the presence of allergenic proteins like tropomyosin has been a critical determinant of the quality of crustacean food products. In shrimp, protein-profiling study showed that high-pressure steaming reduced the level of tropomyosin due to protein degradation, therefore reducing the allergenicity of tropomyosin compared with other heat treatment methods (Lasekan and Nayak 2016). Foodborne pathogen found in seafood is also studied by proteomics.

8.3 Welfare in Aquaculture

It is defined by Ashley as the freedom from hunger and thirst, discomfort, pain, injury, disease, distress, and fear and the freedom to express normal behavior (Ashley 2007). Usually, welfare is strictly related to stress in intensive culture organisms, where proteomics focused either on the liver (due to its central role in most key metabolic processes) or bodily fluids like blood plasma (due to its value in providing nonlethal proteomics-based diagnostic tools); some of them also target the brain, skeletal muscle, osmoregulatory and immune-related organs, and tissues and existing proteomic studies within aquaculture. Studies on the whole larvae, liver (Martin et al. 2001; Vilhelmsson et al. 2004), heart (Lee et al. 2006), kidney (Martin et al. 2007), and muscle (Wang 2009) have been reported using 2-DE-based proteome analyses. Indeed, proteomics can be a powerful tool to investigate the evolution, biodiversity, and physiological adaptations of zebrafish and other fish species. For welfare, fish health aspects have taken a major part, with special focus on viral infection, bacterial diseases, parasites, vaccine development, and hepatic tumors, and skeletal deformities in fish have been studied (Booy et al. 2005; Dios et al. 2008; Xiong et al. 2010; Yan et al. 2011). The molecular metabolic indicator of chronic stress has been proposed in gilthead sea bream (*Sparus aurata*), using a comparative proteomic approach (Alves et al. 2010).

8.4 Immunoproteomics

It is the study of large sets of proteins involved in the immune response; in European sea, bass mucus (*Dicentrarchus labrax*) has been presented as a strategy that could be applied in disease diagnosis (Cordero et al. 2015). The major concern in aquaculture is disease, so both hosts and pathogens are highly concerned, where the hosts include fish, shrimp, crab, and shellfish by using immunoproteomic approach to identify broad cross-immunogens (Li et al. 2009, 2010). To investigate serum profile of yellow croaker infected with *A. hydrophila,* a combined differential proteomic approach was used (EST Library) after post-immunization challenge and to characterize the targeted molecules (Chen et al. 2010). *A. hydrophila*-infected ayu (*Plecoglossus altivelis*) liver proteins were identified using 2-DE-based proteomics, and an observed natural killer cell-enhancing factor-B (NKEF-B), an upregulated protein in the infected fish, was identified (Chen et al. 2011).

8.5 Nutriproteomics

In aqua products, mainly the emphasis is on the contact of bioactive food ingredients with proteins, and they are scrutinized following two different ways: (1) gene expression and (2) examining its posttranslational modifications or for small molecule/protein interactions (Piñeiro et al. 2010). The nutritional differences between the cultured and wild stock of fish can also be monitored by nutriproteomics (Erickson 2005; Monti et al. 2005; Reddish et al. 2008). Another challenge in sustainable aquaculture is the diet composition by replacing a fish meal or oil and the addition of new ingredients to mitigate these issues; nutrioproteomics is a promising field for actual understanding of fish metabolism.

8.6 Climate Change and Toxicology

Applicable for estimating the effect of toxicity in the environment as well as to assessing climate change and its effects on aquatic organisms, toxicology allows a better understanding of the mechanisms behind changes in fish and invertebrate protein brought by variations due to biotic or abiotic factors. Ecological protein biomarkers and stress responses define the specific issues in the aquatic environment (Amelina et al. 2007; Nesatyy et al. 2006). Some proteins such as metallothionein and cytochrome p450 in "sentinel" marine species act as excellent indicators for stress in the environment and potential health threats for humans (Stewart et al. 2008). Algal toxins are major contaminants of the marine, a widespread phenomenon during climate change, posing significant threats to aquatic as well as humans and among them, and microcystins are major algal toxins in the aquatic environment. Therefore, mollusks of the genus *Mytilus* have been used as major natural sentinels in seawater for aquatic toxicity. Species, such as medaka fish (*Oryzias latipes*), zebrafish (*Danio rerio*), mosquitofish (*Gambusia affinis*), and fathead minnow (*Pimephales promelas*), have been among the most commonly used models for various ecotoxicology and biomedical research (Forné et al. 2010). After the recent invention of iTRAQ which has been used in new biomarkers and toxicity signatures, the discovery of early markers in drug toxicity, the target organ analysis, and the curiosity for ecotoxicology is that many proteins usually utilized as bioindicators of toxicity or stress are quantifiable using iTRAQ on a larger scale, providing a global baseline of biological effect from which to assess changes in the proteome (Martyniuk et al. 2012). In zebrafish ovaries, 14 days of BPA exposure cause premature of oocytes and proteins altered in oxidative stress, metabolic shifts, and degradome perturbations (Molina et al. 2021).

Microplastic (MP) is a severe issue in the aquatic system, and fish are expected to experience chronic exposure and to bioaccumulate the plastic particles potentially. In invertebrates such as crustaceans, barnacles, polychaete worms, mussels, and amphipods have ingested MP fragments in controlled studies (Browne et al. 2008; Graham and Thompson 2009). Advanced techniques in proteomics that may hold promise for upcoming reviews in the area include pyrolysis combined with chromatography, mass spectrometry, infrared spectroscopy, scanning electron microscopy with energy dispersive X-ray spectroscopy, field flow fractionation with pyrolysis, and multiangle light scattering (Bouwmeester et al. 2015).

9 Conclusion

Proteins are an essential component of cells, and fish proteomes differ from xenobiotics and normal proteins, making these proteins ideal targets for identifying eco-monitoring markers. Proteomics is an essential tool for verifying gene products by proteomic methods. It is also the first step in “annotating the genome.” However, in many aquatic organisms, whole-genome maps are not available, making it difficult to identify appropriate biomarkers. Understanding fish and shellfish physiological functions is heavily dependent on proteomics. It has been shown that xenobiotics modulate fish reproductive and developmental stages. To improve our understanding of physiological responses to stress in fish, as well as fish welfare management, it is crucial to acquire the major proteins that affect testis maturation, sperm quality, and ovarian regulation, which are crucial to identifying reliable biomarkers of fish welfare.

References

Abdallah C, Dumas-Gaudot E, Renaut J, Sergeant K (2012) Gel-based and gel-free quantitative proteomics approaches at a glance. Int J Plant Genom 2012:494572

Al Shweiki MR, Mönchgesang S, Majovsky P, Thieme D, Trutschel D, Hoehenwarter W (2017) Assessment of label-free quantification in discovery proteomics and impact of technological factors and natural variability of protein abundance. J Proteome Res 16:1410–1424

Alves RN, Cordeiro O, Silva TS, Richard N, De Vareilles M, Marino G, Di Marco P, Rodrigues PM, Conceição LE (2010) Metabolic molecular indicators of chronic stress in gilthead seabream (Sparus aurata) using comparative proteomics. Aquaculture 299:57–66

Amatruda JF, Patton EE (2008) Genetic models of cancer in zebrafish. Int Rev Cell Mol Biol 271:1–34

Amelina H, Apraiz I, Sun W, Cristobal S (2007) Proteomics-based method for the assessment of marine pollution using liquid chromatography coupled with two-dimensional electrophoresis. J Proteome Res 6:2094–2104

Anderson NL, Anderson NG (1998) Proteome and proteomics: new technologies, new concepts, and new words. Electrophoresis 19:1853–1861

Apraiz I, Mi J, Cristobal S (2006) Identification of proteomic signatures of exposure to marine pollutants in mussels (Mytilus edulis). Mol Cell Proteomics 5:1274–1285

Ashley PJ (2007) Fish welfare: current issues in aquaculture. Appl Anim Behav Sci 104:199–235

Bagchi D, Swaroop A, Bagchi M (2015) Genomics, proteomics and metabolomics in nutraceuticals and functional foods. Wiley, Hoboken, NJ

Bantscheff M, Schirle M, Sweetman G, Rick J, Kuster B (2007) Quantitative mass spectrometry in proteomics: a critical review. Anal Bioanal Chem 389:1017–1031

Berndt P, Hobohm U, Langen H (1999) Reliable automatic protein identification from matrix-assisted laser desorption/ionization mass spectrometric peptide fingerprints. Electrophoresis 20: 3521–3526

Bhat IA, Rather MA, Saha R, Pathakota GB, Pavan-Kumar A, Sharma R (2016) Expression analysis of Sox9 genes during annual reproductive cycles in gonads and after nanodelivery of LHRH in Clarias batrachus. Res Vet Sci 106:100–106

Bohne-Kjersem A, Skadsheim A, Goksøyr A, Grøsvik BE (2009) Candidate biomarker discovery in plasma of juvenile cod (*Gadus morhua*) exposed to crude North Sea oil, alkyl phenols and polycyclic aromatic hydrocarbons (PAHs). Mar Environ Res 68(5):268–277

Booy A, Haddow J, Ohlund L, Hardie D, Olafson R (2005) Application of isotope coded affinity tag (ICAT) analysis for the identification of differentially expressed proteins following infection of Atlantic Salmon (Salmo salar) with infectious hematopoietic necrosis virus (IHNV) or Renibacterium salmoninarum (BKD). J Proteome Res 4:325–334

Borcier E, Artigaud S, Gaillard JC, Armengaud J, Charrier G, Couteau J, Receveur J, Ouddane B, Diop M, Amara R, Laroche J (2019) Coupling caging and proteomics on the European flounder (*Platichthys flesus*) to assess the estuarine water quality at micro scale. Sci Total Environ 695: 133760

Bouwmeester H, Hollman PC, Peters RJ (2015) Potential health impact of environmentally released micro-and nanoplastics in the human food production chain: experiences from nanotoxicology. Environ Sci Technol 49:8932–8947

Braceland M, Bickerdike R, Tinsley J, Cockerill D, Mcloughlin M, Graham D, Burchmore R, Weir W, Wallace C, eckersall, P. (2013) The serum proteome of Atlantic salmon, Salmo salar, during pancreas disease (PD) following infection with salmonid alphavirus subtype 3 (SAV3). J Proteomics 94:423–436

Brewis IA, Brennan P (2010) Proteomics technologies for the global identification and quantification of proteins. In: Advances in protein chemistry and structural biology. Elsevier, Amsterdam

Browne MA, Dissanayake A, Galloway TS, Lowe DM, Thompson RC (2008) Ingested microscopic plastic translocates to the circulatory system of the mussel, Mytilus edulis (L.). Environ Sci Technol 42:5026–5031

Cajaraville MP, Bebianno MJ, Blasco J, Porte C, Sarasquete C, Viarengo A (2000) The use of biomarkers to assess the impact of pollution in coastal environments of the Iberian Peninsula: a practical approach. Sci Total Environ 247(2–3):295–311

Campos AM, De Almeida AM (2016) Top-down proteomics and farm animal and aquatic sciences. Proteomes 4:38

Campos A, Tedesco S, Vasconcelos V, Cristobal S (2012) Proteomic research in bivalves: towards the identification of molecular markers of aquatic pollution. J Proteomics 75:4346–4359

Chalkley RJ, Baker PR, Huang L, Hansen KC, Allen NP, Rexach M, Burlingame AL (2005) Comprehensive analysis of a multidimensional liquid chromatography mass spectrometry dataset acquired on a quadrupole selecting, quadrupole collision cell, time-of-flight mass

spectrometer II. New developments in protein prospector allow for reliable and comprehensive automatic analysis of large datasets. Mol Cell Proteomics 4:1194–1204
Chelius D, Bondarenko PV (2002) Quantitative profiling of proteins in complex mixtures using liquid chromatography and mass spectrometry. J Proteome Res 1:317–323
Chen X, Wu Z, Yu S, Wang S, Peng X (2010) Beta2-microglobulin is involved in the immune response of large yellow croaker to Aeromonas hydrophila: a proteomic based study. Fish Shellfish Immunol 28:151–158
Chen G, Zhang C, Li C, Wang C, Xu Z, Yan P (2011) Haemocyte protein expression profiling of scallop Chlamys farreri response to acute viral necrosis virus (AVNV) infection. Dev Comp Immunol 35:1135–1145
Ciereszko A, Dietrich M, Nynca J (2012) The identification of seminal proteins in fish: from a traditional approach to proteomics. J Appl Ichthyol 28:865–872
Cordero H, Brinchmann MF, Cuesta A, Meseguer J, Esteban MA (2015) Skin mucus proteome map of European sea bass (Dicentrarchus labrax). Proteomics 15:4007–4020
Craig R, Cortens JP, Beavis RC (2004) Open source system for analyzing, validating, and storing protein identification data. J Proteome Res 3:1234–1242
Deracinois B, Flahaut C, Duban-Deweer S, Karamanos Y (2013) Comparative and quantitative global proteomics approaches: an overview. Proteomes 1:180–218
Deutsch EW, Lam H, Aebersold R (2008) Peptide Atlas: a resource for target selection for emerging targeted proteomics workflows. EMBO Rep 9:429–434
Dietrich MA, Judycka S, Żarski D, Malinowska A, Świderska B, Palińska-Żarska K, Błażejewski M, Ciereszko A (2021) Proteomic analysis of pikeperch seminal plasma provides novel insight into the testicular development of domesticated fish stocks. Animal 15(7):100279
Dios S, Novoa B, Buonocore F, Scapigliati G, Figueras A (2008) Genomic resources for immunology and disease of salmonid and non-salmonid fish. Rev Fish Sci 16:119–132
Eng JK, Mccormack AL, YATES, J. R. (1994) An approach to correlate tandem mass spectral data of peptides with amino acid sequences in a protein database. J Am Soc Mass Spectrom 5:976–989
Erickson BE (2005) Government and society: EuPO changes name to EuPA| proteomics reveals effects of aquaculture on fish. ACS Publications, Washington, DC
Etienne M, Jérôme M, Fleurence J, Rehbein H, Kündiger R, Mendes R, Costa H, Pérez-Martín R, Piñeiro-González C, Craig A (2000) Identification of fish species after cooking by SDS–PAGE and urea IEF: a collaborative study. J Agric Food Chem 48:2653–2658
Forné I, Abián J, Cerdà J (2010) Fish proteome analysis: model organisms and non-sequenced species. Proteomics 10:858–872
Forne I, Castellana B, Marín-Juez R, Cerda J, Abián J, Planas JV (2011) Transcriptional and proteomic profiling of flatfish (Solea senegalensis) spermatogenesis. Proteomics 11:2195–2211
Gowda H, Chatterjee A, Prasad TK (2015) Protein databases from India. Proteom Res India 26
Graham ER, Thompson JT (2009) Deposit-and suspension-feeding sea cucumbers (Echinodermata) ingest plastic fragments. J Exp Mar Biol Ecol 368:22–29
Graves PR, Haystead TA (2002) Molecular biologist's guide to proteomics. Microbiol Mol Biol Rev 66:39–63
Griffiths J (2008) A brief history of mass spectrometry. Anal Chem 80:5678–5683
Henzel WJ, Billeci TM, Stults JT, Wong SC, Grimley C, Watanabe C (1993) Identifying proteins from two-dimensional gels by molecular mass searching of peptide fragments in protein sequence databases. Proc Natl Acad Sci 90:5011–5015
Hernandez-Castellano LE, Almeida AM, Castro N, Arguello A (2014) The colostrum proteome, ruminant nutrition and immunity: a review. Curr Protein Pept Sci 15:64–74
Jensen ON, Mortensen P, Vorm O, Mann M (1997) Automation of matrix-assisted laser desorption/ ionization mass spectrometry using fuzzy logic feedback control. Anal Chem 69:1706–1714
Jung E, Heller M, Sanchez JC, Hochstrasser DF (2000) Proteomics meets cell biology: the establishment of subcellular proteomes. Electrophoresis 21:3369–3377

Karlsen OA, Bjørneklett S, Berg K, Brattås M, Bohne-Kjersem A, Grøsvik BE, Goksøyr A (2011) Integrative environmental genomics of cod (Gadus morhua): the proteomics approach. J Toxic Environ Health A 74:494–507
Keyvanshokooh S, Kalbassi MR, Hosseinkhani S, Vaziri B (2009) Comparative proteomics analysis of male and female Persian sturgeon (Acipenser persicus) gonads. Anim Reprod Sci 111:361–368
Kim M-S, Pinto SM, Getnet D, Nirujogi RS, Manda SS, Chaerkady R, Madugundu AK, Kelkar DS, Isserlin R, Jain S (2014) A draft map of the human proteome. Nature 509:575
Klinbunga S, Petkorn S, Kittisenachai S, Phaonakrop N, Roytrakul S, Khamnamtong B, Menasveta P (2012) Identification of reproduction-related proteins and characterization of proteasome alpha 3 and proteasome beta 6 cDNAs in testes of the giant tiger shrimp Penaeus monodon. Mol Cell Endocrinol 355:143–152
Lasekan AO, Nayak B (2016) Effects of buffer additives and thermal processing methods on the solubility of shrimp (Penaeus monodon) proteins and the immunoreactivity of its major allergen. Food Chem 200:146–153
Lee J, Valkova N, White MP, Kültz D (2006) Proteomic identification of processes and pathways characteristic of osmoregulatory tissues in spiny dogfish shark (Squalus acanthias). Comp Biochem Physiol Pt D Genom Proteom 1(3):328–343
Li H, Xiong X-P, Peng B, Xu C-X, Ye M-Z, Yang T-C, Wang S-Y, PENG, X.-X. (2009) Identification of broad cross-protective immunogens using heterogeneous antiserum-based immunoproteomic approach. J Proteome Res 8:4342–4349
Li H, Ye M-Z, Peng B, Wu H-K, Xu C-X, Xiong X-P, Wang C, Wang S-Y, Peng X-X (2010) Immunoproteomic identification of polyvalent vaccine candidates from Vibrio parahaemolyticus outer membrane proteins. J Proteome Res 9:2573–2583
Link V, Shevchenko A, Heisenberg C-P (2006) Proteomics of early zebrafish embryos. BMC Dev Biol 6:1
Lundgren DH, Hwang S-I, Wu L, Han DK (2010) Role of spectral counting in quantitative proteomics. Expert Rev Proteomics 7:39–53
Marcelino I, De Almeida AM, Ventosa M, Pruneau L, Meyer DF, Martinez D, Lefrançois T, Vachiéry N, Coelho AV (2012) Tick-borne diseases in cattle: applications of proteomics to develop new generation vaccines. J Proteomics 75:4232–4250
Marco-Ramell A, De Almeida AM, Cristobal S, Rodrigues P, Roncada P, Bassols A (2016) Proteomics and the search for welfare and stress biomarkers in animal production in the one-health context. Mol Biosyst 12:2024–2035
Marie B, Zanella-Cléon I, Guichard N, Becchi M, Marin F (2011) Novel proteins from the calcifying shell matrix of the Pacific oyster Crassostrea gigas. Marine Biotechnol 13:1159–1168
Martens L, Hermjakob H, Jones P, Adamski M, Taylor C, Gevaert K, Vandekerckhove J, Apweiler R (2005) PRIDE: the proteomics identifications database. Proteomics 5:3537–3545
Martin SA, Cash P, Blaney S, Houlihan DF (2001) Proteome analysis of rainbow trout (Oncorhynchus mykiss) liver proteins during short term starvation. Fish Physiol Biochem 24 (3):259–270
Martin SAM, Mohanty BP, Cash P, Houlihan DF, Secombes CJ (2007) Proteome analysis of the Atlantic salmon (Salmo salar) cell line SHK-1 following recombinant IFN-gamma stimulation. Proteomics 7:2275–2286
Martyniuk CJ, Denslow ND (2009) Towards functional genomics in fish using quantitative proteomics. Gen Comp Endocrinol 164:135–141
Martyniuk CJ, Alvarez S, Denslow ND (2012) DIGE and iTRAQ as biomarker discovery tools in aquatic toxicology. Ecotoxicol Environ Saf 76:3–10
Molina AM, Abril N, Lora AJ, Huertas-Abril PV, Ayala N, Blanco C, Moyano MR (2021) Proteomic profile of the effects of low-dose bisphenol A on zebrafish ovaries. Food Chem Toxicol 156:112435
Monti G, De Napoli L, Mainolfi P, Barone R, Guida M, Marino G, Amoresano A (2005) Monitoring food quality by microfluidic electrophoresis, gas chromatography, and mass

spectrometry techniques: effects of aquaculture on the sea bass (Dicentrarchus labrax). Anal Chem 77:2587–2594

Natale M, Maresca B, Abrescia P, Bucci EM (2011) Image analysis workflow for 2-D electrophoresis gels based on ImageJ. Proteoms Insights 4:S7971

Nesatyy VJ, Ammann AA, Rutishauser BV, Suter MJ-F (2006) Effect of cadmium on the interaction of 17β-estradiol with the rainbow trout estrogen receptor. Environ Sci Technol 40:1358–1363

Nissa MU, Pinto N, Mukherjee A, Reddy PJ, Ghosh B, Sun Z, Ghantasala S, Chetanya C, Shenoy SV, Moritz RL, Goswami M (2021) Organ-based proteome and post-translational modification profiling of a widely cultivated tropical water fish, *Labeo rohita*. J Proteome Res 21(2):420–437

Nuez-Ortín WG, Carter CG, Nichols PD, Cooke IR, Wilson R (2018) Liver proteome response of pre-harvest Atlantic salmon following exposure to elevated temperature. BMC Genomics 19(1): 1–13

O'Farrell PH (1975) High resolution two-dimensional electrophoresis of proteins. J Biol Chem 250: 4007–4021

Pandey A, Mann M (2000) Proteomics to study genes and genomes. Nature 405:837

Paredi G, Raboni S, Bendixen E, De Almeida AM, Mozzarelli A (2012) "Muscle to meat" molecular events and technological transformations: the proteomics insight. J Proteomics 75: 4275–4289

Paredi G, Sentandreu M-A, Mozzarelli A, Fadda S, Hollung K, De Almeida AM (2013) Muscle and meat: new horizons and applications for proteomics on a farm to fork perspective. J Proteomics 88:58–82

Patterson SD, Aebersold RH (2003) Proteomics: the first decade and beyond. Nat Genet 33:311

Perkins DN, Pappin DJ, Creasy DM, Cottrell JS (1999) Probability-based protein identification by searching sequence databases using mass spectrometry data. Electrophoresis 20:3551–3567

Petricoin E, Wulfkuhle J, Espina V, Liotta LA (2004) Clinical proteomics: revolutionizing disease detection and patient tailoring therapy. J Proteome Res 3:209–217

Piñeiro C, Barros-Velázquez J, Pérez-MARTÍN RI, Gallardo JM (2000) Specific enzyme detection following isoelectric focusing as a complimentary tool for the differentiation of related gadoid fish species. Food Chem 70:241–245

Piñeiro C, Cañas B, Carrera M (2010) The role of proteomics in the study of the influence of climate change on seafood products. Food Res Int 43:1791–1802

Provan F, Jensen L, Uleberg K, Larssen E, Rajalahti T, Mullins J, Obach A (2013) Proteomic analysis of epidermal mucus from sea lice–infected Atlantic salmon, Salmo Salar L. J Fish Dis 36:311–321

Reddish JM, St-Pierre N, Nichols A, Green-Church K, Wick M (2008) Proteomic analysis of proteins associated with body mass and length in yellow perch, Perca flavescens. Proteomics 8: 2333–2343

Rehbein H (1992) Parvalbumins as marker proteins for the fish species in fishery products. In: Developments in food science, pp 399–405

Rehbein H, Kündiger R, Pineiro C, Perez-Martin RI (2000) Fish muscle parvalbumins as marker proteins for native and urea isoelectric focusing. Electrophoresis 21:1458–1463

Rime H, Guitton N, Pineau C, Bonnet E, Bobe J, Jalabert B (2004) Post-ovulatory ageing and egg quality: a proteomic analysis of rainbow trout coelomic fluid. Reprod Biol Endocrinol 2:26

Rodrigues PM, Silva TS, Dias J, Jessen F (2012) Proteomics in aquaculture: applications and trends. J Proteomics 75:4325–4345

Schultz IR, Nagler JJ, Swanson P, Wunschel D, Skillman AD, Burnett V, Smith D, Barry R (2013) Toxicokinetic, toxicodynamic, and toxicoproteomic aspects of short-term exposure to trenbolone in female fish. Toxicol Sci 136:413–429

Silvestre F, Tu HT, Bernard A, Dorts J, Dieu M, Raes M, Phuong NT, Kestemont P (2010) A differential proteomic approach to assess the effects of chemotherapeutics and production management strategy on giant tiger shrimp Penaeus monodon. Comp Biochem Physiol Pt D Genom Proteom 5:227–233

Sirdeshmukh R (2006) Fostering proteomics in India. ACS Publications, Washington, DC

Soares R, Franco C, Pires E, Ventosa M, Palhinhas R, Koci K, De Almeida AM, Coelho AV (2012) Mass spectrometry and animal science: protein identification strategies and particularities of farm animal species. J Proteomics 75:4190–4206

Solé M, Rodríguez S, Papiol V, Maynou F, Cartes JE (2009) Xenobiotic metabolism markers in marine fish with different trophic strategies and their relationship to ecological variables. Comp Biochem Physiol Pt C Toxicol Pharmacol 149(1):83–89

Stewart JR, Gast RJ, Fujioka RS, Solo-Gabriele HM, Meschke JS, Amaral-Zettler LA, Del Castillo E, Polz MF, Collier TK, Strom MS (2008) The coastal environment and human health: microbial indicators, pathogens, sentinels and reservoirs. Environ Health 7 Suppl 2:S3

Sundaray JK, Rather MA, Kumar S, Agarwal D (2021) Recent updates in molecular endocrinology and reproductive physiology of fish. Springer, Singapore

Talakhun W, Phaonakrop N, Roytrakul S, Klinbunga S, Menasveta P, Khamnamtong B (2014) Proteomic analysis of ovarian proteins and characterization of thymosin-β and RAC-GTPase activating protein 1 of the giant tiger shrimp Penaeus monodon. Comp Biochem Physiol Pt D Genom Proteom 11:9–19

Tepedino V, Berrini A, Borromeo V, Gaggioli D, Cantoni C, Manzoni P, Secchi C (2001) Identification of commercial fish species belonging to the orders pleuronectiformes and gadiformes: library of isoelectric focusing patterns. J AOAC Int 84:1600–1607

Thomson JJ (1913) Bakerian lecture: rays of positive electricity. Proc R Soc Lond A 89:1–20

Tognoli C (2010) Molecular biomarkers for fish welfare and species authentication. Università degli Studi dell'Insubria

Tyers M, Mann M (2003) From genomics to proteomics. Nature 422:193

Urdaneta V, Camafeita E, Poleo G, Guerrero H, Bernal C, Galindo-Castro I, Diez N (2018) Proteomic characterization of vitellogenins from three species of south American fresh water fish. Lat Am J Aquat Res 46(1):187–196

Vieira LR, Hissa DC, De Souza TM, Sá CA, Evaristo JA, Nogueira FC, Carvalho AF, Farias DF (2020) Proteomics analysis of zebrafish larvae exposed to 3, 4-dichloroaniline using the fish embryo acute toxicity test. Environ Toxicol 35(8):849–860

Vilhelmsson OT, Martin SA, Médale F, Kaushik SJ, Houlihan DF (2004) Dietary plant-protein substitution affects hepatic metabolism in rainbow trout (Oncorhynchus mykiss). Br J Nutr 92 (1):71–80

Wang Y (2009) Use of proteomic technique in the quality and safety of aquatic product. Feed Ind 30:58–60. (in Chinese)

Wilkins MR, Sanchez J-C, Gooley AA, Appel RD, Humphery-Smith I, Hochstrasser DF, Williams KL (1996) Progress with proteome projects: why all proteins expressed by a genome should be identified and how to do it. Biotechnol Genet Eng Rev 13:19–50

Xiong X-P, Zhang B-W, Yang M-J, Ye M-Z, Peng X-X, Li H (2010) Identification of vaccine candidates from differentially expressed outer membrane proteins of vibrio alginolyticus in response to NaCl and iron limitation. Fish Shellfish Immunol 29:810–816

Yan F, Zhang Y, Jiang R, Zhong M, Hu Z, Du H, Lun J, Chen J, Li Y (2011) Identification and agglutination properties of hemocyanin from the mud crab (Scylla serrata). Fish Shellfish Immunol 30:354–360

Yates JR, Ruse CI, Nakorchevsky A (2009) Proteomics by mass spectrometry: approaches, advances, and applications. Annu Rev Biomed Eng 11:49–79

Zafar I, Pervez MT, Rather MA, Babar ME, Raza MA, Iftikhar R (2020) Genome-wide identification and expression analysis of PPOs and POX gene families in the selected plant species. Biosci Biotechnol Res Asia 17(2):301–318

Zargar S, Gupta N, Mir R, Rai V (2016) Shift from gel based to gel free proteomics to unlock unknown regulatory network in plants: a comprehensive review. J Adv Res Biotechnol 1:19

Zilli L, Schiavone R, Zonno V, Rossano R, Storelli C, Vilella S (2005) Effect of cryopreservation on sea bass sperm proteins. Biol Reprod 72:1262–1267

Role of Cytochrome P450 in Xenobiotic Metabolism in Fishes (Review)

Saima Andleeb, Muddasir Sharief Banday, Sumaira Rashid, Irfan Ahmad, Mehak Hafeez, Oyas Asimi, Mohd Ashraf Rather, Sajad H. Baba, Azra Shah, Nakeer Razak, Anees Fatima, and Ishraq Hussain

Abbreviations

ATR	Atrazine
CPF	Chlorpyrifos
CYP	Cytochrome P450
FAD	Flavin adenine dinucleotide reductase

S. Andleeb · S. Rashid · M. Hafeez · A. Shah · N. Razak
Division of Fish Genetics and Biotechnology, Faculty of Fisheries, Sher-e-Kashmir University of Agricultural Sciences and Technology, Kashmir, Rangil, Ganderbal, Jammu and Kashmir, India

M. S. Banday
Department of Clinical Pharmacology, Sher-e-Kashmir Institute of Medical Sciences, Soura, Srinagar, Jammu and Kashmir, India

I. Ahmad (✉) · M. A. Rather
Division of Fish Genetics and Biotechnology, Sher-e-Kashmir University of Agricultural Sciences and Technology, Kashmir, Rangil, Ganderbal, Jammu and Kashmir, India
e-mail: ahmadirfan@skuastkashmir.ac.in

O. Asimi
Division of Fish Nutrition and Biochemistry, Faculty of Fisheries, Sher-e-Kashmir University of Agricultural Sciences and Technology, Kashmir, Rangil, Ganderbal, Jammu and Kashmir, India

S. H. Baba
Division of Agri Economics and Hotibusiness Management, Sher-e-Kashmir University of Agricultural Sciences and Technology, Kashmir, Shalimar, Jammu and Kashmir, India

A. Fatima
Department of Botany, Govt College for Women, Cluster University-Kashmir, Srinagar, Jammu and Kashmir, India

I. Hussain
Division of Vety Biochemistry, FVSc & AH, Sher-e-Kashmir University of Agricultural Sciences and Technology, Kashmir, Shuhama, Jammu and Kashmir, India

M. A. Rather et al. (eds.), *Xenobiotics in Aquatic Animals*,
https://doi.org/10.1007/978-981-99-1214-8_11

FMN	Flavin mononucleotide
NADH	Nicotinamide adenine dinucleotide
NADPH	Nicotinamide adenine dinucleotide phosphate

1 Introduction

Due to human interventions, huge loads of pollutants enter into aquatic environment through various sources like dumping and disposal, increased industrialization, and direct discharge. Various studies on cytochrome P450 (CYP) have revealed its use as a biomarker for aquatic contamination (Lee et al. 2005). Molecular biomarkers such as CYP have been shown to be very useful for the detection of fatal disturbances in fish (Bucheli and Fent 1995). They respond to a wide variety of xenobiotics and therefore detect the presence of both known and unknown pollutants relevant for organisms (Lemaire et al. 2010). CYP enzymes help in the transformation of environmental contaminants like harmful drugs and carcinogens and in the disintegration of endogenous substrates like prostanoids, steroids, vitamins, and fatty acids (Havelkova et al. 2007).

Cytochrome P450 was first explained by Klingenberg in 1948, and since then, this enzyme system has been studied intensively. In fishes, the first CYP gene was first isolated from rainbow trout followed by some other fishes in the late 1980s (Stegeman 1989; Winston et al. 1988; Uno et al. 2012). Cytochromes are generally most prevalent in the endoplasmic reticulum or mitochondria of the liver which accounts for 1 to 2% mass of hepatocytes (Kilemade et al. 2009). However, cytochromes are also present in other organs like the olfactory system, heart, gonads, kidney, gills, brain, alimentary canal, and placenta (Arukwe 2002; Arellano et al. 2009; Siroka and Dratichova 2004). Cytochrome was discovered as a pigment with maximum absorption at 450 nm, thus got its name as cytochrome P450; however, the inactive form of CYP has maximum absorption at 420 nm, same as other hemoproteins (Schenkman and Jansson 1998).

Based on the transfer of NADPH electrons to the catalytic site, P450 enzymes are classified into four classes (Table 1) (Werck-Reichhart and Feyereisen 2000).

The cytochrome P450 Standardized Nomenclature Committee suggested categorization based on the degree of similarity between amino acid sequences and has classified P450 genes as isoforms, families, and subfamilies (Nelson 1999). A CYP gene is granted in a subfamily when the homology percentage is greater than 55% and in a family when it is greater than 40% (Nelson 1999). But this type of classification has been argued due to the new sequences that are being described. At the VII P450 International Symposium, a different classification based on biological P450 functions was recommended (Kelly et al. 2006). So far, 18 CYP families are identified in fishes, viz., CYP1, CYP2, CYP3, CYP4, CYP5, CYP7, CYP8, CYP11, CYP17, CYP19, CYP20, CYP21, CYP24, CYP26, CYP27, CYP39,

Table 1 Classification of CYP

Class I	Flavin adenine dinucleotide reductase (FAD) and ferric redoxin sulfur are required Commonly found in eukaryotes Helps in detoxification
Class II	Need flavin mononucleotide (FMN) Commonly found in eukaryotes Helps in detoxification
Class III	There is no requirement for any electron donors
Class IV	Accepts electrons from NADPH (nicotinamide adenine dinucleotide phosphate)

CYP46, and CYP51, out of which only 8 families are studied in detail, i.e., CYP1, CYP2, CYP3, CYP4, CYP11, CYP17, CYP19, and CYP26.

The main functions of different CYP families along with the respective species in which they are found are summarized in Table 2.

CYP has been identified from fresh, marine, and brackish-water fish. Some freshwater fish include Atlantic salmon, rainbow trout, catfish, zebrafish, carp, Chinook salmon, crucian carp, pufferfish, rohu, catla, mrigal carp, medaka, Japanese medaka, common whitefish, toad fish, tilapia, killifish, stripey sea perch, winter flounder, mummichog, fathead minnow, bluegill, blue gourami, and guppy. Some marine water fish include Atlantic croaker, mangrove killifish, European sea bass, marine flatfish, southern stingray, and dogfish shark, while Japanese pufferfish and rita are some examples of brackish-water fish.

All CYP families are found in the liver of respective fish species except CYP11. The sites of induction of different CYP families and subfamilies are summarized in Table 3.

NADPH (nicotinamide adenine dinucleotide phosphate)-cytochrome P450 reductase and the phospholipid membrane fraction are the two key factors influencing CYP activity. The general monooxygenase reaction mediated by CYP manifests as:

$$RH + O_2 + NADPH + H^+ \xrightarrow{\text{Cytochrome P450}} ROH + H_2O + NADP^+$$

In the above monooxygenase reaction, due to the insertion of an oxygen atom, one molecule becomes more polar than the other. In actual, the entire reaction is much more complicated because the cytochrome may utilize oxygen from peroxides in addition to molecular oxygen and NADH may also supply electrons (Shalan et al. 2018).

As depicted in the above reaction, NADPH reductase and membrane phospholipids are also required. The function of NADPH reductase is to transfer electrons on cytochrome P450 with the help of FAD (flavin adenine dinucleotide) and FMN (flavin mononucleotide) prosthetic groups. The detailed schematic representation of this reaction is illustrated in Fig. 1.

Table 2 Functions of CYP families in the respective fish species

Family	Functions	Species	References
CYP 1	Hydroxylation of pregnenolone, metabolism of xenobiotics, probable regulator of gas and fluids in gills Helps in embryogenesis, detoxification, and excretion	Atlantic salmon (*Salmo salar*), Atlantic croaker (*Micropogonias undulatus*), Japanese pufferfish (*Takifugu rubripes*), rainbow trout (*Oncorhynchus mykiss*), catfish (*Ancistrus multispinis*), zebrafish (*Danio rerio*), carp (*Cyprinus carpio*), pufferfish (*Takifugu obscures*), Chinook salmon (*Oncorhynchus tshawytscha*), Rita (*Rita rita*), crucian carp (hybridized *Prussian* carp), mangrove killifish (*Rivulus marmoratus*), European sea bass (*Dicentrarchus labrax*)	Lee et al. (2005), Klemz et al. (2010), Tuan et al. (2014), Rahman and Thomas (2012), Sakamoto et al. (2003), Zanette et al. (2009), Meyer et al. (2002), Brammell et al. (2010), Jung et al. (2011), Stien et al. (1998), Arukwe (2002) and Kim et al. (2004, 2008)
CYP 2	Metabolism of nitrosodialkylamines Metabolism of xenobiotic Hydroxylation of lauric acid Epoxylation of arachidonic acid	Rainbow trout (*Oncorhynchus mykiss*), Japanese pufferfish (*Takifugu rubripes*), striped sea perch (*Lutjanus carponotatus*), threadfin butterfly (*Chaetodon auriga*), atoll butterfly (*Chaetodon mertensii*), zebrafish (*Danio rerio*), graysby sea bass (*Cephalopholis cruentata*), tomtate grunt (*Haemulon aurolineatum*), channel catfish (*Ictalurus punctatus*), rohu (*Labeo rohita*), *Catla catla*, mrigal carp (*Cirrhinus mrigala*)	Ruus et al. (2002), Kaplan et al. (1999), Yang et al. (2000), Wang-Buhler et al. (2005), Haasch (2002), Oleksiak et al. (2000, 2003), Schlenk et al. (2002), Buhler et al. (1994) and Yang et al. (1998)
CYP 3	Metabolism of xenobiotics Hydroxylation of testosterone	Japanese pufferfish (*Takifugu rubripes*), toad fish (*Opsanus tau*), zebrafish (*Danio rerio*), rohu (*Labeo rohita*), *Catla catla*, mrigal carp (*Cirrhinus mrigala*), rainbow trout (*Oncorhynchus mykiss*), European sea bass (*Dicentrarchus labrax*), medaka (*Oryzias latipes*),	Lee et al. (2001), Lee and Buhler (2003), Nelson (2003), Barber et al. (2007), Christen et al. (2010), Kullman and Hinton (2001) and Kashiwada et al. (2005)

(continued)

Table 2 (continued)

Family	Functions	Species	References
		fathead minnow (*Pimephales promelas*)	
CYP 4	Metabolism of free fatty acids Hydroxylation of lauric acid	Toad fish (*Opsanus tau*), zebrafish (*Danio rerio*), rare minnow (*Gobiocypris rarus*) Rainbow trout (*Oncorhynchus mykiss*) Japanese pufferfish (*Takifugu rubripes*) European sea bass (*Dicentrarchus labrax*) Bluegill (*Lepomis macrochirus*)	Simpson (1997), Ibabe et al. (2002) and Falckh et al. (1997)
CYP 5	Biosynthesis of thromboxane	Japanese pufferfish (*Takifugu rubripes*) Zebrafish (*Danio rerio*)	Arellano et al. (2009), Siroka and Dratichova (2004) and Uno et al. (2012)
CYP 7	Steroid metabolism	Japanese pufferfish (*Takifugu rubripes*) Zebrafish (*Danio rerio*)	Arellano et al. (2009), Siroka and Dratichova (2004) and Uno et al. (2012)
CYP 8	Biosynthesis of prostacycline	Japanese pufferfish (*Takifugu rubripes*) Zebrafish (*Danio rerio*)	Arellano et al. (2009), Siroka and Dratichova (2004) and Uno et al. (2012)
CYP 11	Steroid biosynthesis and cholesterol hydroxylation	Rainbow trout (*Oncorhynchus mykiss*), Japanese eel (*Anguilla japonica*), European sea bass (*Dicentrarchus labrax*), Nile tilapia (*Oreochromis niloticus*), Japanese pufferfish (*Takifugu rubripes*), zebrafish (*Danio rerio*), southern stingray (*Dasyatis americana*), Black porgy fish (*Acanthopagrus schlegelii*), Atlantic salmon (*Salmo salar*), Medaka (*Oryzias latipes*)	Nunez and Trant (1997), Hsu et al. (2002), Nelson (2003) and Socorro et al. (2007)
CYP 17	Steroid biosynthesis Hydroxylation of pregnenolone, progesterone, and corticosteroids	Rainbow trout (*Oncorhynchus mykiss*), Japanese pufferfish (*Takifugu rubripes*), zebrafish (*Danio rerio*), fathead minnow (*Pimephales promelas*)	Filby et al. (2007), Wang and Ge (2004), Wang-Buhler et al. (2005) and Yu et al. (2003)

(continued)

Table 2 (continued)

Family	Functions	Species	References
		Dogfish shark (*Squalus acanthias*), European perch (*Perca fluviatilis*)	
CYP 19	Steroid biosynthesis Aromatization of androgens and testosterone	Rare minnow (*Gobiocypris rarus*), rainbow trout (*Oncorhynchus mykiss*), carp (*Cyprinus carpio*), channel catfish (*Ictalurus punctatus*), zebrafish (*Danio rerio*), catfish (*Clarias gariepinus*), Nile tilapia (*Oreochromis niloticus*), guppy (*Poecilia reticulate*), rice field eel (*Monopterus albus*)	Simpson et al. (1994), Chang et al. (1997) and Barney et al. (2008)
CYP 20	Unknown	Zebrafish (*Danio rerio*), Japanese pufferfish (*Takifugu rubripes*)	Arellano et al. (2009), Siroka and Dratichova (2004) and Uno et al. (2012)
CYP 21	Steroid biosynthesis	Zebrafish (*Danio rerio*), Japanese pufferfish (*Takifugu rubripes*)	Arellano et al. (2009), Siroka and Dratichova (2004) and Uno et al. (2012)
CYP 24	Vitamin D metabolism	Zebrafish (*Danio rerio*), Japanese pufferfish (*Takifugu rubripes*)	Arellano et al. (2009), Siroka and Dratichova (2004) and Uno et al. (2012)
CYP 26	Retinoid metabolism Hydroxylation of retinoic acid	Japanese pufferfish (*Takifugu rubripes*), zebrafish (*Danio rerio*)	Gu et al. (2005), Zhao et al. (2005), Nelson (2003) and Kudoh et al. (2002)
CYP 27	Bile acid biosynthesis	Zebrafish (*Danio rerio*), Japanese pufferfish (*Takifugu rubripes*)	Arellano et al. (2009), Siroka and Dratichova (2004) and Uno et al. (2012)
CYP 39	Steroid biosynthesis	Zebrafish (*Danio rerio*)	Arellano et al. (2009), Siroka and Dratichova (2004) and Uno et al. (2012)
CYP 46	Steroid biosynthesis	Zebrafish (*Danio rerio*), Japanese pufferfish (*Takifugu rubripes*)	Arellano et al. (2009), Siroka and Dratichova (2004) and Uno et al. (2012)
CYP 51	Fungal isoforms	Zebrafish (*Danio rerio*), Japanese pufferfish (*Takifugu rubripes*)	Arellano et al. (2009), Siroka and Dratichova (2004) and Uno et al. (2012)

Table 3 Induction sites of cytochrome P450 enzymes in fish

CYP family	Subfamily	Liver	Kidney	Gills	Brain	Heart	Gut	Ovaries	Testis	Blood
CYP 1	1A1	I	I	I		I	I+	I		
	1B1	I	I	I+			I			I
	1B2	I		I+			I			
	1C1	I+	I	I		I				
	1C2	I	I	I						
CYP2	2E1	I								
	2 K1	I	I+		I		I			I
	2 K6	I						I		
	2 M1	I+	I		I		I			I
	2 N1	I	ND	ND	I	I	I	ND	ND	
	2 N2	I	I	I	I	I	I			
	2P1	I	ND	ND		ND	I	ND	ND	
	2P2	I					I			
	2P3	I					I			
	2X	I								
CYP3	3A27	I			I		I+			I
	3A38	I								
	3A40	I								
	3A45	I					I+			
	3A56	I+	I	I	I		I+	I		
	3A30	I+	I	I	I		I+	I		
	3A65	I		I	I	I	I			
	3C1	I				I	I	I		
CYP4	4 T1	I								
CYP11	11A1		I					I	I	

(continued)

Table 3 (continued)

CYP family	Subfamily	Liver	Kidney	Gills	Brain	Heart	Gut	Ovaries	Testis	Blood
CYP17	17A1							I	I+	
	17A2							I	I	
	17B							I	I	
	17C	I	I	I	I		I	I	I+	
CYP19	19A1	I			I			I		
	19A2	I			I			ND		
CYP26	26A1									
	26B1	I			I					
	26D1	I			I					

ND nondetected, *I* induction, *I+* high induction

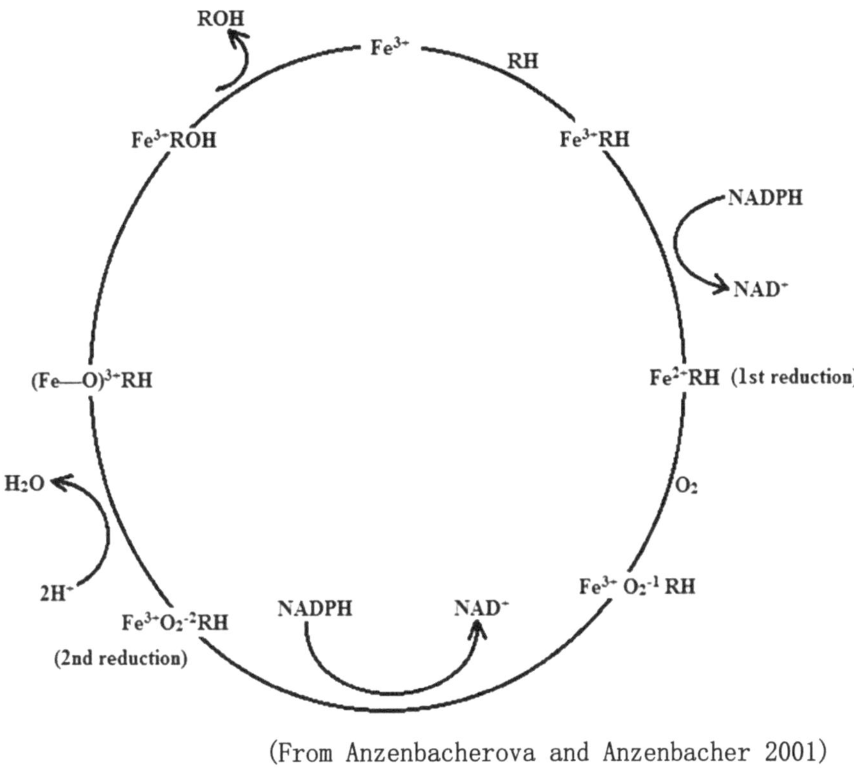

Fig. 1 Metabolic pathway of cytochrome P450. (*Fe* iron atom in P450 heme, *RH* substrate, *ROH* oxidized product)

2 Cytochrome P450 Metabolism

Cytochrome P450 (CYP 450) is recognized to perform a substantial role in the oxidative metabolism/biotransformation of an enormous arraying together the endogenous and exogenous compounds and is thought to be one of the most significant phase I biotransformation enzymes (Siroka and Dratichova 2004). CYP1 to CYP3 are regarded as the most important families of CYP that are accountable for the xenobiotic metabolism and to lesser extent CYP4, while cytochrome P450 enzymes metabolize endogenous substrates (Ioannides and Lewis 2004). Quantifiable reactions to an organism being exposed to xenobiotics are known as biochemical markers. They can react to a set of either similar or extremely diverse xenobiotics because they react to the mechanism of toxic activity rather than the presence of a specific xenobiotic. Biochemical indicators indicate the type of toxicity; for some of them, the strength of the reaction is correlated with the pollution level (Siroka and Dratichova 2004).

P450 enzymes are found mostly in the endoplasmic reticulum of hepatocytes, but their production can also be triggered in organs such as the lungs, colon, kidney, heart, skin, gonads, brain, and placental tissue (Van der Oost et al. 2003). During phase I metabolism, these enzymes regulate oxidation, reduction, and hydrolysis processes, and their function is to biosynthesize substances such as steroids, fatty acids, and prostaglandins (Groves 2005). In fish, CYP1A subfamily plays a significant role in the metabolism and activation of carcinogenesis and is used as a biomarker to estimate contamination of the aquatic environment (Brammell et al. 2010; Jung et al. 2011). Various authors (Rabergh et al. 2000; Morrison et al. 1998; Arukwe 2002; Kim et al. 2004, 2008; Fu et al. 2011) isolated cDNAs encoding CYP1A enzymes from several fish species [rainbow trout (*Oncorhynchus mykiss*), mummichog (*Fundulus heteroclitus*), Atlantic salmon (*Salmo salar*), medaka (*Oryzias latipes*), yellow catfish (*Pelteobagrus fulvidraco*), yellow catfish (*Pelteobagrus fulvidraco*)], respectively, and also from hermaphroditic fish, mangrove killifish (*Rivulus marmoratus*) (Lee et al. 2005). 7-ethoxyresorufin, estradiol, and benzopyrene are all metabolized by CYP1A expressed from zebrafish (*Danio rerio*) cDNA (Scornaienchi et al. 2010). *E. coli* transformed with CYP1A9 cDNA from Japanese eel (*Anguilla japonica*) bioconverts estradiol and flavanone (Uno et al. 2008). Each isoform is involved in the metabolism of a wide variety of substances, and many cytochrome isoforms can metabolize the same substrate. But nearly every isoform has a unique substrate that may be utilized to recognize it (Lewis 2001). However, P450 isoforms are highly substrate-specific in bacterial and mitochondrial cytochrome (Lewis 2001).

3 Effects of Environmental Pollutants on Cytochrome P450 (CYP1A)

Fish CYP1A is induced by a variety of environmental pollutants, and CYP1A has been recognized as a biomarker for the assessment of aquatic pollution. Furthermore, induction of CYP1A has been associated with detrimental outcomes in exposed fish, such as embryonic death and programmed cell death (apoptosis) (Dong et al. 2002). As a result, pharmaceutical substance interactions with the CYP1A enzyme are considered to be toxicologically substantial in fish. By measuring CYP1A mRNA levels, it is possible to track the transcriptional response to the CYP1A induction response caused by pollutants (Rees and Li 2004). There is limited documentation on the toxicity of ATR (atrazine) and CPF (chlorpyrifos) in freshwater fish. It is unclear how CYP1A affects the biotransformation of CPF (chlorpyrifos) and ATR (atrazine) in fish. According to Chang et al. (2005), common carp exposure to 7-ppb ATR (atrazine) could induce CYP1A1 mRNA level after 4 days. According to Xing et al. (2014), CYP1A, which is essential for fish liver antidotal function, was induced in the mRNA expression patterns and EROD activity in carp liver by ATR, CPF, and ATR/CPF combination. Salaberria et al. (2009) revealed a dose-dependent rise in vitellogenin (Vtg) as well as a decrease in CYP1A. Additionally, CYP1A varied in a hormetic manner with testosterone (T) concentrations and was negatively correlated with liver CAT (catalase activity) and 17 beta-estradiol (E2). These results showed the potential for ATR to alter hepatic metabolism, produce estrogenic effects, and induce oxidative stress in vivo, as well as the relationship between these effects. In a previous investigation, a significant alteration in glutathione S-transferase and antioxidant enzymes was found in the liver of the same carp (Xing et al. 2012a, b). These studies (CYP1A, glutathione S-transferase, and antioxidant enzymes) revealed that ATR and CPF, both alone and together, affect the liver of carp. Liver microsomal EROD activity is often used to assess fish CYP1A induction. According to Torre et al. (2011), the effects of musk xylene on EROD activity and CYP1A mRNA levels in PLHC-1 and RTG-2 fish cell lines were distinct. The highest concentration of pesticides used increased EROD activity by about twofold. At the same time, the amount of CYP1A mRNA rose sixfold to sevenfold, as we are all aware that protein is what gives enzymes their chemical makeup. The process of transforming RNA into protein is known as translation, and it can be hampered by a number of reasons. As a result, fluctuations in mRNA levels and enzyme activity are often inconsistent. The results show that pesticides (ATR and CPF) can boost CYP1A expression. However, more research is needed to see if the CYP1A induction has a direct effect on the overall CYP rise.

Cytochrome P450s (CYPs) and heat-shock proteins (HSPs) are key predictors for determining contamination levels in the aquatic system (Yamashita et al. 2004; Alak et al. 2017). Planar constituents of numerous polycyclic aromatic hydrocarbons (PAH), polychlorinated naphthalenes, polychlorinated dibenzodioxins and dibenzofurans (PCDD, PCDF), polychlorinated biphenyls (PCB), and others induce CYP-1A in organisms exposed to a wide spectrum of environmental contaminants (Fent 2001). When a foreign substance binds to a cellular receptor, CYP-1A may be induced (Perdew and Poland 1988). This binding stimulates the CYP-1A gene to express, which enhances RNA transcription (Okey et al. 1994), and thus boost CYP-1A synthesis (Hassanain et al. 2007). As a result, CYP-1A induction is used as a biomarker in fish and fish cell systems to indicate exposure to such contaminants. CYP-1A induction has also been utilized as a biomarker of exposure to different contaminants in a range of vertebrate species, including mammals, in various studies (White et al. 1994), fish (Woodin et al. 1997), reptiles (Rie et al. 2000), and birds (Sanderson et al. 1994). According to previous research, deltamethrin inhibits antioxidant enzymes, increases the expression of heat-shock protein 70, and has negative effects on the expression of IGF-I, IGF-II, and GH (Ceyhun et al. 2010; Aksakal et al. 2010). In fish, cytochrome P450 is essential for the metabolization of a variety of contaminants. In rainbow trout, deltamethrin exposure dramatically increased CYP1A gene expression in a time-dependent way. When a sublethal dose of deltamethrin was used, the pesticide's toxic metabolism was shown to be rapid than in the other groups (Guardiola et al. 2014). The proportion of pesticide or its brain-accumulated metabolites was found to be related to the potential for CYP1A induction to signify neurological toxicity (Johri et al.

2006). Several pyrethroids, particularly DLM, have previously been demonstrated to boost CYP1A activity (Johri et al. 2006; Alak et al. 2017).

It is presently well-established that the activation of xenobiotic metabolism in fish by CYPs is a viable technique for ecotoxicology investigations and environmental pollution biomonitoring (Dong et al. 2009). In recent years, ATR (atrazine) has been related to the induction of CYP isozyme activity in *Chironomus tentans* larvae (Miota et al. 2000). In zebra fish, 3,3,4,4,5-pentachlorobiphenyl can stimulate the expression of cytochrome P4501A, 1B, and 1C genes (Jonsson et al. 2007). Fish have proven to be reliable experimental paradigms for determining how well aquatic ecosystems are doing after being exposed to pollution and biochemical changes. Several research demonstrating the detrimental effects of ATR (atrazine) and CPF (chlorpyrifos) on fish have just recently been published (Wiegand et al. 2001; Kavitha and Venkateswara Rao 2008; De Silva and Samayawardhena 2005). Experiments have demonstrated that exposure to ATR (atrazine), CPF (chlorpyrifos), and mixtures can affect a number of organs, including the liver, kidney, brain, gills, and muscle (Xing et al. 2012a, b; Wang et al. 2011). Because CYPs are the essential enzymes that catalyze the oxidative metabolism of toxicants, including crucial environmental substances, their activity or content is typically altered when the tissues of organisms are damaged by an exogenous toxicant. The gills are involved in gas exchange and come into direct contact with external aquatic chemicals. Furthermore, preliminary studies have shown that benzo(a)pyrene (Bap), indigo, and polyaromatic hydrocarbon (PAH) induction in the gills is more sensitive than that in the liver (Jonsson et al. 2006; Abrahamson et al. 2007).

4 Conclusion

Cytochrome P450 is a biomarker which aids in detoxification in fishes. The maximum expression of this enzyme has been found in the liver. Cytochrome P450 has been identified from many fish families like Salmonidae, Sciaenidae, Tetraodontidae, Siluridae, Cyprinidae, Bagridae, Rivulidae, Moronidae, Fundulidae, Sparidae, Pleuronectidae, Gasterosteidae, Adrianichthyidae, Poeciliidae, Cichlidae, Chaetodontidae, Serranidae, Haemulidae, Centrarchidae, Dasyatidae, Adrianichthyidae, and Squalidae. Several environmental chemicals can inhibit the P450 activity in fish. The list of chemicals comprises chlorinated aromatics (PCB 77, PCB 169), heterocyclic compounds (e.g., piperonyl butoxide), metals (Cd), aromatic hydrocarbons (e.g., benzo[a]pyrene, naphthalene, benzene), and alkylmetals (tributyltin). This system either undergoes direct reduction of molecular dioxygen through peroxide pathway or utilizes electrons from NADPH in order to activate the CYP catalytic pathway.

Acknowledgment The authors are highly indebted to the dean of the Faculty of Fisheries, SKUAST-K, for providing all possible help during the preparation of this manuscript.

Conflict of Interest The authors affirm that they do not have any competing interests.

References

Abrahamson A, Andersson C, Jonsson ME, Fogelberg O, Brunstrom B, Brandt I (2007) Gill EROD in monitoring of CYP1A inducers in fish—a study in rainbow trout (*Oncorhynchus mykiss*) caged in Stockholm and Uppsala waters. Aquat Toxicol 85:1–8

Aksakal E, Ceyhun SB, Erdogan O, Ekinci D (2010) Acute and long-term genotoxicity of deltamethrin to insulin-like growth factors and growth hormone in rainbow trout. Comp Biochem Physiol C 152:451–455

Alak G, Ucar A, Parlak V, Yeltekin AC, Tas IH, Olmez D, Kocaman EM, Yilgin M, Atamanalp M, Yanik T (2017) Assessment of 8-hydroxy-2-deoxyguanosine activity, gene expression and antioxidant enzyme activity on rainbow trout (*Oncorhynchus mykiss*) tissue exposed to biopesticide. Comp Biochem Physiol Pt C Toxicol Pharmacol 203:51–58

Arellano O, Montoya RM, Garcia CM (2009) Endogenous functions and expression of cytochrome P450 in teleost fish: a review. Rev Fish Sci 17:541–556

Arukwe A (2002) Complementary DNA cloning, sequence analysis and differential organ expression of naphthoflavone-inducible cytochrome P4501A in Atlantic salmon (*Salmo salar*). Comp Biochem Physiol 133:613–624

Barber DS, McNally AJ, Garcia-Reyero N, Denslow ND (2007) Exposure to P,P-DDE or dieldrin during the reproductive season alters hepatic CYP expression in largemouth bass (*Micropterus salmoides*). Aquat Toxicol 81:27–35

Barney ML, Patil JG, Gunasekera RM, Carter CG (2008) Distinct cytochrome P450 aromatase isoforms in the common carp (*Cyprinus Carpio*): sexual dimorphism and onset of ontogenic expression. Gen Comp Endocrinol 156:499–508

Brammell BF, McClain JS, Oris JT, Price DJ, Birge WJ, Elskus AA (2010) CYP1A expression in caged rainbow trout discriminates among sites with various degrees of polychlorinated biphenyl contamination. Arch Environ Contam Toxicol 58:772–782

Bucheli TB, Fent K (1995) Induction of cytochrome P450 as a biomarker for environmental contamination in aquatic ecosystems. Crit Rev Environ Sci Technol 25:201–268

Buhler DR, Yang YH, Dreher TW, Miranda CL, Wang JL (1994) Cloning and sequencing of the major rainbow trout constitutive cytochrome P450 (CYP2K1): identification of a new cytochrome P450 gene subfamily and its expression in mature rainbow trout liver and trunk kidney. Arch Biochem Biophys 312:45–51

Ceyhun SB, Snturk M, Ekinci D, Erdogan OC, Iltas A, Kocaman M (2010) Effects of deltamethrin on some enzyme activity and hsp70 expression in rainbow trout (*Oncorhynchus mykiss*). *Comp Biochem Physiol C* 152:215–223

Chang XT, Kobayashi T, Kajiura H, Nakamura M, Nagahama Y (1997) Isolation and characterization of the cDNA encoding the tilapia (*Oreochromis niloticus*) cytochrome P450 aromatase (P450 arom): changes in P450arom mRNA, protein and enzyme activity in ovarian follicles during oogenesis. J Mol Endocrinol 18:57–66

Chang LW, Toth GP, Gordon DA, Graham DW, Meier JR, Knapp CW, DeNoyelles FJ, Campbell S, Lattier DL (2005) Responses of molecular indicators of exposure in mesocosms: common carp (*Cyprinus carpio*) exposed to the herbicides alachlor and atrazine. Environ Toxicol Chem 24:190–197

Christen V, Caminada D, Arand M, Fent K (2010) Identification of a CYP3A form (CYP3A126) in fathead minnow (*Pimephales promelas*) and characterisation of putative CYP3Aenzyme activity. Anal Bioanal Chem 396:585–595

De Silva P, Samayawardhena LA (2005) Effects of chlorpyrifos on reproductive performances of guppy (*Poecilia reticulata*). Chemosphere 58:1293–1299

Dong W, Teraoka H, Yamazaki K, Tsukiyama S, Imani S, Imagawa T, Stegeman JJ, Peterson RE, Hiraga T (2002) 2, 3, 7,8-tetrachlorodibenzo-p-dioxin toxicity in the zebrafish embryo: local circulation failure in the dorsal midbrain is associated with increased apoptosis. Toxicol Sci 69: 191–201

Dong X, Zhu L, Wang J, Wang J, Xie H, Hou X, Jia W (2009) Effects of atrazine on cytochromeP450enzymes of zebrafish (*Danio rerio*). Chemosphere 77:404–412

Falckh JHP, Wu QK, Ahokas JP (1997) CYP4T1-a cytochrome P450 expressed in rainbow trout (*Oncorhynchus mykiss*) liver. Biochem Biophys Res Commun 236:302–305

Fent K (2001) Fish cell lines as versatile tools in ecotoxicology: assessment of cytotoxicity, cytochrome P4501A induction potential and estrogenic activity of chemicals and environmental samples. Toxicol In Vitro 15:477–488

Filby AL, Thorpe KL, Maack G, Tyler CR (2007) Gene expression profiles revealing the mechanisms of anti-androgen- and estrogen-induced feminization in fish. Aquat Toxicol 81:219–231

Fu GH, Yang XL, Zhang HX, Yu WJ, Hu K (2011) Effects of cytochrome P450 1A substrate (difloxacin) on enzyme gene expression and pharmacokinetics in crucian carp (*hybridized Prussian carp*). Environ Toxicol Pharmacol 31(2):307–313

Groves TJ (2005) Models and mechanism of cytochrome P450 action. In: de Montellano RPO (ed) Cytochrome P450: structure, mechanism, and biochemistry, 3rd edn. Kluwer Academic/ Plenum, New York, pp 1–43

Gu X, Xu F, Wang X, Gao X, Zhao Q (2005) Molecular cloning and expression of a novel CYP26 gene (CYP26D1) during zebrafish early development. Gene Expr Patterns 5:733–739

Guardiola FA, Parraga PG, Meseguer J, Cuesta A, Esteban MA (2014) Modulatory effects of deltamethrin-exposure on the immune status, metabolism and oxidative stress in gilthead seabream (*Sparus aurata* L.). Fish Shellfish Immunol 36(1):120–129

Haasch LM (2002) Effects of vehicle, diet, and gender on the expression of PMP70- and CYP2K1/ 2M1-like proteins in the mummichog. Mar Environ Res 54:297–301

Hassanain MA, Abdel-Rahman EH, Abo-Hegab S, Tawfik MAA, Abbas WT (2007) Induction of cytochrome P450 1Al as a biomarker of benzo-apyrene pollution in Egyptian fresh water fish. Pak J Biol Sci 10:1161–1169

Havelkova M, Randak T, Zlabek V, Krijt J, Kropuva H, Pulkrabova J, Svobododova Z (2007) Biochemical markers for the assessment of aquatic environment contamination. Sensors 7: 2599–2611

Hsu HJ, Hsiao P, Kuo MW, Chung BC (2002) Expression of zebrafish CYP11A1 as a maternal transcript and in yolk syncytial layer. Gene Expr Patterns 2:219–222

Ibabe A, Grabenbauer M, Baumgart E, Fahimi HD, Cajaraville MP (2002) Expression of peroxisome proliferator-activated receptors in zebrafish (*Danio rerio*). Histochem Cell Biol 118:231–239

Ioannides C, Lewis DFV (2004) Cytochromes P450 in the bioactivation of chemicals. Curr Top Med Chem 4:1767–1788

Johri A, Yadav S, Singh RL, Dhawan A, Ali M, Parmar D (2006) Long lasting effects of prenatal exposure to deltamethrin on cerebral and hepatic cytochrome P450s and behavioural activity in rat offspring. Eur J Pharmacol 544:58–68

Jonsson EM, Abrahamson A, Brunstrom B, Brandt I (2006) Cytochrome P4501A induction in rainbow trout gills and liver following exposure to water borne indigo, benzo[a] pyreneand3,3′,4,4′,5-pentachlorobiphenyl. Aquat Toxicol 79:226–232

Jonsson ME, Jenny MJ, Woodin BR, Hahn ME, Stegeman JJ (2007) Role of AHR2 in the expression of novel cytochrome P4501 family genes, cell cycle genes, and morphological defects in developing zebra fish exposed to 3,3′, 4,4′, 5- pentachlorobiphenylor2,3,7,8-tetrachlorodibenzo-p-dioxin. Toxicol Sci 100:180–193

Jung JH, Kim M, Yim UH, Ha SY, An JG, Won JH, Han GM, Kim NS, Addison RF, Shim WJ (2011) Biomarker responses in pelagic and benthic fish over 1 year following the Hebei spirit oil spill (Taean, Korea). Mar Pollut Bull 62:1859–1866

Kaplan EAL, Fielding E, Crivello JF (1999) Genetic regulation of liver microsomal CYP2E1 activity among strains of the viviparous fishes *Poeciliopsis occidentalis* and *Poeciliopsis lucida*. Environ Biol Fishes 54:337–343

Kashiwada S, Hinton DE, Kullman SW (2005) Functional characterization of medaka CYP3A38 and CYP3A40: kinetics and catalysis by expression in a recombinant baculo virus system. Comp Biochem Physiol 141:338–348

Kavitha P, Venkateswara Rao J (2008) Toxic effects of chlorpyrifos on antioxidant enzymes and target enzyme acetyl cholinesterase interaction in mosquito fish, Gambusia affinis. Environ Toxicol Pharmacol 26:192–198

Kelly LS, Lamb DC, Kelly DE (2006) Cytochrome P450 biodiversity and biotechnology. Biochem Soc Trans 34:1159–1160

Kilemade M, Hartl MGJ, Ohalloran J, Obrivn NM, Sheehan D, Mothersill C, Van Pelt FNAM (2009) Effects of contaminated sediment from cork harbor, Ireland on the cytochrome P450 system of turbot. Ecotox Environ Safe 72:747–755

Kim IC, Kim YJ, Yoon YD, Kawamura S, Lee YS, Lee JS (2004) Cloning of cytochrome P450 1A (CYP1A) genes from the hermaphrodite fish, *Rivulus marmoratus* and the japanese medaka, *Oryzias latipes*. Mar Environ Res 58:125–129

Kim JH, Raisuddin S, Ki JS, Lee JS, Han KN (2008) Molecular cloning and β naphthoflavone induced expression of a cytochrome P4501A (CYP1A) gene from an anadromous river pufferfish, *Takifugu obscures*. Mar Pollut Bull 57:433–440

Klemz C, Salvo LM, Neto JCB, Bainy ACD, de Assis HCS (2010) Cytochrome p450 detection in liver of catfish *Ancistrus multispinis* (osteichthyes, loricariidae). Braz Arch Biol Technol 2:361–368

Kudoh T, Wilson WS, Dawid BI (2002) Distinct roles for fgf, wnt, and retinoic acid in posteriorizing the neural ectoderm. Development 129:4335–4346

Kullman WS, Hinton DE (2001) Identification, characterization, and ontogeny of a second cytochrome P450 3A gene from the fresh water teleost medaka (*Oryzias latipes*). Mol Reprod Dev 58:149–158

Lee SJ, Buhler DR (2003) Cloning, tissue distribution, and functional studies of a new cytochrome P450 3A subfamily member, CYP3A45, from rainbow trout (*Oncorhynchus mykiss*). Intestinal Ceca Arch Biochem Biophys 412:77–89

Lee SJ, Hedstrom OR, Fischer K, Wang-Buhler JL, Sen A, Cok I, Buhler DR (2001) Immunohistochemical localization and differential expression of cytochrome P450 3A27 in the gastrointestinal tract of rainbow trout. Toxicol Appl Pharmacol 177:94–102

Lee YM, Williams TD, Jung SO, Lee JS (2005) cDNA cloning and expression of a cytochrome P450 1A (CYP1A) gene from the hermaphroditic fish *Rivulus marmoratus*. Mar Pollut Bull 51: 769–775

Lemaire B, Priede IJ, Collins MA, Bailey DM, Schtickzelle N, Thome JP, Rees JF (2010) Effects of organochlorines on cytochrome P450 activity and antioxidant enzymes in liver of roundnose grenadier *Coryphaenoides rupestris*. Aquat Biol 8:161–168

Lewis DFV (2001) Guide to cytochromes P450 structure and function. Taylor & Francis, London, p 215

Meyer JN, Nacci DE, Giulio RTD (2002) Cytochrome P4501A (CYP1A) in killifish (*Fundulus heteroclitus*): heritability of altered expression and relationship to survival in contaminated sediments. Toxicol Sci 68:69–81

Miota F, Siegried BD, Scharf ME, Lydy M (2000) Atrazine induction of cytochromeP450 in *Chironomus tentans* larvae. Chemosphere 40:285–291

Morrison HG, Weil EJ, Karchner SI, Sogin ML, Stegeman JJ (1998) Molecular cloning of CYP1A from the estuarine fish Fundulus heteroclitus and phylogenetic analysis of CYP1 genes: update with new sequences. Comp Biochem Physiol C Pharmacol Toxicol Endocrinol 121(1-3):231–240
Nelson RD (1999) A second CYP26 P450 in human and zebrafish: CYP26B1. Arch Biochem Biophys 371:345–347
Nelson DR (2003) Comparison of P450s from human and *fugu*: 420 million years of vertebrate p450 evolution. Arch Biochem Biophys 409:18–24
Nunez S, Trant JM (1997) Isolation of the putative cDNA encoding cholesterol side chain cleavage cytochrome P450 (CYP11A) of the southern stingray (*Dasyatis americana*). Gene 187:123–129
Okey AB, Riddick DS, Harper PA (1994) The ah receptor: mediator of the toxicity of 2, 3, 7, 8-tetrachlorodibenzo-p-dioxin (TCDD) and related compounds. Toxicol Lett 70:1–22
Oleksiak FM, Wu S, Parker C, Karchner SI, Stegeman JJ, Zeldin DC (2000) Identification, functional characterization, and regulation of a new cytochrome P450 subfamily, the CYP2Ns. J Biol Chem 275:2312–2321
Oleksiak FM, Wu S, Parker C, Qu W, Cox R, Zeldin DC, Stegeman JJ (2003) Identification and regulation of a new vertebrate cytochrome P450 subfamily, the CYP2Ps, and functional characterization of CYP2P3, a conserved arachidonic acid epoxygenase 19- hydroxylase. Arch Biochem Biophys 411:223–234
Perdew GH, Poland A (1988) Purification of the ah receptor from C57BLl6.I mouse liver. J Biol Chem 263:9848–9852
Rabergh CM, Vrolijk NH, Lipsky MM, Chen TT (2000) Differential expression of two CYP1A genes in rainbow trout (*Oncorhynchus mykiss*). Toxicol Appl Pharmacol 165(3):195–205
Rahman MS, Thomas P (2012) Effects of hypoxia exposure on hepatic cytochrome P450 1A (CYP1A) expression in Atlantic croaker: molecular mechanisms of CYP1A down-regulation. PloS One 7:1–14
Rees CB, Li W (2004) Development and application of a real-time quantitative PCR assay for determining CYP1A transcripts in three genera of salmonids. Aquat Toxicol 66:357–368
Rie MT, Lendas KA, Woodin BR, Stegeman JJ, Callard IP (2000) Hepatic biotransformation of enzymes in a sentinel species, the painted turtle (*Chrysemys picta*) from Cape Cod, Massachusetts: seasonal, sex- and location related differences. Biomarkers 5:382–394
Ruus A, Sandvik M, Ugland IK, Skaare UJ (2002) Factors influencing activities of biotransformation enzymes, concentrations, and compositional patters of organochlorine contaminants in members of a marine food web. Aquat Toxicol 61:73–87
Sakamoto KQ, Takahiro A, Yokoyama A, Ushikoshi Y, Hirose H, Ishizuka M, Kazusaka A, Fujita S (2003) Cytochrome P450 induction and gonadal status alteration in common carp (*Cyprinus carpio*) associated with the discharge of dioxin contaminated effluent to the Hikiji river, Kanagawa prefecture, Japan. Chemosphere 51:491–500
Salaberria I, Hansen BH, Asensio V, Olsvik PA, Andersen RA, Jenssen BM (2009) Effects of atrazine on hepatic v metabolism and endocrine homeostasis in rainbow trout (*Oncorhynchus mykiss*). Toxicol Appl Pharmacol 234:98–106
Sanderson JT, Norstrom RJ, Elliott JE, Hart LE, Cheng KM, Bellward GD (1994) Biological effects of polychlorinated dibenzo-p-dioxins, dibenzofurans, and biphenyls in double-crested cormorant chicks (*Phalacrocorax auritus*). J Toxic Environ Health A 41:247–265
Schenkman JB, Jansson I (1998) Spectral analyses of cytochromes P450. In: Phillips IR, Shephard EA (eds) Methods in molecular biology, Cytochrome P450 protocols, vol 107. Humana Press, Totowa, NJ, pp 25–33
Schlenk D, Furnes B, Zhou X, Debusk BC (2002) Cloning and sequencing of cytochrome P450 from channel catfish (*Ictalurus punctatus*). Mar Environ Res 54:391–394
Scornaienchi ML, Thornton C, Willett KL, Wilson JY (2010) Functional differences in the cytochrome P450 1 family enzymes from zebrafish (*Danio rerio*) using heterologously expressed proteins. Arch Biochem Biophys 502(1):17–22

Shalan H, Katoa M, Cheruzela L (2018) Keeping the spotlight on cytochrome P450. BBA Proteins Proteom 1866:80–87

Simpson AE (1997) The cytochrome P4504 (CYP4) family. Gen Pharmacol 28:351–359

Simpson ER, Mahendroo MS, Means GD, Kilgore MW, Hinshelwood MM, Graham-Lorence S, Amarneh B, Ito Y, Fisher CR, Michael MD (1994) Aromatase cytochrome P450, the enzyme responsible for estrogen biosynthesis. Endocr Rev 15(3):342–355

Siroka Z, Dratichova J (2004) Biochemical markers of aquatic environment contamination—cytochrome P450 in fish, a review. Acta Vet Brno 73:123–132

Socorro S, Martins RS, Deloffre L, Mylonas CC, Canario AV (2007) A cDNA for European sea bass (*Dicentrarchus labrax*) 11beta-hydroxylase: gene expression during the thermosensitive period and gonadogenesis. Gen Comp Endocrinol 150:164–173

Stegeman JJ (1989) Cytochrome P450 forms in fish: catalytic, immunological and sequence similarities. Xenobiotica 19:1093–1110

Stien X, Amichot M, Berge JB, Lafaurie M (1998) Molecular cloning of a CYP1A cDNA from the teleost fish *Dicentrarchus labrax*. Comp Biochem Physiol 121:241–248

Torre CD, Monti M, Focardi S, Corsi I (2011) Time-dependent modulation of cyp1a gene transcription and EROD activity by musk xylene in PLHC-1 and RTG-2 fish cell lines. Toxicol In Vitro 25:1575–1580

Tuan T, Kaminishi Y, Funahashi A, Mohamed EAH, Hassanin A, Itakura T (2014) cDNA cloning, characterization and expression of cytochrome P450 family 1 (CYP1A) from Javanese medaka, *Oryzias javanicus* by environmental condition. Afr J Biotechnol 13:1898–1909

Uno T, Ishizuka M, Itakura T (2012) Cytochrome P450 (CYP) in fish. Environ Toxicol Pharmacol 34:1–13

Uno T, Okamoto S, Masuda S, Imaishi H, Nakamura M, Kanamaru K, Yamagata H, Mohamed AH, Kady EL, Kaminishi Y, Itakura T (2008) Bioconversion by functional P450 1A9 and P450 1C1 of Anguilla japonica. 147(3):278–285

Van der Oost R, Beyer RJ, Vermeulen NPE (2003) Fish bioaccumulation and biomarkers in environmental risk assessment: a review. Environ Toxicol Pharmacol 13:57–149

Wang Y, Ge W (2004) Cloning of zebrafish ovarian P450c17 (CYP17, 17alph hydroxylase/17, 20-lyase) and characterization of its expression in gonadal and extra-gonadal tissues. Gen Comp Endocrinol 135:241–249

Wang X, Xing H, Li X, Xu S, Wang X (2011) Effects of atrazine and chlorpyrifos on the RNA levels of IL-1and IFN-g2b in immune organs of common carp. Fish Shellfish Immunol 31:126–133

Wang-Buhler LJ, Lee SJ, Chung WG, Stevens JF, Tseng HP, Hseu TH, Hu CH, Westerfield M, Yang YH, Miranda CL, Buhler DR (2005) CYP2K6 from zebrafish (*Danio rerio*): cloning, mapping, developmental/tissue expression, and aflatoxin B1 activation by baculovirus expressed enzyme. Comp Biochem Physiol 140:207–219

Werck-Reichhart D, Feyereisen R (2000) Cytochromes P450: a success story. Genome Biol 1: 3003.1–3003.9

White RD, Hahn ME, Lockhart WL, Stegeman JJ (1994) Catalytic and immunochemical characterization of hepatic microsomal cytochromes P450 in beluga whale (*Delphinapterus leucas*). Toxicol Appl Pharmacol 126:45–57

Wiegand C, Krause E, Steinberg C, Pflugmacher S (2001) Toxico kinetics of atrazine in embryos of the zebrafish (*Danio rerio*). Ecotoxicol Environ Saf 49:199–205

Winston GW, Shane BS, Henry CB (1988) Hepatic monooxygenase induction and promutagen activation in channel catfish from a contaminated river basin. Ecotoxicol Environ Saf 16:258–271

Woodin BR, Smolowitz RM, Stegeman JJ (1997) Induction of cytochrome P4501A in the intertidal fish *Anoplarchus purpurescens* by Prudhoe Bay crude oil and environmental induction in fish from Prince William sound. Environ Sci Technol 31:1198–1205

Xing H, Li S, Wang Z, Gao X, Xu S, Wang X (2012a) Histopathological changes and antioxidant response in brain and kidney of common carp exposed to atrazine and chlorpyrifos. Chemosphere 88:377–383

Xing H, Wang X, Sun G, Gao X, Xu S (2012b) Effects of atrazine and chlorpyrifos on activity and transcription of glutathione S-transferase in common carp (*Cyprinus carpio* L.). Environ Toxicol Pharmacol 33:233–244

Xing H, Zhang Z, Yao H, Liu T, Wang L, Xu S, Li S (2014) Effects of atrazine and chlorpyrifos on cytochrome P450 in common carp liver. Chemosphere 104(2014):244–250

Yamashita M, Hirayoshi K, Nagata K (2004) Characterization of multiple members of the HSP70 family in platyfish culture cells: molecular evolution of stress protein HSP70 in vertebrates. Gene 336:207–218

Yang YH, Wang JL, Miranda CL, Buhler DR (1998) CYP2M1: cloning, sequencing, and expression of a new cytochrome P450 from rainbow trout liver with fatty acid (Omega-6)-hydroxylation activity. Arch Biochem Biophys 352:271–280

Yang HY, Miranda CL, Henderson MC, Wang-Buhler JL, Buhler DR (2000) Heterologous expression of CYP2K1 and identification of the expressed protein (BV-CYP2K1) as lauric acid (ω-1)-hydroxylase and aflatoxin B1 exo-epoxidase. Drug Metab Dispos 28:1279–1283

Yu H, Cheng H, Gou Y, Xia L, Zhou R (2003) Alternative splicing and differential expression of P450C17 (CYP17) in gonads during sex transformations in the rice field eel. Biochem Biophys Res Commun 307:165–171

Zanette J, Jenny MJ, Goldstone JV, Woodin BR, Watka LA, Bainy ACD, Stegeman JJ (2009) New cytochrome P450 1B1, 1C2 and 1D1 genes in the killifish *Fundulus heteroclitus*: basal expression and response of five killifish CYP1s to the AHR agonist PCB126. Aquat Toxicol 93:234–243

Zhao Q, Dobbs-McAuliffe B, Linney E (2005) Expression of CYP26b1 during zebrafish early development. Gene Expr Patterns 5(3):363–369

Pharmacokinetic-Pharmacodynamic Modeling of Xenobiotics: Fate and Effect in Aquatic Animals

Farah Hanief, Adnan Abubakr, Hafsa Farooq Chashoo, and Monisa Mehboob Malik

1 Introduction

Twenty-first century has so far witnessed an evolution in every sphere of human existence, science, and technology not lagging behind. However, there is a drastic increase in the outbreak of diseases which need proper treatment using appropriate medicines. The aquatic animals have unfortunately become more prone to the diseases and infectious agents due to the pollution. Therefore, it is very important to understand the working mechanisms of available medicines and treatments.

Pharmacokinetics is the branch of medicine which describes the role of absorption of medicine by the body, its distribution throughout the body, and elimination of this medicine from the body of the organisms. On the other hand, how much drug should be given, what is the efficiency of the drug, and for how long does the selected drug remain in the tissue or targeted site are defined as pharmacodynamics (Hoberman et al. 1996). In simple words, the route of the drug throughout the body is termed as pharmacokinetics, while the efficiency of that drug and its impact on the body is termed as pharmacodynamics. The recent research conducted worldwide in the field of drugs has made it of extreme importance to understand the interrelationship and interdependence of pharmacokinetics and pharmacodynamics of drugs. The accurate understanding of pharmacokinetic-pharmacodynamic relationship of drugs will help to predict the efficiency and efficacy of the drugs.

F. Hanief (✉) · A. Abubakr · H. F. Chashoo
Division of Aquatic Environmental Management, Faculty of Fisheries, Rangil, SKUAST-Kashmir, Srinagar, Jammu and Kashmir, India

M. M. Malik
Division of Aquatic Environmental Management, Sher-e-Kashmir University of Agricultural Sciences and Technology, Kashmir, Faculty of Fisheries Rangil, Ganderbal, Jammu and Kashmir, India

M. A. Rather et al. (eds.), *Xenobiotics in Aquatic Animals*,
https://doi.org/10.1007/978-981-99-1214-8_12

Fish being the aquatic animals are cold-blooded in nature which makes them highly susceptible to infections and various diseases. Various pathogens have been reported in fish as bacterial, viral, fungal, protozoans, etc. (Hanief et al. 2021; Rather et al. 2018). All these pathogens have to be treated with different drugs; however, the overall working of the drugs remains somewhat similar. Certain substances termed as xenobiotics have also been reported from aquatic animals. Xenobiotics can be defined as those chemical substances which are present in the body of an organism beyond the normal level. The abnormal levels of these chemicals disrupt the normal metabolic activities and itself become toxic. With respect to the aquatic animals, these xenobiotics have been found to disrupt the normal metabolic activities as development of reproductive organs, gamete formations, hormonal and endocrine disruption, and viability of eggs (Tyrpenou et al. 2003; Pokhrel et al. 2018). The main reason for so much xenobiotic accumulation in aquatic animals can be directly pointed toward the increasing effects of pollution, agricultural runoff, disposal of harmful heavy metals and chemical from various industries, and untreated sewage or partially treated sewage disposed from sewage treatment plants.

2 Pharmacokinetics of Drugs in Aquatic Animals

The drug after administration follows a route of travelling before reaching to its targeted area or site. The steps are as follows:

2.1 Absorption of Drugs

Absorption of drugs can be defined as the movement of drug from the site of administration into the bloodstream. The entry of drug in bloodstream facilitates its flow throughout the body and ultimately will reach to its targeted site. Majority of the administered drugs follow the route of passive transportation. Passive transportation of drugs is also known as the "downhill movement" of the drugs. The flow of administered drugs occurs via cell membranes of the cells. This method works on the principle of concentration gradient.

The major advantages of passive transportation of drugs are that it doesn't require any energy as the flow is across cell membranes. There is no use of any carriers and furthermore no threat of any competitive inhibition.

However, some drugs get absorbed in the bloodstream through active transportation. The term active transportation means that the involvement of certain mediators is required. Hence, this method involves the use of energy also (Tyrpenou et al. 2003). In active transportation, the flow of drugs can be against the concentration gradient as it involves the use of solutes (carriers). The active transportation of drugs can take place through two ways:

2.1.1 Transportation of Drugs

Primary active transportation of drugs: The energy required for drug transportation is derived from the ATP present in the cells.

Secondary active transportation of drugs: The energy required for drug transportation is derived from energy stored as ion gradients as Na^+ and K^+ which are generated during primary active transportation.

2.1.2 Factors Affecting Absorption of Drugs

The absorption of drugs into the body of aquatic animals particularly fish is dependent on various factors as molecular weight of the drug, solubility of lipids, polarity of the drugs within the bloodstream, the reaction of the targeted tissue when the drug comes in contact with it, and the time period within which it shows the required effect.

2.1.3 Route of Drug Administration in Fish

Administration of drugs in fish is usually done through oral cavity (drug administered in pelleted form or mixed with feed), muscle injection (intramuscular), or intravenously (Samuelsen 2006). Intramuscular drug or Intra- venous drug administration is more efficient and result oriented compared to the oral administration. However, intramuscular administration of drugs is not much convenient in aquatic animals.

2.2 Distribution of Drugs

When the drug enters to the targeted site from the bloodstream, it is termed as the distribution of the drug. The transfer of drug from blood to the targeted tissue continues till an equilibrium is achieved. The distribution of drugs depends on several factors as the ionization of the solute molecules, lipid solubility of the drugs, and the binding efficiency of the drugs with the targeted tissues.

2.3 Metabolism of Drugs

The metabolism of drugs can be defined as the biotransformation of drugs from nonpolar lipid-soluble compounds to polar lipid-soluble compounds. The aim of biotransformation of drugs is to prevent the reabsorption of drugs by excretory system so to prevent excretion of the drug with metabolic wastes (Craig and Andes 1996). The biotransformation of drugs also helps to protect the body from accumulation of toxic wastes or xenobiotics.

The biotransformation of drugs is a complex phenomenon. It takes place in several gradual steps, each step aiming in converting the simpler metabolite. The biotransformation of drugs is broadly categorized in two phases, which are as follows:

Phase I: In Phase I, administered drug is subjected to various reactions as oxidation, reduction, hydrolysis, etc. During this phase, the reactions help in activating the drug. Oxidation is the most important reaction in Phase I as it involves using Cytochrome 450 monooxygenase (CYP), NADPH, and oxygen	**Phase II**: In Phase II, glucuronidation of activated drugs from Phase I takes place which helps in generating a conjugated product of the administered drug

The factors which play an important role in the biotransformation of drugs are the exposure to the pollutants and chemicals, the age of the aquatic animals, the habitat, the strength of the immune system, etc. In the overall process of pharmacokinetics, biotransformation plays a key role as it determines the ultimate use of the administered drug.

2.4 Excretion of the Drug

The excretion of the drug helps in eliminating the left-out drug from the body. The excretion of drug holds an important place as it helps in preventing the buildup of toxic substances. The major organs associated with excretion involve the liver and kidney.

The excretion which takes place under the influence of the liver is termed as "hepatic excretion." Liver accomplishes the process of excretion by throwing away the unwanted substances along with bile juices. Hepatic excretion is easily facilitated when the metabolic wastes are in the conjugated form with glucuronic acid.

Renal excretion takes place in kidneys. It is easier to excrete the hydrophilic and lipophilic compounds from the body of the aquatic organisms. The excretion in the kidney takes place in the following steps:

2.4.1 Filtration Mechanism (Steps)

Glomerulus filtration: In glomerulus filtration, the excretion of all the hydrophilic and lipophilic substances takes place. The filtration takes place in the glomerulus part of kidneys. All the drugs which do not have any protein component in them are easily filtered in glomerulus filtration, but drugs with protein presence are not filtered here.

Tubular reabsorption: The excretion of the lipid-insoluble drugs takes place in tubular reabsorption.

Tubular secretion: In this part of excretion, the dissociation of the remaining drug takes place from the bloodstream. The dissociation breaks down the complex drug components into simpler forms which facilitate excretion.

3 Pharmacodynamics of Drugs in Aquatic Animals

The role which a drug plays in the body of an organism or in other words what does a drug do in the body is explained by pharmacodynamics. Pharmacodynamics provides a deeper understanding of what will be the efficiency of the administered drug. The mechanism of pharmacodynamics is well explained by *Clarke* in his *receptor occupation theory* in *1937*.

According to this theory, every drug has a certain number of receptors which are occupied by it during its course of action. The intensity of the effect of the drug is directly proportional to the number of receptors it occupies. The more the receptor, the more the action. Whenever the drug binds itself to a specific receptor, the binding brings in certain structural changes in the framework of the receptor (Maiti et al. 2019). Those drugs which can positively change the structure of the receptor are said to have agonistic effect on the receptor, while those drugs which fail to bring the requisite changes in the receptor are said to have antagonistic effect. After binding itself to the receptor, the pronounced action begins to come into play. This is called as the *action-effect sequence*.

Furthermore, the intensity of the effect also is determined by the concentration of the drug administered. The more the concentration, the more will be the affect and vice versa. This is called as the "*dose-response relationship*." Dose-response relationship plays an important role in determining the fate of pharmacodynamics of the drug or any xenobiotic.

3.1 Drug Dosage

The amount of drug required to generate a specific impact is called as its dosage. The specific concentration of a selected drug can provide the desired results. Therefore, it is very important to determine the exact amount of concentration that should be administered which is known as the dosage of the drug.

Drug dosage can be of the following types:

Standard dose: The concentration of the dug which is almost similar for all the population of the specific organisms.

Regulated dose: The concentration of the drug is measured or determined by frequent administrations in the specific set of targeted species in order to obtain a standard value.

References

Craig WA, Andes D (1996) Pharmacokinetics and pharmacodynamics of antibiotics in otitis media. Pediatr Infect Dis J 15:255–259. 44

Hanief F et al (2021) Immunopathogenesis of haemato-biochemical parameters and haematopoietic organs of *Cyprinus carpio* var. *communis* challenged with *Aeromonas hydrophila*. Aquacult Res 52(12):6280–6287

Hoberman A, Paradise JL, Block S et al (1996) Efficacy of amoxicillin/clavulanate for acute otitis media: relation to Streptococcus pneumoniae susceptibility. Pediatr Infect Dis J 15:955–962

Maiti D, Ansari I, Rather MA, Deepa A (2019) Comprehensive review on wastewater discharged from the coal-related industries–characteristics and treatment strategies. Water Sci Technol *79*(11):2023–2035

Pokhrel H, Baishya S, Phukan B, Pillai D, Rather MA (2018) Occurrence and distribution of multiple antibiotic resistance bacteria of public health significance in backwaters and aquaculture farm. Int J Curr Microbiol App Sci 7(2):975–987

Rather MA, Bhat IA, Sharma N, Sharma R (2018) Molecular and cellular toxicology of nanomaterials with related to aquatic organisms. Cell Mol Toxicol Nanoparticles 1048:263–284

Samuelsen OB (2006) Pharmacokinetics of quinolones in fish: a review. Aquaculture 255(1–4): 55–75

Tyrpenou AE et al (2003) Tissue distribution and depletion of sarafloxacin hydrochloride after in-feed administration in Gilthead Sea bream (*Sparus aurata* L.). Aquaculture 215(1–4): 291–300

Occurrence of Xenoestrogen Alkylphenols (Octylphenols and Nonylphenol) and Its Impact on the Aquatic Ecosystem

Gora Shiva Prasad, Sangram Keshari Rout, Monisa Mehboob Malik, Sutanu Karmakar, Adnan Amin, and Ishtiyaq Ahmad

1 Introduction

The use of chemical substances (natural and synthetic) is increasing exponentially in the present time for diverse purposes. Such substances mostly occur in the day-to-day life as well as in the environment; however, most of these are hazardous and cause immense impact on the environment and the health (EUROSTAT 2019; Gavrilescu et al. 2015; WWF 2020). Increasing urbanization, industrial growth, agriculture, and economic and social development are some of the root causes of high use of these chemicals in personal care products, pharmaceuticals, cosmetics, insecticides, surfactants, endocrine disruptors and hormones, etc. These substances are discharged directly into the freshwater resources due to the lack of advanced technology in wastewater treatment plants (WWTPs) to remove emerging xenobiotic compounds efficiently (Naidu et al. 2016; Van Zijl et al. 2017; Vargas-Berrones et al. 2020a, b). Even though these organic substances may occur in trace levels, they show negative effects on development, reproduction, aquatic life, behavior of animals (Pal et al. 2010), ecology, as well as human health (Vargas-Berrones et al. 2020a, b). Alkylphenol ethoxylates (APEOs) are the type of non-ionic surface-active agents that are used as detergents, emulsifiers, solubilizers, and dispersing agents in products for domestic, agricultural, and industrial application. Ethoxylated

G. S. Prasad · S. K. Rout · S. Karmakar
Faculty of Fishery Sciences, West Bengal University of Animal and Fishery Sciences, Kolkata, India

M. M. Malik (✉) · A. Amin
Division of Aquatic Environmental Management, Sher-e-Kashmir University of Agricultural Sciences and Technology, Kashmir, Faculty of Fisheries Rangil, Ganderbal, Jammu and Kashmir, India

I. Ahmad
Faculty of Fisheries Rangil, SKUAST-Kashmir, Srinagar, India

M. A. Rather et al. (eds.), *Xenobiotics in Aquatic Animals*,
https://doi.org/10.1007/978-981-99-1214-8_13

nonylphenols (NPEOs) account for roughly 80% of all APEOs, while ethoxylated octylphenols (OPEOs) account for about 20% (David et al. 2009; Soares et al. 2008; Sumpter and Jobling 2013). These alkylphenol ethoxylates are degraded to form the alkylphenols in the aquatic compartment (Miyagawa et al. 2016). Alkylphenols (APs) such as nonylphenol (NP) and octylphenol (OP) are endocrine disruptors being mostly used in the textiles, agricultural chemical products, and plastic industry (Salgueiro-González et al. 2012) and as such play a significant role among environmental estrogenic contaminants. Due to their wide use, such compounds end up in aquatic habitats as effluent from urban, industrial discharge, rainfall runoff, etc. Octylphenol concentration in river water worldwide ranges from <0.001 to 1.440 μg L^{-1}(Sharma et al. 2009). Moreover, these compounds show detrimental impact on the hormone-level systems of many organisms (US EPA 2010), like alteration in reproductive, immunological, neurological systems, and may also prove to be carcinogenic to human beings (Uğuz et al. 2009; Thompson et al. 2015). Octylphenol ethoxylates and degradation product octylphenol tend to mimic or block natural hormones in the endocrine system (Jobling et al. 1996) and have been found to have similar characteristic affinity for the estrogen receptors (ERα and ERβ) in reporter cell lines (Paris et al. 2002). Once these compounds bind to ER and AR, they affect endogenous estrogen levels as well as the aromatase activity (conversion of androgen to estrogen) by several cellular pathways (Dong et al. 2014; Dumitrescu et al. 2010). Additionally, the presence of vitellogenin in the male fish of plasma has also been detected (Ali et al. 2017). Previous studies in fish and other aquatic species have recorded that NP exposure causes negative impacts, which include a reduction in gonadosomatic index (Saravanan et al. 2019), a decrease in male spermatogenesis (Cheng et al. 2017), changes in steroids and thyroidal levels (Naderi et al. 2014), destruction of female gonads (Rivero et al. 2008), as well as fecundity reduced to nonylphenol exposure (Hu et al. 2014). When the nonylphenol exposure at environmental-relevant concentration (1 and 10 μg/L) occurs, it induces a wide range of impacts on plasma sex steroids, phosphorus, estradiol-to-testosterone ratio, and intersex among wild populations (Adeogun et al. 2017) and even causes histopathological impairment in different organs in Caspian brown trout during smoltification (Shirdel et al. 2020).

2 Distribution of Alkylphenols

Samples from various countries around the world that have been contaminated by various concentration of alkylphenols (OP and NP) in river water, wastewater treatment plants, drinking water, and sediment samples (Careghini et al. 2015; Wang et al. 2016; Li et al. 2019; Zhao et al. 2021) are presented in Tables 1 and 2. The majority of researchers focused on the identification of octylphenol and nonylphenols in aquatic ecosystems, which is primarily associated with various human activities. The branched alkylphenols (4-tert-octylphenol and nonylphenol)

Table 1 Distribution of alkylphenols (4-nonylphenol (4-NP) and 4-tert-octylphenol (4-t-OP) in different environmental matrix

Country sample sources	Alkylphenol concentration	Observation	Reference
Yong River (China)	4-NP-140–3948 ng L^{-1} 4-t-OP-6–828 ng L^{-1}	Higher concentrations were detected in months with comparatively less precipitation	Cheng et al. (2018)
Pearl River Delta (China)	4-NP-14,540 ng L^{-1} (river water), 3088 ng/g (sediments) 4-t-OP-758 ng L^{-1} (river water)	The mainstream received significant input from tributaries	Zhao et al. (2021)
Canadian MWWTP-associated sites (Wascana)	Median (SD) NP 148 (128.2 ng L^{-1})	NP was identified more frequently in MWWTP-associated sites and urban	Lalonde and Garron (2021)
India, Mumbai, Thane Creek	OP 176 μg kg^{-1} dw (sediment)	Pollutants directly enter from adjacent rivers/outflows containing a larger portion of industrial and municipal wastewater	Tiwari et al. (2016)
Latin America (Mexico)	Surface water (NP), 12.61 μg L^{-1} Wastewater (NP), 12.20 μg L^{-1} Drinking water(NP), 6.08 μg L^{-1}	–	Vargas-Berrones et al. (2020a, 2020b)
Iran	Wastewater sample NP, 2.12 μg L^{-1}	More concentrations of alkylphenols were found in commercial areas and livestock sewages	Bina et al. (2018)

were found in more than 90% of the tests performed, while linear alkylphenols (4-n-octylphenol and 4-n-nonylphenol) were just identified in 12% of the samples. Wastewater treatment plant discharges, as well as fishing and shipping operations, were determined to be the main sources of alkylphenol pollutants (Salgueiro-González et al. 2016). Recent studies have documented that NP concentration in flounder and eel livers were 3–20 times higher than that in muscles, while in all cod liver samples, they were below the limit of quantitation (LOQ). The average NP concentration in the liver of flounder caught in the Gulf of Gdansk was 222 μg kg^{-1} ww (Ruczyńska et al. 2020).

Table 2 Concentration of alkylphenols in aquatic species samples

Country	Species	Sample and 4-tOP	Sample and NP	Reference
Spanish Atlantic coast and Bay of Biscay	*Mytilus galloprovincialis*	9.3–44.4 (ng g^{-1} dw)	9.0–2441 (ng g^{-1} dw)	Salgueiro-González et al. (2016)
Gulf of Gdansk (Southern Baltic)	*M. edulis trossulus*	<0.8–6.4 (ng g^{-1} dw)	18.8–75.6 (ng g^{-1} dw)	Staniszewska et al. (2014)
Atlantic coast, Portugal and Spain	Sea bass	–	Bile 360–898 ng g^{-1}	Fernandes et al. (2008)
China, Xi'an River	Wild fish	–	Bile ND–803 ng ml^{-1}	Wu et al. (2016)
Dan Shui River, Taiwan	Wild tilapia	–	Liver 2028 mg kg^{-1} ww	Chen et al. (2014)
Mediterranean Sea, Sicily	Red mullet	–	10.1–33.7 mg kg^{-1} ww	Errico et al. (2017)
Gulf of Gdańsk	Flounder	–	Liver mean concentration—222 μg kg^{-1} ww	Ruczyńska et al. (2020)
Pulau Kukup, Johor	Golden pomfret (*Trachinotus blochii*)	Mean concentration (0.124 ng/g)	Mean concentration (0.023 ng/g)	Ismail et al. (2018)

3 Effect on the Aquatic Species

3.1 Male Reproductive System

There is serious concern that pollutants are affecting aquatic wild male reproduction. NPs are noticed to have effects on sex differentiation, implying that they directly stimulate gonad development into male or female (Demska-Zakes and Zakes 2006; Maack and Segner 2004; Zafar et al. 2021). It has been seen that such estrogenic pollutants stimulated to decrease the concentration of some circulating hormones like TSH, T_3, T_4, FSH, and testosterone, while other hormones like LH and 17-β-estradiol had increased in the serum of the fish (Sayed et al. 2012). Similarly, other studies also observed significant increase in the levels of estradiol and decrease of testosterone and FSH levels in the plasma of both sexes of smolts (*Salmo trutta caspius*) after 21 days of exposure to NP (Shirdel et al. 2020).

It induces the synthesis of vitellogenin in male fish, alters the testis structure, causes degeneration of spermatogonia (Shirdel et al. 2020; Sundaray et al. 2021), and reduces sperm numbers, resulting in the formation of hermaphrodite individuals and shifts in population sex ratios (Ackermann et al. 2002; Christensen et al. 1999). Gonadosomatic index (GSI) reduced significantly during the spawning season of the male fishes (catfish) of both lower (64 μg.L^{-1}) and higher (160 μg.L^{-1}NP) concentration treatments than control and later progressed to reproductive dysfunction (Suman and Jiwatram 2021). Exposure to Nile tilapia fingerlings at 40 μg NP/L causes decrease in males significantly as compared to control group, whereas at higher concentration (60 and 100 μg NP/L) exposure leads to intersex (testis-ova) condition in male fingerlings which may be due to sex reversal (Ali et al. 2017). In another study, exposure of 4-tert-octylphenol on early life stages of the South American cichlid fish exhibited 15–38.5% testicular oocytes (TOs) in males, but didn't show significant difference from the control group. Further, TOs did not appear to affect male gonad development and functionality, as normal spermatogenesis has been noticed in OP-treated fish testes, which may be due to weaker xenoestrogen effect of octylphenol (Meijide et al. 2016),whereas other fish such as Neotropical cichlid fish exhibited interstitial fibrosis at higher concentration of 150-μg g/L OP (became apparent), but the lobular structure of the testes is preserved (Rey Vázquez et al. 2016).

4 Female Reproductive System

A 2019 study conducted by Adeogun et al. (year) investigated the severity of five endocrine disruptive compounds on *Sarotherodon melanotheron* from Lagos Lagoon (Nigeria). In this study, endocrine-disrupting compounds (including alkylphenols) had inhibited histopathological alteration in fish gonads and revealed a 27.4% prevalence of intersex in the sampled fish, out of which 78% are males (tests-ova) and 22% are females (ova-testis). Further, the severity of intersex was observed in contaminated site compared to control, whereas female fish showed variable levels of atretic oocytes, higher *zrp* mRNA, and alteration in body weight, gonad weight, plasma VTG, and GSI were observed at contaminated site (Adeogun et al. 2019; Rather and Dhandare 2019). Further toxicity studies of nonylphenol effect on female fish treated with 10- and 50-μg 4-NP/g BW had significantly higher plasma VTG concentrations compared to control. In addition to that, increase of E2 and progesterone was noticed in Koi carp (*Cyprinus carpio*) (Amaninejad et al. 2018; Rather et al. 2017). Interestingly, the mean diameter of the oocyte increased in fish (Caspian trout) exposed to nonylphenol concentrations of 10 and 100 μg/L (Shirdel et al. 2020).

5 Effect on Growth of Microalgae

Microalgae species exposed to alkylphenols caused alteration in their metabolism, photosynthesis activity, and raise in basic photosynthesis pigments (phycobiliproteins and chlorophyll a) due to xenobiotic exposure. In this study, for all types of algae investigated, nonylphenols (NPs) have demonstrated more toxicity than octylphenols (OPs): the doses of NPs (EC_{50}) that cause 50% inhibition of microalgal development are 1.5–3.3 times lower than EC_{50} for OP (depending on the culture). They also concluded that AP exposure to microalgae showed that green algae (*Oocystis parva* and *Scenedesmus quadricauda*) are more resistant, as compared to blue-green algae. The presence of low concentration of alkylphenols (NP and OP) stimulated the growth of cyanobacteria of the genus *Microcystis* by 20–30% (Zaytseva et al. 2015). Similarly, other toxicity studies also reported that NP and OP hampered the cell growth and photosynthetic activity of *Chlorella pyrenoidosa* and *Scenedesmus obliquus*. Furthermore, the toxic effects of alkylphenols to microalgae are chemical- and species-specific (Yang et al. 2021).

6 Effect on the Mollusk

Endocrine disruption in invertebrates has been less studied than in fish, mainly due to a lack of understanding or expertise concerning invertebrate endocrinology. The invertebrate VTG as a biomonitor for environmental estrogens has gained much attention in recent time. Molluskan VTG looks promising as an invertebrate biomarker for both synthetic and natural estrogenic compounds. Natural estrogens, such as estrone (E1), 17-estradiol (E2), and estriol (E3), are endocrine-disrupting chemicals (EDCs) that are regularly and directly released into water bodies via wastewater treatment plant effluents, disposal sludges, and rainwater runoff. However, E2 constitutes the most widespread and effective natural estrogen that induces estrogen activity (Nazari and Suja 2016).

M. edulis exposed to 17b-estradiol (E2) (at 5 ng/L, 50 ng/L for females, and 50 ng/L for males) during their early gametogenesis stage significantly increased VTG and ER2 mRNA expression levels in gonad tissues (Table 3). In the same study, at elevated doses (200 ng/L E2), there was no significant change in VTG and ER2 mRNA expression levels in both immature and mature *M. edulis* (Ciocan et al. 2010). Japanese clam was exposed for 7 days to nominal 4-NP concentrations of 0.1 and 0.2 mg/L, which resulted in elevated VTG-like protein levels in male hemolymph and digestive glands (Matozzo and Marin 2005).

7 Conclusion

Xenoestrogens, such as alkylphenols (NP, OP), have an immense effect on various aquatic organisms, resulting in abnormality in male and female reproductive organs, reduction in spermatozoa count, impact on oocyte development, decreased male proportionality, etc. Such pollution can lead to significant economic losses to fishery ecosystem. APs can accumulate along the food chain and ultimately be consumed by human beings. However, in order to safeguard both humans and other aquatic animals, there is an urgent need for public awareness and concern about such pollutants in order to monitor and minimize the use of certain alkyl phenolic compounds.

Table 3 Toxicity of alkylphenols to aquatic organisms

Name of the compounds	Concentration of exposure (range of dose at which the risk was observed)	Type of risks involved	Reference
Alkylphenols (NP and OP)	50- and 100-mg/L tNP or 4-tert-OP was added at the start of the cell culture	Suppression of the cell growth, changes in mitochondria, vacuoles, and cell walls of *Aspergillus tubingensis*	Kuzikova et al. (2020) Gulf of Finland
Alkylphenols (NP and OP)	Nonylphenols in concentrations from 5×10^{-4} to 5×10^{-2} mg/L and octylphenol in concentrations from 5×10^{-3} to 0.25 mg/L)	Inducing the development of toxigenic cyanobacteria *Microcystis aeruginosa*	Zaytseva et al. (2015)
NP	Low dose 64 $\mu g.L^{-1}$ NP	Discontinuous of interlobular connective tissue	Suman and Jiwatram (2021)
	High dose 160 $\mu g.L^{-1}$	Disintegration of germ cells	
4-nonylphenol	0.08 mg/L	Few number of Sertoli cell nuclei hypertrophy was observed	Sayed et al. (2012)
	0.1 mg/L	Follicular atresia, breakdown of yolk granules	
4-NP	100 μg 4-NP	Follicular atretic increased	Amaninejad et al. (2018)

References

Ackermann GE, Schwaiger J, Negele RD, Fent K (2002) Effects of long-term nonylphenol exposure on gonadal development and biomarkers of estrogenicity in juvenile dam (Awba Dam) in Ibadan, Nigeria. Aquat Toxicol 174:10–21

Adeogun AO, Onibonoje K, Ibor OR, Omiwole RA, Chukwuka AV, Ugwumba AO, Ali T, Abdel-Aziz SH, El-Sayed AF, Zeid S (2017) Effects of nonylphenol on plasma steroids, vitellogenin synthesis and sex reversal in Nile tilapia (*Oreochromis niloticus*) fingerlings. NISCAIR-CSIR, pp 953–960

Adeogun AO, Ibor OR, Chukwuka AV, Regoli F, Arukwe A (2019) The intersex phenomenon in Sarotherodon melanotheron from Lagos lagoon (Nigeria): occurrence and severity in relation to contaminants burden in sediment. Environ Pollut 244:747–756

Ali TE, Aziz SHA, El-Sayed AFM, Zeid S (2017) Effects of nonylphenol on plasma steroids, vitellogenin synthesis and sex reversal in Nile tilapia (Oreochromis niloticus) fingerlings. Indian J Geo Mar Sci 46(03):521–528

Amaninejad P, Hosseinzadeh Sahafi H, Soltani M, Hosseini Shekarabi SP (2018) Endocrine disrupting effects of 4-nonylphenol on plasma vitellogenin, reproductive system and histology in koi carp (*Cyprinus carpio*). Int Aquat Res 10(3):263–274

Bina B, Mohammadi F, Amin MM, Pourzamani HR, Yavari Z (2018) Determination of 4-nonylphenol and 4-tert-octylphenol compounds in various types of wastewater and their removal rates in different treatment processes in nine wastewater treatment plants of Iran. Chin J Chem Eng 26:183–190

Careghini A, Mastorgio AF, Saponaro S, Sezenna E (2015) Bisphenol A, nonylphenols, benzophenones, and benzotriazoles in soils, groundwater, surface water, sediments, and food: a review. Environ Sci Pollut Res 22(8):5711–5741

Chen WL, Gwo JC, Wang GS, Chen CY (2014) Distribution of feminizing compounds in the aquatic environment and bioaccumulation in wild tilapia tissues. Environ Sci Pollut Res Int 21(19):11349–11360

Cheng Y, Shan Z, Zhou J, Bu Y, Li P, Lu S (2017) Effects of 4-nonylphenol in drinking water on the reproductive capacity of Japanese quails (*Coturnix japonica*). Chemosphere 175:219–227

Cheng JR, Wang K, Yu J, Yu ZX, Yu XB, Zhang ZZ (2018) Distribution and fate modeling of 4-nonylphenol, 4-t-octylphenol, and bisphenol a in the Yong River of China. Chemosphere 195: 594–605

Christensen LJ, Korsgaard B, Bjerregaard P (1999) The effect of 4-nonylphenol on the synthesis of vitellogenin in the flounder Platichthys flesus. Aquat Toxicol 46:211–219

Ciocan CM, Cubero-Leon E, Puinean AM, Hill EM, Minier C, Osada M, Rotchell JM (2010) Effects of estrogen exposure in mussels, Mytilus edulis, at different stages of gametogenesis. Environ Pollut 158(9):2977–2984

David D, Fenet H, Gomez E (2009) Alkylphenols in marine environments: distribution monitoring strategies and detection considerations. Mar Pollut Bull 58(7):256–262

Demska-Zakes K, Zakes Z (2006) Induction of testis-ova in pikeperch (*Sander lucioperca* (L.)) exposed to 4- nonylphenol. Arch Pol Fish 14(1):29–39

Dong RR, Yang SJ, Feng RJ, Fang LL, Sun YL, Zhang YG, Wang DS (2014) Complete feminization of catfish by feeding Limnodilus, an annelid worm collected in contaminated streams. Environ Res 133:371–379

Dumitrescu G, Ciochina LP, Voia S, Dronca D, Boca L (2010) Histological changes induced in gonads, liver and kidney of zebra fish (Danio rerio) under the effect octylphenol (OP). Sci Papers Anim Sci Biotechnol 43(1):484–489

Errico S, Nicolucci C, Migliaccio M, Micale V, Mita DG, Diano N (2017) Analysis and occurrence of some phenol endocrine disruptors in two marine sites of the northern coast of Sicily (Italy). Mar Pollut Bull 120(1–2):68–74

EUROSTAT (2019) Chemicals production and consumption statistics. https://ec.europa.eu/eurostat/documents/3217494/10165279/KS-DK-19-001-EN-N.pdf/76651a29-b817-eed4-f9f2-92bf692e1ed9

Fernandes D, Zanuy S, Bebianno MJ, Porte C (2008) Chemical and biochemical tools to assess pollution exposure in cultured fish. Environ Pollut 152(1):728–735

Gavrilescu M, Demnerova K, Aamand J, Agathos S, Fava F (2015) Emerging pollutants in the environment: present and future challenges in biomonitoring, ecological risks and bioremediation. New Biotechnol 32:147–156

Hu X-L, Sun Z-W, Wang J-J, An M, Duan S-S (2014) Sublethal toxic effects of nonylphenol ethoxylate and nonylphenol to Moina macrocopa. Bull Environ Contam Toxicol 93:204–208

Ismail NAH, Wee SY, Aris AZ (2018) Bisphenol A and alkylphenols concentrations in selected mariculture fish species from Pulau Kukup, Johor, Malaysia. Mar Pollut Bull 127:536–540

Jobling S, Sheahan D, Osborne A, Matthiessen P, Sumpter JP (1996) Inhibition of testicular growth in rainbow trout (Oncorhynchus mykiss) exposed to estrogenic alkylphenolic chemicals. Environ Toxicol Chem 343:143–155

Kuzikova I, Rybalchenko O, Kurashov E, Krylova Y, Safronova V, Medvedeva N (2020) Defense responses of the marine-derived fungus Aspergillus tubingensis to alkylphenols stress. Water Air Soil Pollut 231(6):1–18

Lalonde B, Garron C (2021) Nonylphenol, octylphenol, and nonylphenol ethoxylates dissemination in the Canadian freshwater environment. Arch Environ Contam Toxicol 80(2):319–330

Li Z, Zhang W, Shan B (2019) The effects of urbanization and rainfall on the distribution of, and risks from, phenolic environmental estrogens in river sediment. Environ Pollut 250:1010–1018

Maack G, Segner H (2004) Life-stage-dependent sensitivity of zebrafish (Danio rerio) to estrogen exposure. Comp Biochem Physiol C Toxicol Pharmacol 139(1–3):47–55. https://doi.org/10.1016/j.cca.2004.09.004

Matozzo V, Marin MG (2005) Can 4-nonylphenol induce vitellogenin-like proteins in the clam *Tapes philippinarum*? Environ Res 97:43–49

Meijide FJ, Vázquez GR, Piazza YG, Babay PA, Itria RF, Nostro FLL (2016) Effects of waterborne exposure to 17β-estradiol and 4-tert-octylphenol on early life stages of the South American cichlid fish Cichlasoma dimerus. Ecotoxicol Environ Saf 124:82–90

Miyagawa S, Sato T, Iguchi T (2016) Handbook of hormones, comparative endocrinology for basic and clinical research. Subchapter 101A – Nonylphenol

Naderi M, Mousavi SM, Safahieh A, Ghatrami ER, Zargham D (2014) Effects of 4- nonylphenol on balance of steroid and thyroid hormones in sexually immature male yellowfin seabream (Acanthopagrus latus). Environ Toxicol 29:459–465

Naidu R, Arias Espana VA, Liu Y, Jit J (2016) Emerging contaminants in the environment: risk-based analysis for better management. Chemosphere 154:350–357

Nazari E, Suja F (2016) Effects of 17β-estradiol (E2) on aqueous organisms and its treatment problem: a review. Rev Environ Health 31(4):465–491

Pal A, Gin KYH, Lin AYC, Reinhard M (2010) Impacts of emerging organic contaminants on freshwater resources: review of recent occurrences, sources, fate and effects. Sci Total Environ 408(24):6062–6069

Paris F, Balaguer P, Terouanne B, Servant N, Lacoste C, Cravedi JP, Nicolas JC, Sultan C (2002) Phenylphenols, biphenols, bisphenol-A and 4-tert-octylphenol exhibit alpha and beta estrogen activities and antiandrogen activity in reporter cell lines. Mol Cell Endocrinol 193:43–49

Rather MA, Dhandare BC (2019) Genome-wide identification of doublesex and Mab-3-related transcription factor (DMRT) genes in Nile tilapia (Oreochromis niloticus). Biotechnology reports 24:e00398

Rather MA, Basha SH, Bhat IA, Sharma N, Nandanpawar P, Badhe M, Sharma R (2017) Characterization, molecular docking, dynamics simulation and metadynamics of kisspeptin receptor with kisspeptin. Int J Biol Macromol 101:241–253

Rey Vázquez G, Meijide FJ, Lo Nostro FL (2016) Recovery of the reproductive capability following exposure to 4-tert-octylphenol in the neotropical cichlid fish Cichlasoma dimerus. Bull Environ Contam Toxicol 96(5):585–590

Rivero CLG, Barbosa AC, Ferreira MFN, Dorea JG, Grisolia CK (2008) Evaluation of genotoxicity and effects on reproduction of nonylphenol in Oreochromis niloticus (Pisces: Cichlidae). Ecotoxicology 17:732–737

Ruczyńska W, Szlinder-Richert J, Nermer T (2020) The occurrence and distribution of nonylphenols and nonylphenol ethoxylates in different species of fish. Environ Sci: Processes Impacts 22(4):1057–1070

Salgueiro-González N, Turnes-Carou I, Muniategui-Lorenzo S, López-Mahía P, Prada-Rodríguez D (2012) Fast and selective pressurized liquid extraction with simultaneous in cell clean up for the analysis of alkylphenols and bisphenol A in bivalve molluscs. J Chromatogr A 1270:80–87

Salgueiro-González N, Turnes-Carou I, Viñas L, Besada V, Muniategui-Lorenzo S, López-Mahía P, Prada-Rodríguez D (2016) Occurrence of alkylphenols and bisphenol a in wild mussel samples from the Spanish Atlantic coast and Bay of Biscay. Mar Pollut Bull 106(1–2):360–365

Saravanan M, Nam SE, Eom HJ, Lee DH, Rhee JS (2019) Long-term exposure to waterborne nonylphenol alters reproductive physiological parameters in economically important marine fish. Comp Biochem Physiol C Toxicol Pharmacol 216:10–18

Sayed AEDH, Mahmoud UM, Mekkawy IA (2012) Reproductive biomarkers to identify endocrine disruption in Clarias gariepinus exposed to 4-nonylphenol. Ecotoxicol Environ Saf 78:310–319

Sharma VK, Anquandah GA, Yngard RA, Kim H, Fekete J, Bouzek K et al (2009) Nonylphenol, octylphenol, and bisphenol-a in the aquatic environment: a review on occurrence, fate, and treatment. J Environ Sci Health A 44(5):423–442

Shirdel I, Kalbassi MR, Esmaeilbeigi M, Tinoush B (2020) Disruptive effects of nonylphenol on reproductive hormones, antioxidant enzymes, and histology of liver, kidney and gonads in Caspian trout smolts. Comp Biochem Physiol C Toxicol Pharmacol 232:108756

Soares A, Guieysse B, Jefferson B, Cartmell E, Lester JN (2008) Nonylphenol in the environment: a critical review on occurrence, fate, toxicity and treatment in wastewaters. Environ Int 34(7): 1033–1049

Staniszewska M, Falkowska L, Grabowski P, Kwaśniak J, Mudrak-Cegiołka S, Reindl AR, Zgrundo A (2014) Bisphenol A, 4-tert-octylphenol, and 4-nonylphenol in the Gulf of Gdańsk (Southern Baltic). Arch Environ Contam Toxicol 67(3):335–347

Suman RC, Jiwatram GG (2021) 4-Nonylphenol affects the structure and function of testis in catfish H. fossilis and C. batrachus. J Sci Res 65(5):155–163

Sumpter JP, Jobling S (2013) The occurrence, causes, and consequences of estrogens in the aquatic environment. Environ Toxicol Chem 32(2):249–251

Sundaray JK, Rather MA, Kumar S, Agarwal D (2021) Recent updates in molecular endocrinology and reproductive physiology of fish. Springer, Singapore

Thompson PA, Khatami M, Baglole CJ, Sun J, Harris SA, Moon EY (2015) Environmental immune disruptors, inflammation and cancer risk. Carcinogenesis, Chemosphere 256–266

Tiwari M, Sahu SK, Pandit GG (2016) Distribution and estrogenic potential of endocrine disrupting chemicals (EDCs) in estuarine sediments from Mumbai, India. Environ Sci Pollut Res 23(18): 18789–18799

Uğuz C, İşcan M, Togan İ (2009) Alkylphenols in the environment and their adverse effects on living organisms. Kocatepe Vet J 2(1):49–58

US EPA (2010) Nonylphenol (NP) and nonylphenol ethoxylates (NPEs) action plan

Van Zijl MC, Aneck-Hahn NH, Swart P, Hayward S, Genthe B, De Jager C (2017) Estrogenic activity, chemical levels and health risk assessment of municipal distribution point water from Pretoria and Cape Town, South Africa. Chemosphere 186:305–313

Vargas-Berrones K, Bernal-Jácome L, de León-Martínez LD, Flores-Ramírez R (2020a) Emerging pollutants (EPs) in Latin América: a critical review of under-studied EPs, case of study-nonylphenol. Sci Total Environ 726:138493

Vargas-Berrones K, de León-Martínez LD, Bernal-Jácome L, Rodriguez-Aguilar M, Ávila-Galarza A, Flores-Ramírez R (2020b) Rapid analysis of 4-nonylphenol by solid phase microextraction in water samples. Talanta 209:120546

Wang, B., Dong, F., Chen, S., Chen, M., Bai, Y., Tan, J., & Wang, Q. (2016). Phenolic endocrine disrupting chemicals in an urban receiving river (Panlong river) of Yunnan–Guizhou plateau: occurrence, bioaccumulation and sources. Ecotoxicol Environ Saf, 128, 133–142

Wu M, Pan C, Yang M, Xu B, Le X, Ma J, Cai L, Chen J (2016) Chemical analysis of fish bile extracts for monitoring endocrine disrupting chemical exposure in water: bisphenol A, alkylphenols, and norethindrone. Environ Toxicol Chem 35(1):182–190. https://doi.org/10.1002/etc.3176

WWF (2020) Nature. In: WWF (ed) Toxic chemicals, pp 462–473

Yang W, Gao X, Wu Y, Wan L, Lu C, Huang J, Zhang W (2021) Chemical-and species-specific toxicity of nonylphenol and octylphenol to microalgae Chlorella pyrenoidosa and Scenedesmus obliquus. Environ Toxicol Pharmacol 81:103517

Zafar I, Iftikhar R, Ahmad SU, Rather MA (2021) Genome wide identification, phylogeny, and synteny analysis of sox gene family in common carp (*Cyprinus carpio*). Biotechnol Rep 30: e00607

Zaytseva TB, Medvedeva NG, Mamontova VN (2015) Peculiarities of the effect of octyl-and nonylphenols on the growth and development of microalgae. Inland Water Biol 8(4):406–413

Zhao JL, Huang Z, Zhang QQ, Ying-He L, Wang TT, Yang YY, Ying GG (2021) Distribution and mass loads of xenoestrogens bisphenol a, 4-nonylphenol, and 4-tert-octylphenol in rainfall runoff from highly urbanized regions: a comparison with point sources of wastewater. J Hazard Mater 401:123747

Xenobiotics' Effect on Fish Reproduction and Development

Jham Lal, Soibam Khogen Singh, Pradyut Biswas, and Reshmi Debbarma

1 Introduction

Xenobiotics are elements that are not naturally occurring inside the body but possess the potential to interact with it and cause teratogenic, mutagenic, or perhaps even cancerous consequences. Examples include air pollution, toxic metals, plastic materials, medications, pesticide, as well as petroleum products (Lang and Pelkonen 1999). Nevertheless, xenobiotics are regarded as hazardous agents because they upset the natural equilibrium, which means these are compounds that alter the body's natural equilibrium after ingesting them (Larini 1997). Fish can absorb xenobiotics via their gills, skin, digestive juices, as well as other organs (Júnior 2019). If it is considered that man-made xenobiotic doses do not affect fish abundance and diversity, so a chronic toxicity study must evaluate how these compounds affect development and reproduction because these processes are crucial population controllers (Bresch 1982).

The ecosystem encompasses everything that occurs naturally in our immediate vicinity and has an impact on how we live on Earth. The continuation of all life on Earth depends on a secure and healthy ecosystem. Nevertheless, in the age of increased industry as well as urbanization, numerous human activities are substantially to blame for the introduction of both dangerous and poisonous contaminants including environmental xenobiotics (Embrandiri et al. 2016; Malla et al. 2018; Bhatt et al. 2022; Rodríguez et al. 2020). Chemicals known as xenobiotic compounds are those that are neither anticipated nor generated normally by living things. Typically, when discussing environmental contaminants, the word "xenobiotic" refers to synthetic substances manufactured in huge quantities for industrialized, agriculture, as well as home usage (Embrandiri et al. 2016; Atashgahi et al. 2018;

J. Lal · S. K. Singh (✉) · P. Biswas · R. Debbarma
Department of Aquaculture, College of Fisheries, Lembucherra, Central Agricultural University, Imphal, Manipur, India

M. A. Rather et al. (eds.), *Xenobiotics in Aquatic Animals*,
https://doi.org/10.1007/978-981-99-1214-8_14

Dinka 2018). The enormous variety of xenobiotic substances that are being released into the environment—either by purpose or by accident—poses a considerable risk to both humans and other animals, which is causing increasing public concern (Jacob and Cherian 2013; Hashmi et al. 2017; Zhu et al. 2017; Dinka 2018). A few examples of environmental xenobiotic compounds include pesticides, polycyclic aromatic hydrocarbons (PAHs), pharmaceutical active compounds (PhACs), personal care products (PCPs), polyphenolic compounds, chlorinated substances, as well as other toxic materials. Concerns regarding these possible negative consequences have grown as its occurrence has increased in many environmental contexts (Crinnion 2010; Kim et al. 2013; Embrandiri et al. 2016; Tsaboula et al. 2016; Dhakal et al. 2018). Their toxicity poses previously unheard-of health dangers as well as threats to the security as well as safety of the natural environment (Godheja et al. 2016; Dvořák et al. 2017; Burgos-Aceves et al. 2018). Because xenobiotics have a strong attraction for organic molecules, when they are discharged into the environment, chemicals can accumulate in the body and cause harm to humans, organisms, as well as the entire environment (Pedersen et al. 2003; Iovdijová and Bencko 2010; Maurya and Malik 2016).

2 Xenobiotics and Toxicity

It has been demonstrated that a large variety of man-made chemical compounds utilized in a variety of industrial and domestic processes interferes with natural endocrinology as well as physiology in living creatures. Neurophysiological, reproduction, and behavioral consequences are the three major categories within which the impacts of xenobiotics mostly on the entire organism are examined. These impacts are frequently interconnected; for example, neurological changes can influence behavior, and behavioral changes can influence reproduction. A substance does not necessarily have an impact on the target animal or group. This always varies depending on the amount of that substance as well as the duration of the exposure. In the end, these impacts may be either immediate or persistent. Acute toxicity is characterised by its rapid onset, characteristic manifestations, frequent fatal outcomes, and rarity of reversibility. Long-term exposure at low concentrations or longer subsequent exposure results in chronic consequences that can eventually result in mortality. A xenobiotic is considered lethal when it either directly causes death or has the potential to do so. Additionally, a toxin is sublethal when its potency is below that which ultimately results in death. Following it, the individual's behavioral and physiological processes regress, which lowers their overall health. The likelihood that such contamination will have an irreversible impact on the environment is only present in the situation of radioactive material (Table 1).

The impact of pollutants on freshwater fish species is reflected in the extinction of certain animals, with potential financial gain for others. The relative importance of processes like predators, competition, and substance cycling changes as the variety generally declines, though not necessarily the total number of distinct organisms.

Table 1 Effects of major heavy metals on fish reproduction

Heavy metal	Fishes	Effects on fish	Reference
Lead (Pb)	Freshwater fishes	Genotoxic, cytotoxic damage in gill and fin epithelial cells in some fishes; in other fishes, it delays embryonic development, obstructs growth, suppresses reproduction, causes kidney dysfunction	Ibemenuga (2013)
Selenium	Zebrafish (*Danio rerio*)	It can negatively influence fish reproductive capability	Penglase et al. (2014)
Selenium and mercury	Zebrafish (*Danio rerio*)	Se and Hg had a synergistic adverse impact on all aspects of fish reproduction	Penglase et al. (2014)
Copper oxide nanoparticles (CuO-NPs)	Guppy (*Poecilia reticulata*)	Chronic exposure affected reproductive traits of fish	ForouharVajargah et al. (2020)
Mercury (Hg), cadmium(Cd), and lead (Pb)	*Tilapia nilotica*	The gonadosomatic index (GSI) decreased in females, retarded gonadal development in the juvenile fish	Ahmed et al. (2010)
Zinc (Zn)	Zebrafish (*Brachydanio rerio*)	The adverse influences on zebrafish fish exposure to zinc can be reversed	Speranza et al. (1977)
Environmental pollutants (heavy metals, drugs, and products)	Fish	The particular component of the reproductive strategy that is most susceptible to such disruption and that this may be at levels well below that which induces mortality or apparent signs of stress	Kime et al. (1999)

Due to the intricacy of contamination, various characteristics of the pollutants also influence how it affects aquatic life. When two or more toxins are existing within the same environment, they may have an additional, antagonistic, as well as cumulative influence on the organism. Zinc and cadmium together are poisonous to fish, which is an illustration of synergistic interactions. In contrast to the hostile elements calcium, zinc, and aluminium, mercury, zinc, cadmium, and chlorine can significantly lessen the severity of cyanide and copper. Phenol and ammonia are additively poisonous to the mayfly *Baetis rhodani* with low doses, but then, at larger quantities, the impact is greater than the additive (Zaki et al. 2014) (Fig. 1).

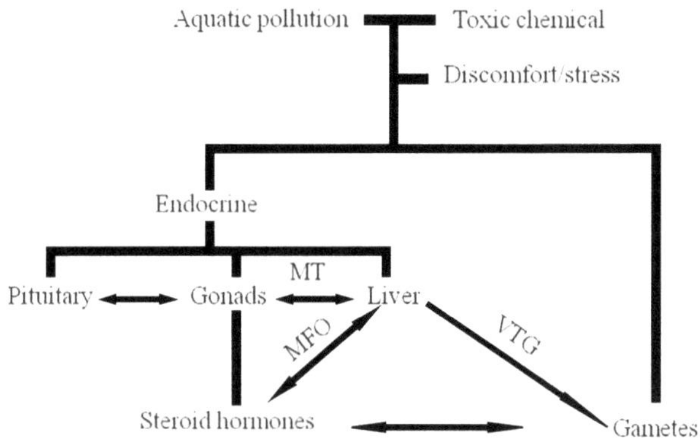

Fig. 1 Components of the reproductive system affected by low-level pollution (Kime 1999)

3 Absorption of Xenobiotics in Fish

Fish can absorb xenobiotics via their gills, stomach juices, skin, as well as other organs. Xenobiotics are then expelled from the body by a variety of discharges, including urine, bile, and feces, as well as gas exchange, either in their original state or after being chemically changed. The kinetic behavior of each chemical has a significant impact on elimination, whether it is unmodified as well as chemically altered. Therefore, lipophilic compounds are quickly absorbed, for instance, through the digestive system, but their expulsion, either through bile or urine, is challenging due to the ease with which they acquire reuptake after passing through the cellular membrane. Therefore, these compounds have a propensity to build up inside the body. Although hydrophilic compounds have worse absorption, they are primarily excreted through the renal pathway (Larini 1997; Oga 1996).

4 Biotransformation of Xenobiotics in Fish

The body possesses a metabolic system that converts the minor polar and lipid-soluble molecules into more significant polar as well as hydrosoluble elements to aid in the elimination of lipid-soluble xenobiotic compounds (Oga 1996). This mechanism is known as biotransformation. A variety of chemical reactions involving liver enzymes makes up the biotransformation pathways. A xenobiotic component may go through one or more changes before derivatives with such a genuine chance of elimination are created. In the second scenario, the initial reaction serves as a prelude, resulting in an intermediate chemical (Phase 1) that will continue to go through a further reaction, ultimately yielding either active or inactive metabolites (phase 2). Phase 1 and 2 processes can occur independent of one another, so a xenobiotic molecule can go through either phase I, then phase II, or both phases simultaneously (Van der Oost et al. 2003). The xenobiotic material is subjected to chemical processes during biotransformation, which typically involves enzymes and results in the creation of a different compound than the one that was first supplied. The process of biotransformation, which is primarily carried out within the liver, involves electrically charging the toxic compound's chemical to prevent its absorption when it passes through into the renal tubules.

The process of biotransformation, which is primarily carried out in the liver, entails electrically charged xenobiotic material to prevent its absorption whenever it passes through into the renal tubules. Because of the basic structural changes that are made throughout this procedure and the decreased likelihood that the chemical will reach the vulnerable tissues, the chemical is typically rendered inactive. Biotransformation is for these materials synonymous with elimination. But occasionally, active metabolites—or even more active ones—are produced in addition to the chemical that was delivered.

5 Stress Signal Transduction by Xenobiotics in Fish

Prolonged exposure to xenobiotic compounds causes stress as well, and hormonal, neuronal, as well as other physiological processes have a role in how we respond to this stress. The study attempts to summarize the different assessment processes elicited by certain candidate pollutants. Xenobiotics have recently attracted attention as just environmental stressors. Intoxication and detoxication are the two main methods by which the toxic compounds' stress response is elicited. Xenobiotic poisoning may manifest itself along any one of the numerous physiological mechanisms. Any poisoning mechanism must include a targeted region of action that is reliant just on chemical components of both the xenobiotic, either organic or inorganic, and other concerns (Bhattacharya 2000).

6 Xenobiotic Action on Steroid Hormones

Periodic reproductive cycles among bony fish, which also are frequently governed by light/dark cycles as well as temperatures, are controlled by hypothalamic and pituitary hormones, which interact on the gonads to increase sex hormones biosynthesis. Therefore, fish reproduction depends on the perfect orchestrating as well as the timing of multiple interconnected activities. The disturbance of the endocrine glands and the restriction of hormone secretion caused by xenobiotics might lead to a reduction in fertility, breeding, or growth (Bhattacharya 2000).

Stress, as regulated by the adrenal, has well-known impacts on the reproductive organs and may be responsible for important unintended consequences that reduce hormone production (Kime et al. 1980; Pickering et al. 1987; Carragher et al. 1989). Growth, reproduction, and immune response are negatively impacted by prolonged increased blood cortisol levels during stressful situations (Barton and Iwama 1991).

Fish exposed to Cd continuously have been found to develop a stress response habitually, as seen by a drop in their earlier increased plasma cortisol level (Fu et al. 1990). There are very few publications accessible regarding either the intrinsic cortisol production or even the mechanisms of action of xenobiotic compounds (Balm et al. 1994; Sumpter et al. 1994; Pottinger et al. 1995; Balm and Pottinger 1995).

7 Xenobiotic Interactions with Steroidogenesis

The effects of xenobiotics on steroid production, including the effects of sterol-ester hydrolase on testosterone hydroxylase. (Civen et al. 1977; Donovan et al. 1978; Clement 1985; Mohapatra et al. 2021). It has been demonstrated that 3B-hydroxysteroid dehydrogenase (HSD) is among the targeted areas of action of xenobiotic compounds (Bhattacharya and Pandey 1989; Bagchi et al. 1990). Molecular or pharmacological mechanisms that are still being studied underlie the responses to stressors, and xenobiotics can interact straightforwardly with protein at locations like steroid hormone receptors to either start or speed up gene transcription (Kime et al. 1980). There have been claims that OP toxicity promotes hormone-mediated protein production; nonetheless, it has been suggested that such a direct influence of a xenobiotic on protein synthesis cannot be excluded (Cisson and Wilson 1981).

8 Vitellogenin as an Indicator of Estrogen Exposure

Vitellogenesis is the process by which ovarian estradiol stimulates the liver to create vitellogenin, which is then integrated into the yolks of forming oocytes (Sundaray et al. 2021). Estrogenic organic pollutants can also stimulate vitellogenin production by acting on hepatic receptors (Pelissero et al. 1993). Although fishes of both sexes, and premature juveniles, contain hepatic estrogen receptors, only the livers of female fish are typically exposed to estrogens. Therefore, the development of vitellogenin through males, juveniles, as well as non-vitellogenic females may serve as bioindicators of exposure to natural estrogens.

However, abnormally high vitellogenin independently may not always be a sign of exposure to environmental estrogen, so care should be used when evaluating vitellogenin data on its own. Female flounder exposure to Rotterdam Port soil in mesocosms had greater vitellogenin levels than control fish. Furthermore, neither treated nor control male fishes had plasma vitellogenin that could be detected, indicating that this impact was not brought on by exogenous estrogen. In contrast to controls, the exposed females' gonads revealed that they possessed vitellogenic oocytes in their ovaries (Janssen et al. 1995).Vitellogenin output was thus most likely a consequence of the increased estradiol secreted by all these prematurely innovative oocytes, possibly as a result of diminished catabolism, because benzopyrene, a primary component of the sediment, hindered hepatic steroid catabolism through monooxygenase enzymes, despite inducing action of such an enzyme as assessed by EROD action (Janssen et al. 1997). Although the specific mechanism of interruption is unknown, these findings indicate that vitellogenin in juveniles or men would be a strong indicator of exogenous estrogen exposure. Nevertheless, if females had not been employed in these tests, there would have been no indications of reproductive failure.

9 Vitellogenin Indicative of Reproductive Dysfunction

Natural estrogens, ingredients in birth control pills, alkyl phenolics, phthalates, certain organochlorine insecticides, and PCBs are among the environmental pollutants thought to stimulate the production of vitellogenin in fish. Although vitellogenesis can be utilized as a biomonitor of estrogenic pollution, it is additionally important to consider regardless of whether this has a bearing on the reproduction or wellness of the fish, whether greater vitellogenin affects the quantity or quality of the eggs generated, as well as what influence it has on the males and juveniles who generally don't ever start producing vitellogenin. Since the mechanism of vitellogenesis includes a complicated interaction between the various glands and hormones of a reproduction endocrine system, estrogen exposure, as demonstrated through vitellogenin synthesis, may very likely have detrimental effects on some other components of fish reproduction (Kime 1998).

Gonadotropin-I (GtH-I), which is released by the pituitary mostly during the vitellogenic phase of the female process, helps to stimulate the ovarian follicle cells to generate estradiol. Estradiol then is transported into the liver through the circulation which performs on receptor sites inside the hepatocytes to generate vitellogenin. Intriguingly, the liver is simultaneously a targeted tissue as well as a key location of estradiol inactivation, which raises the concern about how it "decides" whether it should deactivate the steroids or respond to them by making vitellogenin. The liver's vitellogenin then returns to the ovary's bloodstream wherein, in the presence of GtH-I, this is synthesized into the oocytes (Tyler et al. 1991). It then needs to be converted inside the oocytes into the various molecules that make up yolks. An evaluation system including estradiol, testosterone, or other mediators like inhibins allows communication between both the oocytes and the pituitary. Once oocytes have assimilated a specified amount of vitellogenin as well as reached a species-dependent size, subsequent incorporating stops and aromatase enzymes are turned off. As vitellogenin inclusion is finished for each oocyte, there occurs a gradual shift between estradiol-to-testosterone output and hepatic vitellogenesis declines. Increasing estrogen has little effect on egg size because integration is GtH-I-dependent rather than vitellogenin-dependent.

Reasonable incorporation of vitellogenin into the oocyte follows its withdrawal, but because incorporation is governed by GtH-I (Tyler et al. 1991), the percentage of inclusion cannot be accelerated. As a result, an excess amount of VTG will result in plasma concentrations that are higher than those of normal fish. Therefore, if the xenobiotic can outcompete natural estradiol for pituitary-hypothalamic feedback receptors, it could be able to outcompete this for hepatic receptor sites as well. This may act in an agonistic or antagonistic manner, just like the cardiac receptor, causing a rise or drop in GtH-I. Alterations in GtH generation may occur if xenoestrogen-induced vitellogenin causes a drop in ovary estradiol through a short feedback system. Variations in GtH-I secretions can have an impact on overall oocyte size because VTG must be incorporated into the oocytes for this to happen. Furthermore, it's crucial to keep in mind that GtH-I is influenced by elements other than E2, including such inhibins, etc. (Kime et al. 1999).

Due to the high energy cost of protein synthesis, somatic development must always be sacrificed to produce vitellogenin. There are a few indications indicating estrogen therapy-induced kidney impairment may result in increased plasma vitellogenin levels, which can promote death rates (Herman and Kincaid 1988). In the male population, the response of testis androgens could encompass brain aromatization of testosterone to estradiol; as such an outcome, pituitary GtH formation may be impacted by xenobiotic estrogens. This can result in impaired sperm quality, diminished sperm performance and productivity, failure of GtH-II, as well as inability to produce the last phases of sperm development. These impacts have previously been documented in male fish exposure by alkylphenolics, which also had lower spermatocytes as well as more spermatogonia compared to reference fishes (Jobling et al. 1996). There haven't been any investigations on how estrogenic exposure affects sperm quality, but we're looking into using computer-assisted sperm testing as just a quantitative technique to see if large amounts of vitellogenin in the male population are a sign of low sperm motility (Table 2).

10 Energetic Adaptation

Since an organism could only obtain a certain level of energy toward which multiple processes immediately compete, an enhancement inside the energy assignment through one procedure should be accompanied by a reduction in the energetic distribution to others (Ware 1980, 1982; Sibly and Calow 1983). The ideas of life history as well as optimum foraging give the physiological underpinnings for what happens to the dietary energy that animals consume. Female fish generally develop later than male fishes due to the energy-intensive nature of the reproductive maturing process (Thorpe 1994). In general, the development of the ovary as well as eggs takes more energy than the development of the testes and sperm. Fishes should convert food particles into total (accessible) energy to support their normal metabolism (maintain) as well as operational expenses (Ware 1980).

Table 2 Pollutant-induced reproduction disturbances in fish

Pollutant	Species	Effect	Reference
Alkylphenols	Rainbow trout	Vitellogenesis	Purdom et al. (1994), Harries et al. (1996, 1997)
PCBb	Baltic flounder and herring	Reduced viable hatch	Hansen et al. (1985)
PCB, DDT	Charr	High egg mortality	Monod (1985)
Organic contaminants	Starry flounder	Reduced viable hatch and fertilization	Spies and Rice (1988)
Crude oil	Pacific herring	Premature hatch, egg and larval mortality, morphological deformities, cytogenetic abnormalities	Hose et al. (1996), Kocan et al. (1996), Norcross et al. (1996)
Crude petroleum	Atlantic salmon, flounder	Decreased plasma androgens	Truscott et al. (1983)
Contaminants	Coho salmon	Overripe eggs, fry deformity, low fertilization rate	Flett et al. (1991)
Textile mill effluent	Air-breathing fish	Oocyte atresia	Murugesan and Haniffa (1992)
Vegetable oil factory effluent	Freshwater murrel	Retarded ovarian growth	Saxena and Bhatia (1983)
Vegetable oil factory effluent	Air-breathing fish	Decreased gonadal lipid, increased fatty acids	Kondal et al. (1989)

11 Non-vitellogenic Responses

An organism's life cycle is significantly influenced by the liver, which is a crucial organ (Hinton 1990). These functions include the biotransformation of xenobiotics as well as the gonadal hormone, in addition to the synthesis of Zrp and Vtg for the ovaries (metabolic activities). The impact of xenobiotic compounds inside the liver, however, can be modified by a wide range of additional variables.

The primary site for protein biosynthesis (including Zrp and Vtg) as well as metabolism transformation of both exogenous drugs (xenobiotics) and intrinsic molecules like steroid hormones is indeed the liver cells (Goksøyr and Förlin 1992; Zimniak and Waxman 1993). By the use of certain enzymes from the cytochrome P450 (CYP) superfamily as well as several transferase enzymatic families, environmental pollutants including steroids are metabolically converted or biotransformed inside the liver (Nelson et al. 1996; Lewis 1996; Oritz de Montellano 1995). Isoforms of the CYP1A subfamily have already been widely researched as well as recognized as being the most significant subfamily inside the metabolic activation of many carcinogenic agents (Goksøyr 1995; Ioannides 1990), whereas the CYP3A subfamily (also CYP2K in fishes) has been connected toward steroid metabolism (Zimniak and Waxman 1993; Celander et al. 1996). Extensive study has demonstrated that PAHs, HAHs including 2,3,7,8-tetrachlorodibenzo-p-dioxin (TCDD), as well as similar substances can activate the CYP1A genes via aryl hydrocarbon (Ah) receptors (Okey et al. 1994; Swanson and Bradfield 1993; Hahn and Chandran 1996; Schmidt and Bradfield 1996).

12 Ecological Consequences

Through all stages of ontogeny, reproduction generation is ongoing. As a result, it is vulnerable to the impacts of xenoestrogens and/or environmental pollutants during all phases of the life cycle, involving fertilization, embryonic development, sexual differentiation, oocyte maturation or sperm production, ultimate maturity, and ovulatory. Consequently, the susceptibility to a particular substance may fluctuate according to the phase of reproductive stages (Donaldson 1990). The representation of a sequential manner of reactions to pollution pressures inside a biological system (Bayne et al. 1985). It is always possible to generate emergency alert biomarker indicators since impacts at greater hierarchical levels are usually anticipated via alterations in "earlier" biological systems. The early biological reactions' relevance just at the population as well as ecosystem levels is challenging to determine, though. The challenge is due, among many other things, to the possibility that variations between fish populations and ecosystem processes may result from a wide range of many other factors besides xenobiotic compounds, such as seasonal variations in temperature, saltiness, availability of food, harvesting activity, etc. Acknowledging the basic principles through which chemical compounds, as well as abroad substances (xenobiotics), interrupt fish reproduction is especially essential for achieving the long-term goals of underwater breeding toxicity because it is unattainable to interpret the living organisms' specificity or how each substance influences the reproductive life-history strategic plan of each organism.

Recent risk assessments regarding xenobiotics' reproduction impacts on aquatic creatures rely on in vitro and in vivo laboratory investigations, either explicitly or indirectly. The environmental implications for xenobiotic-induced Zrp as well as Vtg production, on the other hand, are unknown. There is currently considerable evidence indicating aquatic creatures living in bioaccumulating toxic compounds are harmed. Given the various roles that endogenous estrogens perform in healthy physiology, including throughout adult sexual development as well as sex distinction at the earliest stages of life (egg and embryo), there is an obvious need for care in some of these regards (Hunter and Donaldson 1983; Piferrer and Donaldson 1989). In addition, a wide range of additional elements that can modify the impact of xenobiotics may be challenging or even impossible to evaluate.

13 Types of Endocrine Disruptive Chemicals (EDCs)

EDCs come in a variety of kinds and chemical compositions. Diethylstilbestrol, an artificial steroid, is regarded as the initial EDC of significance as well as a perfect example of prolonged poisoning, including impacts in the second and even third generations. Since 1945, a considerable quantity of chemical insecticides for pest management has been produced. Its perspective endocrine-disrupting effects were discovered only during the 1970s. Once discharged into the environment, those resistant hazardous chemicals in water are nearly impossible to remove. When EDCs interact with wastewater cleaning products, metabolic ends are created which attach to steroid hormone receptors, change the expression of multiple genes, as well as ultimately cause endothelial dysfunction (Kar et al. 2021). Additionally, they have an impact on the hypothalamic hypophyseal gonadal (HHG) pathway, which is crucial for regulating response (Senthilkumaran 2015). The likelihood of metal deposition at different trophic levels in aquatic freshwater environments has risen as a consequence of increasing the utilization of metallic nanoparticles. However, research addressing nanoparticle (NP) lethality in the endocrine glands of freshwater fish is surprisingly lacking. Juvenile fishes exposed to Cu-NPs for a brief period of time experienced oxidative stress that hampered their development and progress (Gupta et al. 2016). Cu-NPs' effect on male *Clarias batrachus* was investigated by Murugananthkumar et al. (2016). It's interesting that while there was a significant rise in androgen levels, there was no histological correlation between the transcripts of numerous genes associated with the testis. It was hypothesized that catfish testicular regeneration would be negatively impacted by small-dose exposures to Cu-NPs. In order to establish recommendations as well as regulate the danger of nanoparticles in various organisms, it is crucial to understand the harmful pathways of nanoparticles in aquatic species (Malhotra et al. 2020).

14 Target of endocrine disrupting chemicals in Fish

14.1 Thyroid

The thyroid hormone is a mysterious hormone that is associated with a wide range of physiological procedures. They play an important role in the development of fish larvae into their adult stages (Dickhoff et al. 1990; Galton 1992). Thyroid hormone function in vertebrates is controlled by the hypothalamic-pituitary-thyroid (HPT) pathway. These thyroid tissues of vertebrates are not structured in a compacted gland; instead, the thyroid follicle is scattered into connective tissue in the pharynx region as well as the head kidneys. Aside from controlling metabolic activity, thyroid-stimulating hormones exhibit considerable synergistic effects with several other hormones including synthetic hormones, sex steroids, as well as interrenal steroids (Norris et al. 1999).

14.2 Interrenal Gland

In fishes, the hypothalamic-pituitary-interrenal (HPI) pathway controls the physiological response to stress (Wendelaar Bonga 1997). Prolonged increasing levels of cortisol (the major corticosteroids in fishes) impair reproduction activities, immunological function, as well as developmental activity, much as they do in that other vertebrates. On deposition, high cortisol discharge was reported among mercury, Cd, met acid 50, as well as endosulfan; however, mainly pesticide-treated fishes demonstrated a reverse in levels of cortisol (Mondal 1997).

14.3 Gonads

There are some variations between the sex steroid mechanisms involved in vertebrates and invertebrates, including the existence of the third kind of estrogen receptor (ER)—ERc in addition to ERa as well as b in fish (Hawkins et al. 2000). Another distinction is the variety of different kinds of EDCs that attach to the progestin receptors in fishes as opposed to the mammalian progesterone receptor (Pinter and Thomas 1997). Plasma estrogen concentrations have been employed by numerous researchers as a trustworthy biomarker of toxic compound exposures (Choudhury et al. 1993; Kime 1995; Thomas 1990). It has been shown that several xenobiotics and their metabolites, such as lesser oestrogens, act in the waterways (Sumpter and Jobling 1995).

14.4 Fish Reproduction

Fish are frequently used as modeling techniques in studies of a pattern of intervention imposition by endocrine-disrupting chemicals on fish embryo development considering that they are different season breeders, are sensitive to minor changing environments that can alter the normal reproductive cycle, and are most influenced by toxic elements throughout their reproductive cycle. It is now generally recognized that even low amounts of pollution can cause problems with fish's endocrine systems, as well as that fish can act as an alert system for those other wildlife species (Bhattacharya and Munshi 2021).

15 Conclusion

An increasing number of widely used chemicals and their degradation products are found to be estrogenic in animal and human systems. These effects are observed throughout the trophic levels in the aquatic environment, ranging from zooplankton to top predators. Severe impacts can occur at the level of steroidogenesis, biotransformation, gametogenesis, oogenesis, and spermatogenesis. Many environmental chemicals have been shown to have specific effects on components of the reproductive system of fish. Such components include the pituitary and gonads and their hormones, the liver catabolism of steroids and synthesis of vitellogenin, sperm motility, fertilization rate, and larval survival and rate of abnormality. Vitellogenin production can provide a valuable biomarker of reproductive disruption in fish. Abnormally high or low levels are indicative of disruption at multiple sites of the reproductive endocrine system or of early sexual development and differentiation.

References

Ahmed AM, Gomaa AH, El-Afifi TM (2010) Effect of pollution by mercury, cadmium and lead on reproduction of Tilapia nilotica fish. J Anim Poult Prod 1(9):395–407

Atashgahi S, Shetty SA, Smidt H, De Vos WM (2018) Flux, impact, and fate of halogenated xenobiotic compounds in the gut. Front Physiol 9:888

Bagchi P, Chatterjee S, Ray A, Deb C (1990) Effect of quinalphos, organophosphorus insecticide, on testicular steroidogenesis in fish, Clarias batrachus. Bull Environ Contam Toxicol 44(6): 871–875

Balm PHM, Pottinger TG (1995) Corticotrope and melanotrope POMC-derived peptides in relation to interrenal function during stress in raibow trout (Oncorhynchus mykiss). Gen Comp Endocrinol 98(3):279–288

Balm PHM, Pepels PPLM, Helfrich S, Hovens MLM, Bonga SW (1994) Adrenocorticotropic hormone in relation to interrenal function during stress in tilapia (Oreochromis mossambicus). Gen Comp Endocrinol 96(3):347–360

Barton BA, Iwama GK (1991) Physiological changes in fish from stress in aquaculture with emphasis on the response and effects of corticosteroids. Annu Rev Fish Dis 1:3–26

Bayne BL, Brown DA, Burns K, Dixon DR, Ivanovici A (1985) The effects of stress and pollution on marine animals. Praeger, New York

Bhatt P, Sethi K, Gangola S, Bhandari G, Verma A, Adnan M, Chaube S (2022) Modeling and simulation of atrazine biodegradation in bacteria and its effect in other living systems. J Biomol Struct Dyn 40(7):3285–3295

Bhattacharya S (2000) Signal transduction by xenobiotics in fish. Indian J Exp Biol 38:753–761

Bhattacharya S, Munshi C (2021) Endocrine disruptors in freshwater: impact on teleost reproduction. Proc Zool Soc 74(4):369–377. https://doi.org/10.1007/s12595-021-00397-z

Bhattacharya L, Pandey AK (1989) Inhibition of steroidogenesis and pattern of recovery in testes of DDT exposed cichlid—Oreochromis mossambicus. Bangladesh J Zool 17:1–14

Bresch H (1982) Investigation of the long-term action of xenobiotics on fish with special regard to reproduction. Ecotoxicol Environ Saf 6(1):102–112

Burgos-Aceves MA, Cohen A, Smith Y, Faggio C (2018) MicroRNAs and their role on fish oxidative stress during xenobiotic environmental exposures. Ecotoxicol Environ Saf 148:995–1000. https://doi.org/10.1016/j.ecoenv.2017.12.001

Carragher JF, Sumpter JP, Pottinger TG, Pickering AD (1989) The deleterious effects of cortisol implantation on reproductive function in two species of trout, Salmo trutta L. and Salmo gairdneri Richardson. Gen Comp Endocrinol 76(2):310–321
Celander M, Buhler DR, Förlin L, Goksøyr A, Miranda CL, Woodin BR, Stegeman JJ (1996) Immunochemical relationships of cytochrome P4503A-like proteins in teleost fish. Fish Physiol Biochem 15(4):323–332
Choudhury C, Ray AK, Bhattacharya S, Bhattacharya S (1993) Non lethal concentrations of pesticide impair ovarian function in the freshwater perch, Anabas testudineus. Environ Biol Fish 36(3):319–324
Cisson CM, Wilson BW (1981) Paraoxon increases the rate of synthesis of acetylcholinesterase in cultured muscle. Toxicol Lett 9(2):131–135
Civen M, Lifrak E, Brown CB (1977) Studies on the mechanism of inhibition of adrenal steroidogenesis by organophosphate and carbamate compounds. Pestic Biochem Physiol 7(2):169–182
Clement JG (1985) Hormonal consequences of organophosphate poisoning. Toxicol Sci 5(6part2): 61–77
Crinnion WJ (2010) Toxic effects of the easily avoidable phthalates and parabens. Altern Med Rev 15(3):190–196
Dhakal K, Gadupudi GS, Lehmler HJ, Ludewig G, Duffel MW, Robertson LW (2018) Sources and toxicities of phenolic polychlorinated biphenyls (OH-PCBs). Environ Sci Pollut Res 25(17): 16277–16290. https://doi.org/10.1007/s11356-017-9694-x
Dickhoff WW, Brown CL, Sullivan CV, Bern HA (1990) Fish and amphibian models for developmental endocrinology. J Exp Zool 256(S4):90–97
Dinka DD (2018) Environmental xenobiotics and their adverse health impacts-a general review. J Environ Pollut Hum Health 6:77–88. https://doi.org/10.12691/jephh-6-3-1
Donaldson EM (1990) Reproductive indices as measures of the effects of environmental stressors in fish. Am Fish Soc Symp 8:109–122
Donovan MP, Schein LG, Thomas JA (1978) Effects of pesticides on metabolism of steroid hormone by rodent liver microsomes. J Environ Pathol Toxicol 2(2):447–454
Dvořák P, Nikel PI, Damborský J, de Lorenzo V (2017) Bioremediation 3.0: engineering pollutant-removing bacteria in the times of systemic biology. Biotechnol Adv 35(7):845–866. https://doi.org/10.1016/j.biotechadv.2017.08.001
Embrandiri A, Kiyasudeen SK, Rupani PF, Ibrahim MH (2016) Environmental xenobiotics and its effects on natural ecosystem. In: Plant responses to xenobiotics. Springer, Singapore, pp 1–18. https://doi.org/10.1007/978-981-10-2860-1_1
Flett PA, Munkittrick KR, Kraak G, Leatherland JF (1991) Reproductive problems in Lake Erie coho salmon. In: Scott AP, Sumpter JP, Kime DE, Rolfe MS (eds) Proceedings of the 4th international symposium on the reproductive physiology of fish, fish symposium 91, Sheffield, pp 151–153
ForouharVajargah M, MohamadiYalsuyi A, Sattari M, Prokić MD, Faggio C (2020) Effects of copper oxide nanoparticles (CuO-NPs) on parturition time, survival rate and reproductive success of guppy fish, Poecilia reticulata. J Clust Sci 31(2):499–506
Fu H, Steinebach OM, Van den Hamer CJA, Balm PHM, Lock RAC (1990) Involvement of cortisol and metallothionein-like proteins in the physiological responses of tilapia (*Oreochromis mossambicus*) to sublethal cadmium stress. Aquat Toxicol 16(4):257–269
Galton VA (1992) The role of thyroid hormone in amphibian metamorphosis. Trends Endocrinol Metab 3(3):96–100. https://doi.org/10.1016/1043-2760(92)90020-2
Godheja J, Shekhar SK, Siddiqui SA, Modi DR (2016) Xenobiotic compounds present in soil and water: a review on remediation strategies. J Environ Analyt Toxicol 6(392):2161–0525. https://doi.org/10.4172/2161-0525.1000392
Goksøyr A (1995) Use of cytochrome P450 1A (CYP1A) in fish as a biomarker of aquatic pollution. Toxicol Transit 17:80–95
Goksøyr A, Förlin L (1992) The cytochrome P-450 system in fish, aquatic toxicology and environmental monitoring. Aquat Toxicol 22(4):287–311

Gupta YR, Sellegounder D, Kannan M, Deepa S, Senthilkumaran B, Basavaraju Y (2016) Effect of copper nanoparticles exposure in the physiology of the common carp (Cyprinus carpio): biochemical, histological and proteomic approaches. Aquacult Fisher 1:15–23

Hahn ME, Chandran K (1996) Uroporphyrin accumulation associated with cytochrome P4501A induction in fish hepatoma cells exposed to aryl hydrocarbon receptor agonists, including 2, 3, 7, 8-tetrachlorodibenzo-p-dioxin and planar chlorobiphenyls. Arch Biochem Biophys 329(2): 163–174

Hansen PD, Von Westernhagen H, Rosenthal H (1985) Chlorinated hydrocarbons and hatching success in Baltic herring spring spawners. Mar Environ Res 15(1):59–76

Harries JE, Sheahan DA, Matthiessen P, Neall P, Rycroft R, Tylor T, Sumpter JP (1996) A survey of estrogenic activity in United Kingdom inland waters. Environ Toxicol Chem Int J 15(11): 1993–2002

Harries JE, Sheahan DA, Jobling S, Matthiessen P, Neall P, Sumpter JP, Zaman N (1997) Estrogenic activity in five United Kingdom Rivers detected by measurement of vitellogenesis in caged male trout. Environ Toxicol Chem Int J 16(3):534–542

Hashmi MZ, Kumar V, Varma A (eds) (2017) Xenobiotics in the soil environment: monitoring, toxicity and management, vol 49. Springer. https://doi.org/10.1007/978-3-319-47744-2

Hawkins MB, Thornton JW, Crews D, Skipper JK, Dotte A, Thomas P (2000) Identification of a third distinct estrogen receptor and reclassification of estrogen receptors in teleosts. Proc Natl Acad Sci 97(20):10751–10756. https://doi.org/10.1073/pnas.97.20.10751

Herman RL, Kincaid HL (1988) Pathological effects of orally administered estradiol to rainbow trout. Aquaculture 72(1–2):165–172

Hinton DE (1990) Integrative histopathological effects of environmental stressor on fishes. Am Fish Soc Symp 8:51–66

Hose JE, McGurk MD, Marty GD, Hinton DE, Brown ED, Baker TT (1996) Sublethal effects of the (Exxon Valdez) oil spill on herring embryos and larvae: morphological, cytogenetic, and histopathological assessments, 1989 1991. Can J Fish Aquat Sci 53(10):2355–2365

Hunter GA, Donaldson EM (1983) 5 Hormonal sex control and its application to fish culture. In: Fish physiology, vol 9. Academic Press, pp 223–303

Ibemenuga KN (2013) Bioaccumulation and toxic effects of some heavy metals in freshwater fishes. Anim Res Int 10(3):1792–1798

Ioannides C (1990) Induction of cytochrome P450 I and its influences in chemical carcinogenesis. Biochem Soc Trans 18(1):32–34

Iovdijová A, Bencko V (2010) Potential risk of exposure to selected xenobiotic residues and their fate in the food chain-Part. I: classification of xenobiotics. Ann Agric Environ Med 17(2): 183–192

Jacob J, Cherian J (2013) Review of environmental and human exposure to persistent organic pollutants. Asian Soc Sci 9:11. https://doi.org/10.5539/ass.v9n11P107

Janssen PAH, Lambert JGD, Goos HT (1995) The annual ovarian cycle and the influence of pollution on vitellogenesis in the flounder, Pleuronectes flesus. J Fish Biol 47(3):509–523

Janssen PAH, Lambert JGD, Vethaak AD, Goos HT (1997) Environmental pollution caused elevated concentrations of oestradiol and vitellogenin in the female flounder, Platichthys flesus (L.). Aquat Toxicol 39(3–4):195–214

Jobling S, Sumpter JP, Sheahan D, Osborne JA, Matthiessen P (1996) Inhibition of testicular growth in rainbow trout (Oncorhynchus mykiss) exposed to estrogenic alkylphenolic chemicals. Environ Toxicol Chem Int J 15(2):194–202

Júnior ZR (2019) Biodegradation of xenobiotics in fish: a synthesis. Salão do Conhecimento

Kar S, Sangem P, Anusha N, Senthilkumaran B (2021) Endocrine disruptors in teleosts: evaluating environmental risks and biomarkers. Aquacult Fisher 6(1):1–26

Kim KH, Jahan SA, Kabir E, Brown RJ (2013) A review of airborne polycyclic aromatic hydrocarbons (PAHs) and their human health effects. Environ Int 60:71–80. https://doi.org/10.1016/j.envint.2013.07.019

Kime DE (1995) The effects of pollution on reproduction in fish. Rev Fish Biol Fish 5(1):52–95

Kime DE (1998) Endocrine disruption in fish. Springer
Kime DE (1999) A strategy for assessing the effects of xenobiotics on fish reproduction. Sci Total Environ 225(1–2):3–11
Kime DE, Vinson GP, Major PW, Kilpatrick R (1980) Adrenal-gonad relationships. General, comparative and clinical endocrinology of the adrenal cortex, vol 3, pp 183–264
Kime DE, Nash JP, Scott AP (1999) Vitellogenesis as a biomarker of reproductive disruption by xenobiotics. Aquaculture 177(1–4):345–352
Kocan RM, Hose JE, Brown ED, Baker TT (1996) Pacific herring (Clupea pallasi) embryo sensitivity to Prudhoe Bay petroleum hydrocarbons: laboratory evaluation and in situ exposure at oiled and unoiled sites in Prince William Sound. Can J Fish Aquat Sci 53(10):2366–2375
Kondal JK, Saxena PK, Soni GL (1989) Effect of vegetable oil factory effluent on the gonads of a freshwater teleost, Heteropneustes fossilis (Bl.)—a biochemical study. Environ Pollut 57(2): 117–125
Lang M, Pelkonen O (1999) Metabolism of xenobiotics and chemical carcinogenesis. IARC Sci Publ 148:13–22
Larini L (1997) Toxicologia, 3rd edn. Manole, São Paulo
Lewis DF (1996) Guide to cytochromes P450: structure and function. CRC Press
Malhotra N, Ger TR, Uapipatanakul B, Huang JC, Chen KHC, Hsiao CD (2020) Review of copper and copper nanoparticle toxicity in fish. Nano 10(6):1126
Malla MA, Dubey A, Yadav S, Kumar A, Hashem A, Abd_Allah EF (2018) Understanding and designing the strategies for the microbe-mediated remediation of environmental contaminants using omics approaches. Front Microbiol 9:1132. https://doi.org/10.3389/fmicb.2018.01132
Maurya PK, Malik DS (2016) Bioaccumulation of xenobiotics compound of pesticides in riverine system and its control technique: a critical review. J Ind Pollut Control 32(2):580–594
Mohapatra S, Kumar R, Sundaray JK, Patnaik ST, Mishra CSK, Rather MA (2021) Structural damage in liver, gonads, and reduction in spawning performance and alteration in the haematological parameter of Anabas testudineus by glyphosate-a herbicide. Aquac Res 52(3): 1150–1159
Mondal S (1997) Pesticides and heavy metals influence steroidogenic activity in fish gonad and interrenal. PhD Dissertation, Visva Bharati University, Santiniketan
Monod G (1985) Egg mortality of Lake Geneva charr (Salvelinus alpinus L.) contaminated by PCB and DDT derivatives. Bull Environ Contam Toxicol 35:531–536
Murugananthkumar R, Rajesh D, Senthilkumaran B (2016) Copper nanoparticles differentially target testis of the catfish, Clarias batrachus: in vivo and in vitro study. Front Environ Sci 4:67
Murugesan AG, Haniffa MA (1992) Histopathological and histochemical changes in the oocytes of the air-breathing fish Heteropneustes fossilis (Bloch) exposed to textile-mill effluent. Bull Environ Contam Toxicol 48(6):929–936
Nelson DR, Koymans L, Kamataki T, Stegeman JJ, Feyereisen R, Waxman DJ, Nebert DW (1996) P450 superfamily: update on new sequences, gene mapping, accession numbers and nomenclature. Pharmacogenetics 6(1):1–42
Norcross BL, Hose JE, Frandsen M, Brown ED (1996) Distribution, abundance, morphological condition, and cytogenetic abnormalities of larval herring in Prince William sound, Alaska, following the (Exxon Valdez) oil spill. Can J Fish Aquat Sci 53(10):2376–2387
Norris DO, Knobil E, Neill JD (1999) Thyroid hormones, in subavian vertebrates. Encyclop Reprod 4:807–812
Oga S (1996) Fundamentos de Toxicologia, d Atheneu. São Paulo
Okey AB, Riddick DS, Harper PA (1994) Molecular biology of the aromatic hydrocarbon (dioxin) receptor. Trends Pharmacol Sci 15(7):226–232
Oritz de Montellano PR (1995) Cytochrome P-450. Structure, mechanism and biochemistry
Pedersen JA, Yeager MA, Suffet IH (2003) Xenobiotic organic compounds in runoff from fields irrigated with treated wastewater. J Agric Food Chem 51(5):1360–1372. https://doi.org/10.1021/jf025953q

Pelissero C, Flouriot G, Foucher JL, Bennetau B, Dunogues JFLG, Le Gac F, Sumpter JP (1993) Vitellogenin synthesis in cultured hepatocytes; an in vitro test for the estrogenic potency of chemicals. J Steroid Biochem Mol Biol 44(3):263–272

Penglase S, Hamre K, Ellingsen S (2014) Selenium and mercury have a synergistic negative effect on fish reproduction. Aquat Toxicol 149:16–24

Pickering AD, Pottinger TG, Carragher J, Sumpter JP (1987) The effects of acute and chronic stress on the levels of reproductive hormones in the plasma of mature male brown trout Salmotrutta L. Gen Comp Endocrinol 68(2):249–259

Piferrer F, Donaldson EM (1989) Gonadal differentiation in coho salmon, Oncorhynchus kisutch, after a single treatment with androgen or estrogen at different stages during ontogenesis. Aquaculture 77(2–3):251–262

Pinter J, Thomas P (1997) The ovarian progestogen receptor in the spotted seatrout, Cynoscion nebulosus, demonstrates steroid specificity different from progesterone receptors in other vertebrates. J Steroid Biochem Mol Biol 60(1–2):113–119. https://doi.org/10.1016/s0960-0760(96)00177-x

Pottinger TG, Balm PHM, Pickering AD (1995) Sexual maturity modifies the responsiveness of the pituitary-interrenal axis to stress in male rainbow trout. Gen Comp Endocrinol 98(3):311–320

Purdom CE, Hardiman PA, Bye VVJ, Eno NC, Tyler CR, Sumpter JP (1994) Estrogenic effects of effluents from sewage treatment works. Chem Ecol 8(4):275–285

Rodríguez A, Castrejón-Godínez ML, Salazar-Bustamante E, Gama-Martínez Y, Sánchez-Salinas E, Mussali-Galante P, Ortiz-Hernández M (2020) Omics approaches to pesticide biodegradation. Curr Microbiol 77(4):545–563. https://doi.org/10.1007/s00284-020-01916-5

Saxena PK, Bhatia R (1983) Effect of vegetable oil factory effluent on ovarian recrudescence in the fresh water teleost, Channa punctatus (BI.). Water Air Soil Pollut 20(1):55–61

Schmidt JV, Bradfield CA (1996) Ah receptor signaling pathways. Annu Rev Cell Dev Biol 12(1): 55–89

Senthilkumaran B (2015) Pesticide-and sex steroid analogue-induced endocrine disruption differentially targets hypothalamo–hypophyseal–gonadal system during gametogenesis in teleosts–a review. Gen Comp Endocrinol 219:136–142. https://doi.org/10.1016/j.ygcen.2015.01.010

Sibly R, Calow P (1983) An integrated approach to life-cycle evolution using selective landscapes. J Theor Biol 102(4):527–547

Speranza AW, Seeley RJ, Seeley VA, Perlmutter A (1977) The effect of sublethal concentrations of zinc on reproduction in the zebrafish, Brachydanio rerio Hamilton-Buchanan. Environ Pollut 12(3):217–222

Spies RB, Rice DW (1988) Effects of organic contaminants on reproduction of the starry flounder Platichthys stellatus in San Francisco Bay. Mar Biol 98(2):191–200

Sumpter JP, Jobling S (1995) Vitellogenesis as a biomarker for estrogenic contamination of the aquatic environment. Environ Health Perspect 103(Suppl 7):173–178. https://doi.org/10.1289/ehp.95103s7173

Sumpter JP, Pottinger TG, Rand-Weaver M, Campbell, PM (1994) The wide ranging effects of stress in fish, p 535

Sundaray JK, Rather MA, Kumar S, Agarwal D (2021) Recent updates in molecular endocrinology and reproductive physiology of fish. Springer, Singapore

Swanson HI, Bradfield CA (1993) The AH-receptor: genetics, structure and function. Pharmacogenetics 3(5):213–230

Thomas P (1990) Molecular and biochemical responses of fish to stressors and their potential use in environmental monitoring. Biol Indicat Stress Fish 8:9–28

Thorpe JE (1994) Reproductive strategies in Atlantic salmon, Salmo salar L. Aquac Res 25(1): 77–87

Truscott B, Walsh JM, Burton MP, Payne JF, Idler DR (1983) Effect of acute exposure to crude petroleum on some reproductive hormones in salmon and flounder. Comp Biochem Physiol C Comparat Pharmacol 75(1):121–130

Tsaboula A, Papadakis EN, Vryzas Z, Kotopoulou A, Kintzikoglou K, Papadopoulou-Mourkidou E (2016) Environmental and human risk hierarchy of pesticides: a prioritization method, based on monitoring, hazard assessment and environmental fate. Environ Int 91:78–93. https://doi.org/10.1016/j.envint.2016.02.008

Tyler CR, Sumpter JP, Kawauchi H, Swanson P (1991) Involvement of gonadotropin in the uptake of vitellogenin into vitellogenic oocytes of the rainbow trout, Oncorhynchus mykiss. Gen Comp Endocrinol 84(2):291–299

Van der Oost R, Beyer J, Vermeulen NP (2003) Fish bioaccumulation and biomarkers in environmental risk assessment: a review. Environ Toxicol Pharmacol 13(2):57–149

Ware DM (1980) Bioenergetics of stock and recruitment. Can J Fish Aquat Sci 37(6):1012–1024

Ware DM (1982) Power and evolutionary fitness of teleosts. Can J Fish Aquat Sci 39(1):3–13

Wendelaar Bonga SE (1997) The stress response in fish. Physiol Rev 77(3):591–625. https://doi.org/10.1152/physrev.1997.77.3.591

Zaki MS, Hammam AM, Ata NS (2014) Xenobiotics as stressful factors in aquatic system (in fish). Life Sci J 11(4):188–197

Zhu Y, Boye A, Body-Malapel M, Herkovits J (2017) The toxic effects of xenobiotics on the health of humans and animals. Biomed Res Int 2017:4627872. https://doi.org/10.1155/2017/4627872

Zimniak P, Waxman DJ (1993) Liver cytochrome P450 metabolism of endogenous steroid hormones, bile acids, and fatty acids. In: Cytochrome P450. Springer, Berlin, Heidelberg, pp 123–144

Scientometric Analysis of Ecotoxicological Investigations of Xenobiotics in Aquatic Animals

Bejawada Chanikya Naidu, M. D. Sahana, Muzammal Hoque, and S. Abuthagir Iburahim

1 Introduction

1.1 What Are Xenobiotics?

A chemical substance that is not created naturally by the organism or that was not anticipated to be there is referred to as a xenobiotic (Malchi et al. 2022). They are present in significantly higher concentrations than typical and are frequently discussed in relation to environmental pollutants like polychlorinated biphenyls and dioxins and how they affect the biota (Gupta et al. 2022). They exist in both organic and inorganic forms and can mimic substances produced biologically that are necessary for life. Usually chemicals that are foreign to animal life are referred to as 'xenobiotics', and examples of these include plant elements, medications, insecticides, cosmetics, flavourings, scents, food additives, industrial chemicals, and environmental contaminants (Štefanac et al. 2021). Xenobiotics are of two types: exogenous and endogenous. Exogenous xenobiotics are those that enter an organism through food, medication, or environmental inhalation but are not generally created by the organism itself (e.g. chemicals, pharmaceuticals, pollutants, pesticides, and food additives). Endogenous xenobiotics are physiologically produced substances that resemble exogenous xenobiotic chemicals in some ways (Brandts et al. 2021). These are either created by the human body during various activities or are formed as metabolites (e.g. Eicosanoids, specific fatty acids, bile acids, bilirubin, and steroids). These compounds have the potential to be problematic, and it is important to investigate both their immediate and long-term impacts on people, animals, and the environment (Ortiz et al. 2022). These are either nonbiodegradable or only

B. Chanikya Naidu (✉) · M. D. Sahana · M. Hoque · S. Abuthagir Iburahim
ICAR-Central Institute of Fisheries Education, Mumbai, India
e-mail: chanikya.frmpb006@cife.edu.in

M. A. Rather et al. (eds.), *Xenobiotics in Aquatic Animals*,
https://doi.org/10.1007/978-981-99-1214-8_15

partially biodegradable, have a slow rate of biotransformation, and can last a very long time in the environment (Bhatt et al. 2021).

1.2 Xenobiotics in Various Fields

Since many chemical and pharmaceutical businesses use xenobiotics to make medicines, plastics, detergents, gels, research laboratory chemicals and biochemical kits, perfumes, herbicides, insecticides, and a wide range of other items, xenobiotics are essential for people and society. Many compounds used to enhance daily life (antibiotics, pesticides, dyes, PCPs, additives, etc.) have been developed as a result of technological advancement in the twentieth century; these compounds may not always occur naturally in the environment or may do so at concentrations that are significantly different from those brought on by anthropogenic activity (Mishra et al. 2019). They are categorised as pesticides, pharmaceutical chemicals, personal care items, illegal substances, industrial goods, and nuclear waste and can be found in the air, soil, water, plants, animals, and people (Malchi et al. 2022). Industries, including paper and pulp, fossil fuels, pesticides, explosives, and pharmaceuticals, among others, are important contributors of xenobiotic chemicals to the environment (Faggio et al. 2018).

1.3 World Scenario of Xenobiotics

In their lifetimes, humans are thought to be exposed to one to three million xenobiotics (Esteves et al. 2021). Most of these chemicals undergo a variety of detoxication processes that, in general, make them less poisonous, more polar, and easily excretable before entering the body through nutrition, air, drinking water, drug administration, and lifestyle choices (Collins and Patterson 2020). Xenobiotic pollution of the environment and, consequently, the uptake of these substances by living things have both increased dramatically in recent decades. The introduction of these compounds to ecosystems may result in an increase in allergic reactions, increased mortality of organisms, genetic changes, lowered immune systems, metabolic problems, and disturbances of ecosystem processes all the way up to the biosphere (Kucherenko et al. 2021). There are several immediate and long-term repercussions on natural ecosystems as a result of the diverse medications used, which have caused the release of dangerous chemicals into aquatic areas. Ecosystems are directly impacted by xenobiotics, changing things like community parameters, community structure, diversity, productivity, and energy transfer, as well as succession and population density (Wang et al. 2022). The world develops up to one million new products a year, including about 100,000 chemical compounds. Nearly 15,000 of these are considered possible xenobiotics. Particularly dangerous is the pesticide, hormone, and trans-fatty acid contamination of food (Gan et al. 2022).

1.4 Xenobiotics in Aquatic Environments

In typical sewage treatment facilities, some xenobiotic chemicals are nonbiodegradable and discharged with treated runoff, which could contaminate aquatic systems like rivers, lakes, and estuaries. Common xenobiotic receptors exist in traditional sewage treatment plants and must be treated with municipal wastewater before being released into aquatic systems (Maculewicz et al. 2020). Some trace metals, xenobiotic substances, and synthetic organic chemicals, such as PAHs, phthalates, and pesticides, can be detected in water bodies (Dar et al. 2020; Štefanac et al. 2021). Xenobiotic substances that are released into surface water may leak into groundwater, although this practice is currently severely prohibited since it could compromise the ecological integrity of aquatic ecosystems (Tonelli and Tonelli 2020).

1.5 How It Affects the Aquatic Environment

Within aquatic ecosystems any exposure to xenobiotics in sediment may have negative impacts at lower trophic levels and/or biomagnify and have more severe negative toxic effects at higher trophic levels (Tonelli and Tonelli 2020). Over the past century, there has been a tremendous growth in the variety of synthetic, xenobiotic compounds entering the environment. Their ecological effects in aquatic ecosystems are still poorly known in terms of their nature and severity (Arya and Haq 2019). Long after their original release into the aquatic environment, these hazardous, bioaccumulative, and persistent chemicals still run the danger of having deleterious impacts at all levels of biological organisation. Aquatic species experience oxidative stress and endocrine disruption as a result of the presence of xenobiotic contaminants (Mohapatra et al. 2021; Curpan et al. 2022). Their influence on aquatic ecosystems is well documented, and the manner in which they affect fish and other aquatic creatures can be broadly divided into three categories: behavioural, neurophysiological, and reproductive (Chandana and Kote 2020). It is a global concern that xenobiotic substances, such as poly aromatic hydrocarbons (PAHs), persistent organic pollutants (POPs), polychlorinated biphenyls (PCBs), organochlorine pesticides (OC), heavy metals, tri-butyl tin (TBT), etc., are contaminating the marine environment through various pathways (Gupta et al. 2022). A significant threat to the health of the marine ecosystem is posed by the bioaccumulation of their residues into the tissues of marine creature by considering the importance of the topic we have performed a meta-analysis on xenobiotics to know the trend of research in xenobiotics which will assist the present and future researchers in the particular field.

2 Materials and Methods

In order to guide future research by evaluating knowledge domains, measurable research patterns and intellectual structure, social structures, and emergent research themes, we performed a scientometric study of xenobiotics-related research in this work. The visualisation of data through scientometric analysis allows researchers to examine major changes and developments in the field of study, as well as its overall expansion. In addition, scientometric analysis can spot trends, collaborations, and patterns in publications on a certain subject or field of study. In this paper, the research on xenobiotics was quantitatively displayed and reviewed in a systematic fashion using scientometric techniques. The research made use of the Web of Science (WoS) database (Clarivate Analytics, Philadelphia, Pennsylvania), which includes articles from more than 21,100 high-quality scholarly journals from around the world, covering more than 250 different fields of study (https://clarivate.com/webofsciencegroup/solutions/web-of-science-corecollection/). Microsoft Excel 2019, Vos Viewer and R-studio were used to export the bibliometrics records from the WoS database and conduct the analysis. Average citations per article (ACPP), h-index, and total citations were used to determine how productive institutions and authors were. The h-index is a metric used to evaluate a researcher's productivity in the scientific community by analysing the correlation between their number of citations and the impact of their most highly referenced works (Hirsch 2005). To make bibliometric network maps that anyone can understand, R-studio is utilised. The software package R-studio was utilised to illustrate the hierarchy of command, conduct a co-authorship study, and visualise the citation relationships between individual writers and their respective publications. Different hues were assigned to different clusters (Waltman et al. 2010). The bibliometrics R-package in R version 4.0.1 and R-studio version 1.3.959 was used to examine the retrieved metadata from WoS. Biblioshiny was used for numerous analyses in this scientometrics report, including trend topics, thematic progression, and multiple country publication. Associations between nations, authors, and keywords were used to partition the dataset into multiple color-coded clusters (Van Eck et al. 2010; Waltman et al. 2010). The network visualisation technique was used to depict the connections between them all, with the proximity of two circles representing how closely they are related (Khalil and Crawford 2015; Zhao et al. 2018).

3 Results and Discussion

3.1 Xenobiotics

Scientometric studies examine the development of science in a field that may show the importance of a study, author, or research facility. In the particular subject of science, it is used to compare the influence of a research paper, researcher, journal,

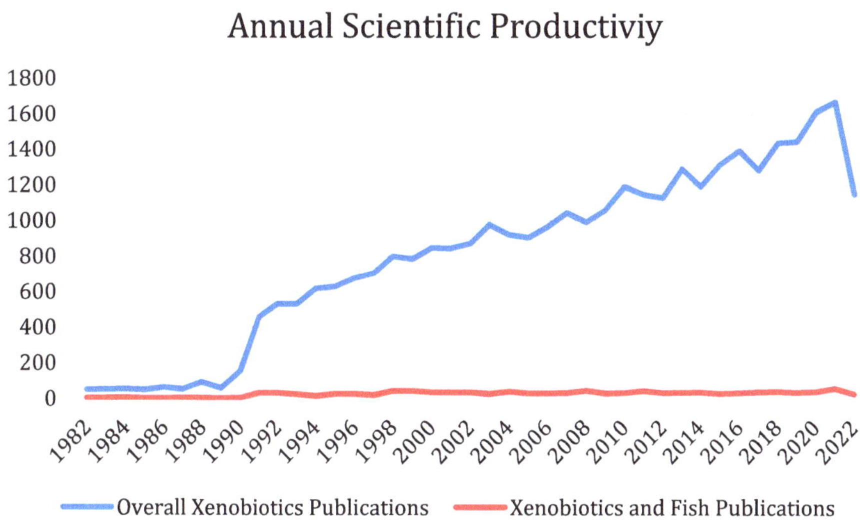

Fig. 1 Annual scientific production trend of research on xenobiotics and xenobiotics in fish

and institutions. Based on a search of the WoS database from 1982 to 2022 using the keywords 'xenobiotics', 'xenobiotics in fish', and 'impact of xenobiotics on fish physiology', this study was conducted to assess the amount of scholarly material on xenobiotics that was available. Between 1982 and 2022, journals included in the xenobiotics WoS category published a total of 33,056 documents on xenobiotics. In 1955, there was the first instance of a scientific paper on xenobiotics that was indexed in the WoS. The metabolism of xenobiotics and other medications was covered in the first publication (Testa 1955). There was a sharp rise in research publications on xenobiotics in the year 1988 and the trend followed till 2020 (Fig. 1). With an average of two authors per document (1.61%), 0.629 documents per author, and a collaboration index of 1.68, there were about 52,574 authors. The average number of citations per document was 37.84, indicating that the documents were of significant academic and research value.

The results of the scientometric analysis show that, between 1982 and 2022, the United States contributed the most publications (30.38%), followed by Germany, China, Japan, and France (Fig. 2).

Pharmacology, toxicology, molecular biology, and ecology were the main areas of concentration for xenobiotics-focused research and scientific publications that took place throughout the past 40 years in many different fields (Fig. 3). Pharmacology leads the field of xenobiotics research with 4070 articles, followed by toxicology (3564), molecular biology (3131), and ecology (2570). Significant xenobiotic research was conducted alongside work in cell biology, biotechnology, genetics, and engineering.

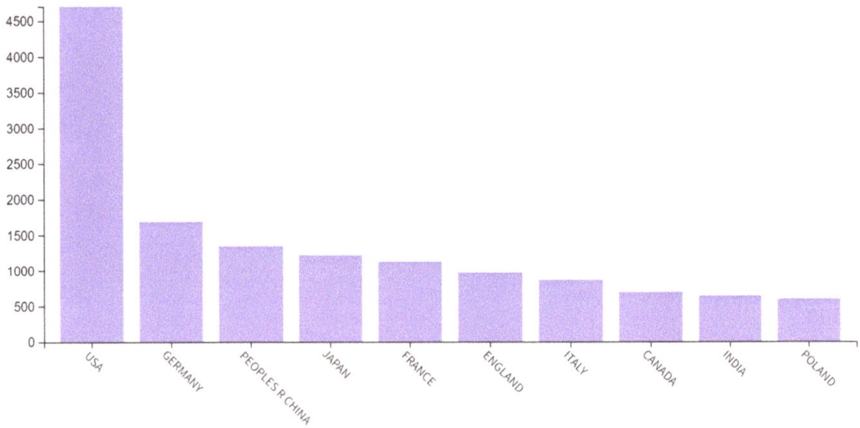

Fig. 2 Publication output of top 10 countries on xenobiotics research

3.2 *Xenobiotics in Fish*

With an average citation per document of 35.71 and a collaboration index of 3.79, about 3876 writers have contributed to 1080 documents about xenobiotics in fish study. According to scientometric study, the top contributing nations for xenobiotics in fish research were the developed nations like the United States, France, Germany, Canada, and China (Fig. 4).

The study of xenobiotics in fish initially fluctuated greatly year to year and peaked in the previous decade. The main institutions that contributed to the research were Oregon State University, the Rudjer Boskovic Institute, the USEPA, the Chinese Academy of Sciences, the University of Aveiro, and the Spanish National Research Council (CSIC). The top researchers in the field of xenobiotics in fish include Andersson, Goksoyr, Schlenk, Cashman, Forlin, James, Pritchard, Monod, and Buhler with a great rate of collaboration (Fig. 5).

3.3 *Effect of Xenobiotics on Fish Physiology*

The effect of xenobiotics on fish physiology was the subject of scientometric analysis, which found a total of 65 published documents by 253 contributing authors, with a collaboration index of 4.13 and an average number of citations per document of 50.71. The publications did not follow any trend over the years with comparatively higher number of publications in 2017 (Fig. 6).

Most research on xenobiotics effects on fish physiology has been done in the domains of toxicology, marine freshwater ecology, endocrine metabolism, and fish pharmacology. Sugiyama, Yamazaki, Gershwin, and Wang are among the authors

Fig. 3 Major focus areas in xenobiotics and its related research during 1982–2022

Fig. 4 Major countries contributing to xenobiotics in fish research during 1982–2022

Fig. 5 Top researchers in the field of xenobiotics in fish from various countries

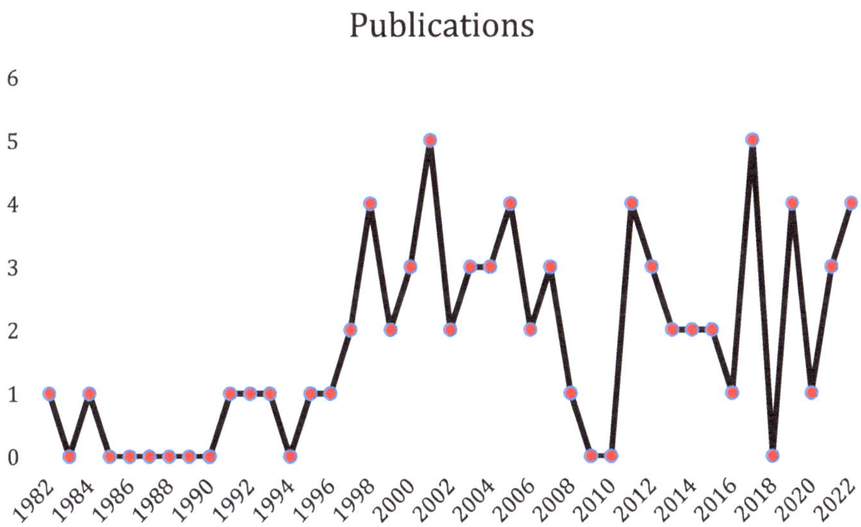

Fig. 6 Annual scientific production trend of research on the effect of xenobiotics on fish physiology during 1982–2022

who have contributed the most to studies on xenobiotics effects on fish physiology, and the most influential countries in the research on xenobiotics effects on fish physiology were the United States, Japan, China, and the United Kingdom (Fig. 7).

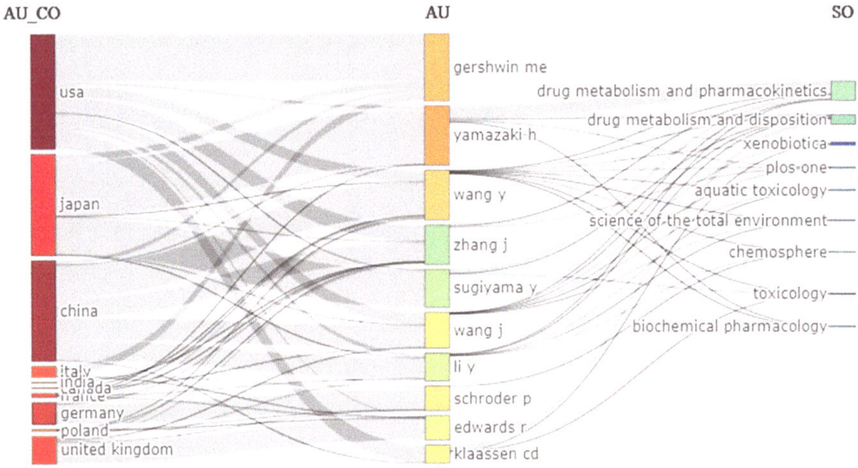

Fig. 7 Highly productive authors of research on the effect of xenobiotics on fish physiology, their networking, and the research source

4 Conclusion

The United States, Germany, and China had more publications in the current study than the other nations combined. China and the United States are large producers of xenobiotics, and perhaps as a result of the negative effects that these xenobiotics have had on their aquatic ecosystems and environment, more study has been conducted in these nations. To have a broad overview of the research investigations conducted in this research field, 33,056 research publications in total were analysed. The research literature produced between 1982 and 2022 was divided and assessed to determine the change in the pattern of authorship and their affiliation. It highlighted the key regions, nations, and authors who are leading the field of research on xenobiotics in fish. In this study, a moderately high level of collaboration was identified both between and within nations. Additionally, it was discovered that developed countries with advanced scientific and technological infrastructure had begun xenobiotics-based research much earlier and produced a large number of publications. However, very recently, the trend has changed as developing countries like India have taken on the difficult studies of xenobiotics in aquatic environments and their effects. The study's scope was constrained by the fact that it was nearly entirely dependent on datasets obtained through WoS. However, the study's general patterns and important measures may be used as a basis for deciding how to strengthen institutions, collaborate more effectively, increase grants, and prioritise research priorities. Similar scientometric analyses of numerous xenobiotics effect on diverse aquatic creatures over time could undoubtedly aid in understanding the direction of the field's study. The scientometrics analysis that was undertaken gives a global picture of research on xenobiotics in developed as well as emerging nations. Policymakers will find the data helpful in identifying the dominant trends in this area. Governments can use it to plan for the future, spending money on new research. Additionally, it has emphasised the key institutions for financial advice and networking, academics for spotting hot-button research subject with the influential and active individuals in the sector, and industries to discover solutions to the xenobiotics in aquatic environment issues. As a result, this research report will aid in identifying potential future line of works.

5 Virtual Reality Application in Various Domains

S. no	Domain	Application	Source
1	Agriculture	VR-based simulation of the natural environment for cows to improve quantity and quality of milk	https://www.agritechtomorrow.com/article/2020/11/smart-farming-is-ready-for-augmented-and-virtual-reality/12516
2	Medicinal science	Application of VR to understand the complex treatment procedure before carrying out the treatment to the patients	https://www.news-medical.net/health/Applications-of-Virtual-Reality-in-Medicine.aspx
3	Remote sensing and GIS	VR- based interaction with real objects using 3D visualisation	Singla (2021)
4	Oceanography	VR- based creation of plankton zoo and 3D visualisation of plankton to learn about phytoplankton	Walcutt et al. (2019)
5	Oceanography	VR application in mimicking natural habitat to elicit behavioural response (e.g. camouflage) in *Loligo opalescens*	Jaffe et al. (2011), Josef (2018)
6	Oceanography	Creation of immersive virtual aquarium with real-walking navigation by using VR	Jung et al. (2013)
7	Fishing areas	Application of VR to map the fishing location present in the US country to know about the different fishing locations	https://www.gmw3.com/2021/12/go-us-west-coast-fishing-in-real-vr-fishings-upcoming-dlc/
8	Engineering and design	VR- based creation of prototype machinery to examineand modify any defects in the system rather than making a physical prototype	https://www.xcubelabs.com/blog/the-applications-of-virtual-reality-in-the-manufacturing-industry/
9	Military	Application of VR for realistic military training to combat high-stress situations and improve skills over handling weapons and range of communications.	https://www.futurevisual.com/blog/uses-vr-military-training/

(continued)

S. no	Domain	Application	Source
10	Sports	VR sports entertainment accelerates the athletes training regimen, giving them a chance to run unlimited reps to perform better in a true event	https://www.strivr.com/solutions/industries/sports/#:~:text=A%20Virtual%20Reality%20(VR)%20sports,best%20when%20it%20truly%20matters
11	Manufacturing	VR- based manufacturing will allow the worker to interact with the inside of an engine to create, repair and to maintain the machinery	https://www.onewatt.eu/post/extended-reality-in-machine-maintenance-and-repair
12	Veterinary medicine	Application of VR to understand the anatomy of dog, cow, and equine	https://guides.lib.vt.edu/vetmed/vranatomy
13	Tourism	VR technology implementation of smart tourism to provide information about destinations and attractions while showing its potential to become a new tourism service	Pestek and Sarvan (2020)
14	Entertainment	VR- based games, theatre, museum, amusement park, gallery, live music concerts, hobby lessons	https://jasoren.com/virtual-reality-for-the-entertainment/
15	Archaeology	A prototype of Pleito Cave was reconstructed by using VR which leads to understanding the real-time information of the caves	Cassidy et al. (2019)
16	Education	VR- based learning is a promising tool to learn the concepts clearly and also engages multiple senses	Christou (2010)
17	Forestry	Application of VR to map the presence of jaguars throughout the Peruvian Amazon for the effective conservation	Bednarz et al. (2016)
18	Environment	Application of VR to know the individual carbon print in their daily activities will make them think in a different way to reduce carbon footprints	https://www.unep.org/news-and-stories/story/experience-your-carbon-footprint-vr
19	Horticulture	VR- based learning of plant physiology can make the individual to grow the plant and can make them to tackle the arising problem effectively	Ai-guo et al. (2011)
20	Medicine	VR technology employed in the drug discovery by engaging different scientist in a place for real-time modification or identification of molecules or drugs	https://www.labcompare.com/10-Featured-Articles/577506-VR-for-Science-Drug-Discovery-and-More-in-the-Virtual-World/

6 Status of Fisheries Education in India

In comparison to veterinary and agricultural education, professional fisheries education in India got established later. The Central Institute of Fisheries Education in Mumbai was started in 1961, and it was in 1969, under the auspices of the University of Agricultural Sciences, Bengaluru, that the first Fisheries College in Mangalore opened, marking a new era in the professional fisheries education in India at the State Agricultural/Veterinary Universities. There are more than 32 fisheries colleges and 3 universities offering fisheries education in India for bachelors, and master and PhD degree programmes. The present annual intake capacity of the B.F.Sc., M.F.Sc., and PhD programme is 1500, 425, and 185, respectively, while the annual out turn is about 80–85% of intake (Kumar et al. 2018). Universities and colleges have always been at the forefront of new technologies, driving innovation, changing industries, and training the next generation of scientists, developers, and business owners. Right now, virtual and augmented reality technologies are at the cutting edge of progress, and things are changing quickly.

7 Possible Virtual Reality Application in Fisheries Education

Immersive VR application allows the students to reach to any possible environment without being actually present.

8 Distance Learning/Virtual Classrooms

Students who are able to think creatively have a better shot at learning the problem-solving, teamwork, and efficiency skills they will need in college and beyond. Teachers can employ exploration-based learning, an active learning strategy that encourages students to learn through curiosity and enquiry, to assist students to acquire confidence and make their education more meaningful (Fig. 8). The concept of distance learning is growing up after the Covid-19 lockdown as people get used to online learning platforms. One of the major problems in online learning platforms is physical engagement, which can be sorted out by virtual reality. Virtual reality classroom in many examples showed significantly better learning motivation, learning outcomes, and positive impacts on learning students' achievement scores (Liou and Chang 2018).

Fig. 8 Virtual reality benefits in education

9 Virtual Field Visits

Visiting a real workplace, especially one that could put students in harm's way, can be prohibitively expensive and time-consuming for those pursuing a practical education. The simulations on underwater environment like trench, coral reef areas; real-time fishing in ocean and freshwater environments, research cruise, aquaculture farms, virtual museums, climate change studies, etc., can be created, which can enable the student to visit these areas from their convenient places and cut down the travelling cost and time. VR creates psychological presence, a sense of being there, due to the immersive VR tracking system that detects the user's body movements and enables them to feel as though their body is moving in the virtual world and the virtual world is reacting to their movement. VR is an effective treatment for some of the phobias like aquaphobia (Morina et al. 2015) and anxiety disorder (Opriş et al. 2012) due to its high level of presence. Virtual reality meetings such as conferences, workshops, and symposia are gaining popularity all over the world. One of the best examples in the fisheries is the thirteenth AFAF (Asian fisheries aquaculture forum) held in virtual mode in 2022.

10 Virtual Fish and Fish Systems

Traditionally, textbooks, live animals, and cadavers are used in fisheries anatomy and biology classes to develop a solid foundation of knowledge. The practical connectivity of students in the anatomy of finfish and shellfish, and biology of finfish and shellfish systems is difficult in the textbook education. Three-dimensional (3D) software programs provide a platform for deeper examination of anatomical structures but are unable to deliver a fully immersive experience. The use of virtual reality (VR) provides a new type of learning environment and avoid sacrificing live organisms (Fig. 9). Dissection and identification of vital organs by the students themselves will be very easy in the VR environment as the users can zoom in and zoom out the parts and they can know the functions of the organs and systems.

11 Virtual Labs

As the accreditation criteria of many accrediting bodies emphasise, acquiring practical skills is a critical component of learning outcomes in any fisheries degree. It will be especially difficult to ensure efficient achievement of these skills with limited physical delivery. One possible solution to this problem is to use virtual lab experiments and simulations to help students understand the concepts, important relationships between variables, and potential impact on experimental rig operation before physically carrying out an experiment in a laboratory within a much shorter

Fig. 9 Glimpse of the Virtual Fish Laboratory developed under NAHEP

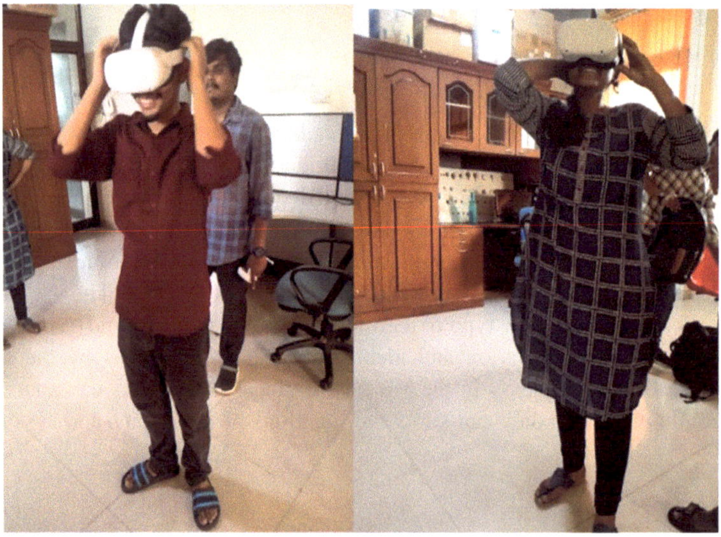

Fig. 10 Students of ICAR-CIFE enjoying the virtual reality programme

timeline. Field of cell biology, toxicology, pharmacology, histology, and pathology studies in aquatic organisms involves hazardous or toxic chemicals in laboratories, which can be avoided in VR education. The wastage of costly chemicals also can be overcome in VR education, and the experiment for teaching can be repeated n number of times (Fig. 10).

12 Conclusion

Those interested in the new teaching technique should not miss this chance to learn more. Virtual reality has the potential to completely transform education at all levels by bringing a new dimension to the learning process. Augmented reality promotes student learning and comprehension through experimentation. We are currently only witnessing the beginnings of a paradigm shift in education brought about by virtual technologies. Increasing accessibility is another key component of the future of VR in the classroom. It benefits both education and technology. Augmented reality is gaining popularity as application development incorporates new technology. As VR hardware and software continue to decrease in price, it will soon permeate all facets of the educational system. Virtual reality (VR) solutions have been proven effective across a variety of educational settings, and they are generally well received by viewers. Augmented reality (AR) can transform human existence due to its many benefits. We must assess unsolved research problems about virtual reality's (VR) future barriers and potential benefits. Recent advances in telecommunication systems, especially the rollout of the 5G network in big economies like China, India, and the United States, will boost VR-based markets. The Web 3.0 is the next big revolution that is happening in the world today. The ongoing Covid-19 pandemic pushed the world to adapt and evolve to challenging situations to keep the economy running. Every aspect of life, be it education, shopping, running business, banking, and attending office, became online. Being online is the new normal these days. This is why VR is going to have a huge impact in all the sectors. VR is increasingly used in education and training in academia, healthcare, tourism, shopping, and in automobile and space industries. From an educational point of view, VR will revolutionise the learning experience. Students can immerse and understand subjects that are challenging to grasp via conventional learning. Increase in educational content in the VR arena will certainly take education to the next level in the coming generations. Medical students can have real-time experience in complex surgeries via pre-recorded real-time surgery videos captured directly through the eyes of the surgeon himself and an engineering student can have hands-on training in 3D modelling of complex machineries. A fishery student in relative terms will get to experience real-time immersive knowledge in a variety of aspects like underwater oceanic observations, craft and gear handling at the sea, fish processing in big industries, aqua feed production in feed manufacturing plants, and hatchery operations to which the students have limited accessibility in many grad schools. In limited resource settings, VR contents can be crafted to aid the students' needs. For example, it is not always necessary for all students to dissect a fish to learn different parts of its body; every time the student will not get exposure to study taxonomy of diverse fish groups within a semester, as fish availability depends on the season and fishing effort which varies substantially across the country. VR can solve these problems by creating interactive contents to give students a real-time immersive learning experience around the year and across the country. Not only does this adaptation make learning wholesome, it cuts down costs and resources associated with it.

References

Ai-guo Z, Ping S, He G (2011) The technology of VR and its application in the experimental teaching of plant physiology. In: 2011 International Conference on E-Business and E-Government (ICEE). IEEE, pp 1–2

Applications of Virtual Reality in Medicine, https://www.news-medical.net/health/Applications-of-Virtual-Reality-in-Medicine.aspx

Arya P, Haq SA (2019) Effects of xenobiotics and their biodegradation in marine life. In: Smart bioremediation technologies. Academic Press, pp 63–81

Augmented Reality and Virtual Reality (AR & VR) Market size is expected to reach USD 571.42 Billion by 2025. Valuates Reports

Bednarz T, Kim J, Brown R, James A, Burrage K, Clifford S, Davis J, Mengersen K, Peterson E, Psaltis S, Vercelloni J (2016) Virtual reality for conservation. In: Proceedings of the 21st international conference on web3d technology, pp 177–178

Bhatt P, Bhandari G, Bhatt K, Maithani D, Mishra S, Gangola S, Bhatt R, Huang Y, Chen S (2021) Plasmid-mediated catabolism for the removal of xenobiotics from the environment. J Hazard Mater 420:126618

Blascovich J, Loomis J, Beall AC, Swinth KR, Hoyt CL, Bailenson JN (2002) Immersive virtual environment technology as a methodological tool for social psychology. Psychol Inq 13(2): 103–124

Bowman DA, McMahan RP (2007) Virtual reality: how much immersion is enough? Computer 40(7):36–43

Brandts I, Solà R, Martins MA, Tvarijonaviciute A, Barreto A, Teles M, Oliveira M (2021) A baseline study on the impact of nanoplastics on the portals of entry of xenobiotics in fish. Mar Pollut Bull 173:113018

Cassidy B, Sim G, Robinson DW, Gandy D (2019) A virtual reality platform for analyzing remote archaeological sites. Interact Comput 31(2):167–176

Chandana GL, Kote NV, Sharath Chandra SP (2020) Recent scenario of impact of xenobiotics on marine fish: an overview. Pharm J 12(6s):1797–1800

Christou C (2010) Virtual reality in education. In: Affective, interactive and cognitive methods for e-learning design: creating an optimal education experience. IGI Global, pp 228–243

Collins SL, Patterson AD (2020) The gut microbiome: an orchestrator of xenobiotic metabolism. Acta Pharm Sin B 10(1):19–32

Cuesta Cambra U, Manas Viniegra L (2016) Integration of immersive virtual reality in communication degrees. Revista Icono 14-Revista Cientifica De Comunicacion Y Tecnologias 14(2)

Curpan AS, Impellitteri F, Plavan G, Ciobica A, Faggio C (2022) Mytilus galloprovincialis: an essential, low-cost model organism for the impact of xenobiotics on oxidative stress and public health. Comp Biochem Physiol C Toxicol Pharmacol 256:109302

Dar SA, Chatterjee A, Rather MA, Chetia D, Srivastava PP, Gupta S (2020) Identification, functional characterization and expression profiling of cytochrome p450 1A (CYP1A) gene in *Labeo rohita* against emamectin benzoate. Int J Biol Macromol 158:1268–1278

Di Natale AF, Repetto C, Riva G, Villani D (2020) Immersive virtual reality in K-12 and higher education: a 10-year systematic review of empirical research. Br J Educ Technol 51(6): 2006–2033

Esteves F, Rueff J, Kranendonk M (2021) The central role of cytochrome P450 in xenobiotic metabolism—a brief review on a fascinating enzyme family. J Xenobiot 11(3):94–114

Experience your carbon footprint in VR, https://www.unep.org/news-and-stories/story/experience-your-carbon-footprint-vr
Extended Reality in Machine Maintenance and Repair, https://www.onewatt.eu/post/extended-reality-in-machine-maintenance-and-repair
Faggio C, Tsarpali V, Dailianis S (2018) Mussel digestive gland as a model tissue for assessing xenobiotics: an overview. Sci Total Environ 636:220–229
Freina L, Ott M (2015) A literature review on immersive virtual reality in education: state of the art and perspectives. In: The international scientific conference elearning and software for education, vol 1, no 133, pp 10–1007
Gan J, Ashraf SS, Bilal M, Iqbal HM (2022) Biodegradation of environmental pollutants using catalase-based biocatalytic systems. Environ Res 214(Pt 2):113914
Gavish N, Gutiérrez T, Webel S, Rodríguez J, Peveri M, Bockholt U, Tecchia F (2015) Evaluating virtual reality and augmented reality training for industrial maintenance and assembly tasks. Interact Learn Environ 23(6):778–798
Global AR and VR Market to Reach USD 165.3 billion by 2027. https://www.vynzresearch.com/ict-media/augmented-reality-and-virtual-reality-market
Go US West Coast Fishing in Real VR Fishing's Upcoming DLC – VRFocus, https://kalem.easterndns.com/news/2021/12/01/go-us-west-coast-fishing-in-real-vr-fishings-upcoming-dlc-vrfocus/
Gupta SK, Singh B, Mungray AK, Bharti R, Nema AK, Pant KK, Mulla SI (2022) Bioelectrochemical technologies for removal of xenobiotics from wastewater. Sustain Energ Technol Assessm 49:101652
Harrington CM, Kavanagh DO, Quinlan JF, Ryan D, Dicker P, O'Keeffe D, Tierney S (2018) Development and evaluation of a trauma decision-making simulator in Oculus virtual reality. Am J Surg 215(1):42–47
Hirsch JE (2005) An index to quantify an individual's scientific research output. Proc Natl Acad Sci 102(46):16569–16572
Ibañez-Etxeberria A, Gómez-Carrasco CJ, Fontal O, García-Ceballos S (2020) Virtual environments and augmented reality applied to heritage education. An evaluative study. Appl Sci 10(7): 2352
Jaffe JS, Laxton B, Zylinski S (2011) The Sub Sea holodeck: a 14-megapixel immersive virtual environment for studying cephalopod camouflage behavior. In: OCEANS 2011 IEEE-Spain. IEEE, pp 1–6
Jensen L, Konradsen F (2018) A review of the use of virtual reality head-mounted displays in education and training. Educ Inf Technol 23(4):1515–1529
Josef N (2018) Cephalopod experimental projected habitat (CEPH): virtual reality for underwater organisms. Front Mar Sci 5:73
Jung S, Choi YS, Choi JS, Koo BK, Lee WH (2013) Immersive virtual aquarium with real-walking navigation. In: Proceedings of the 12th ACM SIGGRAPH international conference on virtual-reality continuum and its applications in industry, pp 291–294
Kalyvioti K, Mikropoulos TA (2014) Virtual environments and dyslexia: a literature review. Proc Comp Sci 27:138–147
Khalil GM, Crawford CAG (2015) A bibliometric analysis of US-based research on the behavioral risk factor surveillance system. Am J Prev Med 48(1):50–57
Kucherenko SV, Ovcharenko AM, Pushenko SL (2021) Xenobiotics: a threat to the health of living organisms. In: E3S web of conferences, vol 285. EDP Sciences, p 03006
Kumar M, Gurjar UR, Keer NR, Kumar S (2018) Professional fisheries education in India: history, current status and future-a review. Int J Curr Microbiol App Sci 7(6):3395–3409
Liou WK, Chang CY (2018) February. Virtual reality classroom applied to science education. In: 2018 23rd international scientific-professional conference on information technology (IT). IEEE, pp 1–4

Maculewicz J, Świacka K, Kowalska D, Stepnowski P, Stolte S, Dołżonek J (2020) In vitro methods for predicting the bioconcentration of xenobiotics in aquatic organisms. Sci Total Environ 739:140261

Malchi T, Eyal S, Czosnek H, Shenker M, Chefetz B (2022) Plant pharmacology: insights into in-planta kinetic and dynamic processes of xenobiotics. Crit Rev Environ Sci Technol 52(19): 3525–3546

Martínez FP (2011) Presente y Futuro de la Tecnología de la Realidad Virtual. Creatividad y sociedad 16:1–39

Mikropoulos TA, Strouboulis V (2004) Factors that influence presence in educational virtual environments. Cyberpsychol Behav 7(5):582–591

Mishra VK, Singh G, Shukla R (2019) Impact of xenobiotics under a changing climate scenario. In: Climate change and agricultural ecosystems. Woodhead Publishing, pp 133–151

Mohapatra S, Kumar R, Sundaray JK, Patnaik ST, Mishra CSK, Rather MA (2021) Structural damage in liver, gonads, and reduction in spawning performance and alteration in the haematological parameter of Anabas testudineus by glyphosate-a herbicide. Aquac Res 52(3): 1150–1159

Morina N, Ijntema H, Meyerbröker K, Emmelkamp PM (2015) Can virtual reality exposure therapy gains be generalized to real-life? A meta-analysis of studies applying behavioral assessments. Behav Res Ther 74:18–24. https://doi.org/10.1016/j.brat.2015.08.010

Opriş D, Pintea S, García-Palacios A, Botella C, Szamosközi Ş, David D (2012) Virtual reality exposure therapy in anxiety disorders: a quantitative meta-analysis. Depress Anxiety 29(2): 85–93. https://doi.org/10.1002/da.20910

Ortiz P, Torres-Sánchez A, López-Moreno A, Cerk K, Ruiz-Moreno Á, Monteoliva-Sánchez M, Ampatzoglou A, Aguilera M, Gruszecka-Kosowska A (2022) Impact of cumulative environmental and dietary xenobiotics on human microbiota: risk assessment for one health. J Xenobiot 12(1):56–63

Panchuk D, Klusemann MJ, Hadlow SM (2018) Exploring the effectiveness of immersive video for training decision-making capability in elite, youth basketball players. Front Psychol 9:2315

Parmaxi A (2020) Virtual reality in language learning: a systematic review and implications for research and practice. Interact Learn Environ 31(3):1–13

Pestek A, Sarvan M (2020) Virtual reality and modern tourism. J Tour Futur 7(2):245–250

Singla JG (2021) Virtual reality based novel use case in remote sensing and GIS. Curr Sci 121(7): 958

Slater M (2003) A note on presence terminology. Presen Conn 3(3):1–5

Smart Farming is Ready for Augmented and Virtual Reality, https://www.agritechtomorrow.com/article/2020/11/smart-farming-is-ready-for-augmented-and-virtual-reality/12516

Sports, https://www.strivr.com/solutions/industries/sports/#:~:text=A%20Virtual%20Reality%20(VR)%20sports,best%20when%20it%20truly%20matters

Štefanac T, Grgas D, Landeka Dragičević T (2021) Xenobiotics—division and methods of detection: a review. J Xenobiot 11(4):130–141

Testa B (1955) The metabolism of drugs and other xenobiotics: biochemistry of redox reactions. Academic Press

Tonelli FCP, Tonelli FMP (2020) Concerns and threats of xenobiotics on aquatic ecosystems. In: Bioremediation and biotechnology, vol 3. Springer, Cham, pp 15–23

Uses of VR in Military Training, https://www.futurevisual.com/blog/uses-vr-military-training/

Van Eck NJ, Waltman L, Dekker R, Van Den Berg J (2010) A comparison of two techniques for bibliometric mapping: multidimensional scaling and VOS. J Am Soc Inf Sci Technol 61(12): 2405–2416

Vet Med Library Information: VR Anatomy, https://guides.lib.vt.edu/vetmed/vranatomy

Virtual Reality for the Entertainment Market: All You Need to Know to Make Investment Decisions, https://jasoren.com/virtual-reality-for-the-entertainment/

VR AR Content Creation Market Ecosystem. Trend, Revenue and growth rate analysis along with decision intelligence – All The Research

VR for Science: Drug Discovery and More in the Virtual World, https://www.labcompare.com/10-Featured-Articles/577506-VR-for-Science-Drug-Discovery-and-More-in-the-Virtual-World/
VR in manufacturing, https://www.xcubelabs.com/blog/the-applications-of-virtual-reality-in-the-manufacturing-industry/
Walcutt NL, Knörlein B, Sgouros T, Cetinić I, Omand MM (2019) Virtual reality and oceanography: overview, applications, and perspective. Front Mar Sci 6:644
Waltman L, Van Eck NJ, Noyons EC (2010) A unified approach to mapping and clustering of bibliometric networks. J Informet 4(4):629–635
Wang X, Sial MU, Bashir MA, Bilal M, Raza QUA, Ali Raza HM, Rehim A, Geng Y (2022) Pesticides xenobiotics in soil ecosystem and their remediation approaches. Sustainability 14(6): 3353
Webster R (2016) Declarative knowledge acquisition in immersive virtual learning environments. Interact Learn Environ 24(6):1319–1333
Zhao L, Deng J, Sun P, Liu J, Ji Y, Nakada N, Qiao Z, Tanaka H, Yang Y (2018) Nanomaterials for treating emerging contaminants in water by adsorption and photocatalysis: systematic review and bibliometric analysis. Sci Total Environ 627:1253–1263

Toxic Effects of Nanomaterials on Aquatic Animals and Their Future Prospective

Imran Zafar, Arfa Safder, Qurat ul Ain, Mouada Hanane, Waqas Yousaf, Ihtesham Arshad, Mohd Ashraf Rather, and Mohammad Amjad Kamal

I. Zafar (✉)
Department of Bioinformatics and Computational Biology, Virtual University, Lahore, Punjab, Pakistan

A. Safder
Institute of Molecular Biology and Biotechnology, The University of Lahore, Lahore, Punjab, Pakistan

Q. u. Ain
Department of Chemistry, Government College Women University, Faisalabad, Pakistan

M. Hanane
VTRS Laboratory, Department of Process Engineering, University Center of Tipaza, El Oued, Algeria
e-mail: mouada-hanane@univ-eloued.dz

W. Yousaf
Department of Botany, Institute of Molecular Biology and Biotechnology (IMBB), The University of Lahore, Lahore, Punjab, Pakistan

I. Arshad
Department of Biotechnology, Faculty of Life Sciences, University of Okara, Okara, Pakistan

M. A. Rather
Division of Fish Genetics and Biotechnology, Sher-e-Kashmir University of Agricultural Sciences and Technology, Kashmir, Rangil, Ganderbal, Jammu and Kashmir, India

M. A. Kamal
Institutes for Systems Genetics, Frontiers Science Center for Disease-Related Molecular Network, West China Hospital, Sichuan University, Chengdu, China

King Fahd Medical Research Center, King Abdulaziz University, Jeddah, Saudi Arabia

Department of Pharmacy, Faculty of Allied Health Sciences, Daffodil International University, Daffodil Smart City, Bangladesh

Novel Global Community Educational Foundation, Hebersham, NSW, Australia

M. A. Rather et al. (eds.), *Xenobiotics in Aquatic Animals*,
https://doi.org/10.1007/978-981-99-1214-8_16

1 Introduction

Recently, nanotechnology in the context of big data for health analytics has been used as one of the most promising fields in wide-ranging domains, and it is also using a new frontier in aquatic animals, aquaculture, fisheries food webs, development, and understanding the remarkable response of aquatic life (Rather et al. 2011; Aklakur et al. 2016; Rayan et al. 2022; Rayan and Zafar 2021). In recent years, nanotechnology has emerged as one of the most promising fields of artificial intelligence, the Internet of Things, and industry 5.0, which are very suitable for a wide range of human activities to improve multiple health responses (Ferosekhan et al. 2014; Bundschuh et al. 2018; Rayan and Zafar 2021). Moreover, advances in nanotechnology are evidenced daily to assess the impact of nanomaterials (NMs) on our environment, especially autotrophs and heterotrophs. Moreover, the currently available evidence is varied and contradictory (Ashraf et al. 2011; Nandanpawar et al. 2013; Kakakhel et al. 2021) for the betterment of therapeutic life (Rayan et al. 2021). The introduction of NMs into the aquatic environment with the help of many intelligent methods like deep learning and machine learning response has many unpredictable consequences, which are most suitable for high-quality accuracy (Sajid et al. 2015; Zafar et al. 2021b). NMs are substances with less than 100 nm in diameter, which possess unique physiochemical characteristics that differ from their surrounding environment (Palmieri et al. 2021). NPs fall into different categories such as natural forms of NP are found in soil, water, or volcanic dust. They are created with the aid of using geological and organic processes (Rai et al. 2018). Many species are able to adapt and evolve in natural NP-rich environments even if they are detrimental (Shokry et al. 2021). Nanoparticles have been produced by companies for many years and used in fields such as agriculture, electronics, medicine, pharmacy, and beautifying materials such as cosmetics (Tijani et al. 2016). Silver NPs, titanium nitride NPs, and zinc oxide NPs from wastewater treatment could be harmful to marine organisms, as per earlier studies in different regions (Yu et al. 2021). Finally, NPs are visible in both (water and earth) environments, are taken up by living things, and build up until they are eliminated by the guard cell or other mechanisms (Selck et al. 2016). NPs are foreign components with unique physical and chemical properties in vivo that can disrupt typical physiological systems. They can occasionally impair embryonic development and cause fatal abnormalities (Rajput et al. 2018). In addition to responding to the known processes, chemicals that makeup NPs also interact with physiochemical attributes of living organisms exhibiting specific unique properties. NPs can readily pass through cell membranes and avoid defence mechanisms because of their tiny size (Cormier et al. 2021). As a result, NMs move about inside the cell, get to organelles like the mitochondria, change the metabolism of the cell, and lead to cell death (Rana et al. 2020). In turn these cause NPs to circulate. If the NPs are too tiny to enter the cell, they could interact with the cell membrane and obstruct processes like signal transduction and ion transport (Medici et al. 2021). NMs can be dangerous due to their chemical composition and physical properties. Positively charged NPs can damage cell membrane. Surface coating of NMs can disrupt cellular structures (Chakraborty et al. 2016). Furthermore, the effect of NMs can be influenced by other chemicals such as impurities. NMs can also absorb elements that are toxic to living organisms (Lei et al. 2018).

Numerous studies have been done throughout the years to identify and comprehend NMs impacts, many of which are still unknown. Understanding any potential negative direct or indirect impact on organisms is essential, given the abundance of NMs in modern society (Deshmukh et al. 2019). NPs have already been shown to be toxic to bacteria, algae, invertebrates, fish, and even humans. Several biological models have been used to evaluate the effects of nanoparticles on living organisms (McClements and Xiao 2017). NMs are detrimental to reproduction and embryonic development in studies on mammals such as mice, teleosts, and model organism zebrafish (Sharma et al. 2016; Okey-Onyesolu et al. 2021). Several studies conducted inside and outside adult tissues have defined Ag-NPs as highly reactive molecules with potential genotoxicity responsible for inducing cell death through oxidative stress (Thines et al. 2017). The inability to detect and quantify engineered nanoparticles in soil, sedimentary rocks, and liquid and other life forms has impeded research on their environmental impact. The outcomes of Co-NPs on *Eisenia fetida*, an earthworm specie, were also investigated using neutron activation (Zhang et al. 2022). Scintillation and autoradiography were employed to identify 4 nm Co NPs containing 59 m^2/g nano powder in spermatogenic cell waste or the environment. Following a literature review, similar findings have previously been discovered in microbes, roundworms, fishes, and cell lines (Ong et al. 2018). Fungicides are usually made from NMs, for example, Ag, ZnO, or CuO, etc. (Al-Bishri 2018). Nontarget species can be adversely affected after being released into the environment, like inhalation of pesticides or exposure to other harmful chemicals. However, our current understanding of the detrimental effects of nanoparticles is incomplete (Kuehr et al. 2021). As a result, earlier researchers have presented many ways to organise this field of research (Muthukumar et al. 2022). Collecting eco toxicological information to assess risk given the type of NP. Experiments with nitrogen dioxide NPs, zinc oxide NPs, copper oxide NPs, silver NPs, single-walled nanotubes (SWNTs) or single-walled carbon nanotubes (SWCNT-NTs), multi-walled nanotubes (MWCNTs), C_{60} fullerenes, are essential (Aruoja et al. 2015). We also raised the question of the experimental environment for understanding the impact and nature of the cytotoxic activity, target cell type, and sample treatment of NPs.

Toxicity can affect organisms living in any environment (air, freshwater, or seawater if inhaled, terrestrial environment) (Cai et al. 2018). Studies and analyses have been performed on different species, including protozoa, various invertebrates, chordates, adults, and embryos to investigate the harmful effects of NMs (Gehrke et al. 2015). This chapter presents some examples of recent research. Rats and fish included in the animal model are widely used in scientific research. On the other hand, rare species such as annelids and molluscs are employed in innovative and informative research. Influence of NMs on aquatic, semiaquatic, and terrestrial organisms is investigated in this chapter. Animal models and their natural habitats

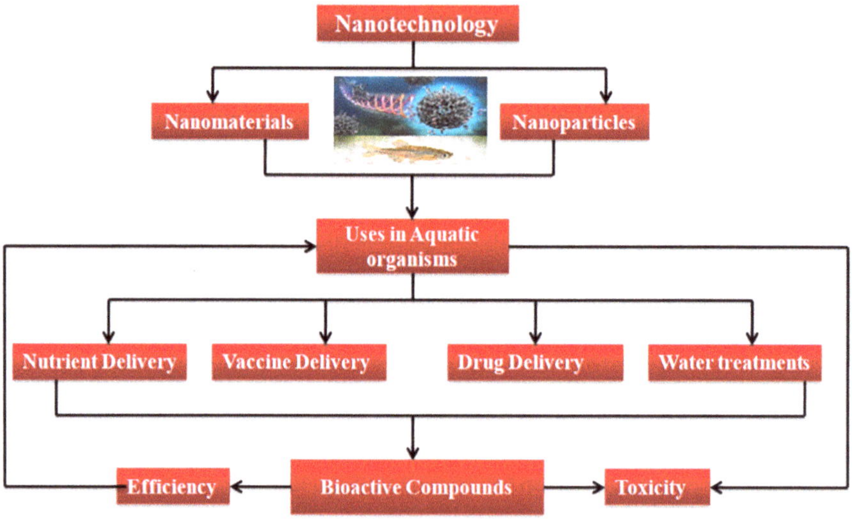

Fig. 1 A detailed overview of the use of nanotechnology in aquaculture

were the starting point for the logic of the text, but NP class may also be the starting point.

The use of nanotechnology enhances the cleaning of aquaculture pools, cure of water, handling and treatment of aquatic disorders, efficient transportation of food and medicine (including hormones and vaccines), and the inability of fish to acquire this matter (Ahmad et al. 2021; Alwash et al. 2022; Sundaray et al. 2022). Many publications are already available that provide detailed overviews of the use of nanotechnology in aquaculture (Fig. 1). However, despite their use, there is a risk of contributing to aquaculture contamination, which is unknown or unnoticed today (Asche et al. 2022; Larsson and Flach 2022). In addition, the excessive use of antibiotics to treat various diseases and other synthetic substances as growth promoters has adverse effects on aquatic ecosystems (Fujita et al. 2023). Worrying scenarios such as developmental and reproductive failure, mortality, and biochemical alterations can lead to enormous economic losses in fisheries (Selck et al. 2016).

On the other hand, they can pose problems regarding human health and environmental safety. It shows gaps, especially for lipophilic bioactive that can be used as natural remedies rather than artificial ones (Zhao et al. 2022). We highlight these innovative potential avenues that are projected to have a significant impact.

2 Aquatic Organisms and Penetration of Nanomaterials

In the field of aquatic animals, the role of bioinformatics and other high technological domains is very remarkable in penetrating reproductions in multiple organisms (Zafar et al. 2021c), and NMs are found to be very impressive domains for undertaking wide-ranging life responses. Furthermore, for domain applications, the infusoria (*Stylonychia mytilus, Tetrahymena pyriformis*), antlers (*Daphnia magna*), and amoeba *(Entamoeba histolytica)* are examples of a few aquatic invertebrates that can ingest C-based NPs. Carbon nanoparticles' capacity to penetrate organs is not entirely understood (Malhotra et al. 2020). *Lumbriculus variegatus* shallow water oligochaete, also known as California black worm, was retained in a similar liquid rich in tagged, unmodified carbon nanotubes at 14 °C, but the tissues were not able to see the nanotubes (Fischer 2015). Earthworms, in particular, and terrestrial oligochaetes, in general, showed similar outcomes (*Eisenia fetida*) (Diez-Ortiz et al. 2015). The structure and changes of carbon nanoparticles may alter their capacity to enter aquatic animals' bodies. Unmodified fullerene C_{60} entered zebrafish *(Danio rerio)* embryos through the chorion but not hydrolysed nanofibers ($C_{60}(OH)_{24}$) (Asil et al. 2020). Unlike fullerenes, the large single-walled nanotube compounds could not cross zebrafish chorion and instead settled on them. Electron microscopy revealed that zebrafish chorionic pores have a diameter of 0.5–0.7 m (Wu et al. 2021).

The tendency of C-based NPs to enter within organs is questionable yet. Only the gastrointestinal tract displayed the signs of nanotubes upon the maintenance of *L. variegatus* in a similar liquid enriched with labelled unmodified carbon nanotubes at 14 °C (Krzyżewska et al. 2016). All terrestrial oligochaetes displayed relatable outcomes. The structural confirmation and variations in NPs may influence their capacity to penetrate aquatic organisms (Thanigaivel et al. 2021). Unaltered and non-hydrolysed fullerene C_{60} penetrated in embryos of zebrafish through chorion. Unlike fullerenes, the large single-walled nanotube compounds could not cross zebrafish chorion and instead settled at their surface. Electron microscopy revealed that zebrafish chorionic pores have a diameter of 0.5–0.7 m (Follmann et al. 2017).

NMs are designed to 'persist as particles in aqueous media', allowing them to cross biological membranes due to their size. In aqueous solutions, NMs potentially generate aggregates, other colloidal suspensions, and colloidal suspensions interacting with aggregates (Yang et al. 2019). This happens because marine habitats are often highly alkaline, have high ionic strength, and already contain 'a wide variety of colloids and natural organic matter'. Due to their proximity to effluents and effluents, they are likely to contain high colloids, organic waste, and NMs (Zheng and Nowack 2022). In freshwater, nanoparticle aggregates slowly sink to the bottom and are more likely to aggregate into the sediment, which can harm benthic animals. In marine ecosystems, 'nanomaterials may concentrate at the boundary between cold and warm currents', but this is unlikely in the freshwater of recycling. This increases the risk for animals that feed in these cold and warm regions, such as tuna (Pulit-Prociak and Banach 2016). Klaine et al. (2008) have a new possibility: accumulation in 'sea surface microlayers' where nanomaterials are confined due to their surface tension and viscous properties (Klaine et al. 2008). This threatens not only seabirds and animals but also species that live in the surface microlayer (Waris et al. 2021). However, no research has been done to look into the different effects of nanomaterial accumulation in the surface microlayers of the ocean.

2.1 Toxicological Effects of Nanomaterials on Aqueous and Terrestrial Ecosystems

Nanotechnology and the excellent incorporation of NMs are part of a multibillion-dollar industry with diverse applications from biological sciences to electronics (Falanga et al. 2020). It is unavoidable that artificial and natural NMs will be released into the atmosphere, impacting water and soil. This might be due to purposeful or inadvertent emissions, which means assessing the expanding sector's possible implications on environmental, human, animal, and plant health (Kaloyianni et al. 2020).

Much research has been published on the transit and fate of putatively engineered nanoparticles (ENMs) availability of reliable toxicological information regarding their exploitation, and safe removal is still scarce. The disparity between ENM synthesis and toxicity data has prompted the scientific community to develop strong, stable, and environmentally acceptable procedures for safe manufacturing and removal (Caixeta et al. 2020). ENMs' inbound properties significantly impact their transit into the environment. Doped ENMs, for example, have a hazardous impact due to their high aggregation stability, low photo bleaching, and delayed photodegradation. Furthermore, recent studies have shown that the transport and toxicological impact of ENMs are primarily due to the dissolved ion concentration rather than the nanomaterial itself or its aggregated form (De Silva et al. 2021) (Table 1).

Although NMs exist naturally in the environment, technological advances have resulted in an abundance of novel and artificial nanomaterials that are not found naturally (Chaukura et al. 2020). Due to a lack of understanding, there is no control over emissions, and many experts are concerned that this might constitute a threat as a new class of environmental hazards. NMs are frequently used in various products due to their limited size and significant surface area, increasing the routes by which they can come in contact with organisms in the surrounding (Yoon et al. 2018). Nanomaterials released into the environment through emissions and industrial and commercial items can significantly impact them, leading them to wind up in wastewater treatment facilities and, in turn, surface water. NMs not screened in wastewater treatment facilities are more likely to 'accumulate in benthic sediments', posing a risk to numerous aquatic organisms (Kahlon et al. 2018). Because of their broad, weak respiratory epithelium, aquatic creatures are 'especially vulnerable to

Table 1 We compiled data on several NPs and target species with potential aquaculture applications, including the key comparative testing settings

Nanomaterials	Functions	Reference
Alginate NPs	Alginate, a naturally occurring polymer, is regularly employed in the food industry to thicken, emulsify and stabilise various products. Recently, successful testing of alginate NPs was successful. Nevertheless, there are significant concerns about its usage due to the absence of accurate toxicity knowledge about these substances	Guo et al. (2013), Khosravi-Katuli et al. (2017), Qi et al. (2015)
Al_2O_3 NPs	Al_2O_3 NPs have good insulating and abrasive characteristics. *Caenorhabditis elegans*, used as live food in aquaculture and aquaria for species larval development, were used to investigate the toxicity of nAl_2O_3. Concentrations greater than 102 mg/L immediately decreased worm development and the number of eggs within worm bodies and progeny, whereas concentrations greater than 203.9 mg/L considerably impeded worm reproduction	Wang et al. (2009)
Ag NPs	Silver nanoparticles can be found in various consumer products, including water purifiers, textiles, pharmaceuticals, and agrochemicals (nAg). nAg has been used in aquaculture to purify water due to its antibacterial properties, and there is a body of research on its toxicity to aquatic species relevant to aquaculture	Márquez et al. (2018)
Au NPs	According to Zhu et al., Au NPs (nAu) are used in various industries' detection. Although it is frequently used, nothing is clear about its ingestion in organisms living in water bodies. Further research observed that Au NPs do not exhibit toxicity (hatching delay)	Khosravi-Katuli et al. (2017), Mohandas et al. (2018)
CeO_2 NPs	CeO_2 nanoparticles are used in various applications, including fuel additives, coatings, electronics, and biomedical devices. There are still many unknowns about how it harms the environment and human health. After 14 days of $nCeO_2$ exposure, zebrafish accumulate only in the liver. During a five-day experimental study, *P. lividus* was exposed to the CeO_2 (50–105 nm) NPs at a concentration of 10 mg/L, causing death later on two days, while the testing model remained alive at 0.1 mg/l	Khosravi-Katuli et al. (2017), Roberta et al. (2021)
Chitosan NPs	Chitosan nanoparticles can cross tight junctions between epithelial cells, potentially	Ahmed et al. (2019), Bhoopathy et al. (2021)

(continued)

Table 1 (continued)

Nanomaterials	Functions	Reference
	threatening humans, animals, and the environment. Zhang (2011) found that *D. rerio* embryos treated with chitosan nanoparticles (200 nm) with high concentration died and deformed at 40 mg/L with nearly 100% mortality	
Cu NPs	Copper NPs (nCu), particularly nCuO, are one of the most prominent metallic nanoparticles (NPS) and display bactericide and antifouling properties, as well as solid heat conductivity, which may influence aquaculture The earlier researcher treated zebrafish juveniles to aquatic nCuO for 48 h and found histological damage, Cu accumulation in the gills, and 82 differentially expressed genes compared to controls	Shah and Mraz (2020), Vicario-Parés et al. (2018)
Fe NPs	The Fe_2O_3 NPs are widely utilised in biological applications such as cellular labelling, drug delivery, tissue regeneration, in vitro bioseparation, and hyperthermia, with additional applications including wastewater purification and as a food additive in aquaculture The researcher revealed both fatal and sub-lethal effects on medaka fish (*Oryzias latipes*) after a 14-day exposure to nFe, stating NPs coated with CMC or cellulose gum were less harmful compared to non-coated forms (ROS production and CAT change)	Mukherjee et al. (2022), Refsnider et al. (2021)
La NPs	According to Mácová et al. (2014), commonly used in water treatment, industry, and medicine. Mácová et al. (2014) exposed boy *D. rerio* and *P. reticulata* for 96 and 144 h, respectively, and reported the following LC50 values: 156.33 5.59 and 128.38 5.29 mg/L, followed by 152.98 8.06 mg/L. As a result, the use of La NPs can have potentially dangerous consequences	Mácová et al. (2014)
Quantum dots	Quantum dots are employed in biosensing, bioimaging, and monitoring the quality of water bodies. According to Khosravi-Katuli, Mykiss treated with O. 2 g/L QDs in 2 days exhibited a rise in overall metallothionein. Lewinsky et al. (2011) treated brine shrimp *(Artemia franciscana)* and crustacean (*Daphnia magna)* with 0.6 mg QD for 24 h. These were then presented to both immature and mature stages of *D. rerio* for 21 days as a	Hébert et al. (2008), Khosravi-Katuli et al. (2017), Wu and Yan (2013)

(continued)

Table 1 (continued)

Nanomaterials	Functions	Reference
	food source. Although zero post-exposure mortality was reported still 4% QD accumulation in young and 8% in adult stages were found. The researchers received comparable outcomes following an in vitro experiment using *O. mykiss* liver cells	
Selenium NPs	Se is a bionutrient product suitable for increasing aquaculture as it is a trace mineral that many species, including fish, need for proper physiological function and growth (Khan et al. 2016). Khan et al. (2016) studied the physiological and biochemical impacts of nSe supplements (0.68 mg/kg diet) on juvenile fish (*Tor putitora*), comparing RBC count, HB level, haematocrit levels, and lytic enzyme activity along conventional diets showed an increase in and other biochemical parameters	Singh and Onuegbu (2020)
Silicon dioxide NPs	According to Babu et al. 2013, they are beneficial for optical imaging and drug delivery, but their use in aquaculture has also been observed, reducing the risk of disease transmission in overcrowded aquaria. Nevertheless, the researcher observed increased mortality and malformations in zebrafish	Babu et al. (2013), Duan et al. (2013), Rahman et al. (2022)
Sn oxide NPs	Owning to the rigidity of low-temperature conductance, tin oxide NPs are crucial for developing optronics, gas sensors, and electrochemical energy storage systems. Regarding $nSnO_2$ toxicity to aquatic organisms, just two life forms are found, and their potential use in aquaculture is currently under investigation. After *P. reticulata* was exposed to 150 mg/L $nSnO_2$ for 5 days, Krysanov et al. (2009) found that tin accumulates in the gonads, spleen, intestine, liver, muscle, and thymus. The results of Falugi et al. (2012) on the effect of sea urchin (*Paracentrotus lividus*) on tin oxide have been mentioned in the 'Fe-NPs' section beforehand	Falugi et al. (2012), Krysanov et al. (2009)
SWCNTs	Carbon nanotubes have been employed in aquaculture setups to improve water treatment and food stability	Khan et al. (2021)
Titanium dioxide NPs	Varnishes, papers, fabric, synthetic polymers, sunblock, makeup, and edible items are some commercially accessible goods that employ $nTiO_2$. Aquaculture may utilise $nTiO_2$ in direct and indirect ways, as discussed in the preceding sections	Khosravi-Katuli et al. (2017), Müller (2007)

(continued)

Table 1 (continued)

Nanomaterials	Functions	Reference
	(Khosravi-Katuli et al. 2017). Investigating its possible toxicity to aquatic creatures is therefore required	
Zinc oxide NPs	According to Rather et al. (2018), ZnO NPs are employed in optoelectronics, cosmetics, catalysts, ceramics, pigments, and aquaculture. Based on concentrations, contact duration, and targeted species, different results have been found regarding the impacts of ZnO	Rather et al. (2018)

contaminants'. Changes in pH, water temperature, and oxygen levels can increase the dangers associated with nanomaterials in aquatic settings and should be considered when assessing risk. Plants are also vulnerable to nanomaterial exposure due to soil contamination or inadvertent discharge (Bakshi 2020).

Live NMs have been found to penetrate live creatures and 'exercise harmful effects' at the cellular level, including membrane rupture, protein inactivation, DNA damage, interruption of energy transmission, and toxic chemical release (Bobori et al. 2020). Due to the significant role of producers and microorganisms in the food chain, it is crucial to comprehend the potential impact that such vast industries may have on biodiversity in upcoming times (Grillo and Fraceto 2022). This study will discuss the toxicological consequences of waste-manufactured NMs in terrestrial and aquatic habitats and their implications for human health and environmental safety. The current investigation will be subjected to the hazardous effects of inappropriately disposed of nanomaterials in the environment and human health.

2.1.1 Uptake of Nanomaterials in Aquatic Ecosystems

Nanoparticles are designed to 'persist as particles in aqueous media', allowing them to cross biological membranes due to their size. In aqueous solutions, nanomaterials potentially generate aggregates, other colloidal suspensions, and colloidal suspensions interacting with aggregates (Wu et al. 2019). This happens because marine habitats are often highly alkaline, have high ionic strength, and already contain 'a wide variety of colloids and natural organic matter'. Due to their proximity to effluents and effluents, they are likely to contain high concentrations of colloids, organic waste, and nanomaterials (Saxena et al. 2020).

In freshwater, NMs' aggregates slowly sink to the bottom and are more likely to aggregate into the sediment, which can harm benthic animals. In marine ecosystems, 'nanomaterials may concentrate at the boundary between cold and warm currents', but this is unlikely in freshwater. In terms of recycling, the concentration of nanomaterials at the boundary between cold and warm currents is a phenomenon observed in marine ecosystems but is unlikely to occur in freshwater (Wu et al.

2017). This increases the risk for animals that feed in these cold and warm regions, such as tuna. Buffle and Leppard (2008) proposed a new possibility: accumulation in 'sea surface microlayers' where nanomaterials are confined due to their surface tension and viscous properties. This endangers not only seabirds and animals but also species that live in the surface microlayer. However, no studies have been performed before to investigate variable implications of nanomaterial accumulation in the surface microlayers of the ocean (Laux et al. 2018).

2.2 *Toxicological Profiling of NMs in the Aquaculture Sector*

Different TNPs are employed in the maritime sector. Several studies are underway to ensure their safety outside the aquaculture industry (Atamanalp et al. 2022). The details of potential NMs in aquaculture applications, including the critical comparative testing settings, are mentioned in Table 2. All their effects on live animals (especially aquatic ones) are unknown, and their use in aquaculture raises public concern. The toxicity of NPs, as mentioned in Fig. 2, might vary based on their delivery method, as well as toxic kinetics and toxic dynamics (Khosravi-Katuli et al. 2017). NP concentrations in feed, on treated surfaces, or in water might be more importantly broad than expected NP ambient levels of up to mcg per litre or more.

Algins, Al_2O_3, Au, Ag, cerium dioxide, CuO, and CsAg nanocomposites area a few examples of NPs along target species with potential aquaculture uses. A summary of crucial relative test parameters. Short-term exposure times have been studied mainly for species of tangential aquaculture relevance (i.e. A. *Salinna annua*).

3 Effects of NPs on Aquatic Animals

Both freshwater and marine environments contain significant amounts of NPs. Finding out how these NPs affect aquatic creatures was made possible by several studies. These results might vary, though (Exbrayat et al. 2015). Recent insights examined nanoparticles as novel contaminants that have variable effects depending on their sizes and are not yet completely understood. Numerous laboratory experiments have revealed that their constant exposure harms fish and invertebrates (Jenifer et al. 2020). The nature of these possible consequences was evaluated using traditional or less conventional animal models. As a result, several works focused on bony fish, specifically the trout *O. mykiss* and the *Danio rerio.* Other research focused on plankton, sea urchins, molluscs, daphnia, and other crustaceans.

Table 2 Profile of toxic nanomaterials

Nanomaterials	Functions	Reference
Carbon nanomaterials	Aggressive behaviour and respiratory disorder have been observed in *O. mykiss* upon nanotubes exposure to water at 0.5–0.1 mg concentration. (Smith et al. 2007). According to the studies, adding fullerene C_{60} induced biochemical alterations in large-mouth bass and fathead minnow, suggesting a detriment impact on the development of both fishes' gills, brain, and liver. A change in behavioural pattern and mortality was observed at 0.25 mL/L of the lowest possible fullerenes concentration in daphnia (*D. magna*). Fullerenes C_{60} and C_{70} are embryotoxic and genotoxic during the early stages of zebrafish embryogenesis by significantly increasing embryo abnormalities and subsequent mortality at 200 h/L maximum doses of fullerene allotrope Additionally, the hydroxylated forms of the fullerenes were less hazardous than the original. However, Petersen et al. (2014) showed that pure fullerene C_{60} dispersed through sonic waves in aquatic environments by 25 mg/l was not poisonous; however, C_{60} dispersed in tetrahydrofuran was toxic and altered the gene expression of larvae. On the other hand, single-walled carbon nanotubes did not affect the continued development of larvae and instead delayed zebrafish hatching Axolotl larvae and Xenopus tadpoles showed no acute or genotoxicity in amphibian tests using carbon nanotubes suspended in water. Xenopus tadpoles were toxic to nanotubes but at high levels. (500 mg/L) only	Haque and Ward (2018), Krysanov et al. (2010), Petersen et al. (2014), Sarasamma et al. (2019), Smith et al. (2007)
Metal oxides nanomaterials	The efficiency of photosynthesis in green algae (*Pseudokirchneriella subcapitata*) was not affected by tests to determine the toxicity of metal oxides (TiO_2, ZrO_2, Al_2O_3, and CeO_2). However, these algae could not grow at 600 g/L concentration.	Basiuk et al. (2011), Hou et al. (2018), Tetu et al. (2017), Wiench et al. (2009)

(continued)

Table 2 (continued)

Nanomaterials	Functions	Reference
	According to a Zn, Al, and TiO toxicity analysis in *D. rerio* juveniles, only ZnO concentration was found harmful for the fish. The zinc oxide nanoparticle lethal dose (LD_{50}) after 96 h was 1.8 mg/l The DAPHIA investigations showed that titanium dioxide nanoparticles are not hazardous, even at concentrations of 100 mg/l in 48-h testing. However, the length of exposure enhanced the toxicity of nanomaterials. For instance, the LD_{50} for titanium dioxide was 2 mg/l during 72 h. In contrast, TiO_2 nanoparticles decreased Daphnia seven's growth and reproduction in the case of prolonged exposure at values of 0.5–5 mg/l	
Titanium dioxide nanomaterials	Rainbow trout exposed to 1 mg/L TiO_2NPs did not experience adverse toxic effects, but sublethal effects such as internal organ disease and biochemical and respiratory abnormalities were found. For two months, TiO_2 nanoparticles were added to the meal of rainbow trout under a year old at doses of 10 and 100 mg/kg, although this did not affect the fish's growth or haematological features. However, it was found that the amounts of copper and zinc ions in the fish nervous system were disturbed, which altered the biochemistry of the gills and gut. The hydrated tin dioxide nanoparticles (SnO_2) in guppies did not have acute or genotoxicity	Handy et al. (2011), Smith et al. (2007)
Metal nanomaterials	Every aquatic organism exhibits different toxicity levels depending on the type of metallic nanoparticles they are exposed to. For instance, the Cu and Ag NPs showed 0.04 and 0.06 mg LD_{50} after 48 h. These nanomaterials, however, were less harmful to fish. Zebrafish died due to gill pathology brought on by copper and silver nanoparticles. For nanomaterials over 48 h, respective LD_{50} concentrations were found to be 7.2 and 0.9 mg/L. Morphological abnormalities in *D. rerio* larvae increased when their	El-Samad et al. (2022), Lacave et al. (2018), Zhao et al. (2013)

(continued)

Table 2 (continued)

Nanomaterials	Functions	Reference
	spawn was incubated with silver nanomaterials, and the scientists also showed a relationship between anomalies and Ag nanomaterial doses. Ag NPs at maximum concentrations accelerated morphological defects and mortality in zebrafish larvae	
Semiconductors	Freshwater muscle (*E. complanata*) reared in aquatic conditions enriched with cadmium telluride and QDs at a concentration range between 1.6 and 8 mg/l showed the vital signs of immunotoxin and Geno toxins presence. The viability and activity of the haemocytes decreased, while oxidative stress and the frequency of DNA fragmentation increased within gills	Giroux et al. (2022), Parolini et al. (2010)
Dendrimers	Sublethal concentrations of fourth-generation polyaminoamide (PAMAM) dendrimer with NH_2 group are found to be more poisonous, interrupting developmental processes and retard the growth of *D. rerio* embryos	Tamayo-Belda et al. (2022)

3.1 In Fish

The immature stage of salmon developed sodium borohydride ($NaBH_4$) after reducing silver NPs when subjected to silver nanoparticle suspension. The size range for colloidal Ag-NPs, both manufactured and purchased, was 3–220 nm (Stanková 2015). In all tests, fish gills collected Ag-NPs, except when NP concentration was least (1 g/L). The effect of response was dose-dependent, which caused a considerable spike in the stress level of gills through HSP70 and plasmatic glucose (Ackerman et al. 2000). Dose-dependent inhibition of Na/K ATPase ubiquitous enzyme exhibits an osmoregulatory default (Mackie et al. 2007). At maximum concentration, 100 g/L silver nanoparticles led to necrosis of gill lamella, resulting in the death of 73% of fish. All these experiments showed how nanoparticle preparation could adversely affect the surrounding organisms (Handy et al. 2008).

Considering their chemical characteristics and behaviour linked to aggregation dynamics and equilibrium of freely available metal ions may cause acute toxicity of metallic nanoparticles. Metallic-NPs can potentially be more toxic to some fish species than their dissolved versions (Barría et al. 2020). Numerous organ pathologies, including those of the gills, liver, gut, and brain, revealed some similarities between the responses to NPs and metal salts. Some consequences for development were also seen (Han et al. 2021). Ag-NPs were applied to the chorion, the egg's

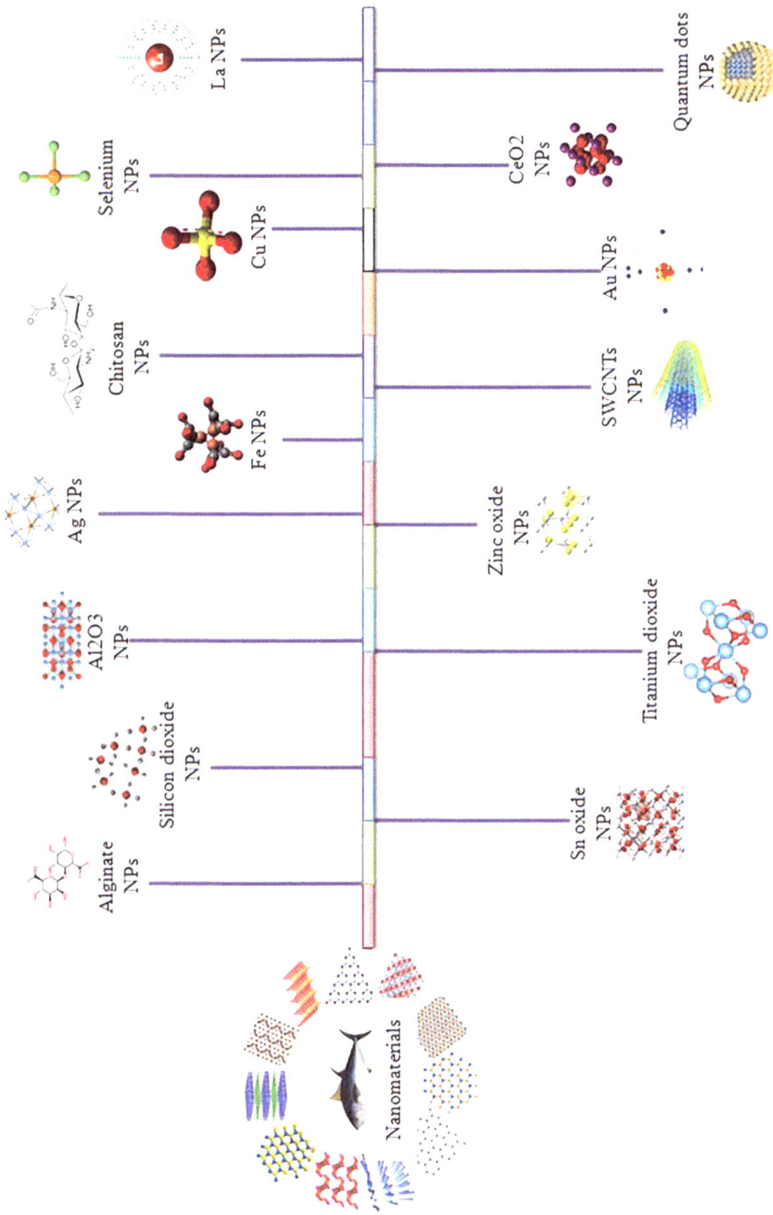

Fig. 2 The influence of NMs and their applications for delivery methods in the context of toxic kinetics and toxic dynamics

membrane, or the growing embryo. Zinc and copper nanoparticles show more adverse consequences for embryos and young animals than the equivalent salt (González-Fernández et al. 2021). It is still feasible that metal-NPs stimulate the relevant stress reactions to obstruct them. Since the end of the 2000s, researchers have been examining how fish react mechanically to NPs and other nanomaterials.

In contrast to other chemical compounds, this research compared the methods of substance absorption, distribution, metabolisation, and excretion in fish (Forouhar Vajargah et al. 2020). The TiO_2-NPs and C_{60} fullerene have been tested for these properties in the gills, digestive system, liver, and adrenals, among other organs. NPs more thoroughly enter the tissues by endocytosis than through diffusion or ionic carriers, such as the equivalent metal ion (Benavides et al. 2016). In fish, NPs might be removed with bile but are seldom expelled through the kidneys. The effects ZnO NPs were investigated in an in vitro experiment employing human and fish liver cell strains. These NPs clumped together, which significantly increased the toxicity of fish cells. The dissolved salts produced by NPs would be hazardous to human cells (George et al. 2014). A comparison of studies using tumour-bearing human hepatocytes Huh7 grown in vitro and in vivo revealed that the 120-nm-diameter Ag-NPs entered the hepatocytes and caused oxidative stress characterised by ROS production, IFN expression, and disruption of the endoplasmic reticulum (Daufresne and Boet 2007).

In vitro studies on *D. rerio* revealed that silver nanoparticles' neurotoxicity is distinct from silver ions. Different forms of Ag nanoparticles, coated with PVP or ionic forms, all have a variable impact on embryonic development (Kumar et al. 2020). Ag+ retards the swim bladder development and is directly connected to multiple deformities in fishes. In response to light stimuli, fish behaviour was also altered. Small-sized Ag nanoparticles coated with PVP induced hyperactivism, whereas large particles induced hypoactivism in affected fishes due to light exposure (Johnston et al. 2010). A thorough examination revealed that the adult nervous system was affected by 1–20 nm Ag–NPs was given to zebrafish embryos. Ag-NPs' production of Ag ions can potentially increase mortality and deformities (Parsai and Kumar 2020). Ag-NPs may affect cell differentiation by inhibiting cells using acetylcholine as an intermediate. In addition, Ni-NPs caused deformities and death in zebrafish embryos. Contrary to the Ni solution, which had no impact, the guts showed thinning upon contact with Ni-coated NPs (Lai et al. 2021). Soluble as well as Ni-NPs of 30, 60, and 100 nm sizes affected skeletal muscles.

A minimal difference between the toxicity of soluble NI and Ni-coated nanoparticles was found. On the other hand, skeletal muscles and the gut were susceptible to 60 nm massive aggregates of dendritic Ni-NPs toxicity (Shaw and Handy 2011). Zebrafish embryos carried 10 nm Au NPs that travelled throughout their whole bodies. However, the impacts on growth were inversely proportional to concentrations; these particles were collected in aggregates with sizes dependent on concentration (Geppert et al. 2021). The observed abnormalities could be the result of being randomly distributed. Ag NPs are more harmful than Au NPs when it comes to toxicity, which depends on chemical characteristics. *D. rerio* embryo can therefore serve as the perfect experimental model for in vivo studies, particularly regarding materials' biocompatibility (Guerrera et al. 2021). A comparison of the impact of $CuSO_4$ and Cu NPs on the gills of *O. mykiss* showed the accumulation of these substances in varying amounts. NP and salt were associated with increased Cu, although the spleen, brain, and muscle showed no signs of product accumulation (Shaw et al. 2012). Finally, at low concentrations, Cu NPs appear to have toxic effects comparable to $CuSO_4$.

Except for Cu and Zn, metallic concentration in tissues remains unaffected by NP accumulation in the brain. Na+/K+ ATPase decreased in gills and intestines. Thiobarbituric acid (TBA) increases in the brain and gills dose-dependently (Tabassum et al. 2016). Ag NPs dose-dependently reduced membrane integrity, and cell metabolism in hepatocytes culture from various species. Au NPs increased ROS without adverse effects. Indeed, the effects of Ag and Au NPs on trout hepatocyte cultures were sometimes contradictory (Singh et al. 2009). *D. rerio* juveniles and embryos were ingested with 25 nm TiO_2 NPs to investigate their impact on developmental processes. During an experimental activity, they were added to commercial food, while in another, fish were given algae that had already been exposed to TiO_2-NPs (Schultz et al. 2014). Hatching occurred early and minimally affected young animals at low concentrations. However, after 14 days of exposure to tainted food, the physiology of the digestive system changed.

4 Toxicity of Nanomaterials

In toxicity studies regarding engineered nanomaterials, numerous research groups have evaluated sublethal and lethal concentrations, cell proliferation, embryonic toxicity, chromosomal abnormalities, and fertility issues in animals.

5 Bio Modification and Migration Along Food Webs

Limited hydrophilicity and instability of aqueous suspensions of NPs are two of the main problems in their use in biology and medicine. Attempts to treat nanomaterials with organic solvents to make them more hydrophilic lead to increased toxicity of the nanomaterials (Wang et al. 2020). However, the ability to create nanoparticle suspensions in water without increasing nanoparticle toxicity has been met with some success. Very shortly, many biologically accessible nanoparticles may enter the environment (Wang et al. 2016). Engineered NMs have been shown to change spontaneously in an aqueous environment, making them more accessible to organisms. In addition, they can be artificially modified to make them more hydrophilic (Souza and Fernando 2016). Naturally present fulvic acid and humic enhance the stability of fullerene and nanotube suspensions. As a result, aggregates do not form or occur in moderate amounts.

In contrast, studies have shown that polysaccharides improve the sedimentation of nanomaterials and reduce their mobility. The ability of nanomaterials to adsorb organic environmental toxins has been shown to increase toxicological impact (Rhim et al. 2013). For example, fullerene C_{60} in water increases the toxicity of phenanthrene to daphnids by order of magnitude. In contrast, titanium dioxide nanomaterials increase the accumulation of arsenic and cadmium in carp organs (Krysanov et al. 2010). It is also recognised that organisms themselves can modify nanomaterials. For example, fullerene C_{60} has been shown to combine with vitamin A to form a molecule in the liver of house mice (Dellinger et al. 2013). Due to their chemical activity, carbon nanotubes can combine with other organic components in the body to produce chemical compounds (proteins, phospholipids, and DNA). Various organic substances bind to nanomaterials, allowing them to enter cells (Mu et al. 2014). Supplementing proteinaceous culture media with nanotubes has shown unexpected results in *Tetrahymena pyriformis* strains.

We found that culture development was stimulated with increasing nanotube concentration. The authors hypothesised that protein–nanotube interactions increase the amount of protein reaching the cell and promote cell development by increasing the amount of protein entering the cell can enter the bodies of aquatic invertebrates, but how far up the food chain it goes is unknown (Naskalska et al. 2021). It has been demonstrated that transportation of QDs from infusoria to rotifers could be due to the food chain. No data on vertebrates is available yet. NMs could potentially move down the food chain and focus on higher consumers. This may influence the severity of toxic effects for species with different trophic levels.

6 Current Nanotechnology in Aquaculture

The use of vaccination in aquaculture is essential as a defence strategy against pathogens to protect hosts against these pathogens. In silico investigations at the genome level (Rather and Dhandare 2019; Rather et al. 2020; Zafar et al. 2021a), oral management, and recent vaccinations in the aquaculture sector are the most reliable and efficient immunisation method (Okeke et al. 2022). The latter is a conventional adjuvant technique requiring an oil-in-water formulation to manufacture the vaccine, with some unfortunate consequences. These compositions and administration techniques occasionally result in fish death (Fajardo et al. 2022). To circumvent these problems, the scientific community has proposed a nanodelivery system as an alternative method of administering vaccines to fish that is believed to be safer and more effective. Among them, alginate particles were selected as the first candidate for oral administration of vaccines to aquatic animals (Sarkar et al. 2022). Alginate particles are often produced by emulsification, one of the most rapid and scalable NP production techniques. Several fish survival, weight, and antigenicity adjuvant researchers have reported alginate and management to improved immunostimulatory responses of carp (*Cyprinus carpio* L.), spotted grouper

(*Epinephelus fuscoguttatus*), and improved protection of flounder (*Scophthalmus maximus* L.) against several diseases (Harikrishnan et al. 2011).

PLG, or PLGA, is a biodegradable copolymer widely employed for encapsulating and transporting various substances into fish. Recent research discovered strong immune-stimulatory and antibody responses in these fish compared to the control group (Wang et al. 2018). Another study team in Japanese flounder also found similar outcomes, with DNA vaccine encased in PLGA demonstrating improved inducing effects on immunological measures against lymphocytes. Compared to the control group, carp (*Cyprinus carpio*) with liposomes encapsulating *Aeromonas salmonicida* antigen had a higher survival rate (83%) and fewer skin ulcers (Shah and Mraz 2020). Hydrophilic antigens significantly increased serum antibody counts, boosting the immunity of common carp.

7 Conclusions

With enormous growth, aquaculture represents one of the fastest-growing sectors. It provides a competent employment platform and significantly impacts the national economy through revenue generation. Integrating nanotechnology in aquaculture promises to reinvent traditional practices and serious challenges. Exceptional physiochemical properties of NMs have extensive applications in aquaculture research and development and promote aquatic organisms and life. NMs are employed in drug delivery, broad-spectrum antimicrobial activity, biomaterials engineering, and ecological remediation. In nanotechnology, NPs have become a significant source of economic and scientific innovation to provide remarkable results.

On the other hand, the widespread use leading to the uncontrolled discharge of these particles and other toxic effluents has negatively impacted multiple modes of life. Various poorly designed, newly engineered NMs and NPs are coming to light every other day, turning them into potential environmental pollutants. They come in sizes from 1 to 100, and their ecological toxicity has been suggested to be linked to their physiochemical properties. As aquatic organisms are directly exposed to distinct metallic nanoparticles via food, water, and sediments, thus, their low concentrations can induce toxicity in fishes and other aquatic organisms. Biomagnification of their mixtures in aquatic life harm terrestrial biodiversity by accumulating in the food chains. The intrinsic chemical reactivity of nanoparticles causes inflammation, oxidative stress, and genotoxicity of biological systems. It has adverse impacts on human health, increasing animal disease rates and degrading the ecosystem.

A sustainable approach regarding the employment of NMs in aquaculture and fisheries demands a deeper insight into their accumulation into ecological systems and impacts on contemporary life forms. Public engagement is crucial in terms of food and safety to preserve confidence in nanotechnology. Moreover, multidisciplinary collaboration between researchers public and private sectors is mandatory for the future success of aquaculture resulting in global benefit.

References

Ackerman P, Forsyth R, Mazur C, Iwama G (2000) Stress hormones and the cellular stress response in salmonids. Fish Physiol Biochem 23(4):327–336

Ahmad A, Abdullah SRS, Hasan HA, Othman AR, Ismail NI (2021) Aquaculture industry: supply and demand, best practices, effluent and its current issues and treatment technology. J Environ Manag 287:112271

Ahmed F, Soliman FM, Adly MA, Soliman HA, El-Matbouli M, Saleh M (2019) Recent progress in biomedical applications of chitosan and its nanocomposites in aquaculture: a review. Res Vet Sci 126:68–82

Aklakur M, Asharf Rather M, Kumar N (2016) Nanodelivery: an emerging avenue for nutraceuticals and drug delivery. Crit Rev Food Sci Nutr 56(14):2352–2361

Al-Bishri WM (2018) Toxicity study of gold and silver nanoparticles on experimental animals. Pharmacophore 1:48–55

Alwash SW, Al-Faragi J, Al-Saadi TM (2022) Effeciency of copper nanoparticles coated mint as anti-fungal against saprolegniasis disease in common carp. Iraqi J Agric Sci 53(5):1129–1137

Aruoja V, Pokhrel S, Sihtmäe M, Mortimer M, Mädler L, Kahru A (2015) Toxicity of 12 metal-based nanoparticles to algae, bacteria and protozoa. Environ Sci Nano 2(6):630–644

Asche F, Eggert H, Oglend A, Roheim CA, Smith MD (2022) Aquaculture: externalities and policy options. Rev Environ Econ Policy 16(2):282–305

Ashraf M, Aklakur M, Sharma R, Ahmad S, Khan M (2011) Nanotechnology as a novel tool in fisheries and aquaculture development: a review. Iran J Energy Environ 2(3):258–261

Asil SM, Ahlawat J, Barroso GG, Narayan M (2020) Nano-material based drug delivery systems for the treatment of neurodegenerative diseases. Biomater Sci 8(15):4109–4128

Atamanalp M, Ucar A, Alak G (2022) Catfish as an ecotoxicological model for assessment of nanoparticle toxicity profiling. Environ Sci Pollut Res 29(3):2843–2856

Babu A, Templeton AK, Munshi A, Ramesh R (2013) Nanoparticle-based drug delivery for therapy of lung cancer: progress and challenges. J Nanomater 2013:863951

Bakshi MS (2020) Impact of nanomaterials on ecosystems: mechanistic aspects in vivo. Environ Res 182:109099

Barría C, Brandts I, Tort L, Oliveira M, Teles M (2020) Effect of nanoplastics on fish health and performance: a review. Mar Pollut Bull 151:110791

Basiuk EV, Ochoa-Olmos OE, De la Mora-Estrada LF (2011) Ecotoxicological effects of carbon nanomaterials on algae, fungi and plants. J Nanosci Nanotechnol 11(4):3016–3038

Benavides M, Fernández-Lodeiro J, Coelho P, Lodeiro C, Diniz MS (2016) Single and combined effects of aluminum (Al_2O_3) and zinc (ZnO) oxide nanoparticles in a freshwater fish, Carassius auratus. Environ Sci Pollut Res 23(24):24578–24591

Bhoopathy S, Inbakandan D, Thirugnanasambandam R, Kumar C, Sampath P, Bethunaickan R et al (2021) A comparative study on chitosan nanoparticle synthesis methodologies for application in aquaculture through toxicity studies. IET Nanobiotechnol 15(4):418–426

Bobori D, Dimitriadi A, Karasiali S, Tsoumaki-Tsouroufli P, Mastora M, Kastrinaki G et al (2020) Common mechanisms activated in the tissues of aquatic and terrestrial animal models after TiO_2 nanoparticles exposure. Environ Int 138:105611

Buffle J, Leppard GG (2008) Sea surface microlayers as an extreme habitat for hydrophobic organic compounds and nanoparticles. Water Air Soil Pollut 191(1–4):371–383

Bundschuh M, Filser J, Lüderwald S, McKee MS, Metreveli G, Schaumann GE et al (2018) Nanoparticles in the environment: where do we come from, where do we go to? Environ Sci Eur 30(1):1–17

Cai Z, Dwivedi AD, Lee W-N, Zhao X, Liu W, Sillanpää M et al (2018) Application of nanotechnologies for removing pharmaceutically active compounds from water: development and future trends. Environ Sci Nano 5(1):27–47

Caixeta MB, Araújo PS, Gonçalves BB, Silva LD, Grano-Maldonado MI, Rocha TL (2020) Toxicity of engineered nanomaterials to aquatic and land snails: a scientometric and systematic review. Chemosphere 260:127654

Chakraborty C, Sharma AR, Sharma G, Lee S-S (2016) Zebrafish: a complete animal model to enumerate the nanoparticle toxicity. J Nanobiotechnol 14(1):1–13

Chaukura N, Madzokere TC, Mugocheki N, Masilompane TM (2020) The impact of nanomaterials in aquatic systems. In: The ELSI handbook of nanotechnology: risk, safety, elsi and commercialization, pp 205–222

Cormier B, Gambardella C, Tato T, Perdriat Q, Costa E, Veclin C et al (2021) Chemicals sorbed to environmental microplastics are toxic to early life stages of aquatic organisms. Ecotoxicol Environ Saf 208:111665

Daufresne M, Boet P (2007) Climate change impacts on structure and diversity of fish communities in rivers. Glob Chang Biol 13(12):2467–2478

De Silva Y, Rajagopalan U, Kadono H (2021) Microplastics on the growth of plants and seed germination in aquatic and terrestrial ecosystems. Glob J Environ Sci Manag 7(3):347–368

Dellinger A, Zhou Z, Connor J, Madhankumar A, Pamujula S, Sayes CM, Kepley CL (2013) Application of fullerenes in nanomedicine: an update. Nanomedicine 8(7):1191–1208

Deshmukh SP, Patil S, Mullani S, Delekar S (2019) Silver nanoparticles as an effective disinfectant: a review. Mater Sci Eng C 97:954–965

Diez-Ortiz M, Lahive E, George S, Ter Schure A, Van Gestel CA, Jurkschat K, Spurgeon DJ (2015) Short-term soil bioassays may not reveal the full toxicity potential for nanomaterials; bioavailability and toxicity of silver ions ($AgNO_3$) and silver nanoparticles to earthworm Eisenia fetida in long-term aged soils. Environ Pollut 203:191–198

Duan J, Yu Y, Shi H, Tian L, Guo C, Huang P, Sun Z (2013) Toxic effects of silica nanoparticles on zebrafish embryos and larvae. PLoS One 8(9):e74606

El-Samad LM, El-Gendy AH, Abdel-Moneim AM, El-Ashram S, Augustyniak M (2022) CuO NPs-induced damage to testes and deregulation of the antioxidant system in wild terrestrial organism Blaps sulcata (Coleoptera: Tenebrionidae). Environ Nanotechnol Monitor Manag 18: 100751

Exbrayat J-M, Moudilou EN, Lapied E (2015) Harmful effects of nanoparticles on animals. J Nanotechnol 2015:861092

Fajardo C, Martinez-Rodriguez G, Blasco J, Mancera JM, Thomas B, De Donato M (2022) Nanotechnology in aquaculture: applications, perspectives and regulatory challenges. Aquacult Fisher 7(2):185–200

Falanga A, Siciliano A, Vitiello M, Franci G, Del Genio V, Galdiero S et al (2020) Ecotoxicity evaluation of pristine and indolicidin-coated silver nanoparticles in aquatic and terrestrial ecosystem. Int J Nanomedicine 15:8097

Falugi C, Aluigi M, Chiantore M, Privitera D, Ramoino P, Gatti M et al (2012) Toxicity of metal oxide nanoparticles in immune cells of the sea urchin. Mar Environ Res 76:114–121

Ferosekhan S, Gupta S, Singh AR, Rather MA, Kumari R, Kothari DC et al (2014) RNA-loaded chitosan nanoparticles for enhanced growth, immunostimulation and disease resistance in fish. Curr Nanosci 10(3):453–464

Fischer J (2015) Pure carbon nanotube toxicity to moth species plutella xylostella and consumption of carbon nanotube material. LOGOS A J Undergrad Res 8

Follmann HDM, Naves AF, Araujo RA, Dubovoy V, Huang X, Asefa T, Oliveira ON (2017) Hybrid materials and nanocomposites as multifunctional biomaterials. Curr Pharm Des 23(26): 3794–3813

Forouhar Vajargah M, Mohamadi Yalsuyi A, Sattari M, Prokić MD, Faggio C (2020) Effects of copper oxide nanoparticles (CuO-NPs) on parturition time, survival rate and reproductive success of guppy fish, Poecilia reticulata. J Clust Sci 31(2):499–506
Fujita R, Brittingham P, Cao L, Froehlich H, Thompson M, Voorhees T (2023) Toward an environmentally responsible offshore aquaculture industry in the United States: ecological risks, remedies, and knowledge gaps. Mar Policy 147:105351
Gehrke I, Geiser A, Somborn-Schulz A (2015) Innovations in nanotechnology for water treatment. Nanotechnol Sci Appl 8:1
George S, Gardner H, Seng EK, Chang H, Wang C, Yu Fang CH et al (2014) Differential effect of solar light in increasing the toxicity of silver and titanium dioxide nanoparticles to a fish cell line and zebrafish embryos. Environ Sci Technol 48(11):6374–6382
Geppert M, Sigg L, Schirmer K (2021) Toxicity and translocation of Ag, CuO, ZnO and TiO 2 nanoparticles upon exposure to fish intestinal epithelial cells. Environ Sci Nano 8(8): 2249–2260
Giroux MS, Zahra Z, Salawu OA, Burgess RM, Ho KT, Adeleye AS (2022) Assessing the environmental effects related to quantum dot structure, function, synthesis and exposure. Environ Sci Nano 9(3):867–910
González-Fernández C, Díaz Baños FG, Esteban MÁ, Cuesta A (2021) Functionalized nanoplastics (NPs) increase the toxicity of metals in fish cell lines. Int J Mol Sci 22(13):7141
Grillo R, Fraceto LF (2022) Impacts of magnetic iron oxide nanoparticles in terrestrial and aquatic environments. In: Toxicology of nanoparticles and nanomaterials in human, terrestrial and aquatic systems, pp 147–164
Guerrera MC, Aragona M, Porcino C, Fazio F, Laurà R, Levanti M et al (2021) Micro and nano plastics distribution in fish as model organisms: histopathology, blood response and bioaccumulation in different organs. Appl Sci 11(13):5768
Guo H, Lai Q, Wang W, Wu Y, Zhang C, Liu Y, Yuan Z (2013) Functional alginate nanoparticles for efficient intracellular release of doxorubicin and hepatoma carcinoma cell targeting therapy. Int J Pharm 451(1–2):1–11
Han Y, Lian F, Xiao Z, Gu S, Cao X, Wang Z, Xing B (2021) Potential toxicity of nanoplastics to fish and aquatic invertebrates: current understanding, mechanistic interpretation, and meta-analysis. J Hazard Mater 427:127870
Handy RD, Henry TB, Scown TM, Johnston BD, Tyler CR (2008) Manufactured nanoparticles: their uptake and effects on fish—a mechanistic analysis. Ecotoxicology 17(5):396–409
Handy R, Al-Bairuty G, Al-Jubory A, Ramsden C, Boyle D, Shaw B, Henry T (2011) Effects of manufactured nanomaterials on fishes: a target organ and body systems physiology approach. J Fish Biol 79(4):821–853
Haque E, Ward AC (2018) Zebrafish as a model to evaluate nanoparticle toxicity. Nano 8(7):561
Harikrishnan R, Kim M-C, Kim J-S, Han Y-J, Jang I-S, Balasundaram C, Heo M-S (2011) Immunomodulatory effect of sodium alginate enriched diet in kelp grouper Epinephelus brneus against streptococcus iniae. Fish Shellfish Immunol 30(2):543–549
Hébert N, Gagné F, Cejka P, Cyr D, Marcogliese D, Blaise C et al (2008) The effects of a primary-treated municipal effluent on the immune system of rainbow trout (Oncorhynchus mykiss): exposure duration and contribution of suspended particles. Comparat Biochem Physiol C Toxicol Pharmacol 148(3):258–264
Hou J, Wu Y, Li X, Wei B, Li S, Wang X (2018) Toxic effects of different types of zinc oxide nanoparticles on algae, plants, invertebrates, vertebrates and microorganisms. Chemosphere 193:852–860
Jenifer AA, Malaikozhundan B, Vijayakumar S, Anjugam M, Iswarya A, Vaseeharan B (2020) Green synthesis and characterisation of silver nanoparticles (AgNPs) using leaf extract of Solanum nigrum and assessment of toxicity in vertebrate and invertebrate aquatic animals. J Clust Sci 31(5):989–1002

Johnston BD, Scown TM, Moger J, Cumberland SA, Baalousha M, Linge K et al (2010) Bioavailability of nanoscale metal oxides TiO_2, CeO_2, and ZnO to fish. Environ Sci Technol 44(3): 1144–1151

Kahlon SK, Sharma G, Julka J, Kumar A, Sharma S, Stadler FJ (2018) Impact of heavy metals and nanoparticles on aquatic biota. Environ Chem Lett 16(3):919–946

Kakakhel MA, Wu F, Sajjad W, Zhang Q, Khan I, Ullah K, Wang W (2021) Long-term exposure to high-concentration silver nanoparticles induced toxicity, fatality, bioaccumulation, and histological alteration in fish (Cyprinus carpio). Environ Sci Eur 33(1):1–11

Kaloyianni M, Dimitriadi A, Ovezik M, Stamkopoulou D, Feidantsis K, Kastrinaki G et al (2020) Magnetite nanoparticles effects on adverse responses of aquatic and terrestrial animal models. J Hazard Mater 383:121204

Khan MA, Khan HA, Ali Z, Asiri AM, Al-Misned FA (2016) Physiological and biochemical impacts of nSe supplements (0.68 mg/kg diet) on juvenile fish (Tor putitora). Biol Trace Elem Res 173(2):415–422

Khan FSA, Mubarak NM, Tan YH, Khalid M, Karri RR, Walvekar R et al (2021) A comprehensive review on magnetic carbon nanotubes and carbon nanotube-based buckypaper for removal of heavy metals and dyes. J Hazard Mater 413:125375

Khosravi-Katuli K, Prato E, Lofrano G, Guida M, Vale G, Libralato G (2017) Effects of nanoparticles in species of aquaculture interest. Environ Sci Pollut Res 24(21):17326–17346

Klaine SJ, Alvarez PJ, Batley GE, Fernandes TF, Handy RD, Lyon DY et al (2008) Nanomaterials in the environment: behavior, fate, bioavailability, and effects. Environ Toxicol Chem Int J 27(9):1825–1851

Krysanov EY, Demidova T, Pel'gunova L, Badalyan S, Rumyantseva M, Gas'kov, A. (2009) Effect of hydrated tin dioxide ($SnO_2 \cdot xH_2O$) nanoparticles on guppy (Poecilia reticulata Peters, 1860). Dokl Biol Sci:426, 288–429

Krysanov EY, Pavlov D, Demidova T, Dgebuadze Y (2010) Effect of nanoparticles on aquatic organisms. Biol Bull 37(4):406–412

Krzyżewska I, Kyzioł-Komosińska J, Rosik-Dulewska C, Czupioł J, Antoszczyszyn-Szpicka P (2016) Inorganic nanomaterials in the aquatic environment: behavior, toxicity, and interaction with environmental elements. Arch Environ Protect 42(1):87–101

Kuehr S, Kosfeld V, Schlechtriem C (2021) Bioaccumulation assessment of nanomaterials using freshwater invertebrate species. Environ Sci Eur 33(1):1–36

Kumar N, Chandan NK, Wakchaure G, Singh NP (2020) Synergistic effect of zinc nanoparticles and temperature on acute toxicity with response to biochemical markers and histopathological attributes in fish. Comp Biochem Physiol C Toxicol Pharmacol 229:108678

Lacave JM, Vicario-Parés U, Bilbao E, Gilliland D, Mura F, Dini L et al (2018) Waterborne exposure of adult zebrafish to silver nanoparticles and to ionic silver results in differential silver accumulation and effects at cellular and molecular levels. Sci Total Environ 642:1209–1220

Lai W, Xu D, Li J, Wang Z, Ding Y, Wang X et al (2021) Dietary polystyrene nanoplastics exposure alters liver lipid metabolism and muscle nutritional quality in carnivorous marine fish large yellow croaker (Larimichthys crocea). J Hazard Mater 419:126454

Larsson D, Flach C-F (2022) Antibiotic resistance in the environment. Nat Rev Microbiol 20(5): 257–269

Laux P, Riebeling C, Booth AM, Brain JD, Brunner J, Cerrillo C et al (2018) Challenges in characterising the environmental fate and effects of carbon nanotubes and inorganic nanomaterials in aquatic systems. Environ Sci Nano 5(1):48–63

Lei C, Sun Y, Tsang DC, Lin D (2018) Environmental transformations and ecological effects of iron-based nanoparticles. Environ Pollut 232:10–30

Lewinsky HA, Rosenkranz P, Hare L (2011) Toxicity of nanoparticles to the brine shrimp Artemia franciscana and the cladoceran Daphnia magna. J Nanopart Res 13(2):643–651

Mackie PM, Gharbi K, Ballantyne JS, McCormick SD, Wright PA (2007) Na+/K+/2Cl– cotransporter and CFTR gill expression after seawater transfer in smolts (0+) of different Atlantic salmon (Salmo salar) families. Aquaculture 272(1–4):625–635

Mácová S, Babula P, Plhalová L, Žáčková K, Kizek R, Svobodová Z (2014) Acute toxicity of lanthanum to fish Danio rerio and Poecilia reticulata. J Biochem Technol 2(5):46–47
Malhotra N, Villaflores OB, Audira G, Siregar P, Lee J-S, Ger T-R, Hsiao C-D (2020) Toxicity studies on graphene-based nanomaterials in aquatic organisms: current understanding. Molecules 25(16):3618
Márquez JCM, Partida AH, del Carmen M, Dosta M, Mejía JC, Martínez JAB (2018) Silver nanoparticles applications (AgNPS) in aquaculture. Int J Fisher Aquatic Stud 6(2):05–11
McClements DJ, Xiao H (2017) Is nano safe in foods? Establishing the factors impacting the gastrointestinal fate and toxicity of organic and inorganic food-grade nanoparticles. npj Sci Food 1(1):1–13
Medici S, Peana M, Pelucelli A, Zoroddu MA (2021) An updated overview on metal nanoparticles toxicity. Semin Cancer Biol 76:17–26
Mohandas A, Deepthi S, Biswas R, Jayakumar R (2018) Chitosan based metallic nanocomposite scaffolds as antimicrobial wound dressings. Bioact Mater 3(3):267–277
Mu Q, Jiang G, Chen L, Zhou H, Fourches D, Tropsha A, Yan B (2014) Chemical basis of interactions between engineered nanoparticles and biological systems. Chem Rev 114(15): 7740–7781
Mukherjee K, Saha A, Bhattacharjee S, Mallick A, Sarkar B (2022) Nanoscale iron for sustainable aquaculture and beyond. Biocatal Agric Biotechnol 102440:44
Müller N (2007) Nanoparticles in the environment: what concentrations of nano titanium dioxide, carbon nanotubes and nano silver are we exposed to? ETH Zurich, Department of Environmental Sciences
Muthukumar H, Palanirajan SK, Shanmugam MK, Arivalagan P, Gummadi SN (2022) Photocatalytic degradation of caffeine and E. coli inactivation using silver oxide nanoparticles obtained by a facile green co-reduction method. Clean Techn Environ Policy 24(4):1087–1098
Nandanpawar P, Badhe M, Rather MA, Sharma R (2013) Chitosan nanoparticles for gene delivery in Macrobrachium rosenbergii (DE MAN 1879). J Cell Tissue Res 13(2):3619
Naskalska A, Borzęcka-Solarz K, Rozycki J, Stupka I, Bochenek M, Pyza E, Heddle JG (2021) Artificial protein cage delivers active protein cargos to the cell interior. Biomacromolecules 22(10):4146–4154
Okeke ES, Chukwudozie KI, Nyaruaba R, Ita RE, Oladipo A, Ejeromedoghene O et al (2022) Antibiotic resistance in aquaculture and aquatic organisms: a review of current nanotechnology applications for sustainable management. Environ Sci Pollut Res:1–34
Okey-Onyesolu CF, Hassanisaadi M, Bilal M, Barani M, Rahdar A, Iqbal J, Kyzas GZ (2021) Nanomaterials as nanofertilizers and nanopesticides: an overview. ChemistrySelect 6(33): 8645–8663
Ong CB, Ng LY, Mohammad AW (2018) A review of ZnO nanoparticles as solar photocatalysts: synthesis, mechanisms and applications. Renew Sust Energ Rev 81:536–551
Palmieri V, De Maio F, De Spirito M, Papi M (2021) Face masks and nanotechnology: keep the blue side up. Nano Today 37:101077
Parolini M, Binelli A, Cogni D, Provini A (2010) Multi-biomarker approach for the evaluation of the cyto-genotoxicity of paracetamol on the zebra mussel (Dreissena polymorpha). Chemosphere 79(5):489–498
Parsai T, Kumar A (2020) Tradeoff between risks through ingestion of nanoparticle contaminated water or fish: human health perspective. Sci Total Environ 740:140140
Petersen EJ, Henry TB, Zhao J, MacCuspie RI, Kirschling TL, Dobrovolskaia MA et al (2014) Identification and avoidance of potential artifacts and misinterpretations in nanomaterial ecotoxicity measurements. Environ Sci Technol 48(8):4226–4246
Pulit-Prociak J, Banach M (2016) Silver nanoparticles–a material of the future…? Open Chem 14(1):76–91
Qi W-W, Yu H-Y, Guo H, Lou J, Wang Z-M, Liu P et al (2015) Doxorubicin-loaded glycyrrhetinic acid modified recombinant human serum albumin nanoparticles for targeting liver tumor chemotherapy. Mol Pharm 12(3):675–683

Rahman ANA, Shakweer MS, Algharib SA, Abdelaty AI, Kamel S, Ismail TA et al (2022) Silica nanoparticles acute toxicity alters ethology, neuro-stress indices, and physiological status of African catfish (Clarias gariepinus). Aquacult Rep 23:101034

Rai PK, Kumar V, Lee S, Raza N, Kim K-H, Ok YS, Tsang DC (2018) Nanoparticle-plant interaction: implications in energy, environment, and agriculture. Environ Int 119:1–19

Rajput VD, Minkina TM, Behal A, Sushkova SN, Mandzhieva S, Singh R et al (2018) Effects of zinc-oxide nanoparticles on soil, plants, animals and soil organisms: a review. Environ Nanotechnol Monitor Manag 9:76–84

Rana A, Yadav K, Jagadevan S (2020) A comprehensive review on green synthesis of nature-inspired metal nanoparticles: mechanism, application and toxicity. J Clean Prod 272:122880

Rather MA, Dhandare BC (2019) Genome-wide identification of doublesex and Mab-3-related transcription factor (DMRT) genes in nile tilapia (Oreochromis niloticus). Biotechnol Rep 24: e00398

Rather MA, Sharma R, Aklakur M, Ahmad S, Kumar N, Khan M, Ramya VL (2011) Nanotechnology: a novel tool for aquaculture and fisheries development. A prospective mini-review. Fish Aquac J 16(1–5):3

Rather MA, Bhat IA, Sharma N, Sharma R (2018) Molecular and cellular toxicology of nanomaterials with related to aquatic organisms. Cell Mol Toxicol Nanopart 1048:263–284

Rather MA, Dutta S, Guttula PK, Dhandare BC, Yusufzai S, Zafar MI (2020) Structural analysis, molecular docking and molecular dynamics simulations of G-protein-coupled receptor (kisspeptin) in fish. J Biomol Struct Dyn 38(8):2422–2439

Rayan RA, Zafar I (2021) Monitoring technologies for precision health. In: The smart cyber ecosystem for sustainable development, pp 251–260

Rayan RA, Zafar I, Tsagkaris C (2021) Artificial intelligence and big data solutions for COVID-19. In: Intelligent data analysis for COVID-19 pandemic. Springer, pp 115–127

Rayan RA, Tsagkaris C, Zafar I, Moysidis DV, Papazoglou AS (2022) Big data analytics for health: a comprehensive review of techniques and applications. In: Big data analytics for healthcare, pp 83–92

Refsnider JM, Garcia JA, Holliker B, Hulbert AC, Nunez A, Streby HM (2021) Effects of harmful algal blooms on stress levels and immune functioning in wetland-associated songbirds and reptiles. Sci Total Environ 788:147790

Rhim J-W, Park H-M, Ha C-S (2013) Bio-nanocomposites for food packaging applications. Prog Polym Sci 38(10–11):1629–1652

Roberta P, Elena Maria S, Carmelo I, Fabiano C, Maria Teresa R, Sara I et al (2021) Toxicological assessment of CeO_2 nanoparticles on early development of zebrafish. Toxicol Res 10(3): 570–578

Sajid M, Ilyas M, Basheer C, Tariq M, Daud M, Baig N, Shehzad F (2015) Impact of nanoparticles on human and environment: review of toxicity factors, exposures, control strategies, and future prospects. Environ Sci Pollut Res 22(6):4122–4143

Sarasamma S, Audira G, Samikannu P, Juniardi S, Siregar P, Hao E et al (2019) Behavioral impairments and oxidative stress in the brain, muscle, and gill caused by chronic exposure of C70 nanoparticles on adult zebrafish. Int J Mol Sci 20(22):5795

Sarkar B, Mahanty A, Gupta SK, Choudhury AR, Daware A, Bhattacharjee S (2022) Nanotechnology: a next-generation tool for sustainable aquaculture. Aquaculture 546:737330

Saxena P, Sangela V, Ranjan S, Dutta V, Dasgupta N, Phulwaria M, Rathore DS (2020) Aquatic nanotoxicology: impact of carbon nanomaterials on algal flora. Energy Ecol Environ 5(4): 240–252

Schultz AG, Boyle D, Chamot D, Ong KJ, Wilkinson KJ, McGeer JC et al (2014) Aquatic toxicity of manufactured nanomaterials: challenges and recommendations for future toxicity testing. Environ Chem 11(3):207–226

Selck H, Handy RD, Fernandes TF, Klaine SJ, Petersen EJ (2016) Nanomaterials in the aquatic environment: a European Union–United States perspective on the status of ecotoxicity testing, research priorities, and challenges ahead. Environ Toxicol Chem 35(5):1055–1067

Shah BR, Mraz J (2020) Advances in nanotechnology for sustainable aquaculture and fisheries. Rev Aquac 12(2):925–942

Sharma N, Rather MA, Ajima MN, Gireesh-Babu P, Kumar K, Sharma R (2016) Assessment of DNA damage and molecular responses in Labeo rohita (Hamilton, 1822) following short-term exposure to silver nanoparticles. Food Chem Toxicol 96:122–132

Shaw BJ, Handy RD (2011) Physiological effects of nanoparticles on fish: a comparison of nanometals versus metal ions. Environ Int 37(6):1083–1097

Shaw BJ, Al-Bairuty G, Handy RD (2012) Effects of waterborne copper nanoparticles and copper sulphate on rainbow trout, (Oncorhynchus mykiss): physiology and accumulation. Aquat Toxicol 116:90–101

Shokry A, Khalil M, Ibrahim H, Soliman M, Ebrahim S (2021) Acute toxicity assessment of polyaniline/Ag nanoparticles/graphene oxide quantum dots on Cypridopsis vidua and Artemia Salina. Sci Rep 11(1):1–9

Singh N, Onuegbu CU (2020) Nano-minerals as livestock feed additives. In: 21st century nanoscience–a handbook. CRC Press, p 20-21-20-12

Singh N, Manshian B, Jenkins GJ, Griffiths SM, Williams PM, Maffeis TG et al (2009) NanoGenotoxicology: the DNA damaging potential of engineered nanomaterials. Biomaterials 30(23–24):3891–3914

Smith CJ, Shaw BJ, Handy RD (2007) Toxicity of single walled carbon nanotubes to rainbow trout, (Oncorhynchus mykiss): respiratory toxicity, organ pathologies, and other physiological effects. Aquat Toxicol 82(2):94–109

Souza VGL, Fernando AL (2016) Nanoparticles in food packaging: biodegradability and potential migration to food—A review. Food Packag Shelf Life 8:63–70

Stanková R (2015) Combined effects of silver nanoparticles and 17α-ethinyl estradiol on turbot (Scophthalmus maximus) NTNU

Sundaray JK, Dixit S, Rather A, Rasal KD, Sahoo L (2022) Aquaculture omics: an update on the current status of research and data analysis. Mar Genomics 64:100967

Tabassum H, Ashafaq M, Khan J, Shah MZ, Raisuddin S, Parvez S (2016) Short term exposure of pendimethalin induces biochemical and histological perturbations in liver, kidney and gill of freshwater fish. Ecol Indic 63:29–36

Tamayo-Belda M, Pulido-Reyes G, Rosal R, Fernández-Piñas F (2022) Nanoplastic toxicity towards freshwater organisms. Water Emerg Contamin Nanoplast 1(4):19

Tetu S, Paulsen IT, Gillings MR (2017) Pangsatabam Kamala Devi

Thanigaivel S, Vickram A, Anbarasu K, Gulothungan G, Nanmaran R, Vignesh D et al (2021) Ecotoxicological assessment and dermal layer interactions of nanoparticle and its routes of penetrations. Saudi J Biolog Sci 28(9):5168–5174

Thines R, Mubarak N, Nizamuddin S, Sahu J, Abdullah E, Ganesan P (2017) Application potential of carbon nanomaterials in water and wastewater treatment: a review. J Taiwan Inst Chem Eng 72:116–133

Tijani JO, Fatoba OO, Babajide OO, Petrik LF (2016) Pharmaceuticals, endocrine disruptors, personal care products, nanomaterials and perfluorinated pollutants: a review. Environ Chem Lett 14(1):27–49

Vicario-Parés U, Lacave JM, Reip P, Cajaraville MP, Orbea A (2018) Cellular and molecular responses of adult zebrafish after exposure to CuO nanoparticles or ionic copper. Ecotoxicology 27(1):89–101

Wang H, Wick RL, Xing B (2009) Toxicity of nanoparticulate and bulk ZnO, Al_2O_3 and TiO_2 to the nematode Caenorhabditis elegans. Environ Pollut 157(4):1171–1177

Wang Z, Yin L, Zhao J, Xing B (2016) Trophic transfer and accumulation of TiO_2 nanoparticles from clamworm (Perinereis aibuhitensis) to juvenile turbot (Scophthalmus maximus) along a marine benthic food chain. Water Res 95:250–259

Wang Q, Wang X, Wang X, Feng R, Luo Q, Huang J (2018) Generation of a novel Streptococcus agalactiae ghost vaccine and examination of its immunogenicity against virulent challenge in tilapia. Fish Shellfish Immunol 81:49–56

Wang Y-L, Lee Y-H, Chiu I-J, Lin Y-F, Chiu H-W (2020) Potent impact of plastic nanomaterials and micromaterials on the food chain and human health. Int J Mol Sci 21(5):1727

Waris AA, Athar T, Fatima H, Nisar M (2021) Nanotoxicology-toxicology of nanomaterials and incidental nanomaterials. In: Nano-materials: synthesis, characterisation, hazards and safety. Elsevier, pp 123–143

Wiench K, Wohlleben W, Hisgen V, Radke K, Salinas E, Zok S, Landsiedel R (2009) Acute and chronic effects of nano-and non-nano-scale TiO_2 and ZnO particles on mobility and reproduction of the freshwater invertebrate Daphnia magna. Chemosphere 76(10):1356–1365

Wu P, Yan X-P (2013) Doped quantum dots for chemo/biosensing and bioimaging. Chem Soc Rev 42(12):5489–5521

Wu F, Bortvedt A, Harper BJ, Crandon LE, Harper SL (2017) Uptake and toxicity of CuO nanoparticles to daphnia magna varies between indirect dietary and direct waterborne exposures. Aquat Toxicol 190:78–86

Wu F, Harper BJ, Harper SL (2019) Comparative dissolution, uptake, and toxicity of zinc oxide particles in individual aquatic species and mixed populations. Environ Toxicol Chem 38(3): 591–602

Wu J, Cui X, Ke PC, Mortimer M, Wang X, Bao L, Chen C (2021) Nanomaterials as novel agents for amelioration of Parkinson’s disease. Nano Today 41:101328

Yang L, Luo X-B, Luo S-L (2019) Assessment on toxicity of nanomaterials. In: Nano-materials for the removal of pollutants and resource reutilization. Elsevier, pp 273–292

Yoon H, Pangging M, Jang M-H, Hwang YS, Chang Y-S (2018) Impact of surface modification on the toxicity of zerovalent iron nanoparticles in aquatic and terrestrial organisms. Ecotoxicol Environ Saf 163:436–443

Yu G, Wang X, Liu J, Jiang P, You S, Ding N et al (2021) Applications of nanomaterials for heavy metal removal from water and soil: a review. Sustainability 13(2):713

Zafar I, Iftikhar R, Ahmad SU, Rather MA (2021a) Genome wide identification, phylogeny, and synteny analysis of sox gene family in common carp (Cyprinus carpio). Biotechnol Rep 30: e00607

Zafar I, Rather M, Ain Q, Rayan R (2021b) Precision medicine and omics in the context of deep learning. In: Deep learning for biomedical applications. CRC Press, pp 1–19

Zafar I, Saba M, Raza MA, Rather MA, Rayan RA, Fatima M, Ishneiwra RM (2021c) Roles of bioinformatics in reproductive science. In: Recent updates in molecular endocrinology and reproductive physiology of fish. Springer, pp 203–228

Zhang W (2011) Characterizing, imaging, and quantifying the environmental behavior and biological interactions of metal-based nanoparticles. Georgia Institute of Technology

Zhang S, Ren S, Pei L, Sun Y, Wang F (2022) Ecotoxicological effects of polyethylene microplastics and ZnO nanoparticles on earthworm Eisenia fetida. Appl Soil Ecol 176:104469

Zhao X, Wang S, Wu Y, You H, Lv L (2013) Acute ZnO nanoparticles exposure induces developmental toxicity, oxidative stress and DNA damage in embryo-larval zebrafish. Aquat Toxicol 136:49–59

Zhao Y, Xue B, Bi C, Ren X, Liu Y (2022) Influence mechanisms of macro-infrastructure on micro-environments in the recirculating aquaculture system and biofloc technology system. Rev Aquac

Zheng Y, Nowack B (2022) Meta-analysis of bioaccumulation data for nondissolvable engineered nanomaterials in freshwater aquatic organisms. Environ Toxicol Chem 41(5):1202–1214

Microplastic Contamination in Aquatic Organisms: An Ecotoxicological Perspective

Udai Ram Gurjar, Suman Takar, Adnan Amin, and K. A. Martin Xavier

1 Introduction

The majority of the plastic debris present in aquatic environments are microplastics (Eriksen et al. 2013). Plastics have become the prime marine pollutants, contributing to approximately 80% of all global marine debris. Several studies have estimated that eight million tonnes of plastic products enter the sea each year. Plastic waste is expected to increase from 50 million tonnes in 2015 to 150 million tonnes by 2025 (Jambeck et al. 2015). With the rising global population and changes in lifestyle, humans are producing huge amounts of plastic waste. Just 192 coastal countries produced approximately 275 million tonnes of plastic discards in 2010. This figure is expected to rise by an order of magnitude by 2025 (Jambeck et al. 2015). In litter, a variety of different sizes and shapes of plastic debris have been identified. Macroplastics (>200 mm) are easily visible to the human eye, including large items like discarded water bottles, plastic bags, and food containers. Mesoplastics are slightly smaller in size (5–200 mm), though typically still visible with relative

U. R. Gurjar (✉)
DRPCAU-Krishi Vigyan Kendra, East Champaran, Bihar, India

Fisheries Resource Harvest and Post-Harvest Management Division, ICAR-Central Institute of Fisheries Education, Mumbai, Maharashtra, India

S. Takar
The Neotia University, Kolkata, West Bengal, India

A. Amin
Division of Aquatic Environmental Management, Sher-e-Kashmir University of Agricultural Sciences and Technology, Kashmir, Faculty of Fisheries Rangil, Ganderbal, Jammu and Kashmir, India

K. A. Martin Xavier
Fisheries Resource Harvest and Post-Harvest Management Division, ICAR-Central Institute of Fisheries Education, Mumbai, Maharashtra, India

M. A. Rather et al. (eds.), *Xenobiotics in Aquatic Animals*,
https://doi.org/10.1007/978-981-99-1214-8_17

ease (Eriksen et al. 2013). Microplastics are defined as plastic particles less than five millimetres in size or their largest dimension (Arthur et al. 2009). It is estimated that roughly 5.25 trillion floating particles of microplastics were present on ocean surfaces globally in 2013 (Eriksen et al. 2013). Humans are using the rivers and oceans as garbage cans or dumping grounds; therefore, global freshwater and coastal water resources are becoming polluted (Sharma and Chatterjee 2017). Microplastics are a global concern, affecting aquatic (freshwater and marine), terrestrial, and distant arctic habitats, consequently influencing numerous lifeforms (de Sá et al. 2018; Gurjar et al. 2021a). Plastic debris is found in all over the global environment due to its widespread use and durability of synthetic polymers (Duis and Coors 2016). Globally, several commercially important aquatic organisms ingested microplastic litter (GESAMP 2016; UNEP 2016). Therefore, it is essential to know the microplastic interactions with flora and fauna at an individual level. Numerous studies have observed the availability of microplastics in the gastrointestinal tracts of different fish species out of which many are commercially important (Rochman et al. 2015; Karuppasamy et al. 2020; Robin et al. 2020; Gurjar et al. 2022). The ingestion of MP items by organisms leads to reduce food consumption and fitness. In the case of fish, they can injure gills and block the intestinal tracts. The current studies show that additives such as phthalates, BPA, PBDEs, nonylphenol, and antioxidants can leach out from microrubber or plastic products into the aquatic resources (Amelia et al. 2021), which poses serious health issues to aquatic animals such as endocrine-disrupting, gene alteration, toxicity, and DNA damage (Chen et al. 2019). Microplastics pose threats to aquatic organisms as well as human life; therefore, it is necessary to stop the overuse of plastic products and implement new policies to monitor the plastic litter sources (Amelia et al. 2021; Prata 2018). Chronic biological effects are observed in various aquatic biota due to accumulation of MPs in their tissues (Hamed et al. 2020; Chen et al. 2019). MPs can be loaded with a wide range of substances to meet industry and consumer demand (preservatives, additives). Furthermore, these MPs act as vectors and have the ability to adsorb toxic chemicals in the environment. As a result, this chapter discusses some of the most important aspects of microplastics such as sources, eco-toxicological effects on aquatic organisms, and their management.

2 Sources of Microplastics

Microplastics in the environment are a very diverse group of particles that vary in shape, size, chemical composition, colours, and density. Microplastics in the environment come from various sources. Land-based sources of MPs contributed nearly 80% of the total plastic waste in the oceanic environments. Industrial, domestic, and coastal activities are the primary paths for entering the plastic items in the aquatic ecosystem (Derraik 2002). MP pollution in the aquatic environment is contributed mostly by the plastics being disposed into the river system, which ultimately makes its way to the ocean water and thus degrades within a given time period. It has been assessed that rivers transported 70–80% of plastic to marine ecosystems. The creation of plastic items from the manufacturing units (Lechner et al. 2014; Sadri and Thompson 2014), leaking of microbeads and plastic powders produced by air-blasting techniques (Claessens et al. 2011; Zbyszewski et al. 2014), fishing activities, aquaculture industries, coastal tourism, and careless usage of fishing gears are further sources of microplastic contamination in the marine and coastal habitat (Desforges et al. 2014; Gurjar et al. 2022). Hence, marine resources are mostly contaminated by plastics that enter the coastal waters through wind, rivers, canals, wastewater, and industrial discharge (Thompson 2006; Moore 2008). There are several sources of MPs and nanoplastics in marine environments such as toothpaste, cosmetic products, face wash, washing products, and hand cleansers, which enter the water canals via industrial drainage and domestic discharges (Derraik 2002; Fendall and Sewell 2009; Duis and Coors 2016).

2.1 *Primary Microplastics*

Primary microplastics have been generally described as plastics generated or released into the environment in lengths ranging from a few micrometres to 5 mm (Lassen et al. 2012). MPs are utilised as exfoliants in a variety of personal care products, including toothpaste, and facial and hand cleaners. Primary microplastics can persist in their original form or can break down into secondary microplastics. Based on chemical contents, these primary microplastics are generated by the unintended release of intermediary plastic raw material (nurdles, mermaid tears, pellets) and arise as by-products of various events such as particles released from maintenance of plastic items, industrial manufacture units, fibres, and dust (GESAMP 2015). The primary microplastics' entry route into the environment is primarily determined by their application: microplastic particles from cosmetic products generally enter the environment via wastewater, whereas primary MPs used as raw materials for various products may enter into the environment via unintentional loss during transportation and transhipment or waste runoff from various industrial plants. When primary microplastics are too small to be retained by sewage treatment plants, they either enter the oceans directly or indirectly via rivers and streams.

2.2 *Secondary Microplastics*

Secondary microplastics are formed through the fragmentation of larger plastic materials (Lassen et al. 2012). These microplastics are formed when larger plastic products degrade or break into tiny pieces. It is estimated that approximately 75–90% of the plastic waste in the marine ecosystem comes from land-based sources, with the remaining 10–25% coming from ocean-based sources (Mehlhart and Blepp 2012).

The following are the ways for the entry of secondary microplastics into the natural habitat, such as (1) fibres from the textile industries might arrive through waste discharge following the washing and drying of clothes; (2) weathering of plastic items utilised in the agriculture sector may arrive through the surface soil runoff; (3) microplastics generated during abrasion of tires arrive through surface runoff and air; (4) weathering and breakdown of different substances in landfills by photodegradation may lead microplastics into the rivers, ocean, and atmosphere via surface runoff and wind forces; and (5) fragmentation and breakdown of plastic debris along the coastline may endure in coastal beaches and sediments or be further transferred to offshore waters (Browne et al. 2011; Napper and Thompson 2016).

3 Microplastics in Aquatic Organisms

Inland ecosystems can act as receivers, sinks, and transporter of plastic pollution. Rainfall is a key source of environmental microplastics pollution that brings inland water resources. The marine resources are affected by plastic pollution due to terrestrial runoff, industrial waste, domestic and laundry waste through rivers, or direct discharge. Anthropogenic activities in the coastal area, such as recreation, fishing, and the marine industry, contribute significantly to MP pollution in marine resources. Marine snow is an important mechanism for transporting microplastics from the ocean surface to deep pelagic and mesopelagic zones, and it may also increase their bioavailability to benthic biota.

Microplastics enter the terrestrial environment, and soils have already been assumed to be a sink for MPs. Most consumer plastics are initially buoyant and abundant on the surface of water, or the top 20 cm of the water column, which has been addressed in numerous studies (Duis and Coors 2016). MP concentrations were maximum in the most populated and tourism parts of the lake-friendly area. A parallel macroplastic survey revealed that household plastics items (plastic bags and bottles) and fishing gear were the most prevalent macroplastic substances at shore areas of the lakes.

Microplastics in the ocean can absorb and release toxic substances and are ingested by marine biotas (Amelia et al. 2021). Worldwide, more than 220 aquatic species have been recorded to ingest microplastic litter; out of this, 58% are commercially important species (GESAMP 2016; UNEP 2016). Therefore, it is essential to know the microplastic interactions with flora and fauna at an individual level. Globally, numerous authors have observed the availability of MPs in the gastrointestinal tracts of wild and commercially important fish and shellfishes species from the North Sea (Foekema et al. 2013), English Channel (Lusher et al. 2013), Indo-Pacific Ocean (Rochman et al. 2015), the Baltic Sea (Rummel et al. 2016), the Adriatic Sea (Anastasopoulou et al. 2018; Pellini et al. 2018), the Mediterranean Sea (Nadal et al. 2016; Güven et al. 2017), the East China Sea and

South China Sea (Jabeen et al. 2017); the North Eastern Atlantic (Neves et al. 2015), Bay of Bengal (Hossain et al. 2019), and Indian waters (Karuppasamy et al. 2020; Robin et al. 2020; Saha et al. 2021; Gurjar et al. 2021b, 2022). The percentage occurrence of MPs in fish varied from 2.6% in the North Sea (Foekema et al. 2013) to 8.95% in the southeast Bay of Bengal (Karuppasamy et al. 2020), 18% in the Central Mediterranean (Romeo et al. 2015), 21.40% in the southwest coast of India (Robin et al. 2020), 35% in the North Pacific Central Gyre (Boerger et al. 2010), 41.1% in Kerala waters (Daniel et al. 2020), 58% in the Mediterranean Sea (Nadal et al. 2016; Güven et al. 2017), 77% in Tokyo Bay (Tanaka and Takada 2016), 95% in the Adriatic sea (Pellini et al. 2018), and 100% in the East China Sea and South China Sea, R'ıo de la Plata, Adriatic Sea, Mumbai (Arabian sea) waters (Jabeen et al. 2017; Pazos et al. 2017; Anastasopoulou et al. 2018; Gurjar et al. 2022). Remarkably, microplastics were found in samples collected from fish markets, including 25% and 28% of fish caught off the west coast of the United States and Indonesian waters, respectively (Rochman et al. 2015).

Similarly, MPs have been found within 20% of individual fish species (*Oreochromis niloticus* and *Lates niloticus*) bought from the Tanzania fish market originating from Lake Victoria (Biginagwa et al. 2016). Jabeen et al. (2017) collected 27 fish species from Shanghai fish markets originating from freshwater Lake Taihu and marine waters (East China Sea, South China Sea, and Yangtze Estuary) and found that microplastic ingestion varied from 1.1 to 7.2 particles/individual. Wild fish larvae of commercially valued species harvested from the English Channel have also been observed with microplastic ingestion (Steer et al. 2017). However, many commercial fish species are ingested with microplastics, still limited information about their impact on consumption is available. Microplastics can be retained in the digestive tract or egested with faecal matter.

Various studies have revealed the connections between newly introduced plastic materials and aquatic animals (Wright et al. 2013). Different shapes and sizes of microplastics have been ingested by filter feeders, detritus feeders, corals, as well as predators (Browne et al. 2008, 2013; Cole et al. 2013; Lusher et al. 2013; Setala et al. 2014; Carlos de Sa et al. 2015; Hall et al. 2015; Sussarellu et al. 2016; Gurjar et al. 2022). Hence, there is no doubt that several aquatic organisms are consuming MPs. Fish are the most vulnerable aquatic faunas to accidental microplastic ingestion. In general, MPs are consumed directly at the lower trophic level and then pass through the food chain to reach higher trophic levels (Nelms et al. 2018).

In nature, there are numerous ways through which incidental ingestion of MPs happens such as trophic transfer, being swallowed with natural food, or predation (Cedervall et al. 2012; Peters and Bratton 2016). Deliberate ingestion occurs when fish falsely ingest plastic as food (do Sul and Costa 2007). MPs' incidence or ingestion was noticed in several internal organs of aquatic organisms, such as guts, gills, muscles, livers, brain, and gonads (Ding et al. 2018; Su et al. 2019; Wang et al. 2019). Some fishes regularly feed on microplastic items existing in the natural environment (Carson et al. 2013). Welden and Cowie (2016) have observed microfibres in the GI tract of lobsters from North and West Scotland. Microplastic ingestion is also observed in pelagic fishes such as European sardines and anchovy

(Pennino et al. 2020). MPs have been recorded in killer whales (top marine predators) from the North Pacific (Harlacher 2020) and demersal sharks of the North-East Atlantic (Parton et al. 2020). Therefore, it might have a greater chance of trophic transfer of microplastics due to the predatory species feeding upon plastic-contaminated prey items (Lusher et al. 2016). But still there is no conclusive proof of MP bio-magnification in situ at the higher trophic species, although bioaccumulation of MPs was observed within trophic levels (Miller et al. 2020).

4 Eco-Toxicological Effects of Microplastics

Microplastics have various environmental effects, including physical (mechanical) effects on organisms, acting as vectors for hydrophobic pollutants and substrates for organisms, and influencing sediment (soil) properties (Duis and Coors 2016). Aquatic animals impacted by the microplastic particles trigger global awareness and concern adequately. Consumption of MP items may reduce the amount of food consumed as well as the organism's health. They can also cause intestinal blockages and damage to gill tissues and intestinal tracts. Recent studies show that additives such as phthalates, PBDEs, BPA, nonylphenol, and antioxidants can leach out from plastic or rubber items into the aquatic resources (Amelia et al. 2021), which poses serious health issues (endocrine-disrupting, DNA damage, toxicity, gene alteration) to aquatic animals (Chen et al. 2019). Recognising that images of marine animals impacted by the negative effects of plastic particles raise global awareness and concern, a growing number of researchers think that plastic items pose a more significant threat to aquatic environments and, potentially, human health (Derraik 2002; Thompson et al. 2004; Ng and Obbard 2006; Barnes et al. 2009; Fendall and Sewell 2009; Lozano and Mouat 2009). As the abundance and pervasiveness of microplastics in both pelagic and benthic ecosystems increase, the probability of encountering and interacting with these small particles with aquatic biota will increase. Globally, in different studies authors noticed the ecotoxicological effects of microplastics on different aquatic organisms such as fish, crustaceans, molluscs, Annelid, Echinoderms, and rotifers. These groups play a significant role in aquatic food chain and webs (de Sá et al. 2018). Because of their feeding behaviour and habitat, these organisms are likely to be affected by plastic pollution. Also, several of these organisms are extensively used for human food; therefore, they are a probable source of MPs' contaminants for humans. MPs have been shown to cause oxidative stress, reduce feeding activity, neurotoxicity, reduce growth, reduce reproductive fitness, genotoxicity, and ultimately cause death in aquatic organisms (Table 1).

In laboratories studies, authors have observed the ecotoxicological effects of nanoplastics and microplastics on fishes, such as disturbing organ and tissue damage, intestinal permeability, disorders of the intestinal microbiome, immune and behavioural functions, reproductive potentials, metabolism activities, and even brain (Gu et al. 2020; Qiao et al. 2019; Pitt et al. 2018; Lei et al. 2018; Jacob et al. 2020). Yang et al. (2020) described that polystyrene microplastics with the size of 50 μm

Table 1 Effects of microplastics on aquatic organisms in various ecosystems

Habitat	Plastic polymer	Organisms	Findings	References
Marine water	Polyethylene	*Mytilus galloprovincialis*	Immune response, genotoxicity, and oxidative stress-related toxic effects	Avio et al. (2015)
Marine water	Polyethylene	*Pomatoschistus microps*	Reduced predatory performance and efficiency	de Sá et al. (2015)
Estuarine water	Polyethylene	*Pomatoschistus microps*	Neurotoxicity	Luis et al. (2015)
Marine water	Polyethylene	*Dicentrarchus labrax*	Mortality and induction of the cytochrome P450	Mazurais et al. (2015)
Freshwater (lab experiment)	Polystyrene and polyethylene	*Danio rerio*	Immunity, behavioural activities, and transcriptional processes are all altered	Limonta et al. (2019)
Freshwater	*Polyester*	*Gambusia holbrooki*	MPs in head and body parts	Su et al. (2019)
Marine water	Polystyrene	*Sebastes schlegelii*	Increased oxygen consumption and ammonia; reduced growth and swimming speed	Yin et al. (2019)
Marine water (lab experiment)	Polystyrene	*Girella laevifrons*	Leukocyte diffusion, hyperaemia, and crypt cell degeneration	Ahrendt et al. (2020)
Marine water	Polyester and polyethylene	*Dicentrarchus labrax, Scomber colias, Trachurus treacherous*	Neurotoxicity as well as oxidative damage	Barboza et al. (2020)
Freshwater (laboratory experiment)	Microplastics	*Oreochromis niloticus*	Oxidative stress and DNA damage, extreme production of reactive oxygen species and change in the antioxidant parameters	Hamed et al. (2020)
Freshwater (reservoir)	Microplastics	*Micropterus salmoides, Dorosoma cepedianum*	Alteration in physiological functions and feeding	Hurt et al. (2020)
Freshwater (lab experiment)	Polystyrene	*Salmo trutta*	Changes in resting habits	Schmieg et al. (2020)
Freshwater (lab experiment)	6PPD and 6PPD quinone	*Danio rerio*	Induce developmental behavioural, and cardiotoxicity, oxidative stress, locomotor activity	Varshney et al. (2022)

deposited in the intestinal tracts of *Carassius auratus* larvae could lead to oxidative stress, guts, liver damage, gills damage, and obstruct the growth, and nanoplastics (NPs) can enter into the muscle tissues through the epidermis of larvae, consequently having adverse effects. In crucian carp, even the NP materials pass through blood to

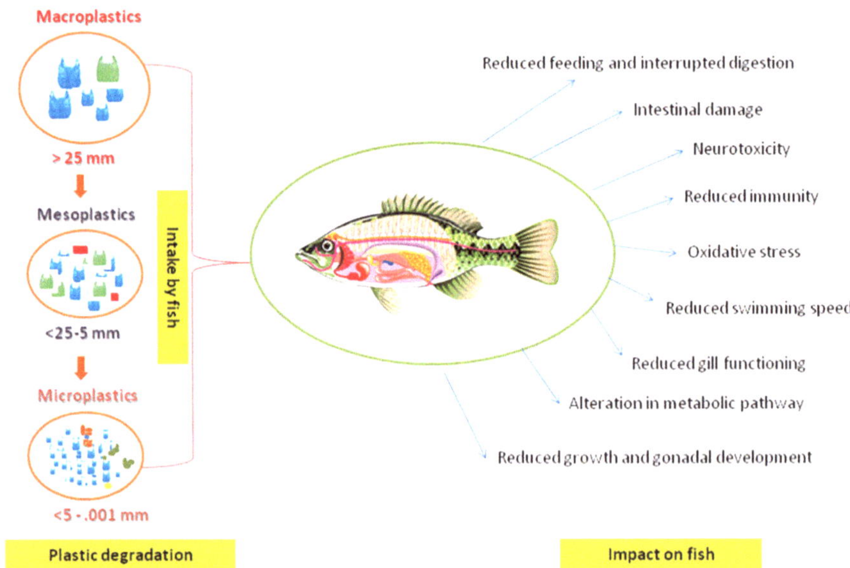

Fig. 1 Impact of microplastic on fish health

the brain, which can cause brain injury and behaviour dysfunction (Kashiwada 2006; Mattsson et al. 2017). The collective effects of MPs and other associated stressors (e.g. absorbed chemicals, temperature, heavy metals, and nanoparticles) on fish were too reported (Ferreira et al. 2016). Furthermore, microplastics are an issue for filter–feeder megafaunas such as baleen whales and sharks since they would like to filter a huge quantity of water per day to fulfil their nutrition and food requirements (Germanov et al. 2018). In studies, authors have observed the incidence of microplastics with their associated pollutants (such as persistent organic pollutants, and additive phthalates) in baleen whales, *Balaenoptera physalus*, implying direct MP ingestion and feeding on polluted prey items through filter-feeding (Fossi et al. 2012, 2014, 2016). Recent studies have evidenced that additives like phthalates, PBDEs, BPA, nonylphenol, and antioxidants can leach out from microrubber or plastic materials into the aquatic ecosystems (Liu et al. 2019, Chen et al. 2019, Khaled et al. 2018, Paluselli et al. 2018, Turner et al. 2020), which poses serious health issues (endocrine-disrupting, DNA damage, toxicity, gene alteration) to aquatic animals (Chen et al. 2019; Boyle et al. 2020; Kolomijeca et al. 2020; Capolupo et al. 2020; Roda et al. 2020; Oliviero et al. 2019; Pikuda et al. 2018). Microplastics can provide a possible route for the transfer of absorbed contaminants into the tissue of aquatic organisms, causing a health risk (Fig. 1). Though the combined risks of MPs and allied pollutants may be species-specific and chemical-specific (Teuten et al. 2009; Browne et al. 2013; Bakir et al. 2016; Campanale et al. 2020; Amelia et al. 2021).

5 Management of Microplastics

The impact of three comprehensive management strategies, such as plastic waste management, waste reduction, and environmental recovery, has been evaluated at various levels of effort to estimate plastic emissions in 173 countries by 2030 (Borrelle et al. 2020). Stopping overuse, increasing awareness and behavioural changes in people, improving waste separation technology, expanding the measures of recycling, and reusing plastic products are the simplest and superb ways to reduce or mitigate plastic pollution. People need to develop household-based systems that prevent the entry of microplastics into the sewer lines. Pollution caused by microplastics can be reduced by reducing the sources of pollution, implementing appropriate waste management plans, and integrating plastic into the circular economy. To reduce MPs in the environment, we need to develop infrastructure facilities, waste valorisation, and cost-effective plastic waste management techniques and possible substitutes. Non-recyclable plastics can produce energy, transforming plastic waste into a usable and sustainable source of energy. Plastic litter discharge into environments can be considerably reduced by implementing appropriate waste management activities, extending the shelf life of plastic items, and raising awareness among people, allowing the ocean life to be restored (Mallik et al. 2021).

6 Conclusions

Microplastics are a significant problem in the aquatic environment. A large amount of research in this field, along with the large volume of results describing the problem of MPs and their effects on fish and aquatic life, has shed some light on this issue. MPs can cause stress, intestinal obstruction, and health problems, and more research is being conducted to determine the full potential risks of MPs in aquatic organisms. Plastics contain a large number of substances that can bioaccumulate throughout the trophic food chain. More studies are required to know the MPs' threshold level and its effects throughout the food chain or web, as well as studies that better manage these issues. More attention needs to be given to knowing the human health risks of consuming MP-contaminated seafood, and more emphasis should be placed on fish processing interventions to reduce MP contamination in seafood prior to consumption. From the above-discussed issues, it can be suggested that proper mitigation measures and enforcement of new laws or guidelines are required to control microplastic pollution in the aquatic environment. Increased awareness and education on marine litter by incorporating these aspects into educational curricula and outreach materials targeted to different age groups to promote behavioural change is needed.

References

Ahrendt C, Perez-Venegas DJ, Urbina M, Gonzalez C, Echeveste P, Aldana M, Pulgar J, Galbán-Malagón C (2020) Microplastic ingestion cause intestinal lesions in the intertidal fish *Girella laevifrons*. Mar Pollut Bull 151:110795

Amelia TSM, Khalik WMAWM, Ong MC, Shao YT, Pan HJ, Bhubalan K (2021) Marine microplastics as vectors of major ocean pollutants and its hazards to the marine ecosystem and humans. Prog Earth Planet Sci 8(1):1–26

Anastasopoulou A, Viršek MK, Varezić DB, Digka N, Fortibuoni T, Koren Š, Mandić M, Mytilineou C, Pešić A, Ronchi F, Šiljić J (2018) Assessment on marine litter ingested by fish in the Adriatic and NE Ionian Sea macro-region (Mediterranean). Mar Pollut Bull 133:841–851

Arthur C, Baker JE, Bamford HA (2009) Proceedings of the international research workshop on the occurrence, effects, and fate of microplastic marine debris, September 9–11, 2008. University of Washington Tacoma, Tacoma, WA

Avio CG, Gorbi S, Milan M, Benedetti M, Fattorini D, D'Errico G, Pauletto M, Bargelloni L, Regoli F (2015) Pollutants bioavailability and toxicological risk from microplastics to marine mussels. Environ Pollut 198:211–222

Bakir A, O'Connor IA, Rowland SJ, Hendriks AJ, Thompson RC (2016) Relative importance of microplastics as a pathway for the transfer of hydrophobic organic chemicals to marine life. Environ Pollut 219:56–65

Barboza LGA, Lopes C, Oliveira P, Bessa F, Otero V, Henriques B, Raimundo J, Caetano M, Vale C, Guilhermino L (2020) Microplastics in wild fish from north East Atlantic Ocean and its potential for causing neurotoxic effects, lipid oxidative damage, and human health risks associated with ingestion exposure. Sci Total Environ 717:134625

Barnes DK, Galgani F, Thompson RC, Barlaz M (2009) Accumulation and fragmentation of plastic debris in global environments. Philos Trans R Soc Lond B Biol Sci 364(1526):1985–1998

Biginagwa FJ, Mayoma BS, Shashoua Y, Syberg K, Khan FR (2016) First evidence of microplastics in the African Great Lakes: recovery from Lake Victoria Nile perch and Nile tilapia. J Great Lakes Res 42(1):146–149

Boerger CM, Lattin GL, Moore SL, Moore CJ (2010) Plastic ingestion by planktivorous fishes in the North Pacific central gyre. Mar Pollut Bull 60(12):2275–2278

Borrelle SB, Ringma J, Law KL, Monnahan CC, Lebreton L, McGivern A, Murphy E, Jambeck J, Leonard GH, Hilleary MA, Eriksen M (2020) Predicted growth in plastic waste exceeds efforts to mitigate plastic pollution. Science 369(6510):1515–1518

Boyle D, Catarino AI, Clark NJ, Henry TB (2020) Polyvinyl chloride (PVC) plastic fragments release Pb additives that are bioavailable in zebrafish. Environ Pollut 263:114422

Browne MA, Dissanayake A, Galloway TS, Lowe DM, Thompson RC (2008) Ingested microscopic plastic translocates to the circulatory system of the mussel, *Mytilus edulis* (L.). Environ Sci Technol 42(13):5026–5031

Browne MA, Crump P, Niven SJ, Teuten E, Tonkin A, Galloway T, Thompson R (2011) Accumulation of microplastic on shorelines woldwide: sources and sinks. Environ Sci Technol 45(21):9175–9179

Browne MA, Niven SJ, Galloway TS, Rowland SJ, Thompson RC (2013) Microplastic moves pollutants and additives to worms, reducing functions linked to health and biodiversity. Curr Biol 23(23):2388–2392

Campanale C, Massarelli C, Savino I, Locaputo V, Uricchio VF (2020) A detailed review study on potential effects of microplastics and additives of concern on human health. Int J Environ Res Public Health 17(4):1212

Capolupo M, Sørensen L, Jayasena KDR, Booth AM, Fabbri E (2020) Chemical composition and ecotoxicity of plastic and car tire rubber leachates to aquatic organisms. Water Res 169:115270

Carlos de Sa L, Luis LG, Guilhermino L (2015) Effects of microplastics on juveniles of the common goby (*Pomatoschistus microps*): confusion with prey, reduction of the predatory performance and efficiency, and possible influence of developmental conditions. Environ Pollut 196:359–362

Carson HS, Nerheim MS, Carroll KA, Eriksen M (2013) The plastic-associated microorganisms of the North Pacific gyre. Mar Pollut Bull 75(1–2):126–132

Cedervall T, Hansson LA, Lard M, Frohm B, Linse S (2012) Food chain transport of nanoparticles affects behaviour and fat metabolism in fish. PLoS One 7(2):e32254

Chen Q, Allgeier A, Yin D, Hollert H (2019) Leaching of endocrine disrupting chemicals from marine microplastics and mesoplastics under common life stress conditions. Environ Int 130: 104938

Claessens M, De Meester S, Van Landuyt L, De Clerck K, Janssen CR (2011) Occurrence and distribution of microplastics in marine sediments along the Belgian coast. Mar Pollut Bull 62(10):2199–2204

Cole M, Lindeque P, Fileman E, Halsband C, Goodhead R, Moger J, Galloway TS (2013) Microplastic ingestion by zooplankton. Environ Sci Technol 47(12):6646–6655

Daniel DB, Thomas SN, Thomson KT (2020) Assessment of fishing-related plastic debris along the beaches in Kerala coast, India. Mar Pollut Bull 150:110696

de Sá LC, Luís LG, Guilhermino L (2015) Effects of microplastics on juveniles of the common goby (*Pomatoschistus microps*): confusion with prey, reduction of the predatory performance and efficiency, and possible influence of developmental conditions. Environ Pollut 196:359–362

de Sá LC, Oliveira M, Ribeiro F, Rocha TL, Futter MN (2018) Studies of the effects of microplastics on aquatic organisms: what do we know and where should we focus our efforts in the future? Sci Total Environ 645:1029–1039

Derraik JG (2002) The pollution of the marine environment by plastic debris: a review. Mar Pollut Bull 44:842–852

Desforges JPW, Galbraith M, Dangerfield N, Ross PS (2014) Widespread distribution of microplastics in subsurface seawater in the NE Pacific Ocean. Mar Pollut Bull 79(1–2):94–99

Ding J, Zhang S, Razanajatovo RM, Zou H, Zhu W (2018) Accumulation, tissue distribution, and biochemical effects of polystyrene microplastics in the freshwater fish red tilapia (*Oreochromis niloticus*). Environ Pollut 238:1–9

do Sul JAI, Costa MF (2007) Marine debris review for Latin America and the wider Caribbean region: from the 1970s until now, and where do we go from here? Mar Pollut Bull 54(8): 1087–1104

Duis K, Coors A (2016) Microplastics in the aquatic and terrestrial environment: sources (with a specific focus on personal care products), fate and effects. Environ Sci Eur 28(1):1–25

Eriksen M, Mason S, Wilson S, Box C, Zellers A, Edwards W, Farley H, Amato S (2013) Microplastic pollution in the surface waters of the Laurentian Great Lakes. Mar Pollut Bull 77(1–2):177–182

Fendall LS, Sewell MA (2009) Contributing to marine pollution by washing your face: microplastics in facial cleansers. Mar Pollut Bull 58(8):1225–1228

Ferreira P, Fonte E, Soares ME, Carvalho F, Guilhermino L (2016) Effects of multi-stressors on juveniles of the marine fish *Pomatoschistus microps*: gold nanoparticles, microplastics and temperature. Aquat Toxicol 170:89–103

Foekema EM, De Gruijter C, Mergia MT, van Franeker JA, Murk AJ, Koelmans AA (2013) Plastic in North Sea fish. Environ Sci Technol 47(15):8818–8824

Fossi MC, Panti C, Guerranti C, Coppola D, Giannetti M, Marsili L, Minutoli R (2012) Are baleen whales exposed to the threat of microplastics? A case study of the Mediterranean fin whale (*Balaenoptera physalus*). Mar Pollut Bull 64(11):2374–2379

Fossi MC, Coppola D, Baini M, Giannetti M, Guerranti C, Marsili L, Panti C, de Sabata E, Clò S (2014) Large filter feeding marine organisms as indicators of microplastic in the pelagic environment: the case studies of the Mediterranean basking shark (*Cetorhinus maximus*) and fin whale (*Balaenoptera physalus*). Mar Environ Res 100:17–24

Fossi MC, Marsili L, Baini M, Giannetti M, Coppola D, Guerranti C, Caliani I, Minutoli R, Lauriano G, Finoia MG, Rubegni F (2016) Fin whales and microplastics: the Mediterranean Sea and the sea of Cortez scenarios. Environ Pollut 209:68–78

Germanov ES, Marshall AD, Bejder L, Fossi MC, Loneragan NR (2018) Microplastics: no small problem for filter-feeding megafauna. Trends Ecol Evol 33(4):227–232

GESAMP (2015) Sources, fate and effects of microplastics in the marine environment: part 2 of a global assessment. In: Kershaw PJ, Rochman CM (eds) Report study, vol 90. GESAMP, London, p 96

GESAMP (2016) Sources, fate and effects of microplastics in the marine environment: part 2 of a global assessment. Report Study, vol 93. IMO/FAO/UNESCO-IOC/ÚNIDO/WMO/IAEA/UN/UNEP/UNDP, London

Gu W, Liu S, Chen L, Liu Y, Gu C, Ren HQ, Wu B (2020) Single-cell RNA sequencing reveals size-dependent effects of polystyrene microplastics on immune and secretory cell populations from zebrafish intestines. Environ Sci Technol 54(6):3417–3427

Gurjar UR, Xavier KM, Shukla SP, Deshmukhe G, Jaiswar AK, Nayak BB (2021a) Incidence of microplastics in gastrointestinal tract of golden anchovy (*Coilia dussumieri*) from north east coast of Arabian Sea: the ecological perspective. Mar Pollut Bull 169:112518

Gurjar UR, Xavier M, Nayak BB, Ramteke K, Deshmukhe G, Jaiswar AK, Shukla SP (2021b) Microplastics in shrimps: a study from the trawling grounds of North Eastern part of Arabian Sea. Environ Sci Pollut Res 28(35):48494–48504

Gurjar UR, Xavier KM, Shukla SP, Jaiswar AK, Deshmukhe G, Nayak BB (2022) Microplastic pollution in coastal ecosystem off Mumbai coast, India. Chemosphere 288:132484

Güven O, Gökdağ K, Jovanović B, Kıdeyş AE (2017) Microplastic litter composition of the Turkish territorial waters of the Mediterranean Sea, and its occurrence in the gastrointestinal tract of fish. Environ Pollut 223:286–294

Hall J, Trent S, Thomas KL, O'Donovan MC, Owen MJ (2015) Genetic risk for schizophrenia: convergence on synaptic pathways involved in plasticity. Biol Psychiatry 77(1):52–58

Hamed M, Soliman HA, Osman AG, Sayed AE-DH (2020) Antioxidants and molecular damage in Nile tilapia (*Oreochromis niloticus*) after exposure to microplastics. Environ Sci Pollut Res 27: 1–8

Harlacher J (2020) Whale, what do we have here? Evidence of microplastics in top predators: analysis of two populations of resident killer whale fecal samples. University of Washington, Seattle

Hossain MS, Sobhan F, Uddin MN, Sharifuzzaman SM, Chowdhury SR, Sarker S, Chowdhury MSN (2019) Microplastics in fishes from the Northern Bay of Bengal. Sci Total Environ 690: 821–830

Hurt R, O'Reilly CM, Perry WL (2020) Microplastic prevalence in two fish species in two US reservoirs. Limnol Oceanogr Lett 5:147–153

Jabeen K, Su L, Li J, Yang D, Tong C, Mu J, Shi H (2017) Microplastics and mesoplastics in fish from coastal and fresh waters of China. Environ Pollut 221:141–149

Jacob H, Besson M, Swarzenski PW, Lecchini D, Metian M (2020) Effects of virgin micro-and nanoplastics on fish: trends, meta-analysis, and perspectives. Environ Sci Technol 54(8): 4733–4745

Jambeck JR, Geyer R, Wilcox C, Siegler TR, Perryman M, Andrady A, Narayan R, Law KL (2015) Plastic waste inputs from land into the ocean. Science 347(6223):768–771

Karuppasamy PK, Ravi A, Vasudevan L, Elangovan MP, Mary PD, Vincent SG, Palanisami T (2020) Baseline survey of micro and mesoplastics in the gastro-intestinal tract of commercial fish from southeast coast of the bay of Bengal. Mar Pollut Bull 153:110974

Kashiwada S (2006) Distribution of nanoparticles in the see-through medaka (*Oryzias latipes*). Environ Health Perspect 114(11):1697–1702

Khaled A, Rivaton A, Richard C, Jaber F, Sleiman M (2018) Phototransformation of plastic containing brominated flame retardants: enhanced fragmentation and release of photoproducts to water and air. Environ Sci Technol 52(19):11123–11131

Kolomijeca A, Parrott J, Khan H, Shires K, Clarence S, Sullivan C, Chibwe L, Sinton D, Rochman CM (2020) Increased temperature and turbulence alter the effects of leachates from tire particles on fathead minnow (*Pimephales promelas*). Environ Sci Technol 54(3):1750–1759

Lassen C, Hansen SF, Magnusson K, Norén F, Hartmann NIB, Jensen PR, Nielsen TG, Brinch A (2012) Microplastics-occurrence, effects and sources of releases to the environment in Denmark. Significance 2(2)

Lechner A, Keckeis H, Lumesberger-Loisl F, Zens B, Krusch R, Tritthart M, Glas M, Schludermann E (2014) The Danube so colourful: a potpourri of plastic litter outnumbers fish larvae in Europe's second largest river. Environ Pollut 188:177–181

Lei L, Wu S, Lu S, Liu M, Song Y, Fu Z, Shi H, Raley-Susman KM, He D (2018) Microplastic particles cause intestinal damage and other adverse effects in zebrafish *Danio rerio* and nematode *Caenorhabditis elegans*. Sci Total Environ 619:1–8

Limonta G, Mancia A, Benkhalqui A, Bertolucci C, Abelli L, Fossi MC, Panti C (2019) Microplastics induce transcriptional changes, immune response and behavioral alterations in adult zebrafish. Sci Rep 9:1–11

Liu X, Shi H, Xie B, Dionysiou DD, Zhao Y (2019) Microplastics as both a sink and a source of bisphenol a in the marine environment. Environ Sci Technol 53(17):10188–10196

Lozano LR, Mouat J (2009) Marine litter in the Northeast Atlantic region: assessment and priorities for response. OSPAR, London

Luis L, Ferreira P, Fonte E, Oliveira M, Guilhermino L (2015) Does the presence of microplastics influence the acute toxicity of chromium (VI) to early juveniles of the common goby (*Pomatoschistus microps*)? A study with juveniles from two wild estuarine populations. Aquat Toxicol 164:163–174

Lusher AL, Mchugh M, Thompson RC (2013) Occurrence of microplastics in the gastrointestinal tract of pelagic and demersal fish from the English Channel. Mar Pollut Bull 67(1–2):94–99

Lusher AL, O'Donnell C, Officer R, O'Connor I (2016) Microplastic interactions with North Atlantic mesopelagic fish. ICES J Mar Sci 73(4):1214–1225

Mallik A, Xavier KM, Naidu BC, Nayak BB (2021) Ecotoxicological and physiological risks of microplastics on fish and their possible mitigation measures. Sci Total Environ 779:146433

Mattsson K, Johnson EV, Malmendal A, Linse S, Hansson LA, Cedervall T (2017) Brain damage and behavioural disorders in fish induced by plastic nanoparticles delivered through the food chain. Sci Rep 7(1):1–7

Mazurais D, Emande B, Quazuguel P, Severe A, Huelvan C, Madec L, Mouchel O, Soudant P, Robbens J, Huvet A, Zambonino-Indante J (2015) Evaluation of the impact of polyethylene microbeads ingestion in European sea bass (*Dicentrarchus labrax*) larvae. Mar Pollut Bull 112 (Part A):78–85

Mehlhart G, Blepp M (2012) Study on land-sourced litter (LSL) in the marine environment: review of sources and literature. Institut für Angewandte Ökologie, Darmstadt

Miller ME, Hamann M, Kroon FJ (2020) Bioaccumulation and biomagnification of microplastics in marine organisms: a review and meta-analysis of current data. PLoS One 15(10):0240792

Moore CJ (2008) Synthetic polymers in the marine environment: a rapidly increasing, long-term threat. Environ Res 108(2):131–139

Nadal MA, Alomar C, Deudero S (2016) High levels of microplastic ingestion by the semipelagic fish Bogue *Boops boops* (L.) around the Balearic Islands. Environ Pollut 214:517–523

Napper IE, Thompson RC (2016) Release of synthetic microplastic plastic fibres from domestic washing machines: effects of fabric type and washing conditions. Mar Pollut Bull 112(1–2): 39–45

Nelms SE, Galloway TS, Godley BJ, Jarvis DS, Lindeque PK (2018) Investigating microplastic trophic transfer in marine top predators. Environ Pollut 238:999–1007

Neves D, Sobral P, Ferreira JL, Pereira T (2015) Ingestion of microplastics by commercial fish off the Portuguese coast. Mar Pollut Bull 101(1):119–126

Ng KL, Obbard JP (2006) Prevalence of microplastics in Singapore's coastal marine environment. Mar Pollut Bull 52(7):761–767

Oliviero M, Tato T, Schiavo S, Fernández V, Manzo S, Beiras R (2019) Leachates of micronized plastic toys provoke embryotoxic effects upon sea urchin *Paracentrotus lividus*. Environ Pollut 247:706–715

Paluselli A, Fauvelle V, Galgani F, Sempéré R (2018) Phthalate release from plastic fragments and degradation in seawater. Environ Sci Technol 53(1):166–175

Parton RG, McMahon KA, Wu Y (2020) Caveolae: formation, dynamics, and function. Curr Opin Cell Biol 65:8–16

Pazos RS, Maiztegui T, Colautti DC, Paracampo AH, Gomez N (2017) Microplastics in gut contents of coastal freshwater fish from Río de la Plata estuary. Mar Pollut Bull 122(1–2):85–90

Pellini G, Gomiero A, Fortibuoni T, Fabi G, Grati F, Tassetti AN, Polidori P, Vega CF, Scarcella G (2018) lastic soles: microplastic litter in the gastrointestinal tract of *Solea solea* from the Adriatic Sea. In: Proceedings of the international conference on microplastic pollution in the Mediterranean Sea. Springer, Cham, pp 137–149

Pennino MG, Bachiller E, Lloret-Lloret E, Albo-Puigserver M, Esteban A, Jadaud A, Bellido JM, Coll M (2020) Ingestion of microplastics and occurrence of parasite association in Mediterranean anchovy and sardine. Mar Pollut Bull 158:111399

Peters CA, Bratton SP (2016) Urbanization is a major influence on microplastic ingestion by sunfish in the Brazos river basin, Central Texas, USA. Environ Pollut 210:380–387

Pikuda O, Xu EG, Berk D, Tufenkji N (2018) Toxicity assessments of micro-and nanoplastics can be confounded by preservatives in commercial formulations. Environ Sci Technol Lett 6(1): 21–25

Pitt JA, Kozal JS, Jayasundara N, Massarsky A, Trevisan R, Geitner N, Wiesner M, Levin ED, Di Giulio RT (2018) Uptake, tissue distribution, and toxicity of polystyrene nanoparticles in developing zebrafish (*Danio rerio*). Aquat Toxicol 194:185–194

Prata JC (2018) Airborne microplastics: consequences to human health? Environ Pollut 234:115–126

Qiao R, Sheng C, Lu Y, Zhang Y, Ren H, Lemos B (2019) Microplastics induce intestinal inflammation, oxidative stress, and disorders of metabolome and microbiome in zebra fish. Sci Total Environ 662:246–253

Robin RS, Karthik R, Purvaja R, Ganguly D, Anandavelu I, Mugilarasan M, Ramesh R (2020) Holistic assessment of microplastics in various coastal environmental matrices, southwest coast of India. Sci Total Environ 703:134947

Rochman CM, Tahir A, Williams SL, Baxa DV, Lam R, Miller JT, Teh FC, Werorilangi S, Teh SJ (2015) Anthropogenic debris in seafood: plastic debris and fibers from textiles in fish and bivalves sold for human consumption. Sci Rep 5:14340

Roda JFB, Lauer MM, Risso WE, dos Reis Martinez CB (2020) Microplastics and copper effects on the neotropical teleost *Prochilodus lineatus*: is there any interaction? Comp Biochem Physiol A Mol Integr Physiol 242:110659

Romeo T, Pietro B, Pedà C, Consoli P, Andaloro F, Fossi MC (2015) First evidence of presence of plastic debris in stomach of large pelagic fish in the Mediterranean Sea. Mar Pollut Bull 95(1): 358–361

Rummel CD, Löder MG, Fricke NF, Lang T, Griebeler EM, Janke M, Gerdts G (2016) Plastic ingestion by pelagic and demersal fish from the North Sea and Baltic Sea. Mar Pollut Bull 102(1):134–141

Sadri SS, Thompson RC (2014) On the quantity and composition of floating plastic debris entering and leaving the Tamar estuary, Southwest England. Mar Pollut Bull 81(1):55–60

Saha M, Naik A, Desai A, Nanajkar M, Rathore C, Kumar M, Gupta P (2021) Microplastics in seafood as an emerging threat to marine environment: a case study in Goa, west coast of India. Chemosphere 270:129359

Schmieg H, Burmester JKY, Krais S, Ruhl AS, Tisler S, Zwiener C, Köhler H-R, Triebskorn R (2020) Interacting effects of polystyrene microplastics and the antidepressant amitriptyline on early life stages of Brown trout (*Salmo trutta f. fario*). Water 12:2361. https://doi.org/10.3390/w12092361

Setala O, Fleming-Lehtinen V, Lehtiniemi M (2014) Ingestion and transfer of microplastics in the planktonic food web. Environ Pollut 185:77–83

Sharma S, Chatterjee S (2017) Microplastic pollution, a threat to marine ecosystem and human health: a short review. Environ Sci Pollut Res 24(27):21530–21547

Steer M, Cole M, Thompson RC, Lindeque PK (2017) Microplastic ingestion in fish larvae in the western English Channel. Environ Pollut 226:250–259

Su L, Nan B, Hassell KL, Craig NJ, Pettigrove V (2019) Microplastics biomonitoring in Australian urban wetlands using a common noxious fish (*Gambusia holbrooki*). Chemosphere 228:65–74. https://doi.org/10.1016/j.chemosphere.2019.04.114

Sussarellu R, Suquet M, Thomas Y, Lambert C, Fabioux C, Pernet MEJ, Le Goic N, Quillien V, Mingant C, Epelboin Y, Corporeau C (2016) Oyster reproduction is affected by exposure to polystyrene microplastics. Proc Natl Acad Sci 113(9):2430–2435

Tanaka K, Takada H (2016) Microplastic fragments and microbeads in digestive tracts of planktivorous fish from urban coastal waters. Sci Rep 6(1):1–8

Teuten EL, Saquing JM, Knappe DR, Barlaz MA, Jonsson S, Björn A, Rowland SJ, Thompson RC, Galloway TS, Yamashita R, Ochi D (2009) Transport and release of chemicals from plastics to the environment and to wildlife. Philos Trans R Soc Lond B Biol Sci 364(1526):2027–2045

Thompson RC (2006) Plastic debris in the marine environment: consequences and solutions. Marine Nature Conservation in Europe, vol 193. Federal Agency for Nature Conservation, Bonn, pp 107–115

Thompson RC, Olsen Y, Mitchell RP, Davis A, Rowland SJ, John AW, McGonigle D, Russell AE (2004) Lost at sea: where is all the plastic? Science 304(5672):838

Turner A, Holmes L, Thompson RC, Fisher AS (2020) Metals and marine microplastics: adsorption from the environment versus addition during manufacture, exemplified with lead. Water Res 173:115577

UNEP (2016) Marine plastic debris and microplastics–global lessons and research to inspire action and guide policy change. United Nations Environment Programme, Nairobi

Varshney S, Gora AH, Siriyappagouder P, Kiron V, Olsvik PA (2022) Toxicological effects of 6PPD and 6PPD quinone in zebrafish larvae. J Hazard Mater 424:127623

Wang J, Liu X, Li Y, Powell T, Wang X, Wang G, Zhang P (2019) Microplastics as contaminants in the soil environment: a mini-review. Sci Total Environ 691:848–857

Welden NA, Cowie PR (2016) Environment and gut morphology influence microplastic retention in langoustine, *Nephrops norvegicus*. Environ Pollut 214:859–865

Wright SL, Rowe D, Thompson RC, Galloway TS (2013) Microplastic ingestion decreases energy reserves in marine worms. Curr Biol 23(23):1031–1033

Yang H, Xiong H, Mi K, Xue W, Wei W, Zhang Y (2020) Toxicity comparison of nano-sized and micron-sized microplastics to goldfish *Carassius auratus* larvae. J Hazard Mater 388:122058

Yin L, Liu H, Cui H, Chen B, Li L, Wu F (2019) Impacts of polystyrene microplastics on the behavior and metabolism in a marine demersal teleost, black rockfish (*Sebastes schlegelii*). J Hazard Mater 380:120861. https://doi.org/10.1016/j.jhazmat.2019.120861

Zbyszewski M, Corcoran PL, Hockin A (2014) Comparison of the distribution and degradation of plastic debris along shorelines of the Great Lakes, North America. J Great Lakes Res 40(2): 288–299

Effects of Xenobiotics and Their Degradation in Aquatic Life

Iqram ul Haq, Umer Aziz Mir, Sabeehah Rehman, Rasy Fayaz Choh Wani, Mir Shabir Hussain, and Ishteyaq Majeed Shah

1 Introduction

With advancements in scientific technologies, industrialisation, or globalisation, human lives are affected in both ways, be it positive or negative. Since advancements in technologies are definitely bringing change in our lives, but nonetheless there are growing concerns about the fact that globalisation or industrialisation has impacted heavily and negatively on aquatic bodies. Among the negative aspects, one major concern is the introduction of new substances into these waterbodies. The sources of these new substances are medicines, by-products of various personal care products, among others, and finally ending up in waterbodies. The term xenobiotics is actually a combination of two words with xenos meaning foreign (something which is not natural) and bios standing for life. So, xenobiotics represent a class of chemicals or substances that are not naturally present in any given water ecosystem. The non-biodegradable nature of these substance makes it impossible for wastewater treatment plants to effectively remove them from waste water, making it possible to end up in various food chains and infecting and affecting human lives.

Water-soluble environmental chemicals can enter the body via the gills of aquatic creatures, whereas hydrophobic xenobiotics can enter the body via contaminated food. Respiratory work of aquatic organisms is generally carried by gills (Carvalho

Iqram ul Haq
Division of Wildlife Sciences, SKAUST-Kashmir, Srinagar, India

U. A. Mir · R. F. C. Wani · I. M. Shah (✉)
Advanced Research Laboratory, Department of Zoology, University of Kashmir, Srinagar, India
e-mail: ishteyaq.scholar@kashmiruniversity.net

S. Rehman
Centre of Research for Development, University of Kashmir, Srinagar, India

M. S. Hussain
Wildlife Research Lab, Department of Zoology, University of Kashmir, Srinagar, India

M. A. Rather et al. (eds.), *Xenobiotics in Aquatic Animals*,
https://doi.org/10.1007/978-981-99-1214-8_18

2011; Ray and Ringø 2014). The functions of gills other than those of respiration include waste excretion, maintenance of homeostasis, including pH, hormone production, etc. (Foyle et al. 2020; Zhang et al. 2021). Even though the gill structure of most bivalves has underwent secondary evolution to serve as feeding appendages, however, the prime function remains osmoregulation. Less emphasis has been placed on ion transport (Riisgård et al. 2015; Moreira et al. 2015). Whether we talk of fish or mollusc gill, gill structure represents an important and vital link between aquatic fauna and environment as it is the structure that comes in contact (frequently) with aquatic environment. Since gills are lined with selective barriers, controlling uptake of nutrients as well as removal of toxic xenobiotics is needed (Armitage et al. 2013; Wang and Wang 2015). The presence of such selective barriers throughout the digestive tract confers fishes the property to control absorption of many small molecules. According to the study carried out by Collinder et al. (2009), and Karasov and Douglas (2013), the digestive structure of both terrestrial and aquatic animals is essentially the same. However, the variation observed in relative length and volumes of various regions like oesophagus, stomach, with an oesophagus, stomach, midgut, and hindgut is attributed to food and nutrient extraction and absorption. In case of both fish and humans, the epithelia of small and large intestines is largely responsible for nutrient uptake, including water and ions (Sundell and Rønnestad 2011; Kiela and Ghishan 2016), and this process (i.e. nutrient uptake whether micro- or macronutrients) is mediated by simple diffusion, kind of enhanced permeability, or the involvement of secondary active transport coupled with electrical potentiality (Sundell and Rønnestad 2011). According to the studies carried out by Müller et al. (2017) and Nicklisch and Hamdoun (2020), an absolute integration of solute carrier (SLC) proteins and ABC transporters, found on the apical and basolateral membranes of enterocytes, on the other hand, inhibits the absorption of dietary toxins and contaminants, including xenobiotics and biotoxins.

Despite the differences in gastrointestinal macro- and microanatomy between aquatic and terrestrial species, there is limited knowledge regarding the chemical composition, amounts of food intake, and xenobiotic defence mechanisms in the digestive tracts of fish and like organisms. Furthermore, knowledge of the molecular interactions of water and foodborne pathogens with these transportation networks, as well as how these interactions may change nutritional homeostasis and hazardous contaminant bioaccumulation, is lacking.

2 Epithelial Transport in Aquatic Organisms' Digestive System

Since the prime function associated with digestive system is absorption of nutrients, water and minerals form the food to produce energy used for growth and development. In addition to this function, this digestive system serves as an environmental barrier, preventing xenobiotic absorption and its accumulation. As already discussed, integration of several SLC and ATP binding cassette transporters present in the gills as well as intestines of many aquatic organisms, including fishes, performs a key role in xenobiotic efflux, ion flux, cell signalling processes, and absorption of nutrient material.

2.1 Transporters of Nutrients and Endogenous Substrates

Secondary active transporters of the solute carrier (SLC) family typically mediate transepithelial transfer of nutrients in the gastrointestinal system. SLCs are the most common type of secondary active membrane transporter in humans (Höglund et al. 2011; Hediger et al. 2013). SLCs can facilitate bidirectional transport (Kottra et al. 2002; Winter et al. 2011); however, they are mostly involved in nutrient and ion uptake (Zhang et al. 2018; Felmlee et al. 2020; Song et al. 2020). In spite of their role in nutrient absorption and metabolic equilibrium, these transporters are, however, infamously understudied both in humans and aquatic creatures (César-Razquin et al. 2015; Barat et al. 2016).

3 Impact of Xenobiotics on Aquatic Life/Fauna

Advancements in the system of industrialisation as well as urbanisation have given birth to a wide range of pollutants of which xenobiotics have found a top-notch place in the toxic list of pollutants. Azole, phenolic, polycyclic aromatic hydrocarbon (PAH), halogenated, personal care product (PCP), pharmaceutical active ingredient (PhAC), pesticide, nitroaromatic, triazine, and chlorinated chemicals have an adverse effect on the environment due to their long-term durability and sluggish to nonexistent biodegradation in ecosystems. Anthropogenic sources such as urbanisation and population expansion are producing xenobiotic contamination in the environment. Massive volumes of toxic substances discharged into the environment pollute whole ecosystems. The list includes sediments, aromatic hydrocarbons, pesticides, fertilisers, herbicides, among others.

Anthropogenic activities, including urban transportation, spraying housing, industrial production, and building construction, are large contributors of both ground and surface water pollution in urban environments through diffusive and point contributions. Numerous investigations have revealed the presence of various chemicals and signs of human intervention in urban water systems (Strauch et al. 2008). Mishra et al. (2019) examined various trace metals, xenobiotic pollutants, and synthetic organic pollutants with the likes of phthalates, PAHs, and pesticides, among others, in diverse bodies of water. Xenobiotic chemicals can infiltrate water bodies via a variety of routes. These include (a) continual inputs from commercial and fossil fuel products, as well as sewage effluents; (b) surface water runoff from highways and land surfaces; (c) particle deposition in the air; and (d) solid waste burning (Essumang 2010). Also, through the leaching process, xenobiotic chemicals also reach the water table, affecting the very ecology of various aquatic ecosystems (Fent et al. 2006). In the presence of xenobiotic contaminants, aquatic organisms experience oxidative stress. Recently, research carried out by Ibor et al. (2019) at the artificial Eleyele lake in Nigeria showed elevation in oxidative stress response in fish fauna in the presence of xenobiotic contamination.

Xenobiotics negatively imparts the metabolism of marine organisms, particularly growing fish embryos, leading to morphological deformities, functional abnormalities, stunned growth, and eventual death. Additionally, fish with altered body forms, physiological abnormalities, delayed hatching, and mortality have been seen (Arya and Haq 2019). Dyes and paints are xenobiotics even in trace doses because they obstruct sunlight penetration and gas exchange (Abdelkader et al. 2011). The major xenobiotic pollutants of marine life include pesticides and herbicides. In agricultural and everyday life, chemicals like organophosphates, nitrophenols, morpholine, pyrethroids, and carbamates are routinely used; these chemicals eventually wind up in many bodies of water, such as the sea and ocean. Insecticides like -cypermethrin are very dangerous to marine life and invertebrates (Zhang et al. 2011). Environmental xenobiotics are any manufactured substances that are not ordinarily anticipated to occur in any organism. The presence of various environmental xenobiotics greatly and negatively affects both ecosystems and humans. Pesticides, polychlorinated biphenyls, persistent organic pollutants, dangerous heavy metals, etc., are some fine examples of environmental xenobiotics, and there is a wealth of scientific data demonstrating the adverse health effects that these substances are having on people. Among the various health concerns connected with environmental xenobiotics, neurotoxicity, immunotoxicity, nephrotoxicity, hepatotoxicity, and cancer are the most commonly highlighted. However, the general population is less aware of their potential toxicity and different routes of exposure. As a result, an attempt was made in this research to critically examine current literature in order to help future investigations on the health effects of xenobiotics so that the findings of such studies can address the situation on the ground. As a result, this chapter in nutshell discusses a number of particular xenobiotic compounds, as well as the ways in which they can be ingested and the long-term repercussions of doing so.

4 Impacts of Xenobiotics on the Ecosystem

Around 24% of world illnesses and some 13 million deaths are attributed to environmental pollutants. Today, detectable quantities of pharmaceutical preparations can be detected in water and foods, including rivers and oceans, as either the original drug or a metabolite (Banjoko 2014). The impact of medication on people and animals goes beyond the fundamental goals of conventional medical care. The majority of APIs (active pharmaceutical ingredients) derived from medications, whose by-products may contaminate the environment, are contributed by the pharmaceutical sector.

Pollution is detected physiologically by aquatic creatures. Fent et al. (2006) examined the presence and effects of pharmaceuticals in the aquatic environment, discussed putative mechanisms of action based on research on mammals, and assessed the acute and long-term effects on species of ecotoxicity. Pharmaceuticals find their way into the environment either in original shape or in the form of metabolites. Humans normally eliminate chemicals through digestion, excretion, and wastewater disposal. Human drugs are the most prevalent type of medication discovered in municipal wastewater. Pharmaceuticals can be found in high amounts in hospital wastewater, industrial wastewater, and landfill leachates, finding their ways into rivers, lakes, etc., and maybe in drinking water and ending up in damaging ecosystems and associated fauna. Sewage used for agricultural purposes also poses a high risk of contamination. Medication that has an environmental impact typically has a high manufacturing volume, long-term environmental persistence, and biological activity. According to recent research, rising levels of pharmaceuticals identified in surface waterways throughout the world have raised concerns, notably concerning their impact on aquatic vegetation and animals. Fish are the aquatic organisms with whom humans have the most pharmaceutical targets. Medicines' long-term impacts on aquatic species are little known. According to the study carried out by Cuklev et al. (2012), fish when subjected to a dose of l g/L diclofenac, both gene expression and organ histology are found to be altered. NSAIDs like diclofenac, ibuprofen, among others, were detected at 27 sites along the Kaveri velar and Tami rapini rivers of south India, posing the greatest toxicity risk for all those depending on these rivers for water purposes (Shanmugam et al. 2013). Several therapeutic concoctions have been found in both streams and rivers. It has been discovered that antidiabetic and antihistamines diphenhydramine greatly discombobulate the biofilm population, which is vital to the ecology. Microbe aggregates known as biofilms are actually the cells that adhere to one another and/or a surface and are usually encased in a self-produced matrix of extracellular polymeric polymers. Impacts of pollutants like diphenylamine on biofilm can be determined by the fact that organisms in food web such as fish and insects are largely impacted because these biofilms are food suppliers for vertebrates, which in turn are easily accessed by higher food web organisms (Rosi-Marshall 2013). Some mussels spawn prematurely as a result of antidepressants, disturbing aquatic homeostasis. Furthermore, fluoxetine and propranolol have been found to be toxic to many zooplankton and benthic species. Microbes produce biofilms in response to cellular identification of specific or nonspecific attachment sites, nutritional signals, and planktonic cell exposure to subinhibitory antibiotic dosages (Karatan and Watrick 2009). In female sea snails exposed to tributyltin, imposex (masculinisation) was seen. Dichloro-diphenyldichloroethylene (DDE) -induced eggshell thinning in birds is one of the best examples of reproductive harm leading to significant population declines in a variety of European and North American raptor species. Over time, DDT exposure in male western gulls has been linked to ovo-testis. The majority of cleaning products contain the broad-spectrum antibacterial chemical triclosan (TCS) to prevent the growth of bacteria, fungi, and mildew. Domestic wastewater, leaking septic systems, and sewage overflows all cause triclosan to infiltrate streams. Long-term usage of these antibiotics breeds germs that are resistant to them, potentially diminishing the effectiveness of crucial treatments (Drury et al. 2013).

The most well-known effects of EDC in aquatic animals include reduced reproduction and development (Kid et al. 2007). Recent studies discovered many brain targets for EDC that are present in meaningful quantities in surface waters. PCBs have been demonstrated to reduce reproductive and immunological function in Wadden Sea harbour seals, as well as Baltic grey and ringed seals in field studies (polychlorinated biphenyls). Other food-chain animals that may be harmed include the guinea pigs, polar bears, and rabbits. A chemical spill in Florida resulted in altered genital development in alligators. Furthermore, DDT-complicated experimental research employing alligator eggs has been linked to the reported androgenic and oestrogenic effects.

Although urbanisation, population growth, commercialisation, and globalisation have both positive and bad consequences on our lives, they are undeniably causing change (Bulucea et al. 2012). International links, technical innovation, and market expansion intensify global concerns such as economic centralisation and the relaxation and ease of the movement of commodities and services. Nonetheless, despite the benefits, globalisation has a detrimental impact on the environment from an economic and political aspect, and a healthy environment is a need for a good level of life (Banerjee et al. 2008). Technology development, increased longevity, better access to medical care (for humans and animals), routine use of personal care items, and/or pesticides all result in new substances being released into the environment (Jaffe 2005; Eugine and Vincent 2016). A thorough study should be done on these substances' immediate and long-term impacts on people, animals, and the ecosystem because they have the potential to be hazardous either alone or in combination (air, water, and soil). Environmental problems have been caused by poor usage of education and drug disposal, along with the disregard for the environment demonstrated by some businesses, despite the fact that knowledge of the issue is growing (Wu 1999). Pollution develops as a result of man-made toxins that either do not dissolve or disintegrate extremely slowly in the environment (El-Saad and Elgerbed 2010). Science has yet to discover a good and practical artificial deterioration approach that meets all requirements. The term 'xenobiotics' refers to foreign materials in living form and is derived from the Greek terms 'xenos' (foreign) and 'bios' (life) (life). Wastewater treatment plants and rainwater runoff are primarily responsible for the occurrence of xenobiotics in freshwater (Ahlborg et al. 1992; Anetor et al. 2008; Cataudella et al. 2012). The removal of xenobiotics from wastewater by wastewater treatment facilities is frequently insufficient, allowing xenobiotics to infiltrate public sewers, enter the food chain, and directly damage people (Rosas and Eskenazi 2008; Neal and Guilarte 2012), as well as contributing to micropollutant pollution of aquatic bodies (Julvez and Grandjean 2009; Soderland et al. 2010; Descamps and Deschamps 2012). Even though colonies of bacteria and other microorganisms have been discovered to be successful in breaking down particular xenobiotics, activated sludge is typically insufficient for this task.

Communities would have to adapt to xenobiotics and operating parameters in wastewater (Garcia et al. 2012; Singh et al. 2017) that are financially unviable in typical plants. A lot of effort is being put into developing and refining biological or physicochemical mechanisms that are more effective in removing xenobiotics from water; these processes will be discussed later. Pharmaceuticals and personal care products, for example, were shown to have a substantial impact on removal efficiency due to technical developments (PPCPs), and secondary treatment procedures were found to be varied (and inefficient) in eliminating pharmaceutical pollutants (Ozaydin 2017). Various international organisations like US EPA, EMA, and EEA, among others, have been carrying out the high-end research to lessen the harmful nature/toxic effects of various xenobiotics and to look for the ones (pollutants) that need urgent readdressing. However, in order to prevent or minimise the negative impacts of xenobiotics and identify the most pressing pollutants, it is required to know the amounts of contaminants in the environment that are affecting both people and animals. Xenobiotics are distinct, have an impact on both the environment and public health, and their potential for harm is not fully recognised, claims the US EPA. Many directives and laws which are in place aim to enhance environmental quality by continuously looking and monitoring a list of harmful compounds. It is critical to identify pollution sources and implement the most cost-effective and ecologically friendly strategies to reduce pollutant emissions at their source. (Kim et al. 2015), and environmental quality standards (EQS) for those priority compounds were established by Directive 2008/105/EC. This list is being continuously updated based on data sets on toxicological impacts. In the sphere of water policy, Directive 2013/60/EC includes a list of 45 priority compounds. The most latest scientific and technological information is provided in the EQS for those substances. The research 'Modes of action of the current Priority Chemicals list under the Water Framework Directive and other chemicals of interest' is one of several papers on the modes of action (MoA) and impacts of priority substances and other compounds on the WFD's Watch List (WL). The research includes information on assessing these chemicals using effect-based methodologies (biomarkers and bioassays), with an emphasis on combinations of drugs and their possible interactions in the aquatic environment. Second, chemicals on the priority list are divided into 17 groups, while those on the watch list are divided into 8 groups. The European Medicines Agency (https://www.ema.europa.eu/en, accessed on 5 June 2021) offers scientific advice on the best strategy for meeting regulatory requirements that apply to medical goods in the European Union.

5 Biodegradation of Xenobiotics

The degradation of non-biodegradable complex materials into products that are acceptable in the environment includes carbon dioxide, water, and biomass. These compounds are redistributed into the environment through ecological cycles like the

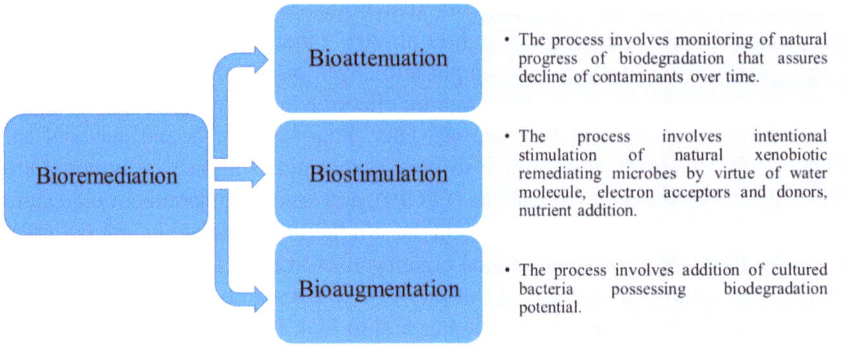

Fig. 1 General methods employed by microbes in the bioremediation process

sulphur cycle, nitrogen cycle, and carbon cycle. The degradation process is accomplished by the metabolic action of certain microorganisms including bacteria in natural environmental conditions (Karak 2012). Microorganisms play a key role in biodegradation, with the advancement in technology and molecular biology genetically engineered microorganisms are being extensively used to counter environmental problems such as the addition and accretion of xenobiotics into the biosphere. Over the decades the acceleration in environmental pollution by xenobiotics has emerged as a serious concern (Rathore et al. 2022). The microbes employ different method for the bioremediation of xenobiotics. some of the methods are shown in Fig. 1.

Xenobiotics like phenolics, azodyes, personal care products, halogenated compounds, pharmaceuticals active compounds, polycyclic aromatic hydrocarbon, nitroaromatic compounds, triazines, pesticides, antibiotics, and chlorinated compounds negatively impact the environment owing to their non- or slow biodegradable nature and their ability of persistence in the environment. The most serious threat of xenobiotics is biomagnification apart from causing adverse effects at each tropic level by making their way into the food chain (Zhou et al. 2022). All life forms, plants, and animals, including human and environmental health, are adversely affected by major xenobiotic compounds.

The aquatic habitat forms the pool of accumulation of xenobiotics and has emerged as a sink for the hazardous complex polymers that are usually non-biodegradable. Efforts have been made to eliminate these compounds from the ecosystem; these methods include degradation by coagulation, adsorption, filtration, electrolysis, chemical precipitation, and ozonation. However, the microbial degradation of xenobiotics has evolved promptly as a reliable approach for being eco-friendly and cost-effective.

Among the different strategies, the research of microbial enzymes for bioremediation is growing in significance, nevertheless, on a worldwide scale. The advancement and use of cutting-edge molecular approaches are providing new insights into the structural and functional characteristics of complex microorganisms. These techniques include proteomics, metagenomics, transcriptomics, metabolomics, etc.

For bioremediation of xenobiotics methods like phytoremediation, bioaugmentation, landfarming, rhizo-filtration, bio-stimulation, composting and bioreactors have been used widely (Azubuike et al. 2016). Bioremediation uses the service of various microorganisms for destruction, eradication, immobilisation, or detoxification of a wide range of chemical wastes and other harmful chemicals from the ecosystem/ environment. Bioremediation involves the systems of living organisms, most importantly, bacteria, fungi, plants, and their enzymes (Ijoma and Tekere 2017). Especially microbes, bacteria, and fungi possess the capability to degrade xenobiotics by means of endo- and exo-enzymes systems (Singh 2014). Microorganisms involve two basic mechanisms for the biodegradation process; aerobic and anaerobic biodegradation (Sharma and Fulekar 2009). The basic equations for the aerobic and anaerobic bioremediation of xenobiotics can be mentioned as follows):

$$\text{Xenobiotics} + O_2 \ {}^{(\text{Microbes})} \rightarrow CO_2 + H_2O + \text{residue (Aerobic)}$$

$$\text{Xenobiotics} \ {}^{(\text{Microbes})} \rightarrow CO_2 + CH_4 + H_2O + \text{residue (Anaerobic)}$$

6 Degradation of Xenobiotics Through Bacteria

Xenobiotics find their way into environments as a result of anthropogenic activities, resulting in ecosystem damages and environmental pollutions. Opposed to this, certain metabolites of bacteria have xenobiotic degrading capabilities. Bacterial strains from several genera, including *Burkholderia, Bacillus, Pseudomonas, Sphingomonas, Kocuria, Chromohalobacter* and *Achromobacter*, are known to degrade xenobiotics completely or mineralise when subjected to axenic and anoxic environments (Zhang et al. 2020). Microorganisms have a remarkable capacity for catabolism involved in the biodegradation process because of a wide range of genes and enzymes. The capability of bacteria to multiply rapidly and adapt to diverse environmental conditions is important. Certain bacterial species have been identified and cultured for the bioremediation process (Table 1); however, such microbes are very limited though with incomparable capabilities to degrade xenobiotic compounds. Apart from the culture-dependent technique, more modern techniques like genomics/metagenomics and transcriptomics have led to the identification of specific genes that actually impart the biodegradation character to a microbe. These techniques have also led to the characterisation of a wider community of microbes that were otherwise uncultured and left unidentified. Genome investigations of bacterial strains that break down xenobiotics have revealed that these strains evolved recently by integrating genes for xenobiotic degradation, with mobile genetic components being essential for gene acquisition (Nagata et al. 2019). However, the origin and evolution of such genes in the microbiome are yet not clear. Below, Fig. 2 represents the different degrading enzymes produced by bacteria against xenobiotics.

Table 1 Promising microbes against different xenobiotics

S. no.	Bacteria	Promising against	Reference
1	*Pseudomonas*	Aliphatic hydrocarbon degradation	
2	*Microbacterium D-2*	Pesticide degradation (dicofol)	Lu et al. (2019)
3	*Enterobacter*	Polyethylene, plastic	Ren et al. (2019)
4	*Micrococcus*		
5	*Alcaligenes*		
6	Sphingopyxis,		Russell et al. (2021)
7	*Hyphomicrobiaceae*		
8	*Achromobacter*		
9	*Purpureocillium*		
10	Mesorhizobium,		
11	*Aeromonas*		
12	*Gordonia*	Oil degradation	
13	*Rhodococcus*		
14	*Sphingobium*		
15	*Penicillium*		
16	*Candida*		
17	*Sphingopyxis*		Russell et al. (2021)
18	*Trichoderma*		
19	*Rhodotorula*		
20	*Rhodopseudomonas,*		
21	*Thalassolituus*	Oil-degradation	
22	*Afipia*		Russell et al. (2021)
23	*Bacillus amyloliquefaciens*	Organophosphorous pesticide (phoxim) degradation	
24	*Stenotrophomonas*		Russell et al. (2021)
25	*Oligotropha*		Russell et al. (2021)
26	*Mesorhizobium*		
27	*Rhodopseudomonas palustris*	Hexabromocyclododecane degradation	
28	*Oleispira*	Oil-degradation	
29	*Trichoderma hamatum*	DDT-degradation	
30	*Burkholderia sp. strain C3*	N-methyl carbamates pesticides	Seo et al. (2013)
31	*Photobacterium ganghwense*	Cyfluthrin degradation	Singh et al. (2018)

(continued)

Table 1 (continued)

S. no.	Bacteria	Promising against	Reference
32	*Mycobacterium sp.* DBP42	Phthalate and plasticisers	Wright et al. (2020)
33	*Halomonas sp.* ATBC 28	Phthalate and plasticisers	
34	*Drechslera sp. 678*	Methyl tertiary-butyl ether (MtBE), an additive used in gasoline	d'Errico et al. (2021)
35	*Fusarium verticillioides*	Lactam and lactone xenobiotic degradation	Gao et al. (2022)
36	*Sphingobium chungbukense*	PAH-degrading	

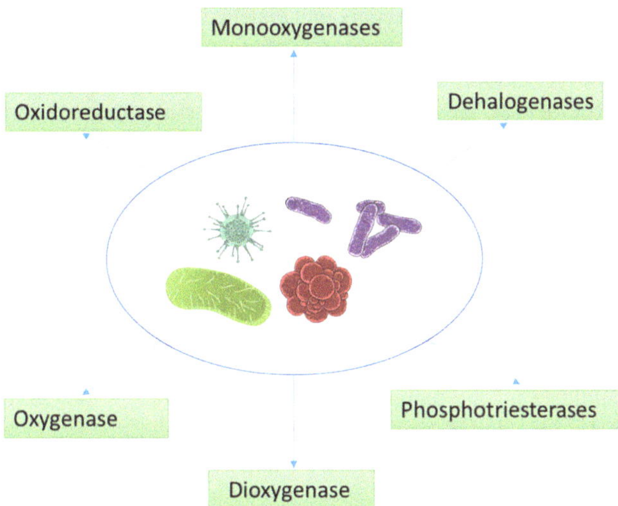

Fig. 2 Xenobiotic degrading enzymes produced by bacteria

Many bacterial metabolic enzymes like cytochrome P450s, cellulase, laccases, proteases, phytase, lipase, among others, are believed to play a vital role in the management and breakdown of a good number of xenobiotics by degrading the dyes, aromatic hydrocarbons, and halogenated compounds. The first step in enzymatic biodegradation is identifying an appropriate enzyme for bioremediation application; this enzyme needs to be able to convert the target contaminants into less-toxic by-products (Gangola et al. 2019). Aliphatic hydrocarbons are broken down either by mono- or dioxygenases, resulting in the creation of peroxide, which is then transformed into fatty acids (Okolafor and Ekhaise 2022). The fatty acid molecule oxidises to create intermediates in the TCA cycle, which are ultimately broken down into carbon dioxide and water.

7 Degradation of Xenobiotics Through Fungi

Apart from bacteria, fungi are top-notch players for bioremediation of xenobiotic compounds and the process is also referred to as 'mycoremediation', wherein fungi are utilised in the bioremediation of hazardous contaminants including hazardous phenolics, dyes, polycyclic aromatic hydrocarbons, polythene, among others. The degradation of xenobiotics via fungal metabolism involves the adsorption of the compounds to the chitinous cell wall of the fungi and as such has highlighted the significance of xenobiotic breakdown by an intracellular enzymatic process (Mishra et al. 2021).

Many members of the group have been recognised to possess biodegradation capabilities, including *Aspergillus, Trichoderma, Penicillium, Fusarium, Cryptococcus, Rhodotorula, Pichia, Candida, Exophiala,* and *Aureobasidium* (Bhatt et al. 2020). Fungi, because of their diversity and genetic functionality, are responsible for the evolution of novel traits for bioremediation of xenobiotics. According to Bosshard (2011), various members (species) of genus *Aspergillus, A. niger, A. flavus,* and *A. oryzae,* are frequently utilised in the management (breakdown) of low-density polyethylene (LDPE) because of their intrinsic potential to grow easily and extensively at soil and other waste sites, and are thought to have longer incubation period compared with other fungal members (species). A study by Sangale et al. (2019), identified *Aspergillus sydowii* strain PNPF15/TS and *Aspergillus terreus* strain MANGF1/WL effective against polythene degradation. *Metarhizium brunneum* ARSEF has been characterised as metabolic biodegrading fungi against herbicides (ametryn and s-triazene) (Szewczyk et al. 2018). Among other fungi used in bioremediation, white rot fungi (WRF) have evolved as an important candidate for the bioremediation of xenobiotics by producing numerous enzymes for the degradation process.

However, research has demonstrated that fungal consortium yields a better outcome than the usage of single species for the bioremediation process (Saroj et al. 2015). Saroj et al. (2015) developed a fungal consortium by combining three fungal strains, *Aspergillus niger* SAR-6, *Penicillium oxalicum* SAR-3, and *Aspergillus flavus* SAB-3 and the consortium had a comparatively elevated capability to break down azo dyes. Another consortium developed by Wang et al. (2022) consisted of *Trametes hirsuta* BYL-3, *T. versicolor* BYL-7, and *T. hirsuta* BYL-8, the consortium exhibited enhanced lignin degradation. The degradation capability is measurable, and different techniques employed are (Table 2).

8 Conclusion

Advancements in the techniques and methods in molecular biology and bioinformatics have provided new perceptions of bioremediation. The bioremediation process can be enhanced with the application of techniques like genome editing, which

Table 2 Techniques employed for measuring degradability

Polymer deterioration	Scanning electron microscopy	Zahra et al. (2010)
Bio-fragmentation	Size-exclusion chromatography	Sangale et al. (2019)
	High-performance liquid chromatography	
	Fourier-transform infrared spectroscopy	
Plastic degradation by fungi	Spectroscopic methods	
	Fourier-transform infrared spectroscopy)	

enables the modification of microbial strains with the boosted capability of degrading many xenobiotics simultaneously and/or with a rapid rate of degradation (Janssen and Stucki 2020). Breakthroughs in the very advancement of genetic modification technologies have opened the doors of knowledge and information, thus providing a platform for exploring the potential of highly competent microorganisms in the breakdown of xenobiotics and their biodegradation.

References

Abdelkader E, Nadjia L, Ahmed B (2011) Degradation study of phenazin neutral red from aqueous suspension by paper sludge. J Chem Eng Process Technol 2:109–114

Ahlborg U, Hanberg A, Kenne K (1992) Risk assessment of polychlorinated biphenyls (PCBs). Institute of Environmental Medicine, Stockholm

Anetor JI, Anetor GO, Iyanda AA, Adeniyi FAA (2008) Environmental chemicals and human neurotoxicity: magnitude, prognosis and markers. Afr J Biomed Res 11:1–12

Armitage JM, Arnot JA, Wania F, Mackay D (2013) Development and evaluation of a mechanistic bioconcentration model for ionogenic organic chemicals in fish. Environ Toxicol Chem 32:115–128. https://doi.org/10.1002/etc.2020

Arya P, Haq SA (2019) Effects of xenobiotics and their biodegradation in marine life. In: Smart bioremediation technologies. Academic, Cambridge, MA, pp 63–81

Azubuike CC, Chikere CB, Okpokwasili GC (2016) Bioremediation techniques–classification based on site of application: principles, advantages, limitations and prospects. World J Microbiol Biotechnol 32(11):1–18

Banerjee BD, Chakraboti A, Suke SG, Ahmed RS, Tripathi AK (2008) Xenobiotic-induced immune alterations: implications in health and disease. Indian J Biochem Biophys 45:7–15

Banjoko B (2014) Environmental pharmacology–an overview. In: Pharmacology and therapeutics. IntechOpen, London

Barat A, Sahoo PK, Kumar R, Pande V (2016) Solute carriers (SLCs) identified and characterized from kidney transcriptome of golden mahseer (tor putitora) (fam: Cyprinidae). Comp Biochem Physiol B Biochem Mol Biol 200:54–61. https://doi.org/10.1016/j.cbpb.2016.06.003

Bhatt P, Rene ER, Kumar AJ, Zhang W, Chen S (2020) Binding interaction of allethrin with esterase: bioremediation potential and mechanism. Bioresour Technol 315:123845

Bosshard PP (2011) Incubation of fungal cultures: how long is long enough? Mycoses 54(5):e539–e545

Bulucea CA, Rosen MA, Mastorakis NE, Bulucea CA, Brindusa CC (2012) Approaching resonant absorption of environmental xenobiotics harmonic oscillation by linear structures. Sustainability 4:561–573

Carvalho O (2011) Comparative physiology of the respiratory system in the animal kingdom. Open Biol J 4:35–46. https://doi.org/10.2174/1874196701104010035

Cataudella E, Malaguarnera G, Gagliano C, Condorelli G, Antic T, Rampello L, Erdogan O, Rampello L, Malaguarnera M (2012) Pesticides exposure and the management of acute hepatic injury. Acta Med Mediterr 28:245–252

César-Razquin A, Snijder B, Frappier-Brinton T, Isserlin R, Gyimesi G, Bai X, Superti-Furga G (2015) A call for systematic research on solute carriers. Cell 162(3):478–487

Collinder E, Björnhag G, Cardona M, Norin E, Rehbinder C, Midtvedt T (2009) Gastrointestinal host-microbial interactions in mammals and fish: comparative studies in man, mice, rats, pigs, horses, cows, elk, reindeer, salmon and cod. Microb Ecol Health Dis 15:66–78. https://doi.org/10.1080/08910600310014980

Cuklev F, Fick J, Cvijovic M, Kristiansson E, Förlin L, Larsson DJ (2012) Does ketoprofen or diclofenac pose the lowest risk to fish? J Hazard Mater 229:100–106

d'Errico G, Aloj V, Flematti GR, Sivasithamparam K, Worth CM, Lombardi N, Ritieni A, Marra R, Lorito M, Vinale F (2021) Metabolites of a Drechslera sp. endophyte with potential as biocontrol and bioremediation agent. Nat Prod Res 35(22):4508–4516

Descamps CM, Deschamps F (2012) Nervous system disorders induced by occupational and environmental toxic exposure. Open J Prev Med 2(3):272–278

Drury R, Scott J, Rosi-Marshall EJ, Kelly JJ (2013) Triclosan exposure increases triclosan resistance and influences taxonomic composition of benthic bacterial communities. Environ Sci Technol 47(15):8923–8930

El-Saad AA, Elgerbed AS (2010) Dimethoate-induced hepatotoxicity in rats and the protective roles of vitamin E and N-acetylcysteine. Egyp J Exp Biol 6(2):219–230

Essumang DK (2010) Distribution, levels, and risk assessment of polycyclic aromatic hydrocarbons (PAHs) in some water bodies along the coastal belt of Ghana. Sci World J 10:972–985

Eugine M, Vincent T (2016) Prevalence of persistent organic pollutants in Blantyre—Malawi. Am J Environ Prot 4(3):61–66

Felmlee MA, Jones RS, Rodriguez-Cruz V, Follman KE, Morris ME (2020) Monocarboxylate transporters (SLC16): function, regulation, and role in health and disease. Pharmacol Rev 72(2): 466–485

Fent K, Escher C, Caminada D (2006) Estrogenic activity of pharmaceuticals and pharmaceutical mixtures in a yeast reporter gene system. Reprod Toxicol 22(2):175–185

Foyle KL, Hess S, Powell MD, Herbert NA (2020) What is gill health and what is its role in marine finfish aquaculture in the face of a changing climate? Front Mar Sci 7:400. https://doi.org/10.3389/fmars.2020.00400

Gangola S, Joshi S, Kumar S, Pandey SC (2019) Comparative analysis of fungal and bacterial enzymes in biodegradation of xenobiotic compounds. In: Smart bioremediation technologies. Academic, Cambridge, MA, pp 169–189

Gao M, Gu X, Satterlee T, Duke MV, Scheffler BE, Gold SE, Glenn AE (2022) Transcriptomic responses of fusarium verticillioides to lactam and lactone xenobiotics. Front Fungal Biol 3: 923112

Garcia FP, Ascencio SC, Oyarzun JG, Hernandez AC, Alavarado PV (2012) Pesticides: classification, uses and toxicity Measures of exposure and genotoxic risks. J Res Environ Sci Toxicol 1(11):279–293

Hediger MA, Clémençon B, Burrier RE, Bruford EA (2013) The ABCs of membrane transporters in health and disease (SLC series): introduction. Mol Asp Med 34(2–3):95–107

Höglund C, Sneitz N, Radominska-Pandya A, Laakonen L, Finel M (2011) Phenylalanine 93 of the human UGT1A10 plays a major role in the interactions of the enzyme with estrogens. Steroids 76(13):1465–1473

Ibor OR, Adeogun AO, Regoli F, Arukwe A (2019) Xenobiotic biotransformation, oxidative stress and obesogenic molecular biomarker responses in Tilapia guineensis from Eleyele Lake, Nigeria. Ecotoxicol Environ Saf 169:255–265

Ijoma GN, Tekere M (2017) Potential microbial applications of co-cultures involving ligninolytic fungi in the bioremediation of recalcitrant xenobiotic compounds. Int J Environ Sci Technol 14(8):1787–1806

Jaffe R (2005) Xenobiotics: managing toxic metals, biocides, hormone mimics, solvents and chemical disruptors, DK6006-C029.qxd
Janssen DB, Stucki G (2020) Perspectives of genetically engineered microbes for groundwater bioremediation. Environ Sci Processes Impacts 22(3):487–499
Julvez J, Grandjean P (2009) Neurodevelopmental toxicity risks due to occupational exposure to industrial chemicals during pregnancy. Ind Health 47:459–468
Karak N (2012) Vegetable oil-based polymers: properties, processing and applications. Woodhead Publishing, Sawston
Karasov WH, Douglas AE (2013) Comparative digestive physiology. Compr Physiol 3:741–783. https://doi.org/10.1002/cphy.c110054
Karatan E, Watrick P (2009) Signals, regulatory networks and materials that build and break bacterial biofilms. Microbiol Mol Biol Rev 73(2):310–347
Kid KA, Blanchfield PJ, Mills KH et al (2007) Collapse of a fish population after exposure to a synthetic estrogen. Proc Natl Acad Sci U S A 104(21):8897–8901
Kiela PR, Ghishan FK (2016) Physiology of intestinal absorption and secretion. Best Pract Res Clin Gastroenterol 30(2):145–159
Kim NH, Hyun YY, Lee KB, Rhu S, Kook-Hwan KH, Ahn C (2015) Environmental heavy metal exposure and chronic kidney disease in the general population. J Korean Med Sci 30:272–277
Kottra G, Stamfort A, Daniel H (2002) PEPT1 as a paradigm for membrane carriers that mediate electrogenic bidirectional transport of anionic, cationic, and neutral substrates∗. J Biol Chem 277:32683–32691. https://doi.org/10.1074/jbc.M204192200
Lu P, Liu HM, Liu AM (2019) Biodegradation of dicofol by microbacterium sp. D-2 isolated from pesticide-contaminated agricultural soil. Appl Biol Chem 62(1):1–9
Mishra VK, Singh G, Shukla R (2019) Impact of xenobiotics under a changing climate scenario in climate change and agricultural ecosystems. Woodhead Publishing, Sawston, pp 133–151
Mishra S, Lin Z, Pang S, Zhang W, Bhatt P, Chen S (2021) Recent advanced technologies for the characterization of xenobiotic-degrading microorganisms and microbial communities. Front Bioeng Biotechnol 9:632059
Moreira R, Pereiro P, Canchaya C, Posada D, Figueras A, Novoa B (2015) RNA-Seq in Mytilus galloprovincialis: comparative transcriptomics and expression profiles among different tissues. BMC Genomics 16:728. https://doi.org/10.1186/s12864-015-1817-5
Müller J, Keiser M, Drozdzik M, Oswald S (2017) Expression, regulation and function of intestinal drug transporters: an update. Biol Chem 398:175–192. https://doi.org/10.1515/hsz-2016-0259
Nagata Y, Kato H, Ohtsubo Y, Tsuda M (2019) Lessons from the genomes of lindane-degrading sphingomonads. Environ Microbiol Rep 11(5):630–644
Neal AP, Guilarte TR (2012) Mechanisms of heavy metal neurotoxicity: lead and manganese. J Drug Metab Toxicol S5:002
Nicklisch SC, Hamdoun A (2020) Disruption of small molecule transporter systems by transporter-interfering chemicals (TICs). FEBS Lett 594(23):4158–4185
Okolafor F, Ekhaise FO (2022) Microbial enzyme remediation of poly-aromatic hydrocarbon (PAH’s): a review. J Int Environ Appl Sci 17(1):10–21
Ozaydin FN (2017) Xenobiotics and chronic disease. J Toxicol 1(4):1–2
Rathore S, Varshney A, Mohan S, Dahiya P (2022) An innovative approach of bioremediation in enzymatic degradation of xenobiotics. Biotechnol Genet Eng Rev 38(1):1–32
Ray AK, Ringø E (2014) The gastrointestinal tract of fish. In: Merrifield D, Ringo E (eds) Aquaculture nutrition. Wiley, Hoboken, NJ, pp 1–13
Ren L, Men L, Zhang Z, Guan F, Tian J, Wang B, Wang J, Zhang Y, Zhang W (2019) Biodegradation of polyethylene by Enterobacter sp. D1 from the guts of wax moth galleria mellonella. Int J Environ Res Public Health 16(11):1941
Riisgård HU, Funch P, Larsen PS (2015) The mussel filter–pump—present understanding, with a re-examination of gill preparations. Acta Zool 96:273–282. https://doi.org/10.1111/azo.12110
Rosas LG, Eskenazi B (2008) Pesticides and child neurodevelopment. Curr Opin Pediatr 20:191–197

Rosi-Marshall E (2013) Streams stressed by pharmaceutical pollution. Www. Environmentalchange.Nd.Edu/events/2. Accessed 10 Aug 2013

Russell JN, Perry BJ, Bergsveinson J, Freeman CN, Sheedy C, Nilsson D, Braul L, Yost CK (2021) Metagenomic and metatranscriptomic analysis reveals enrichment for xenobiotic-degrading bacterial specialists and xenobiotic-degrading genes in a Canadian prairie two-cell biobed system. Environ Microbiol Rep 13(5):720–727

Sangale MK, Shahnawaz M, Ade AB (2019) Potential of fungi isolated from the dumping sites mangrove rhizosphere soil to degrade polythene. Sci Rep 9(1):1–11

Saroj S, Dubey S, Agarwal P, Prasad R, Singh RP (2015) Evaluation of the efficacy of a fungal consortium for degradation of azo dyes and simulated textile dye effluents. Sustain Water Resour Manag 1(3):233–243

Seo JS, Keum YS, Li QX (2013) Metabolomic and proteomic insights into carbaryl catabolism by Burkholderia sp. C3 and degradation of ten N-methylcarbamates. Biodegradation 24(6): 795–811

Shanmugam G, Ramasamy K, Selvaraj KK, Sampath S, Ramaswamy BR (2013) Triclosan in fresh water fish Gibelion catla from the Kaveri River, India, and its consumption risk assessment. Environ Forensic 15(3):207–212

Sharma J, Fulekar MH (2009) Potential of Citrobacter freundii for bioaccumulation of heavy metal–copper. Biol Med 1(3):7–14

Singh R (2014) Microorganism as a tool of bioremediation technology for cleaning environment: a review. Proc Int Acad Ecol Environ Sci 4(1):1

Singh N, Gupta VK, Kumar A, Sharma B (2017) Synergistic effects of heavy metals and pesticides in living systems. Front Chem 5:1–9

Singh DP, Prabha R, Gupta VK, Verma MK (2018) Metatranscriptome analysis deciphers multifunctional genes and enzymes linked with the degradation of aromatic compounds and pesticides in the wheat rhizosphere. Front Microbiol 9:1331

Soderland P, Lovekar S, Weiner DE, Brooks DR, Kaufma JS (2010) Chronic kidney disease associated with environmental toxins and exposures. Adv Chronic Kidney Dis 17(3):254–264

Song W, Li D, Tao L, Luo Q, Chen L (2020) Solute carrier transporters: the metabolic gatekeepers of immune cells. Acta Pharm Sin B 10:61–78. https://doi.org/10.1016/j.apsb.2019.12.006

Strauch G, Möder M, Wennrich R, Osenbrück K, Gläser HR, Schladitz T, Muller C, Schirmer K, Reinstorf F, Schirmer M (2008) Indicators for assessing anthropogenic impact on urban surface and groundwater. J Soils Sediments 8:23–33

Sundell KS, Rønnestad I (2011) Integrated function and control of the gutlintestinal absorption. In: Farrell AP (ed) Encyclopedia of fish physiology. Academic, Cambridge, MA, pp 1311–1321

Szewczyk R, Kuśmierska A, Bernat P (2018) Ametryn removal by Metarhizium brunneum: biodegradation pathway proposal and metabolic background revealed. Chemosphere 190: 174–183

Wang X, Wang W-X (2015) Physiologically based pharmacokinetic model for inorganic and methylmercury in a marine fish. Environ Sci Technol 49:10173–10181. https://doi.org/10.1021/acs.est.5b02301

Wang J, Li L, Xu H, Zhang Y, Liu Y, Zhang F, Shen G, Yan L, Wang W, Tang H, Qiu H (2022) Construction of a fungal consortium for effective degradation of rice straw lignin and potential application in bio-pulping. Bioresour Technol 344:126168

Winter TN, Elmquist WF, Fairbanks CA (2011) OCT2 and MATE1 provide bidirectional agmatine transport. Mol Pharm 8:133–142. https://doi.org/10.1021/mp100180a

Wright RJ, Bosch R, Gibson MI, Christie-Oleza JA (2020) Plasticizer degradation by marine bacterial isolates: a proteogenomic and metabolomic characterization. Environ Sci Technol 54(4):2244–2256

Wu RS (1999) Eutrophication, water borne pathogens and xenobiotic compounds: environmental risks and challenges. Mar Pollut Bull 39(11–22):1999

Zahra S, Abbas SS, Mahsa MT, Mohsen N (2010) Biodegradation of low-density polyethylene (LDPE) by isolated fungi in solid waste medium. Waste Manag 30(3):396–401

Zhang C, Wang S, Yan Y (2011) Isomerization and biodegradation of beta-cypermethrin by Pseudomonas aeruginosa CH7 with biosurfactant production. Bioresour Technol 102:7139–7146

Zhang Y, Zhang Y, Sun K, Meng Z, Chen L (2018) The SLC transporter in nutrient and metabolic sensing, regulation, and drug development. J Mol Cell Biol 11:1–13. https://doi.org/10.1093/jmcb/mjy052

Zhang W, Lin Z, Pang S, Bhatt P, Chen S (2020) Insights into the biodegradation of lindane (γ-hexachlorocyclohexane) using a microbial system. Front Microbiol 11:1–12

Zhang L, Yang F, Li Y, Cao H, Huang A, Zhuang Y, Zhang C, Hu G, Mao Y, Luo J, Xing C (2021) The protection of selenium against cadmium-induced mitophagy via modulating nuclear xenobiotic receptors response and oxidative stress in the liver of rabbits. Environ Pollut 285:117301

Zhou Y, Yao L, Pan L, Wang H (2022) Bioaccumulation and function analysis of glutathione S-transferase isoforms in Manila clam Ruditapes philippinarum exposed to different kinds of PAHs. J Environ Sci 112:129–139

Transcriptome Analysis of Aquatic Species Exposed to Endocrine Disruptors

Deepak Agarwal, A. Rathipriya, Mohd Ashraf Rather, and Gulshan Kumar

1 Introduction

Several heavy metals are persistent in the environment mainly in the aquatic environment. These heavy metals are entered into the environment through numerous anthropogenic and natural sources. The major sources of these heavy metals are wastewater from industries, agriculture, and mining that enter the aquatic environment and induce several kinds of stress to the fish and other aquatic animals. Sometimes, the bioavailability and toxicity of these compounds are so high that they can lead to the collapse of the entire ecosystem. Numerous heavy metals have been reported in the aquatic environment through several studies like cadmium, zinc, mercury, polychlorinated biphenyls (PCB), bisphenol A (BPA), linuron, etc. Some powerful heavy metals act as endocrine disruptor compounds (EDCs), for example, arsenic, cadmium, Hg, BPA, etc., which mimic the biological activities or block the receptors of steroid hormones, such as androgens, estrogens, and glucocorticoids (Kim et al. 2016). EDCs are defined as the exogenous compounds that interfere with the hormone-regulated physiological pathways of animals and lead to the elimination of natural hormones required for the maintenance of homeostasis (Kavlock et al. 1996). The EDCs interact and disrupt the hypothalamic-pituitary–gonadal (HPG) axis, resulting in diminished growth, abnormal function, and development of the organism. Thousands of genes are affected by EDCs, requiring new tools to monitor their global effects. Using the high-throughput sequencing for gene transcription

D. Agarwal (✉) · A. Rathipriya
Tamil Nadu Dr. J. Jayalalithaa Fisheries University—IFPGS, OMR Campus, Chennai, India

M. A. Rather
Division of Fish Genetics and Biotechnology, Sher-e-Kashmir University of Agricultural Sciences and Technology, Kashmir, Rangil, Ganderbal, Jammu and Kashmir, India

G. Kumar
College of Fisheries Sciences, Gumla, Birsa Agricultural University, Ranchi, Jharkhand, India

M. A. Rather et al. (eds.), *Xenobiotics in Aquatic Animals*,
https://doi.org/10.1007/978-981-99-1214-8_19

offers a reliable and effective means of monitoring the effects of EDCs on humans and other vertebrates including fish, as well as elucidating the mechanism involved (Baker and Hardiman 2014).

2 Different Endocrine Disruptor Compounds (EDCs)

The role of heavy metals as putative endocrine-disrupting chemicals is due to their chemistry. Their physical, chemical, or in the case of uranium, radioactive properties explain their widespread use in industries (Dyer 2007). In biological processes, certain heavy metals like zinc, copper, manganese, and iron are essential in trace amounts, but their lethal concentrations are usually toxic. Arsenic, cadmium, chromium, lead, and mercury are among the priority metals that are extremely important for human health because of their high level of toxicity.

2.1 Arsenic

Among the most abundant elements on earth, arsenic is widely used in semiconductor manufacture and as a pesticide. Environmental arsenic can also be found in glass and copper smelters, coal combustion, and uranium mining. Drinking water is the most extensive source of environmental exposure. Among the many forms of arsenic, arsenite, and arsenic oxide are the most commonly encountered forms that facilitate the most rapid absorption from the gastrointestinal tract, so they are the most harmful. Some studies suggested that humans and other animals have suffered endocrine disruption caused by arsenic that can produce estrogenic-like effects by direct or indirect stimulation of estrogen receptor-α (Waalkes et al. 2004).

2.2 Cadmium

There is a large amount of cadmium (Cd) dispersed throughout the environment, majorly through mining, smelting, and electroplating. It can also be found in consumer products such as batteries, pigments, etc. Ingestion of food and inhalation of cigarette smoke are the two main ways that people are exposed to cadmium with rare absorption through the skin. Additionally, certain foods such as leafy vegetables, potatoes, and some seafood like crabs, mollusks, and dried seaweed contain trace quantities of cadmium. Measuring the amount of cadmium in blood or urine is a conventional method for detecting exposure. According to studies, cadmium produces reactive oxygen species (ROS), which cause single-strand DNA damage and impair the synthesis of nucleic acids and proteins (Mitra 1984; Stohs and Bagchi 1995). In contrast to other carcinogenic metals, cadmium is not a particularly potent mutagen, yet it can affect male reproduction in mice at a dose of 1 mg/kg body weight (Rossman et al. 1992).

2.3 Chromium

In the crust of the earth, chromium (Cr) is a naturally occurring element. Chromium is released into the air, water, and soil via a range of natural and anthropogenic sources, with industry sectors being the main source of emission. Chromium's environmental burden was mostly increased by the wastewater discharged into the environment by the metallurgical, refractory, and chemical industries (Tchounwou et al. 2012). The amount of chromium in fish varies according to their age, stage of development, and other physiological factors. Additionally, it caused cytotoxicity and had a negative effect on fish behavior, such as gill epithelium hypertrophy and paraplegia and irregular swimming. Numerous studies have found that fish exposed to chromium experience negative hematological consequences such as anemia, thrombocytopenia, and a reduction in hemoglobin and total erythrocyte count (Aslam and Yousafzai 2017).

2.4 Lead

The earth's crust also contains trace amounts of lead, an element that is naturally occurring and bluish-gray in color. Although lead naturally occurs in the environment, human activities like burning fossil fuels, mining, and manufacturing cause significant concentrations to be released. Inhalation of lead-contaminated dust or aerosols, as well as consumption of lead-contaminated food, water, and paints, is the main way that people get exposed to lead (Abadin et al. 2007). Lead compounds appear to cause genetic damage by several indirect pathways, including oxidative damage, interference with DNA-binding proteins, blockage of DNA repair pathways and interactions with tumor suppressor proteins, etc. Being on top of the aquatic food chain, fishes are the most vulnerable to the damaging consequences of lead exposure. Lead induced toxicity in fish through bioaccumulation in specific tissues depending on the habitat (freshwater or seawater) and route (waterborne or dietary exposure) (Lee et al. 2019). In addition to causing neurotoxicity, lead deposition also disrupts neurotransmitter function in fish. Several systems in fish are disrupted by lead toxic exposure, which can be employed as a marker of toxicity in aquatic environments (Ishaque et al. 2020).

2.5 Mercury

Mercury is a heavy metal found in nature. It exists in three forms: elemental mercury vapor (Hg^0), inorganic mercurous (Hg^{+1}), mercuric (Hg^{+2}), and organic mercury compounds. Each of these has its own set of toxicological effects. The most prevalent organic form of mercury found in the environment is methyl mercury, which is created when microorganisms in soil and water methylate the inorganic forms of mercury (Dopp et al. 2004). Mercury that enters streams is methylated by bacteria and algae. Methyl mercury then enters fish, shellfish, and finally humans through the food chain (Sanfeliu et al. 2003). It has also been documented that predatory freshwater fish species, including pike, bass, and walleye, can accumulate high levels of methyl mercury, which can be almost fully absorbed by animals or humans. Mercury causes molecular oxygen to prematurely shed its electrons, which increases the production of reactive oxygen species (ROS). It has been demonstrated that selenium directly binds to mercury or acts as a cofactor for glutathione peroxidase, enhancing its capacity to remove ROS (Ara et al. 2022).

Along with the abovementioned major pollutants, several anthropogenic chemical pollutants that affect aquatic life's endocrine system are ultimately disposed of in water bodies. Some common chemical pollutants found in the aquatic environment are 17-α-ethinylestradiol, citalopram (an SSRI group drug found frequently in sewage and surface waters), azole fungicides, for example, difenconazole, fadrozole, ketoconazole, tebuconazole, cyproconazole, expoxiconazole, imidazole, metoconazole, and nocodazole (Huang et al. 2022), triclosan, bisphenol A, fluorene-9-bisphenol, quercetin, tretinoin, p,p′-DDE isomer (DDT metabolite), atrazine (most common agricultural herbicide used in the United States), propiconazole (inhibitor of CYP51 commonly used in agriculture to control fungal growth on fruit, vegetable, and cereal crops), triclocarban (antimicrobial agent interact with the vertebrate endocrine system), etc. (Dar et al. 2020).

3 The Workflow of Transcriptomics Through Illumina Sequencing

This section describes the overall workflow for transcriptome analysis (Fig. 1). The RNA sequencing process begins with sample collection, which is an important step in transcriptome analysis. The most important step in any type of research is sample collection, that necessitates extreme caution. Later, samples should be stored in RNA and properly frozen in liquid nitrogen. Before sequencing, tissues from different fishes in the same experiment can be pooled in equimolar concentration to reduce the number of samples. Transcripts are the RNA that is expressed inside the cell and must be extracted from the sample using the most appropriate and suitable method, such as the Trizol method. The integrity of purified RNA is an important factor in determining the quality of RNA as measured by Bioanalyser. The samples having

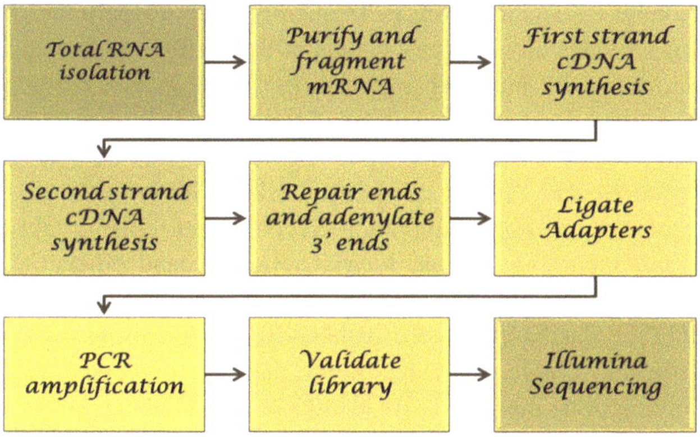

Fig. 1 Flow chart of Illumina RNA-seq

RIN (RNA integrity number) ≥8 are considered best for library preparation (Sundaray et al. 2022). RNA samples are kept at −80 °C until they are needed. Following mRNA purification and fragmentation, random hexamers are used to convert it to cDNA. The library is prepared and validated using the same Bioanalyser instrument with a different chip. The library can be sequenced on any high-throughput sequencer, but Illumina is the preferred and recommended platform for the transcriptomics study.

4 Bioinformatics Analysis

Following sequencing, a large amount of raw data is received, which requires reliable and precise scanning. To analyze the data, numerous bioinformatics tools are freely available in the public domain. Some of the tools are free to use, while others require a subscription fee. The primary tools required for transcriptome data analysis are discussed in this section. FastQC software is used to check the quality of raw reads generated in the FASTQ file format, such as quality score, per base sequence content, per base GC content, sequence length distribution, and duplicate sequences. The raw reads are then subjected to a quality filter, which uses software such as PRINSEQ and Kraken to trim and remove unwanted or low-quality sequences.

Following that, the short quality reads are assembled into a reference transcriptome (Contigs). Depending on the availability of reference genomes, two approaches can be taken. If a reference genome sequence is available, reference-guided assembly is used; otherwise, de novo RNA-seq assembly is used. Trinity, Velvet, and CLC genomics workbench, TopHat-Cufflinks, and other software and packages are available. Once the assembly is complete, the next step is to determine homology in order to determine putative gene descriptions for all of the newly assembled contigs. Local BLAST is frequently used to perform the homology search, which provides all necessary details of the contigs such as gene id, protein id, gene coordinates, hit numbers, expectation (E)-value, etc. All contigs must be functionally annotated as well. The predicted coding sequences were classified using GO (Gene Ontology) assignments based on their functions: biological process, molecular function, and cellular component. Blast2GO is the most common and widely used tool for functional annotation. Several other tools for GO analysis have been developed, including GOstat, EasyGO, Gorilla, and others. The assembled annotated contigs are also compared to the Kyoto Encyclopedia of Genes and Genomes (KEGG) database for gene pathway prediction using the Blast2GO program. Because all contigs in transcriptome data are expressed genes, their EC (Enzyme Commission) number is frequently provided using EC prediction tools such as E-zyme, ECOH, ECPred, and others.

5 Transcriptomics of EDCs-Exposed Aquatic Life

To screen compounds for interactions with the estrogen, androgen, and thyroid pathways, in vivo high-throughput screening (HTS) and computational approaches are still needed. A new and emerging tool for characterizing and quantifying the complete transcriptome is RNA sequencing (RNA-seq). It is accurate and reliable and can determine the amount of RNA present in a sample at any given time. Additionally, it can be used for any species and is not constrained by prior knowledge of the genome because it can find transcripts even when the associated genomic sequence is not available. A promising method used for toxicity monitoring in the recent era is transcriptomics.

Transcriptome analysis was carried out using RNA-seq to explore the underlying factors of the decreased fertility in zebrafish (Porseryd et al. 2018). They used testes from male zebrafish whose reported reduced fertility was linked to developmental exposure to 17α-ethynylestradiol (EE2). Following exposure to 1.2 and 1.6 ng/L EE2, respectively, RNA-seq analysis showed that 249 and 16 genes, respectively, had differential expression. Both exposures changed the expression of 11 genes in the same manner. The hypothesis that the direct effects on estrogen target genes are remediated after recovery of exposure is supported by the finding that the three genes were connected to the GO term response to estrogen in the 1.2 ng/L exposure group and no gene was associated with this term in the 1.6 ng/L exposure group. On exposure to EE2, numerous genes involved in spermatogenesis, for example, cxcl12a, foxc1b, and cox2, were found to be upregulated in zebrafish. The combinatory effects of EE2 and citalopram were also investigated in zebrafish by Porseryd et al. (2017). The study demonstrated that, while having very negligible effects on its own at these low concentrations, citalopram has an impact on the behavior of male zebrafish exposed to very low concentrations of EE2.

Male bivalves may experience reproductive problems and feminization as a result of endocrine-disrupting compounds (EDCs). Blalock et al. (2018) employed a transcriptomics technique to find numerous putative biomarker genes in *Mytilus edulis*, a marine mussel that has been exposed to 17α-ethynylestradiol. Multiple homologs to the cholesterol side chain cleavage complex and other genes in the vertebrate steroidogenesis pathway were identified and used to create the Coastal Biosensor for Endocrine Disruption essay.

In a Saaristo et al. 2010 study, Saaristo et al. examined the effects of 1–4 weeks of EE2 exposure on nest building, courtship, and aggressive behavior in male sand gobies (*Pomatoschistus minutus*) fish. Their research revealed that exposure to EE2 slows nest building and reduces male leading and courtship behavior. Male fish exposed to EE2 were also considerably less aggressive than the control. Additionally, EE2 exposure raised the Vtg and Zrp mRNA levels in males and lowered the hepatosomatic index and ultimately male reproductive success is likely to suffer as a result of these alterations.

To collect crucial information about the potential endocrine activity of propiconazole in fish, Skolness et al. (2013) examined apical reproductive endpoints (fecundity, fertility, hatch) and measured changes in gonadal expression of steroidogenic genes, cholesterol, vitellogenin, and sex steroid concentrations in the plasma of Fathead minnows (*Pimephales promelas*). Females exposed to propiconazole had lower plasma levels of E2 and VTG, and the 500 and 1000 g/L treatment groups had lower egg production. Females also showed signs of a compensatory response, including increased gonad weight and upregulation of genes encoding for important steroidogenic proteins, such as CYP19, CYP17, CYP11A, and steroidogenic acute regulatory protein.

In silico computational analysis was carried out by Basili et al. (2018) to discover several substances that have the potential to disrupt the endocrine system in biological systems. They reported two potential endocrine disruptors quercetin (a flavonoid found in many fruits, vegetables, leaves, and grains used as a dietary supplement) and tretinoin (a retinoic acid used to treat acne and leukemia and acts by forcing APL cells to differentiate and stops them from proliferating) that can alter key biological processes involved in the ovary development of the Largemouth bass (*Micropterus salmoides*). Reproductive targets such as the estrogen receptors, the aromatase (cyp19a), and the vtgr in the ovary were reported to be disrupted by these two identified EDCs.

The disruption of several EDCs in aquatic animals has also been documented in numerous studies using transcriptome analysis (Table 1).

6 Conclusion

According to extensive research in vertebrate species, exposure to anthropogenic chemicals with hormone-like activity has affected numerous genomic pathways over many years. The interactions of the chemicals with nuclear receptors mediate this effect. Furthermore, a number of contaminants have been discovered to obstruct non-genomic (non-classical) pathways, but it is still unknown how this causes

Table 1 Study of different EDCs in fish through transcriptomics

S. no.	EDcs	Fish	Platform	Tissue	Salient findings	Reference
1	EE2	Three-spined stickleback (*Gasterosteus aculeatus*)	Microarray	Liver	Hepatic vitellogenins and choriogenins, along with a number of additional EE2-responsive genes, showed evidence of transcriptional induction.	Katsiadaki et al. (2010)
2	EE2	Guppy	Illumina MiSeq 300	Brain	EE2 altered the expression of 165 transcripts in males, with 88 transcripts downregulated and 77 upregulated, and 120 transcripts in females, with 62 downregulated and 58 upregulated.	Saaristo et al. (2021)
3	Benzo α pyrene (BaP), 17α-ethinylestradiol (EE2)	Polar cod (*Boreogadus saida*)	IluminaNovaSeq 6000	Liver	Several pathways, including the estrogen receptor and aryl hydrocarbon (Ahr) pathways, were found to be enriched. BaP had anti-estrogenic effects as a result of the combination exposure, as evidenced by the reduction in EE2-activated transcription of several estrogen target genes.	Yadetie et al. (2021)
4	17α-Ethinylestradiol (EE2) and progesterone (P4)	Seahorse (*Hippocampus erectus*)	Illumina HiSeq	Ovary, testes, and brood pouch	Both EDCs reduced the male brood pouch growth and lowered the levels of genes linked to spermatogenesis in the testes sometimes resulting in male feminization	Qin et al. (2020)
5	*Tetracapsuloides bryosalmonae* and EE2	Rainbow trout (*Oncorhynchus mykiss*)	Illumina HiSeq 2500	Kidney	Showed a weak immunological reaction that was adversely linked with the overexpression of genes involved in several metabolic and inflammation-resolution pathways	Bailey et al. (2019)

6	Bisphenol a and EE2	Medaka (*Oryzias latipes*)	Illumina Hiseq 2500	Testes	Global variations in gene expression throughout the early stages of gametogenesis in the medaka testes were responsive to chronic BPA and EE2 exposure	Bhandari et al. (2020)
7	17α-Ethinylestradiol (EE2)	Korean rose bitterling (*Rhodeus uyekii*)	Illumina HiSeq 2500	Hepatopancreas and skin	EE2 disturbed the endocrine system and regulated both carcinogenic and apoptotic gene expressions in *R. uyekii*	Kong et al. (2015)
8	Mancozeb	Zebrafish (*Danio rerio*)	Hiseq sequencing platform	Larvae	MZ-induced cardiac developmental toxicity in zebrafish Cyp-related genes (cyp1c2 and cyp3c3) involved in apoptosis of myocardial cells were significantly upregulated after MZ treatment	Wang et al. (2021)
9	17α-Ethinylestradiol (EE2)	Sardine (*Sardinops sagax*) and mackerel (*Scomber japonicus*)	Illumina GAIIx	Liver	EE2 exposure altered expression patterns of key genes involved in important metabolic and physiological processes	Renaud et al. (2019)
10	Triclosan, bisphenol A (BPA) and fluorene-9-bisphenol	Zebrafish (*Danio rerio*)	Illumina Hiseq 4000	Liver	EDCs exposure led to decreased global m^6A (epigenetic marker) level and abnormal expression of m^6A modulators in larvae EDCs exposure disrupted multiple physiological processes including drug Metabolism, sucrose metabolism, fat metabolism and bile secretion	Sun et al. (2020)
11	Methyl-mercury (MeHg)	Largemouth bass (*Micropterus salmoides*)	Microarray	Whole brain	MeHg regulated the expression targets of neuropeptide receptor and steroid signaling, as well as structural components of the cell	Richter et al. (2014)

endocrine disruption. With the advent of new-generation sequencing techniques in recent years, studies on the effect of EDCs on fish physiology and biology at the molecular level have become more time and cost-effective. The data generated by high-throughput sequencers may lay the groundwork for developing new strategies and policies to reduce the use and release of harmful chemicals into the environment.

References

Abadin H, Ashizawa A, Llados F, Stevens YW (2007) Toxicological profile for lead. Agency for Toxic Substances and Disease Registry, Atlanta

Ara A, Jan M, Sofi PA, Rashid M, Lone AA, Dar ZA et al (2022) Reactive oxygen species, oxidative damage and their production, detection in common Bean (Phaseolus vulgaris L.) under water stress conditions. In: Advances in plant defense mechanisms. IntechOpen, London

Aslam S, Yousafzai AM (2017) Chromium toxicity in fish: a review article. J Entomol Zool Stud 5(3):1483–1488

Bailey C, von Siebenthal EW, Rehberger K, Segner H (2019) Transcriptomic analysis of the impacts of ethinylestradiol (EE2) and its consequences for proliferative kidney disease outcome in rainbow trout (*Oncorhynchusmykiss*). Comp Biochem Physiol C Toxicol Pharmacol 222:31–48

Baker ME, Hardiman G (2014) Transcriptional analysis of endocrine disruption using zebrafish and massively parallel sequencing. J Mol Endocrinol 52(3):R241

Basili D, Zhang JL, Herbert J, Kroll K, Denslow ND, Martyniuk CJ et al (2018) In silico computational transcriptomics reveals novel endocrine disruptors in largemouth bass (*Micropterussalmoides*). Environ Sci Technol 52(13):7553–7565

Bhandari RK, Wang X, VomSaal FS, Tillitt DE (2020) Transcriptome analysis of testis reveals the effects of developmental exposure to bisphenola or 17α-ethinylestradiol in medaka (*Oryziaslatipes*). Aquat Toxicol 225:105553

Blalock BJ, Robinson WE, Loguinov A, Vulpe CD, Krick KS, Poynton HC (2018) Transcriptomic and network analyses reveal mechanistic-based biomarkers of endocrine disruption in the marine mussel, Mytilusedulis. Environ Sci Technol 52(16):9419–9430

Dar SA, Chatterjee A, Rather MA, Chetia D, Srivastava PP, Gupta S (2020) Identification, functional characterization and expression profiling of cytochrome p450 1A (CYP1A) gene in Labeo rohita against emamectin benzoate. Int J Biol Macromol 158:1268–1278

Dopp E, Hartmann LM, Florea AM, Rettenmeier AW, Hirner AV (2004) Environmental distribution, analysis, and toxicity of organometal (loid) compounds. Crit Rev Toxicol 34(3):301–333

Dyer CA (2007) Heavy metals as endocrine-disrupting chemicals. In: Endocrine-disrupting chemicals. Humana Press, pp 111–133

Huang T, Zhao Y, He J, Cheng H, Martyniuk CJ (2022) Endocrine disruption by azole fungicides in fish: a review of the evidence. Sci Total Environ 153412:153412

Ishaque A, Ishaque S, Arif A, Abbas HG (2020) Toxic effects of lead on fish and human. Biol Clin Sci Res J 2020(1)

Katsiadaki I, Williams TD, Ball JS, Bean TP, Sanders MB, Wu H et al (2010) Hepatic transcriptomic and metabolomic responses in the stickleback (*Gasterosteusaculeatus*) exposed to ethinyl-estradiol. Aquat Toxicol 97(3):174–187

Kavlock RJ, Daston GP, DeRosa C, Fenner-Crisp P, Gray LE, Kaattari S et al (1996) Research needs for the risk assessment of health and environmental effects of endocrine disruptors: a report of the US EPA-sponsored workshop. Environ Health Perspect 104(suppl 4):715–740

Kim YJ, Lee N, Woo S, Ryu JC, Yum S (2016) Transcriptomic change as evidence for cadmium-induced endocrine disruption in marine fish model of medaka, *Oryziasjavanicus*. Mol Cell Toxicol 12(4):409–420

Kong HJ, Lee IK, Kim J, Kim WJ, Kim HS, Cho WS et al (2015) RNA-Seq-based transcriptome analysis of Korean rose bitterling (*Rhodeusuyekii*) exposed to synthetic estrogen 17-α-ethinylestradiol (EE2). Mar Genomics 24:233–236

Lee JW, Choi H, Hwang UK, Kang JC, Kang YJ, Kim KI, Kim JH (2019) Toxic effects of lead exposure on bioaccumulation, oxidative stress, neurotoxicity, and immune responses in fish: a review. Environ Toxicol Pharmacol 68:101–108

Mitra RS (1984) Protein synthesis in Escherichia coli during recovery from exposure to low levels of Cd2+. Appl Environ Microbiol 47(5):1012–1016

Porseryd T, Kellner M, Caspillo NR, Volkova K, Elabbas L, Ullah S et al (2017) Combinatory effects of low concentrations of 17α-etinylestradiol and citalopram on non-reproductive behavior in adult zebrafish (*Daniorerio*). Aquat Toxicol 193:9–17

Porseryd T, Caspillo NR, Volkova K, Elabbas L, Källman T, Dinnétz P et al (2018) Testis transcriptome alterations in zebrafish (*Daniorerio*) with reduced fertility due to developmental exposure to 17α-ethinylestradiol. Gen Comp Endocrinol 262:44–58

Qin G, Zhang Y, Zhang B, Zhang Y, Liu Y, Lin Q (2020) Environmental estrogens and progestins disturb testis and brood pouch development with modifying transcriptomes in male-pregnancy lined seahorse *Hippocampus erectus*. Sci Total Environ 715:136840

Renaud L, Agarwal N, Richards DJ, Falcinelli S, Hazard ES, Carnevali O et al (2019) Transcriptomic analysis of short-term 17α-ethynylestradiol exposure in two Californian sentinel fish species sardine (*Sardinopssagax*) and mackerel (*Scomberjaponicus*). Environ Pollut 244: 926–937

Richter CA, Martyniuk CJ, Annis ML, Brumbaugh WG, Chasar LC, Denslow ND, Tillitt DE (2014) Methylmercury-induced changes in gene transcription associated with neuroendocrine disruption in largemouth bass (*Micropterussalmoides*). Gen Comp Endocrinol 203:215–224

Rossman TG, Roy NK, Lin WC (1992) Is cadmium genotoxic? IARC Sci Publ 118:367–375

Saaristo M, Craft JA, Lehtonen KK, Lindström K (2010) Exposure to 17α-ethinylestradiol impairs courtship and aggressive behaviour of male sand gobies (*Pomatoschistusminutus*). Chemosphere 79(5):541–546

Saaristo M, Craft JA, Tyagi S, Johnstone CP, Allinson M, Ibrahim KS, Wong BB (2021) Transcriptome-wide changes associated with the reproductive behaviour of male guppies exposed to 17α-ethinylestradiol. Environ Pollut 270:116286

Sanfeliu C, Sebastià J, Cristòfol R, Rodríguez-Farré E (2003) Neurotoxicity of organomercurial compounds. Neurotox Res 5(4):283–305

Skolness SY, Blanksma CA, Cavallin JE, Churchill JJ, Durhan EJ, Jensen KM et al (2013) Propiconazole inhibits steroidogenesis and reproduction in the fathead minnow (*Pimephalespromelas*). Toxicol Sci 132(2):284–297

Stohs SJ, Bagchi D (1995) Oxidative mechanisms in the toxicity of metal ions. Free Radic Biol Med 18(2):321–336

Sun L, Ling Y, Jiang J, Wang D, Wang J, Li J, Wang X, Wang H (2020) Differential mechanisms regarding triclosan vs. bisphenol a and fluorene-9-bisphenol induced zebrafish lipid-metabolism disorders by RNA-Seq. Chemosphere 251:126318

Sundaray JK, Dixit S, Rather A, Rasal KD, Sahoo L (2022) Aquaculture omics: an update on the current status of research and data analysis. Mar Genomics 64:100967

Tchounwou PB, Yedjou CG, Patlolla AK, Sutton DJ (2012) Heavy metal toxicity and the environment. In: Molecular, clinical and environmental toxicology. Springer, Basel, pp 133–164

Waalkes MP, Liu J, Chen H, Xie Y, Achanzar WE, Zhou YS, Cheng M-L, Diwan BA (2004) Estrogen signaling in livers of male mice with hepatocellular carcinoma induced by exposure to arsenic in utero. J Natl Cancer Inst 96(6):466–474

Wang Y, Yu Z, Fan Z, Fang Y, He L, Peng M, Liu C (2021) Cardiac developmental toxicity and transcriptome analyses of zebrafish (*Daniorerio*) embryos exposed to Mancozeb. Ecotoxicol Environ Saf 226:112798

Yadetie F, Brun NR, Vieweg I, Nahrgang J, Karlsen OA, Goksøyr A (2021) Transcriptome responses in polar cod (*Boreogadussaida*) liver slice culture exposed to benzo [a] pyrene and ethynylestradiol: insights into anti-estrogenic effects. Toxicol In Vitro 75:105193

Fundamentals of Genotoxicity and Antiparasitic Drugs Associated with Genotoxicity in Fish

J. Mori Krinal, G. Ad Viral, Mohd Ashraf Rather, and Sujit Kumar

1 Introduction

Fisheries resources are contributing enormously toward providing nutritional as well as livelihood security around the world. These resources are highly vulnerable to the anthropogenic activities such as overexploitation, pollution, etc. Natural waterbodies are receiving significantly high number of chemicals through industrial and sewage discharge, along with runoff water from the agricultural field. Many of such chemicals are affecting the aquatic life by causing direct poisoning or leading to toxic effects at various levels and impacting important physiological pathways.

Over the last few decades, aquatic ecosystems have become more polluted due to industrialization, dense urbanization, and agricultural practices, and have received huge quantities of various physiologically active compounds, including organic and inorganic chemicals.

There are a large number of persistent pollutants that are present in the environment at extremely low levels, yet they may build up in aquatic creatures' tissues through bio-magnification at concentrations that are several folds higher than safety level. Additionally, there is rising worry over the contaminating effects of new, uncontrolled pollutants in water, including pharmaceuticals, diagnostic aids, steroids and hormones, and personal care items (Boleda et al. 2009; Santos et al. 2010; Aklakur et al. 2016). In aquatic habitats, organisms are frequently exposed to a complex combination of chemicals, including parent substances and their transformation products. These pollutants can harm biological diversity, organ function,

J. Mori Krinal · G. Ad Viral · S. Kumar (✉)
Department of Fish Biotechnology, Postgraduate Institute of Fisheries Education and Research, Kamdhenu University, Himmatnagar, India

M. A. Rather
Division of Fish Genetics and Biotechnology, Sher-e-Kashmir University of Agricultural Sciences and Technology, Kashmir, Rangil, Ganderbal, Jammu and Kashmir, India

M. A. Rather et al. (eds.), *Xenobiotics in Aquatic Animals*,
https://doi.org/10.1007/978-981-99-1214-8_20

population size, and ecosystem health (Ginebreda et al. 2014; Vorosmarty et al. 2010; Rather et al. 2018). As a result of their involvement in numerous pathological processes, such as carcinogenesis and reproductive effects, as well as the fact that they cause harm that extends beyond the individual and is present in succeeding generations, a wide range of environmental contaminants that directly or indirectly affect DNA have significant toxicological relevance (Devaux et al. 2011; Lewis and Galloway 2009; Dar et al. 2020).

Disease management is always a major challenge for aquaculturist, and most often farmers face disease outbreaks in their culture system. Parasitic disease is one of the highly prevalent problems that lead to morbidity and mortality and cause serious economic and ecological damage to aquaculture (Shinn et al. 2015). Economic losses are because of mortality of culture animals, expenditure on chemicals and drugs, reduced growth performance, and increased feed conversion ratios. Some of the parasites seriously impacting the aquaculture are argulus, lernaea, white spot, trichodynia, etc.

Various antiparasitic strategies are used in aquaculture, mainly dominated by the use of chemicals and drugs. Different chemicals and drugs such as diflubenzuron, teflubenzuron, emamectin benzoate, cypermethrin, deltamethrin, ivermectin, fenbendazole, chloramine-T, copper sulfate, malachite green, levamisole, metronidazole, potassium permanganate, trichlorfon, etc., are used in aquaculture to control the parasites (Mishra et al. 2017). These chemicals are used extensively, and most of the time higher doses are used for the rapid removal of parasites. Many of these chemicals are known to cause toxicity to the fish as well.

Genotoxicity tests have been used successfully as indicators of toxicity of many drugs and environmental pollutants, and fish is considered a good model for these types of analysis due to its easier handling and monitoring under controlled conditions, and moreover, low cost compared to other models. Genotoxicity studies analyze how the cell's integrity is impacted by the damaging effect on the genetic material (DNA, RNA) of the cell.

Exposure to various types of xenobiotics that include drugs, pollutants, etc., may result in aberrant physiological reactions and have detrimental consequences on the growth, development, behavior, and reproduction of living things (Bistodeau et al. 2006; Giesy et al. 2000; Ginebreda et al. 2014; Eganhouse and Sherblom 2001; Lee and Peart 2000). The exposure to particular kinds of DNA-damaging chemicals (drugs/pesticides/pollutants) has been linked to the development of epizootic neoplasms in a range of ectothermic species, including fish, echinoderms, and shellfish (Mix 1986; Malins et al. 1990; Bolognesi 1990; Myers et al. 1987, 2003, 2008). Genotoxicity is one of the major issues associated with the many chemicals received into the aquatic system either unintentionally or intentionally as therapeutics agents.

1.1 Genotoxicity and Underlying Mechanism

Genotoxicity is one of the major problems associated with damage to the genetic material caused by genotoxin and ultimately affecting DNA integrity and sometimes may lead to carcinogenic. A genotoxin is a substance that has the property to affect DNA either directly or indirectly. Genotoxins can affect organisms by affecting their genetic material through different mechanisms. It includes chemicals as well as radiation, along with some other biological factors. The genotoxic nature of any chemical can be tested using various in vitro biochemical and cell-based assays, namely, bacterial mutagenesis (Ames test), comet assay, chromosomal aberration, mouse lymphoma assay, and the micronucleus test. Genotoxins are classified based on their effects as carcinogenic, mutagenic, and teratogenic or birth defect-causing agents (Mohamed et al. 2017).

The genotoxic substances either interact directly or indirectly with DNA and damage the cellular genetic materials. DNA damage associated with the covalent attachment of a chemical to DNA is known as DNA adduct. Other mechanisms of genotoxicity include chromosomal breakage, deletion of nucleotide from DNA fusion of uni-nuclear cell to form a multinuclear cell, mis-segregation, non-disjunction, etc. (Savale 2018; Nandanpawar et al. 2018). At the level of the individual cell, genotoxins themselves impart the primary mechanisms for genotoxicity. It necessitates the uptake of nanomaterials by cells and frequently follows by interactions with biomolecules. This kind of system has two subclasses (i.e., direct and indirect acting agent).

1.1.1 Direct Acting Agents

They directly harm the genetic material through physical or chemical interaction. This might involve the development of DNA damage at certain locations inside the DNA molecule, which cause mutagenesis because they may render the nucleotide unstable. They cause damage that results in its loss, making it a fundamental location that is frequently the focus of repairs that end in strand breaks (Jenkins et al. 2005; Doak et al. 2007).

1.1.2 Indirect Acting Agents

Direct DNA disorder can only be brought about if the genotoxins can enter the cell nucleus, or if they are allowed to remain free in the cytoplasm and come into contact with DNA directly during mitosis (i.e., cell division) when the nuclear membrane ruptures (Ferguson and Denny 2007; Berry et al. 2007). They cause genetic harm by generating intermediary biomolecules, which are often seen in the cell division cycle (Nativo et al. 2008). Since DNA replication and cell division are complex, multifactorial processes involving several proteins, they are excellent targets for chemicals with indirect genotoxic effects (Fadeel et al. 2012). The production of oxidative stress may result in indirect DNA damage in addition to interfering with DNA replication. (Federico et al. 2007; Karihtala et al. 2009) Interactions with non-DNA targets that result in genotoxic effects are known as indirect mechanisms of genotoxicity. Recent studies on proteins have concentrated on the inhibition of repair enzymes (such as OGG1, XPD, and Ni), cell cycle regulators (such as p53, Rb, and cyclins), apoptosis-related gene products (such as p53, bax, and bcl-2), nuclear lamins, antioxidant defense proteins (such as glutathione), metabolizing enzymes, and tubulins of the mitotic/meiotic spindle.

Overall the direct interaction of genotoxins may cause different types of lesions, including DNA adducts, breaks, and damage that ultimately may lead to mutation by impacting gene, chromosome, or genome, whereas indirect interactions are mediated by means of generation of reactive-free radicals, damage to the enzymes responsible for DNA repair, inhibition of spindle formation, and finally affecting the cell cycle that has ultimate effects on cell proliferation and end with carcinogenesis (Kirsch-Volders et al. 2000).

1.1.3 Genotoxicity Assessment Techniques

As part of the safety evaluation process, regulatory bodies from all over the world need information on the genotoxic potential of various chemicals and drugs either used in fisheries and aquaculture or reaches to the aquatic system along with runoff water or industrial discharge. Preliminary research is required to determine the fundamental toxicological status of these chemicals. The safety aspects are assessed using toxicological data, which will aid in anticipating the chemical's potential risk analysis. Because substances that test positive in these assays have the potential of carcinogens and mutagens, they have mostly been employed to forecast carcinogenicity and genotoxicity (Savale 2018).

The genotoxic potential of any chemicals is analyzed following a genotoxicity assay. Two approaches to genotoxicity assay include in vitro analysis and in vivo analysis. The in vitro test is used to explore the cellular genetic material-damaging potential of the different substrates, products, or environmental components. Cytogenetic studies incorporating different mammalian cells can be employed for such studies. Chromosome and chromatid gaps, chromosome breakage, chromatid deletions, fragmentation, translocation, complex rearrangements, and many other abnormalities can be found in cells exposed to genotoxic substances. MNT test, RAPD, and COMMET assay are examples of frequently used in vitro methods. The goal of in vivo testing is to assess the possibility of DNA damage that can alter the chromosomal structure or disrupt the mitotic apparatus, and may lead to chromosome number change. ADME and DNA repair are two factors that may affect genotoxicity. It can also find genotoxic substances that in vitro assays failed to pick up. As per the publication data on genotoxicity research, the Ames test was featured in the highest percentage followed by comet assay and MN test, whereas other strategies are employed at significantly low level (Turkez et al. 2017).

1.1.4 Micronuclei Test (MNT)

The purpose of this test is to identify chemicals that induce cytogenetic damage, resulting in the production of micronuclei containing lagging chromosomal fragments or complete chromosomes. A micronucleus is a circular cytoplasmic structure (Fig. 1) that includes chromatin mass (DNA). When reproducing cell populations are exposed to chromosomal breakage by clastogens (agents that cause chromosomal breakage) or chromosome loss due to mitotic spindle failure, micronuclei form. Micronuclei have the same morphology as the larger nuclei but are smaller (Fig. 1). It is not related or linked to the primary nuclei. The primary nuclei may be touched but not overlapping. It is also non-retractile and so distinguishable from staining particles. It is a quick approach for determining the mutagenicity of different chemicals (Hosseinimehr et al. 2017).

1.1.5 Comet Assay

The comet assay/single-cell gel electrophoresis is considered a highly sensitive and rapid method for measuring DNA damage and repair in a single eukaryotic cell (Turkez et al. 2017). This assay is based on the fact that DNA with a break-in loop loose supercoiling and freely moves toward an anode during electrophoresis at high pH. The results observed by fluorescence microscopy show a comet containing a distinct head and tail (Fig. 2). The head represents intact DNA, while the tail represents damaged or broken DNA. The intensity and length of the comet tail compared to the head reflects are used to quantify the DNA damage.

1.1.6 Chromosomal Aberrations

Any aberration associated with the shape, size, or structure of chromosomes is called chromosome aberration. Chromosomal aberrations test allows studying the

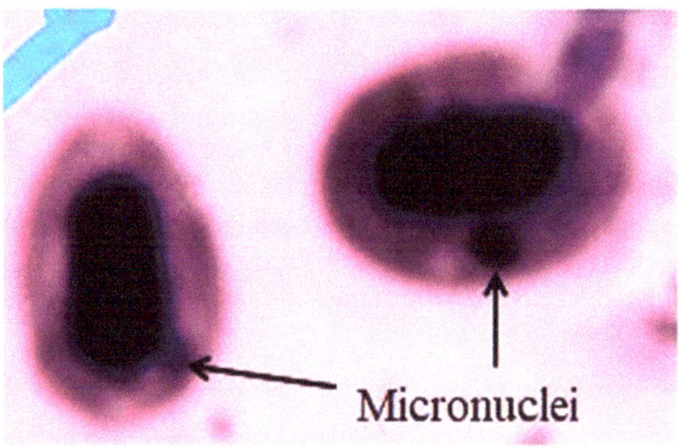

Fig. 1 Image showing micronucleus

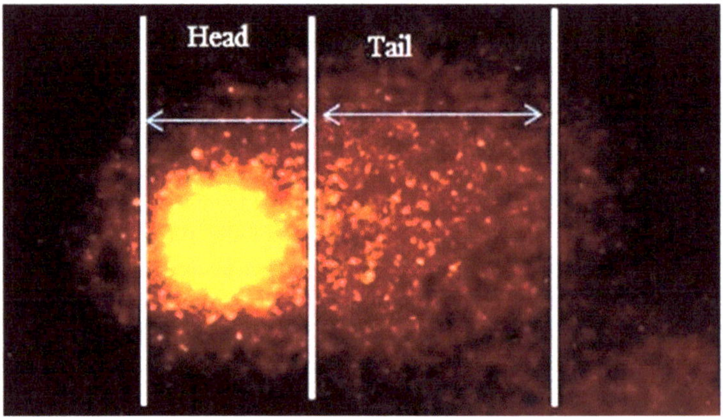

Fig. 2 DNA damage as shown in comet assay (head and tail portion)

structural and numerical abnormality associated with chromosomes. Sometimes errors in repair also lead to chromosomal aberrations. They are translated by chromosome fusion, translocations (exchange of chromosome segments), deletion (losses of a chromosome segment), and duplications (Himri et al. 2018).

1.1.7 Sister Chromatid Exchange

Sister chromatid exchange (SCE) is the exchange of genetic material between two identical sister chromatids. The test of sister chromatid exchange allows revealing the mutual exchanges of genetic elements between sister chromatids during the S phase of the cellular cycle. These breaks in both DNA strands facilitate such exchange. Sister chromatid exchanges are formed by the rupture, exchange, and repair between DNA molecules in homologous regions in the chromatids of duplicating chromosomes. This assay determines reciprocal exchanges of chromatid segments prevailing at low levels in untreated cells (Himri et al. 2018).

1.1.8 Ames Test

The Bacterial Reverse Mutation Assay is also known as the Ames test. It is used in laboratories to test for gene mutation. The technique uses many different bacterial strains to compare the different changes in the genetic material. The result of this test detects the majority of genotoxic carcinogens and genetic changes; the types of mutations detected are frameshift and base substitutions (Savale 2018).

2 Toxicity Associated with Antiparasitic Drugs

Toxicity is a chemical-induced adverse effect on the living organism, and sometimes the extent of damage is very high and may lead to the death of the organism. The toxicity can be classified into acute toxicity (short-term exposure effect), chronic toxicity (long-term exposure effect), local toxicity, and systematic toxicity. Antiparasitic drugs primarily act on parasites but also cause toxicity to the host fish as well, especially at higher concentrations. Several studies with different fish species have been performed to understand the toxicity potential of antiparasitic drugs.

Many antiparasitic drugs used in aquaculture are meant for use in agriculture as a pesticide. The exposure of such chemicals is not limited only to the pest but also reaches the aquatic system and affect many non-target organisms such as fish, gastropods, crustaceans, etc. Exposure to such chemicals at various concentration influences the important hematological parameters such as blood cells, plasma, and serum-level alterations that lead to histological changes involving gill, muscle, liver, kidneys, etc. Different categories of pesticides are also known to damage DNA and induce the various level of genotoxicity. Continuous usage of these chemicals can affect the aquatic systems at a severe level as a result of increased bioaccumulation in the food chain (Tahir et al. 2021).

Patil et al. (2022) comprehensively performed a nationwide survey based on a questionnaire to gather information about the use of chemicals in freshwater and shrimp aquaculture in India. They reported 14 mg PCU^{-1} (population correction unit) use of the antiparasitic drug in India. They reported the application of nine antiparasitic chemicals with the highest use by pangasius farmers followed by Indian major carps and other freshwater fishes. The order of use of various antiparasitic chemicals is deltamethrin (11.14%), quinalphos (5.14%), ivermectin (4.84%), amitraz (4.80%), cypermethrin (4.48%), and albendazole (3.64%). Many aqua farmers use a combination of multiple compounds for parasite management in a crop cycle, whereas around 41.13% of farmers reported use of single antiparasitic compound.

Sujin et al. (2020) investigated the toxicity caused by fipronil in Zebrafish (*Danio rerio*). The zebrafish were treated with 5 μg/L, 25 μg/L, and 50 μg/L of fipronil for 12, 24, and 72 h. To closely observe the toxic effects, 12 h and 24 h of additional time points were set in the exposure test. They reported the adverse effects of fipronil on fish.

Satpathy and Parida (2020) reported dose- and time-dependent increase in the mortality rate, and anxiety signs in the form of behavioral changes in response to different test concentrations of Kandhamalhaladi. They also suggested the potential use of Kandhamalhaladi supplemented with feed for the treatment of fish diseases because of its inhibitory effects on the pathogen and lower toxic effects.

Rhaul et al. (2016) investigated the toxic effects of ivermectin on zebrafish juvenile and adults and 96 h-LC10 values were 14.0 and 55.4 μg/L, respectively. The LC10 of embryo after 96 h was found to be 147.1 μg/L, and increased developmental defects and hatching failure were observed at high concentrations of drug (>400 μg/L), whereas biochemical and behavioral responses such as lethargy were reported at even lower concentrations also (<60 μg/L) in all life stages.

Thiripurasundari et al. (2014) reported ivermectin (7 μg/L)-associated neuronal degeneration and necrosis in brain tissue, along with hepatotoxic effects (hepatic cell degradation, vacuolations) at 24 h post-exposure in *Catla catla*. The results of this study concluded that zebrafish can tolerate even higher concentrations of ivermectin compared to catla for prolonged duration.

Hanqing et al. (2018) analyzed acute toxicity of fipronil (insecticide) on zebrafish embryos after 75 h of fertilization and observed defective embryonic development such as bent spine and shortened body length, and the 96-h LC50 value was found to be 459 μg/L. The result also revealed the presence of 44 differentially expressed genes, 10 GO terms, and 3 KEGG pathways overlapped among the three concentrations.

Wang et al. (2018) reported the LC50 values of KRM (pesticide) following 96 h semi-static test on *Danio rerio* for different life stages such as embryonic, larval, juvenile, and adult stages, and it was found to be in the range of 0.034–0.61 mg/L, which were higher than CYP ranging from 1.05 to 4.42 mg/L. Pesticide mixtures of CYP and KRM exhibited a synergistic effect on embryonic zebrafish.

Malachite green is applied against ectoparasite and external fungus on fish and fish eggs (Culp and Beland 1996) in freshwater. The dye seems to act as an irreversible respiratory enzyme poison (Alderman and Clifton-Hadley 1993). The tissue level of malachite green especially when co-administrated with formalin accumulates in exposed fish to a level greater than the initial exposure concentration (Clifton-Hadley and Alderman 1987). It is also effective against proliferative kidney disease of rainbow trout, *Oncorhynchus mykiss*. It is also found effective against established clinical or subclinical infection caused by the ciliate *Ichthyophthirius multifiliis* exposed under laboratory conditions (Clifton-Hadley and Alderman 1987). It is now banned in many countries mainly due to its higher rate of bioaccumulation in tissue (Alderman and Clifton-Hadley 1993) and associated teratogenic and carcinogenic effects (Meyer and Jorgenson 1983).

Ivermectin (IVM) is a broad-acting anti-helminthic used for treatment in animals and fishes as well. This study reported that concentrations as low as 0.25 μg/L affect the swimming behavior of zebrafish and also affect the feed intake. Further, the effects on weight gain were more in male zebrafish than in females. Fish exposed to 25 μg/L showed darker coloration and mild curvature of the spine (Domingues et al. 2015).

2.1 Genotoxicity Study in Fishes

Obiakor et al. (2021) performed in vivo genotoxicity associated with Sb in the silver perch RBC, and fishes were exposed to sublethal concentrations of 0.4, 0.9, and 1.8 mg/L Sb (III), and 0.9, 2, and 5 mg/LSb (V) for 14 days followed by comet assay and MN test. The comet assay analysis of all Sb (III) concentrations showed non-dose-dependent DNA damage after 48 h, whereas afterward no further increase in DNA damage was witnessed in relation to control. Fishes of all the treatments for Sb (III) died after 14 days. A significant increase in the cytotoxicity index was observed at 1.8 mg/L Sb (III) concentration post 2 days of exposure.

Mehra and Chadha (2021) evaluated the genotoxicity potential of naphthalene-2-sulfonate (2NS) in freshwater fish (*Channa punctatus*). They selected two sublethal concentrations of 2.38 g/15 g body weight (1/4 of LC50), and 4.77 g/15 g body weight (1/2 of LC50) for acute exposure. They analyzed blood sample initially for 24, 48, 72, 96 h., and later after 30 days and 60 days post-exposure to the subchronic effects. They reported dose and time-dependent DNA damage by comet and micronucleus tests after 60 days. Further recovery from damaged DNA was also observed after exposure cessation to 2NS.

Canedo and Rocha (2021) summarized the existing literature on genotoxicity of different pollutants in zebrafish model system. Various types of DNA damage reported against vast categories of pollutants include single-strand breaks, double-strand breaks, adduct formation, and changes in important gene expression involved in DNA damage repair. Very limited studies have been reported related to the capacity of fishes to repair the damage.

Mir et al. (2014) analyzed the genotoxicity potential of a few heavy metal ions and polycyclic hydrocarbons in fishes. These genotoxicants are often disposed of in free-flowing waterbodies, and the fauna, particularly, fishes, are highly affected. These genotoxicants affect the growth as well as the reproductive performance of fish such as fecundity. The liver and kidney are the primary organs that play a pivotal role in detoxification through various enzymatic actions such as metallothionin, SOD, alanine aminotransferase, etc. Genotoxicants damage the liver by affecting its histology and biochemistry.

Pandey et al. (2018) used the comet test and random amplified polymorphic DNA to investigate profenofos (organophosphate)-induced genotoxicity (DNA damage) in freshwater fish *Channa punctatus* following in vivo exposure in a semi-static setup after exposing fish to a sublethal dose of 1.16 ppb (1/2 of LC50), and DNA damage was compared with control fish using erythrocytes. The RAPD profile of fish specimens exposed to PFF showed the appearance/disappearance of bands as well as an increase/decrease in band intensity compared to the control, which indicates the development of DNA break in treatment. Similarly, the comet assay result also indicated increased tail DNA percentage in treatment. Finally, they concluded that the pesticide exerts a genotoxic effect on the fish.

The genotoxic effects of terbufos and fenthion were evaluated individually as well as in combination on HepG2 cells and zebrafish embryos (Wahyuni et al. 2021) by neutral comet and H2AX phosphorylation assay. The result showed the development of double-strand break in DNA of cells received treatment of terbufos and/or fenthion. The toxic as well as the genotoxic effect was significantly less at an equimolar (40 μM) combination of these pesticides without impacting homologous recombination repair activity of DNA compared to individual exposure to terbufos or fenthion. A decreased expression of *Xrcc2* gene (DNA homologous repair gene) was observed in HepG2 cells after exposure to terbufos and/or fenthion. Moreover, the combined pesticides decreased *Xrcc6* gene expression (DNA non-homologous end joining repair genes). Only fenthion reduced the expression of the HR genes (Rad51 and Rad18) at 24 h in zebrafish embryos. The combined pesticides increased the expression of the HR genes (Rad51 and Xrcc2) after 48 h of exposure, but terbufos or fenthion reduced the expression of these four genes (Rad51, Rad18, Xrcc2, Xrcc6). Additionally, fenthion or the combination pesticide dramatically reduced the hatching rate of zebrafish embryos at 72 hpf.

Amaeze et al. (2020) assessed the hematological or genotoxic effects of sublethal concentrations of several agricultural pesticides (deltamethrin, carbofuran, dichlorvos, chlorpyrifos, cypermethrin, dimethoate, fipronil, abamectin, lambda-cyhalothrin, and paraquat) in *Clarias gariepinus*. The LC50–96 h. was initially estimated individually for every pesticide. They were then subjected to sublethal doses (1/100th 96 h LC50) over a 21-day period. In every instance, a control experiment was observed with catfish housed in municipal water that had been dechlorinated. The LC50–96 h. was found to vary widely for different pesticides from 2.043 μg/L (Lambda-cyhalothrin) to 10284.288 μg/L (Paraquat). Further, significant differences ($P < 0.05$) were also reported between various hematological parameters between treatment and control. A significantly higher proportion of micronucleus and nuclear abnormalities were also observed in pesticide-exposed fishes.

Ali et al. (2008) studied CPF (chlorpyrifos)-induced DNA damage in *Channa punctatus* exposed to various concentrations of CPF using MNT and comet assay. The LC50–96 h in a semi-static system was found to be 811.98 μg/L, and concentration-dependent DNA damage was reported based on MNT and comet assay analysis in erythrocytes and gill.

Marco et al. (2016) reported monceren 250 SC fungicide-induced DNA damage based on genotoxicity parameters of the alkaline comet assay such as tail length, tail moment, and tail intensity in zebrafish embryos. They reported a significant rise in these parameters in zebrafish embryos compared with control embryos and also concluded that the tail intensity was the most appropriate parameter for genotoxicity levels assessment in zebrafish embryos.

2.2 Gene Expression Associated with Genotoxicity

Derikvandy et al. (2020) analyzed the toxic effects of different percentages (0, 0.5, 1, and 2) of untreated waste water in zebrafish for 21 days using biochemical indicators, oxidative stress markers, and gene expression analysis involved in detoxification in the liver. They reported upregulation of *sod1, gset-1a,* and *gpx1a* gene in the liver after exposure to 2% sewage for 21 days. Significant upregulation of *Gsr, Ces2* and *Cyp-1a, Mt1,* and *Mt2* genes was also observed in the liver cell compared to control. The levels of lactate dehydrogenase, antioxidants, aspartate aminotransferase, malonaldehyde, and alkaline phosphatase were also increased significantly in fish that received 1 and 2% of sewage water compared to the control group ($P < 0.01$). The alanine aminotransferase activity was increased significantly ($P < 0.01$) in fish that received 2% sewage, whereas a significant reduction in gamma-glutamyl-transferase level was observed in the same treatment. Catalase activity invariably increased in all treatments compared to control.

Verbueken et al. (2018) assessed cytochrome P450 activity in zebrafish embryos and larvae until 14 day post-fertilization (dpf) by using a nonspecific CYP substrate, (benzyloxy-methyl-resorufin (BOMR), CYP1-specific substrate like 7-ethoxyresorufin (ER) and transcript analysis of CYP1A, CYP1B1, CYP1C1, CYP1C2, CYP2K6, CYP3A65, CYP3C1, phase II enzymes uridine di-phosphate glucuronosyltransferase 1A1 (UGT1A1), sulfo-transferase 1st1 (SULT1ST1), and an ATP-binding cassette (ABC) drug transporter like abcb4 gene in zebrafish for 32 dpf. Using quantitative PCR, they reported a low to undetectable role of these proteins in the disposition of xenobiotics before 72 h post-fertilization.

Gaaied et al. 2019 reported upregulation of *Gstp1* gene expression at a lower dose of 2,4-D (0.02 mg/L) and a downregulation at the highest dose (0.8 mg/L) in 96 hpf zebrafish larvae along with a significant increase in various enzymatic activities.

Jin et al. (2010) evaluated the effects of atrazine (ATZ) treatment on female zebrafish and reported a significant change in the level of SOD and CAT, along with GSH and MDA amount in the liver. The transcript levels for these genes encoding antioxidant proteins, such as Cu/Zn–Sod, Mn–Sod, Cat, and Gpx, were also found to be significantly upregulated in the liver of fish exposed to various concentrations of ATZ for 14 days. The levels of *Ucp-2* and *Bcl-2* were also altered significantly in high ATZ treatment groups.

Cong et al. (2020) studied the toxic effects of DMP (dimethyl phthalate) exposure on adult zebrafish liver. The LC50–96 h was found to be 45.8 mg/L. The malondialdehyde level was found to increase at 0.5, 4.6, and 22.9 mg/L concentrations for 96 h. whereas lower antioxidant capacity was observed compared to the control solvent group. The superoxide dismutase level was significantly higher after 24 h exposure at low concentration compared to 0 h and then reduced at high concentrations after exposure for 96 h. The catalase and glutathione S-transferase activities were significantly reduced after 96 h of exposure to high concentrations of DMP, with the up- or downregulation of the related transcriptional expression.

Gonzalez et al. (2006) evaluated the effects of cadmium exposure on selected transcript levels in the gill, liver, skeletal muscles, and brain of the zebrafish. The Cd level was comparatively low in skeletal muscle, but the transcript levels of *mt1,cyt, bax, gadd,* and *rad51* genes were found to be upregulated on the seventh day. The *c-jun, pyc,* and *tap* genes were also found to be upregulated in muscles at day 21, whereas the transcript levels of *bax, gadd,* and *rad51* genes reduced to basal levels. Although the level of Cd was the highest in the liver, only *mt1* and *c-jun* genes displayed a differential expression in the liver after 21 days. The *mt1, mt2,* and *c-jun* gene expression levels were found to be increased in the brain after 21 days. The gill response was significantly higher post 7 days of exposure as evident through differential expression of oxidative stress–response hsp70 and mitochondrial sod genes, along with genes involved in mitochondrial metabolism and metal detoxification.

3 Conclusion

Genotoxicity is one of the major concerns associated with various chemicals present in pollutants and drugs. Fishes exposed to such chemicals are liable to exhibit abnormal behavior and reduced performance. Antiparasitic drugs used in aquaculture are potential xenobiotics that cause genotoxicity and require urgent attention at policy level to regulate its use. Many such antiparasitic drugs are pesticides and reach the natural waterbody with runoff water. Genotoxicity analysis of fish sample collected from different natural waterbody can help in understanding the safety aspects of aquatic animal in the natural waterbody.

References

Aklakur M, Asharf Rather M, Kumar N (2016) Nanodelivery: an emerging avenue for nutraceuticals and drug delivery. Crit Rev Food Sci Nutr 56(14):2352–2361

Alderman DJ, Clifton-Hadley RS (1993) Malachite green: a pharmacokinetic study in rainbow trout, Oncorhynchus mykiss (Walbaum). J Fish Dis 16(4):297–311. https://doi.org/10.1111/j.1365-2761.1993.tb00864.x

Ali D, Nagpure NS, Kumar S, Kumar R, Kushwaha B (2008) Genotoxicity assessment of acute exposure of chlorpyrifos to freshwater fish Channa punctatus (Bloch) using micronucleus assay and alkaline single-cell gel electrophoresis. Chemosphere 71(10):1823–1831. https://doi.org/10.1016/j.chemosphere.2008.02.007

Amaeze NH, Komolafe BO, Salako AF, Akagha KK, Briggs TMD, Olatinwo OO, Femi MA (2020) Comparative assessment of the acute toxicity, haematological and genotoxic effects of ten commonly used pesticides on the African catfish, Clariasgariepinus Burchell 1822. Heliyon 6(8):e04768. https://doi.org/10.1016/j.heliyon.2020.e04768

Berry CC, de la Funente JM, Chu SW, Curtis AS (2007) Nuclear localization of HIV-1 tat functionalized gold nanoparticles. IEEE Trans Nanobioscience 6:262–269

Bistodeau TJ, Barber LB, Bartell SE, Cediel RA, Grove KJ (2006) Larval exposure to environmentally relevant mixtures of alkylphenolethoxylates reduces reproductive competence in male fathead minnows. Aquat Toxicol 79(3):268–277

Boleda MR, Galceran MT, Ventura F (2009) Monitoring of opiates, cannabinoids and their metabolites in wastewater, surface Water and finished water in Catalonia. Spain Water Res 43:126–136

Bolognesi C (1990) Carcinogenic and mutagenic effects of pollutants in marine organisms: a review. In: Grandjean E (ed) Carcinogenic, mutagenic, and teratogenic marine pollutants: impact on human Health and the environment. Portfolio Publishing, London, pp 67–83

Canedo A, Rocha TL (2021) Zebrafish (Danio rerio) using as model for genotoxicity and DNA repair assessments: historical review, current status and trends. Sci Total Environ 762:144084. https://doi.org/10.1016/j.scitotenv.2020.144084

Clifton-Hadley RS, Alderman DJ (1987) The effects of malachite green upon proliferative kidney disease. J Fish Dis 10(2):101–107. https://doi.org/10.1111/j.1365-2761.1987.tb00725.x

Cong B, Liu C, Wang L, Chai Y (2020) The impact on antioxidant enzyme activity and related gene expression following adult zebrafish (Danio rerio) exposure to dimethyl phthalate. Animals 10(4):717. https://doi.org/10.3390/ani10040717

Culp S, Beland FA (1996) Malachite green: a toxicological review. J Am Coll Toxicol 15:219. https://doi.org/10.3109/10915819609008715

Dar SA, Chatterjee A, Rather MA, Chetia D, Srivastava PP, Gupta S (2020) Identification, functional characterization and expression profiling of cytochrome p450 1A (CYP1A) gene in Labeo rohita against emamectin benzoate. Int J Biol Macromol 158:1268–1278

Derikvandy A, Pourkhabbaz HR, Banaee M, Sureda A, Haghi N, Pourkhabbaz AR (2020) Genotoxicity and oxidative damage in zebrafish (Danio rerio) after exposure to effluent from ethyl alcohol industry. Chemosphere 251:126609. https://doi.org/10.1016/j.chemosphere.2020.126609

Devaux A, Fiat L, Gillet C, Bony S (2011) Reproduction impairment following paternal genotoxin exposure in brown trout Salmo trutta and Arctic charr Salvelinusalpinus. Aquat Toxicol 101: 405–411

Doak SH, Johnson G, Quick E, Jenkins GJS, Parry J (2007) Mechanistic influences for mutation induction curves after exposure to DNA-reactive carcinogens. Cancer Res 67:3904–3911

Domingues I, Oliveira R, Soares A, Amorim M (2015) Effects of ivermectin on Danio rerio: a multiple endpoint approach: behaviour, weight and subcellular markers. Ecotoxicology 25:491. https://doi.org/10.1007/s10646-015-1607-5

Eganhouse RP, Sherblom PM (2001) Anthropogenic organic contaminants in the effluent of a combined sewer overflow: impact on Boston harbour. Marine Environ Res 51:51

Fadeel A, Pietroiusti A, Shvedova A (2012) Adverse effects of engineered nanomaterials: exposure, toxicology, and impact on human health. Academic Press, London

Federico A, Morgillo F, Tuccillo C, Ciardiello F, Loguercioc C (2007) Chronic inflammation and oxidative stress in human carcinogenesis. Int J Cancer 121:2381–2386

Ferguson LR, Denny WA (2007) Genotoxicity of non-covalent interactions: DNA intercalators. Mutat Res 623:14–23

Gaaied S, Oliveira M, Le Bihanic F, Cachot J, Banni M (2019) Gene expression patterns and related enzymatic activities of detoxification and oxidative stress systems in zebrafish larvae exposed to the 2, 4-dichlorophenoxyacetic acid herbicide. Chemosphere 224:289–297. https://doi.org/10.1016/j.chemosphere.2019.02.125

Giesy JP, Pierens SL, Snyder EM, Miles-Richardson S, Kramer VJ et al (2000) Effects of 4-nonylphenol on fecundity and biomarkers of estrogenicity in fathead minnows Pimephalespromelas. Environ Toxicol Chem 19:1368–1377

Ginebreda A, Kuzmanovic M, Guasch H, López de Alda M, López-Doval JC et al (2014) Assessment of multi-chemical pollution in aquatic ecosystems using toxic units: compound prioritization, mixture characterization and relationships with biological descriptors. Sci Total Environ 468(469):715–723

Gonzalez P, Baudrimont M, Boudou A et al (2006) Comparative effects of direct cadmium contamination on gene expression in gills, liver, skeletal muscles and brain of the zebrafish (Danio rerio). Biometals 19:225–235. https://doi.org/10.1007/s10534-005-5670-x

Hanqing X, Xingang L, Yang JB, Fengshou D, Jun X, Xiaohu W, Yang Y, Yongquan Z (2018) Fipronil-induced toxic effects in zebrafish (Danio rerio) larvae by using digital gene expression profiling. Sci Total Environ 639(2018):550–559. https://doi.org/10.1016/j.scitotenv.2018.05.159

Himri I, Saboni A, Ouahhoud S, Elyoubi M, Lahmass I, Khoulati A, Saalaoui E (2018) Place of the genotoxicity in the ecotoxicology domain. SMETox J 1(1):10–17. https://revues.imist.ma/index.php/SMETox_Journal/article/view/11778/6658

Hosseinimehr SJ, Fathi M, Ghasemi A, Shiadeh SNR, Pourfallah TA (2017) Celecoxib mitigates genotoxicity induced by ionizing radiation in human blood lymphocytes. Res Pharm Sci 12(1): 82. https://doi.org/10.4103/1735-5362.199051. (in aquatic organisms. J. Toxicol EnViron Health, Part B 12, 65–82. 10.1093/czoolo/60.2.273)

Jenkins GJS, Doak SH, Johnson G, Quick E, Parry J (2005) Do dose response thresholds exist for genotoxic alkylating agents? Mutagenesis 20:389–398

Jin Y, Zhang X, Shu L, Chen L, Sun L, Qian H, Fu Z (2010) Oxidative stress response and gene expression with atrazine exposure in adult female zebrafish (Danio rerio). Chemosphere 78(7): 846–852. https://doi.org/10.1016/j.chemosphere.2009.11.044

Karihtala P, Soini Y, Vaskivuo L, Bloigu R, Puistola U (2009) DNA adduct 8-hydroxydeoxyguanosine, a novel putative marker of prognostic significance in ovarian carcinoma. Int J Gynecol Cancer 19:1047–1051

Kirsch-Volders M, Aardema M, Elhajouji A (2000) Concepts of thresholds in mutagenesis and carcinogenesis. Mutat Res 464(3):11

Lee HB, Peart TE (2000) Bisphenol a contamination in Canadian municipal and industrial wastewater and sludge samples. Water Qual Res J Canada 35:283–298

Lewis C, Galloway T (2009) Reproductive consequences of paternal genotoxin exposure in marine invertebrates. Environ Sci Technol 43:928–933

Malins DC, Ostrander GK, Haimanot R, Williams P (1990) A novel DNA lesion in neoplastic livers of feral fish: 2,6-diamino-4-hydroxy-5-formamidopyrimidine. Carcinogenesis 11:1045–1047

Marco KC, Berenyce G, Calder E, Marıa C, Martınez V, Ernesto M, Caldero ME, S. (2016) DNA damage assessment in zebrafish embryos exposed to Monceren 250 SC fungicide using the alkaline comet assay. Zebrafish 13:442. https://doi.org/10.1089/zeb.2016.1265

Mehra S, Chadha P (2021) Naphthalene-2-sulfonate induced toxicity in blood cells of freshwater fish Channa punctatus using comet assay, micronucleus assay and ATIR-FTIR approach. Chemosphere 265:129147. https://doi.org/10.1016/j.chemosphere.2020.129147

Meyer FP, Jorgenson TA (1983) Teratological and other effects of malachite green on development of rainbow trout and rabbits. Trans Am Fish Soc 112(6):818–824. https://doi.org/10.1577/15488659(1983)112<818:TAOEOM>2.0.CO;2

Mir MI, Khan S, Bhat SA, Reshi AA, Shah FA, Balki MH, Manzoor R (2014) Scenario of genotoxicity in fishes and its impact on fish industry. IOSR-JESTFT 8(6):2319–2402

Mishra SS, Das R, Das BK, Choudhary P, Rathod R, Giri BS, Debbarma J, Sahoo SN, Barua A, Sahu A, Patil PK (2017) Status of aqua-medicines, drugs and chemicals use in India: a survey re-port. J Aquac Fisheries 1(004):1–15

Mix MC (1986) Cancerous disease in aquatic animals and their association with environmental pollutants: a critical literature review. Mar Environ Res 20:1–141

Mohamed S, Sabita U, Rajendra S, Raman D (2017) Genotoxicity: mechanisms, testing guidelines and methods. Glob J Pharm Pharmaceu Sci 1:133–138. https://doi.org/10.19080/GJPPS.2017.02.555575

Myers MS, Rhodes LD, McCain BB (1987) Pathologic anatomy and patterns of occurrence of hepatic neoplasms, putative preneoplastic lesions, and other idiopathic hepatic conditions in English sole Parophrysvetulus from Puget Sound, Washington. J Natl Cancer Inst 78:333–363

Myers MS, Johnson LL, Collier TK (2003) Establishing the causal relationship between polycyclic aromatic hydrocarbon (PAH) exposure and hepatic neoplasms and neoplasia-related liver lesions in English sole Pleuronectesvetulus. Hum Ecol Risk Assess 9:67–94

Myers MS, Anulacion BF, French BL, Reichert WL, Laetz CA et al (2008) Improved flatfish health following remediation of a PAH-contaminated site in Eagle Harbor, Washington. Aquat Toxicol 88:277–288

Nandanpawar PC, Rather MA, Badhe MR, Sharma R (2018) Assessment of DNA damage during gene delivery in freshwater prawn by chitosan reduced gold nanoparticles. Biosci Biotechnol Res Asia 15(1):39–49

Nativo P, Prior IA, Brust M (2008) Uptake and intracellular fate of surface-modified gold nanoparticles. ACS Nano 2:1639–1644

Obiakor MO, Tighe M, Pereg L, Maher W, Taylor AM, Wilson SC (2021) A pilot in vivo evaluation of Sb (III) and Sb (V) genotoxicity using comet assay and micronucleus test on the freshwater fish, silver perch Bidyanusbidyanus (Mitchell, 1838). Environ Adv 5:100109. https://doi.org/10.1016/j.envadv.2021.100109

Pandey AK, Nagpure NS, Trivedi SP (2018) Genotoxicity assessment of pesticide profenofos in freshwater fish Channa punctatus (Bloch) using comet assay and random amplified polymorphic DNA (RAPD). Chemosphere 211:316–323. https://doi.org/10.1016/j.chemosphere.2018.07.182

Patil PK, Mishra SS, Pradhan PK, Manna SK, Abraham JT, Solanki HG, Jena J (2022) Usage pattern of chemicals, biologicals and veterinary medicinal products in Indian aquaculture. Rev Aquac 14:2038. https://doi.org/10.1111/raq.12688

Rather MA, Bhat IA, Sharma N, Sharma R (2018) Molecular and cellular toxicology of nanomaterials with related to aquatic organisms. In: Cellular and molecular toxicology of nanoparticles. Springer, Cham, pp 263–284

Rhaul O, Grisolia CK, Marta SM, Soares AM, Domingues I (2016) Multilevel assessment of ivermectin effects using different zebrafish life stage. Comp Biochem Physiol Part C Toxicol Pharmacol 187:50–61. https://doi.org/10.1016/j.cbpc.2016.04.004

Santos LHMLM, Araujo AN, Fachini A, Pena A, Deleure-Matos C et al (2010) Ecotoxicological aspects related to the presence of pharmaceuticals in the aquatic environment. J Hazard Mat 175: 45–95

Satpathy L, Parida SP (2020) Acute toxicity assessment and behavioral responses induced by KandhamalHaladi in adult zebrafish (Danio rerio). Biointerface Res Appl Chem 11(1): 7368–7381. https://doi.org/10.33263/BRIAC111.73687381

Savale SK (2018) Genotoxicity of drugs: Introduction, prediction and evaluation. Asian J Biomat Res 4(6):1–29

Shinn AP, Pratoomyot J, Bron JE, Paladini G, Brooker EE, Brooker AJ (2015) Economic costs of protistan and metazoan parasites to global mariculture. Parasitology 142:196–270. https://doi.org/10.1017/S0031182014001437

Sujin L, Sangah O, Seonghye K, Wonho L, Juyoung C, Hani L, Yujin L, Suhkmann K (2020) NMR-based Metabolomic responses of zebrafish (danio Rerio) by Fipronil exposure. J Korean Magn Reson Soc 24(4):104–116. https://doi.org/10.6564/JKMRS.2020.24.4.104

Tahir R, Ghaffar A, Abbas G, Turabi TH, Kausar S, Xiaoxia D, Abdelgayed SS (2021) Pesticide induced hematological, biochemical and genotoxic changes in fish: a review. Agrobiol Rec 3: 41–57. https://doi.org/10.47278/journal.abr/2021.005

Thiripurasundari M, Sathya K, Uma A, Srinivasan MR, Rajasekar P (2014) A comparative study on the toxicity of ivermectin in zebra fish and catla fish models. Indo Am J Pharm Res 4:3683–3688

Turkez H, Arslan ME, Ozdemir O (2017) Genotoxicity testing: progress and prospects for the next decade. Expert Opin Drug Metab Toxicol 13(10):1089–1098. https://doi.org/10.1080/17425255.2017.1375097

Verbueken E, Bars C, Ball JS, Periz-Stanacev J, Marei WF, Tochwin A, Van Cruchten SJ (2018) From mRNA expression of drug disposition genes to in vivo assessment of CYP-mediated biotransformation during zebrafish embryonic and larval development. Int J Mol Sci 19(12): 3976. https://doi.org/10.3390/ijms19123976

Vorosmarty CJ, McIntyre PB, Gessner MO, Dudgeon D, Prusevich A et al (2010) Global threats to human water security and river biodiversity. Nature 468:334

Wahyuni EA, Lin HD, Lu CW, Kao CM, Chen SC (2021) The cytotoxicity and genotoxicity of single and combined fenthion and terbufos treatments in human liver cells and zebrafish embryos. Sci Total Environ 758:143597. https://doi.org/10.1016/j.scitotenv.2020.143597

Wang Y, Dai D, Yu Y, Yang G, Shen W, Wang Q, Zhao X (2018) Evaluation of joint effects of cyprodinil and kresoxim-methyl on zebrafish, Danio rerio. J Hazard Mater 352:80–91. https://doi.org/10.1016/j.jhazmat.2018.03.023

Endocrine-Disrupting Compounds (EDCs) as Emerging Aquatic Contaminants: Emphasis on Reproduction and Development

A. Rathipriya, Deepak Agarwal, E. Suresh, and Mohd Ashraf Rather

1 Introduction

Numerous genomic pathways have been found to have been affected by exposure to anthropogenic chemicals with hormone-like activity over the past 25 years, according to considerable studies in vertebrate species. This effect is mediated through the chemicals' interactions with nuclear receptors. Additionally, a variety of contaminants have been found to obstruct non-genomic (non-classical) pathways, but it is yet unclear how this causes endocrine disruption. There has been a significant increase in the number of publications detailing the effects of endocrine-disrupting chemicals on fish reproduction in recent years, particularly those focusing on hypothalamus–pituitary–gonadal axis dysregulation. They may have different effects on the physiology of male or female reproduction according to their ability to resemble endogenous hormones. They may have different effects on the physiology of male or female reproduction depending on their ability to resemble endogenous hormones. The gonadosomatic index changes, intersex gonad development, gametogenesis inhibition, and lower fertility rate have all been extensively studied (Dar et al. 2012; Mohapatra et al. 2021). Males in some wild species have shown changes in sperm quantity, fertility, and motility. Females showed similar negative effects, including decreased maturation and oocyte growth, as well as an increase in apoptotic processes. Gamete viability, one of the key indications of reproductive endocrine disturbance, may be impacted by these routes. Pollutants also cause epigenetic pathways modifications, which result in specific toxicological

A. Rathipriya · D. Agarwal (✉) · E. Suresh
Tamil Nadu Dr. J. Jayalalithaa Fisheries University—IFPGS, OMR Campus, Chennai, India

M. A. Rather
Division of Fish Genetics and Biotechnology, Sher-e-Kashmir University of Agricultural Sciences and Technology, Kashmir, Rangil, Ganderbal, Jammu and Kashmir, India

M. A. Rather et al. (eds.), *Xenobiotics in Aquatic Animals*,
https://doi.org/10.1007/978-981-99-1214-8_21

processes or abnormal cellular reactions that may have an impact on future generations via the germline.

Endocrine-disrupting chemicals are a type of synthetic or natural chemical that has a variety of structural properties and the ability to interact with the endocrine system. EDCs include herbicides and pesticides, plastic pollutants, heavy metals, biocides, heat stabilizers, chemical catalysts, medicines and dietary additives, flame retardants, and other substances. Given their heterogeneity in physicochemical properties and distinct biological effects, it is not surprising that EDCs interact with the endocrine system via a variety of modes of action (Henley and Korach 2006). They disrupt the neuroendocrine and endocrine processes required for embryonic growth and reproduction (Godfrey et al. 2017). Once in the environment, they simulate, reject, and change the amounts of naturally occurring steroids by influencing the rates of their synthesis, metabolism, expression, and activity at receptor sites. EDCs and natural hormones share structural similarities, allowing them to bind to the same receptors and cause similar biological reactions. The three main gonadal steroids that affect the reproductive system are cholesterol-derived oestrogens, androgens, and progesterone. While androstenedione is the starting material for the biosynthesis of testosterone, androstenedione is the starting material for the biosynthesis of oestrogens. In fish, two distinct aromatase enzymes are encoded: CYP19a, which is primarily expressed in the gonads, and CYP19b, which is primarily expressed in the brain (Zhang et al. 2014). Oestrogens were also known to regulate a variety of reproductive processes (Yilmaz et al. 2015). The androgen receptor, which is activated by androgens such as testosterone, is required for male development, as well as secondary sex traits (Schulz et al. 2010). Despite the fact that the majority of EDCs have ecological concentrations that are much lower than the range established by regulatory agencies, it is important to remember that an environmental health issue with EDCs is linked to bioaccumulation, chronic exposures, and the eventual occurrence of antagonistic interaction among them. Furthermore, because EDCs are known to function via non-monotonic dose–response curves, the effects seen at high doses cannot be extrapolated to the low-dose range. It is demonstrated that endocrine disruption affects animals, particularly aquatic species. Dietary absorption caused by contaminated food is a significant source of contamination (Carnevali et al. 2017). Due to the wide dispersion of contaminants in the environment, the current debate is not about whether or not endocrine disruption occurs in wildlife, but about the concentrations at which it occurs, the underlying mechanisms, and whether or not endocrine disruption has any ecologically significant effects.

There have been more recent reports of EDCs' impacts on invertebrates (Cuvillier-Hot and Lenoir 2020). Invertebrates play a significant ecological role in every ecosystem and contribute significantly to global biodiversity (Wilson 1994). Therefore, understanding the effects of EDCs on invertebrate reproduction is crucial. Due to their small size, invertebrates are predicted to be significantly exposed to environmental pollutants, which could indicate greater bioaccumulation (Cuvillier-Hot and Lenoir 2020).

The imposex and intersex phenotypes observed in Mollusca and Arthropoda species are well-established instances of EDC impacts on reproduction in invertebrates (Grilo and Rosa 2017). The majority of these phenotypic conditions are caused by organotin substances such as tributyltin, triphenyltin, and dibutyltin, which are used as crop fungicides and pesticides, slimicides in industrial water systems, and marine antifouling agents. Similar symptoms have also been linked to oestrogenic substances like 17-estradiol and 17-ethynylestradiol exposure (Grilo and Rosa 2017). Male sexual organs are non-functional and are known as the imposex phenotype (De Wolf et al. 2001). Individuals with the intersex condition, in contrast side, have gonadal tissue from the opposite sex, such as spermatocytes in the ovary or oocytes in the testis (Bahamonde et al. 2013; Gomes et al. 2009). Both abnormal morphological and physiological conditions, according to Oehlmann et al. (1998), are irreversible and result in poorer reproduction, biased sex ratios, decreased fertility, and the possibility of a concurrent decline in biodiversity and populations (Cuvillier-Hot and Lenoir 2020).

Understanding the genetic loci responsible for fish sex determination can provide insight into reproduction, which is beneficial to aquaculture and fisheries (Sundaray et al. 2021). Sex determination within fishes is a dynamic process that has been observed in families or genera and can be modulated by external factors (Devlin and Nagahama 2002). Sex can be determined by monogenic or polygenic systems, which can be found on autosomes or sex chromosomes. We can compare sexes using transcriptomics to determine which loci contribute to sex determination (Chen et al. 2015; Sun et al. 2013; Lin et al. 2017). The transcriptome of reproductive tissue can also provide biological insight into sex-related differences or sex differentiation in fish (Zhang et al. 2019; Tao et al. 2018; Agarwal et al. 2020). Dmy/dmrt1Yb, DMW, DMRT1, Sox9, SDy, and Sox3 are just a few of the candidate genes associated with sex determination, gametogenesis, and gonadal differentiation and maturation in fish (Matsuda et al. 2002; Yokoi et al. 2002; Yano et al. 2012; Takehana et al. 2014). Rapid cell division processes occur in the early stages of vertebrate embryo development, whereas synchronous cells divide the zygote in a blastula. Maternal RNA is essential during this time, and these gene transcripts were inherited in growing oocytes and direct embryogenesis (Chapman et al. 2014). In such studies, these transcripts encode regulators or participants in cell cycling, proliferation, growth, and apoptosis, and cytoskeleton (Aegerter et al. 2005; Bonnet et al. 2007).

2 Endocrine-Disrupting Chemicals (EDCs)

EDCs are disrupt hormones that regulate developmental processes and maintain homeostasis in their production, transport, release, metabolism, action, binding, or elimination (EPA Environmental Protection Agency (EPA) 2014). EDCs can have an impact on human, fish, and wildlife health by interfering with the endocrine system. The effects of EDCs are those on the hormone system, which can change how the endocrine system regulates vital physiological processes (Morales et al. 2014). Because EDCs can imitate or disrupt endogenous hormone function, they may have a substantial negative influence on an organism's ability to survive during critical developmental and reproductive stages (Scholz and Mayer 2008). The development of the neuroendocrine system has been markedly impaired in various papers to document the effects of chemical exposure (Weber et al. 2013). By interfering with enzymes or receptors involved in metabolism and steroid synthesis, EDCs may also have an impact on several reproductive processes (Ma et al. 2012). EDCs can either promote or prevent fish oocyte maturation (Tokumoto et al. 2005). Endocrine disruptors can have an effect on many aspects of transcription and transcriptional regulation that control gene expression (Dominguez et al. 2014). The mode of action of EDCs can be classified into disturbance of production, metabolism, secretion or transport of natural hormones, and interruption in hormone receptors during production and function (Rotchell and Ostrander 2003). Based on the properties of the molecule in action, this division distinguishes different and discrete routes of an EDC mechanism (Goksøyr 2006). Despite the fact that only one substance can use all of the aforementioned routes of action, based on the amount administered to the organism (Goksøyr 2006), it was discovered that exposure to EDCs during developmental stages can harm human development and damage aquatic organisms (Wiegand et al. 2001).

There have been reports linking human testicular dysgenesis syndrome development to exposure to EDC (Santos et al. 2007). Furthermore, Wiegand et al. (2001) stated that fish development may be impacted by the fish's inability to detoxify the environment and the absence of EDC degradation. According to Barnhoorn et al. (2004), exposure to EDCs in water is linked to several reproductive effects in fish, including the induction of intersex, a decrease in the levels of various hormones, and a reduction in gamete production (Caballero-Gallardoa et al. 2016). Estrone (El), 17-estradiol (E2), and 17-ethinylestradiol (EE2), which are produced and naturally occurring steroid oestrogens, are among the most potent EDCs present in these effluents. Many manufacturing substances that cause endocrine disruption in animals include bisphenol A (BPA) and nonylphenol (NP) (Xu et al. 2014). Numerous studies have shown that exposure to these contaminants can alter or prevent normal fish development and reproduction. Since l7-ethinylestradiol (EE2) is a potent oestrogen, Kidd et al. (2007) noted that its effect on the environment causes sex reversal in male fish, resulting in a completely female population. According to Baumann et al. (2013), both sexes' gonadal maturation may be impaired. They further clarified that the fish species, developmental stage, and exposure time all played a role in these significant aberrations in gonadal development (Bhat et al. 2016; Rathor et al. 2017). Additionally, it was discovered that zebrafish exposed to chronic oestrogen had negative effects on growth, induced the production of vitellogenin, delayed the commencement of maturation, impaired sex ratio and sexual differentiation, and reduced the success of fecundity and fertilization (Ankley et al. 2009). Concentrations of 1.1 ng/L and 3–5 ng/L of EE2 resulted in complete suppression of fertilization success and failed population recruitment (Fenske et al. 2005; Schäfers et al. 2007). According to Xu et al. (2008), zebrafish exposed to EE2 has negatively impacted reproduction, resulting in impairments in both male and female performance. According to Van der Kraak and Lister (2011), natural 17-estradiol suppresses the oocyte maturation of zebrafish. Additionally, zebrafish treated by the pharmaceutical oestrogen EE2 and anabolic androgen 17-trenbolone showed an increase in vitellogenin levels, which led to fish feminization at a concentration of 10 ng/L EE2. After receiving 50 ng/L of 17-trenbolone, it was also discovered that the production of vtg was masculinized and decreased (Orn et al. 2006). Additionally, among the most often observed negative consequences were a decline in gonadal development, a decrease in fecundity coupled with a decrease in fertility (Xu et al. 2008), and impairments in gonadal differentiation (Fenske and Segner 2004).

Compared to endogenous oestrogen, nonylphenol (NP) can naturally be active in the body for a long time (Caballero-Gallardoa et al. 2016). When NP binds to the oestrogen receptor, it competes with oestrogen and affects fish development and reproduction (Chaube et al. 2013) (Puy-Azurmendi et al. 2014). Since fish are particularly susceptible to EDCs, including triclosan (Shanmugam et al. 2014) and organochlorine insecticides (heptachlor epoxide, dieldrin, and hexachlorobenzene), several studies have connected fish consumption to human risk. As a result, the aquatic environment is regarded as a major basin for chemical disruptors.

3 Effects of EDC Exposure on Fish Reproduction and Development

Antifouling chemical 3, 3″ diindolylmethane (DIM) produced abnormal production of vitellogenin and eggshell proteins in the testis of marine medaka, *Oryzias melastigma*, at low levels, implying oestrogenic action (Chen et al. 2017). When DEHP exposure was tested in zebrafish at concentrations of 0.2 and 20 g/L, it was discovered that DEHP can have a detrimental effect on reproduction by stopping the mitotic process during spermatogenesis and lowering embryo development (up to 90%). These changes were linked to an increase in spermatozoa DNA fragmentation (Corradetti et al. 2013). Zebrafish exposed to environmentally relevant concentrations of triclocarban and inorganic mercury developed severe histological lesions in the testis, resulting in a reduction in size and mature sperm abundance, similar to fish exposed to the chemicals separately (Wang et al. 2016). Similar to this, exposure to DES, flutamide, and their mixture decreased the number of sperm and changed the meiotic and apoptotic processes, affecting spermatogenesis (Yin et al. 2017). A possible delay in spermatogenesis was expected in male *Cyprinus carpio* fish exposed to 2 mg/L PFOA for 56 days, where spermatogonia and spermatocytes were the predominant germ cells. There was also evidence of enhanced Sertoli cell proliferation and interstitial tissue in the testis (Giari et al. 2016). Goldfish exposed to BPA had higher levels of hepatic vitellogenin, but not of gonadosomatic index, hepatosomatic index, or E2. This implies that BPA might affect testicular spermatogenesis, changing sperm maturation (Hatef et al. 2012).

Numerous types of research on fish have been concentrated on reproductive endpoints since abnormalities in sexual development were among the first physiological consequences to be noticed as a result of exposure to EDCs. According to Uren Webster et al.'s (2014) research, zebrafish exposed to glyphosate at high concentration of 10 mg/L experienced changes in the transcript profiles of the gonads for the ovary cyp19a1 and esr1 and the testis hsd3b2, cat, and sod1 genes. Chemicals that disrupt hormones (EDCs) are now widely dispersed in the environment. Due to its potential neurotoxicity and endocrine-disrupting effects, tetrabromobisphenol A bis (2-hydroxyethyl) ether (TBBPA-DHEE) contamination in environmental media poses a serious risk to people and aquatic organisms. However, little is known about how TBBPA-DHEE affects the endocrine system in aquatic organisms. Okeke et al. (2022) observed the neurotoxicity and reproductive endocrine-disruptive effects of TBBPA-DHEE by observing the neurobehavioral changes, vitellogenin, testosterone, 17-estradiol, and gene expression levels in adult male and female zebrafish exposed to TBBPA-DHEE as well as using transcriptomic analysis to uncover additional potential neuroendocrine-disrupting mechanisms. The neuroendocrine disruptive effect of TBBPA-DHEE was detected and is partially explained by the high upregulation of genes.

In breeding colonies of zebrafish, the study identified the genetic processes linked to reproductive disruption after exposure to a model environmental oestrogen (EE2). It was accomplished by examining the gonadal transcriptomes of specific zebrafish exposed to EE2 and comparing these to the impacts on their reproductive physiology to find candidate gene pathways causing the phenotypic changes brought on by EE2.

A ubiquitous and movable metalloid contaminant in the environment is arsenic. Although certain bacterial species utilize arsenic compounds as respiratory metabolites, it is extremely hazardous to the majority of species. The molecular basis of arsenate toxicity has been studied utilizing RNA-SAGE in zebrafish. Iron ion transport, translation, oxidation-reduction, cell redox, and homeostasis were among the primary biological processes highly enriched in differentially expressed genes, according to transcriptional profiles based on zebrafish genome (Xu et al. 2013). In recent years, it has become clear how EE2 affects fish brain development. In one study, the neural transcriptome of male and female guppies was examined using RNA-seq to identify the effects of exposure to 8 ng/L and 38 ng/L EE2 (Saaristo et al. 2017). According to their research, EE2 had a feminizing effect on the male transcriptome and changed transcript abundances in a sex-specific way.

The effects of early exposure to the herbicide atrazine or EE2 on sexual differentiation and gene expression in gonadal tissue were examined by Leet et al. (2020). They exposed largemouth bass (*Micropterus salmoides*) to concentrations of 1, 10, or 100 g atrazine/L or 1 or 10 ng EE2/L from 7 to 80 days post-spawn, and they observed histological growth and transcriptome changes in gonad tissue. They observed an almost 100% female sex ratio in fish exposed to EE2 at a concentration of 10 ng/L, most likely as a result of male sex reversal.

In aquatic studies, 17-methyltestosterone (MT), an artificial androgenic substance used to produce masculinization of both secondary sex characteristics and gonads, was investigated by Wang et al. (2020). The growth and development of fish were retarded by exposure to MT at 200 ng/L in the Stone moroko (*Pseudorasbora parva*). Male and female RNA-seq analysis identified 7758 and 11,543 DEGs, respectively. Males were more clearly disrupted by MT than females were, and this was predominantly shown in the immune system.

The effects of EE2 exposure on Pacific sardines (*Sardinops sagax*) and chub mackerel (*Scomber japonicus*) were studied by Renaud et al. (2019). The liver RNA of wild sardines and mackerel exposed for 5 h in a lab setting to a concentration of 12.5 pM EE2 conducted RNA sequencing (RNA-seq). In male sardine and mackerel, ambient levels of EE2 altered fundamental biological processes and pathways, as shown by the development of molecular markers for metabolic, hormonal, and immunological dysfunction, as well as carcinogenesis in exposed fish.

The physiological effects of chronic exposure to wastewater treatment plants and stormwater effluents were explored by Bertucci et al. (2018) in the Asian clam (*Corbicula fluminea*). They discovered a group of 3181 transcripts whose abundance changed in response to the quality of the water. Clams from the reference clean site and those exposed to wastewater treatment plant effluents showed the greatest changes in transcriptome profiles. The majority of the transcripts with differential expression were connected to signalling pathways for energy metabolism, indicating a lack of nutrients and/or energy as well as hypoxic conditions as a result of the contaminants in the effluents.

Copepods (*Eurytemora affinis*) were subjected to sublethal doses of the pesticide pyriproxyfen and the insecticide chlordecone by Legrand et al. (2016). Males and females (400 each) were separated for RNA extraction after 48 h. After EDC exposures, 2566 distinct genes showed differential expression in comparison to controls with comparable numbers of DE genes with both chemicals. After both exposures, males had more differential gene expression than females.

Guo et al. (2021) investigated the effects of phenolic chemicals in Ba River effluent on the ovary of the sharp belly (*Hemiculter leucisculus*), a freshwater fish, using transcriptome and metabolomic analysis. Compared to upstream and remote sites, oocyte development was stimulated in fish gathered near wastewater discharge. The differentially expressed genes from RNA-seq revealed that fish ovaries had histopathological changes that were most likely caused by increased steroid hormone production.

Salmonid species were subjected to relatively short exposure days. Hepatic vtg and er mRNA levels increased in juvenile rainbow trout exposed to NP for 3 or 6 days, whereas brain gnrh2 mRNA levels decreased in a dose-dependent manner (Vetillard and Bailhache 2006). Furthermore, 4 days of EE2 and NP exposure increased vtg transcript levels in yolk sac larvae, fry, and smolts, as well as plasma levels of VTG and er mRNA in Atlantic salmon (*Salmo salar*) smolts (Duffy et al. 2014; Breves et al. 2018). An appealing experiment that mimicked maternal transfer by exposing rainbow trout oocytes to BPA in the ovarian fluid for three hours resulted in a delay in hatching and yolk absorption in the embryo (Aluru et al.

2010). In zebrafish larvae, low doses of BPA exposure at 2 h post fertilization (hpf) to 120 hpf resulted in earlier hatching time, an increase in GNRH3 neurons, and higher expression of genes linked to reproduction such as gnrh3, lh, fsh, and er (Qiu et al. 2016). After 96 h of exposure to various phthalates, sex steroid hormones (E2 and T) in fathead minnow eggs either increased or decreased (Mankidy et al. 2013). Increases in iNOS (inducible nitric oxide synthase) were seen in zebrafish embryos exposed to BPS, and genes predicted to activate innate immune cells were found in the RNA-seq data (Qiu et al. 2020).

Acute exposure-based experimental methods have been successful in identifying the disrupting effects of NP and EE2 in adult fish. It increased plasma VTG, decreased plasma E2 and T after 24 or 48 h, and altered the expression of genes involved in hormone metabolism, steroid binding, sterol metabolism, and cell development in the liver after 24 and 168 h in gravid female zebrafish (Hoffmann et al. 2006). After a week of exposure to EE2, subadult coho salmon (*Oncorhynchus kisutch*) showed an increase in hepatic vtg and pituitary luteinizing hormone (lh) subunit mRNA, as well as induction of gonadotropin-releasing hormone (gnrh) receptor expression in females (Harding et al. 2013). Three days of NP and BPA exposure resulted in hepatic vtg mRNA expression and reproductive damage in adult male Swordtail fish (*Xiphophorus helleri*) (Kwak et al. 2001).

4 Conclusion

A better comprehension of every aquaculture organisms' reproductive pattern is critical for maintaining reproductive performance and assisting in artificial breeding programmes. Each of the fishes has a well-developed sex-determining system. Thus, identifying relevant transcripts from the tissues will aid in the analysis of spermatogenesis processes. In any method of breeding or culture, reproduction is critical in large-scale commercial industry sectors that use some form of artificial reproduction. When dealing with wild species, species identification is critical because hybridization is feasible. Furthermore, numerous marine species have vast dispersion with probable population structures along their ranges. Transcriptomic analyses can provide insight into individual fitness as well as an understanding of how reproduction works at the molecular level within the species of interest.

References

Aegerter S, Jalabert B, Bobe J (2005) Large scale real-time PCR analysis of mRNA abundance in rainbow trout eggs in relationship with egg quality and post ovulatory ageing. Mol Reprod Dev 72(3):377–385

Agarwal D, Gireesh-Babu P, Pavan-Kumar A, Koringa P, Joshi CG, Chaudhari A (2020) Transcriptome analysis of *Clarias magur* brain and gonads suggests neuro-endocrine inhibition of milt release from captive GnRH-induced males. Genomics 112(6):4041–4052

Aluru N, Leatherland JF, Vijayan MM (2010) Bisphenol a in oocytes leads to growth suppression and altered stress performance in juvenile rainbow trout. PloS One 5(5):e10741

Ankley GT, Bencic DC, Breen MS, Collette TW, Conolly RB et al (2009) Endocrine disrupting chemicals in fish: developing exposure indicators and predictive models of effects based on mechanism of action. Aquat Toxicol 92(3):168–178

Bahamonde PA, Munkittrick KR, Martyniuk CJ (2013) Intersex in teleost fish: are we distinguishing endocrine disruption from natural phenomena? Gen Comp Endocrinol 192:25–35

Barnhoorn I, Bornman M, Pieterse G, Van Vuren J (2004) Histological evidence of intersex in feral sharptooth catfish (*Clarias gariepinus*) from an estrogen-polluted water source in Gauteng, South Africa. Environ Toxicol 19(6):603–608

Baumann L, Holbech H, Keiter S, Kinnberg KL, Knörr S et al (2013) The maturity index as a tool to facilitate the interpretation of changes in vitellogenin production and sex ratio in the fish sexual development test. Aquat Toxicol 128–129:34–42

Bertucci A, Pierron F, Gourves P-Y, Klopp C, Lagarde G, Pereto C, Dufour V, Gonzalez P, Coynel A, Budzinski H et al (2018) Whole-transcriptome response to wastewater treatment plant and stormwater effluents in the Asian clam, Corbicula fluminea. Ecotoxicol Environ Saf 165:96–106

Bhat IA, Rather MA, Saha R, Pathakota GB, Pavan-Kumar A, Sharma R (2016) Expression analysis of Sox9 genes during annual reproductive cycles in gonads and after nanodelivery of LHRH in *Clarias batrachus*. Res Vet Sci 106:100–106

Bonnet E, Fostier A, Bobe J (2007) Microarray-based analysis of fish egg quality after natural or controlled ovulation. BMC Genomics 8(1):55

Breves JP, Duffy TA, Einarsdottir IE, Bjornsson BT, McCormick SD (2018) In vivo effects of 17alpha-ethinylestradiol, 17beta-estradiol and 4-nonylphenol on insulin-like growth-factor binding proteins (igfbps) in Atlantic salmon. Aquat Toxicol 203:28–39

Caballero-Gallardoa K, Olivero-Verbela J, Freeman JL (2016) Toxicogenomics to evaluate endocrine disrupting effects of environmental chemicals using the zebrafish model. Curr Genomics 17:515–527

Carnevali O, Notarstefano V, Olivotto I, Graziano M, Gallo P, Di Marco Pisciottano I, Vaccari L, Mandich A, Giorgini E, Maradonna F (2017) Dietary administration of EDC mixtures: a focus on fish lipid metabolism. Aquat Toxicol 185:95–104

Chapman RW, Reading BJ, Sullivan CV (2014) Ovary transcriptome profiling via artificial intelligence reveals a transcriptomic fingerprint predicting egg quality in striped bass, Morone saxatilis. PLoS One 9(5):e96818

Chaube R, Gautam GJ, Joy KP (2013) Teratogenic effects of 4-nonylphenol on early embryonic and larval development of the catfish *Heteropneustes fossilis*. Arch Environ Contam Toxicol 64(4): 554–561

Chen X, Mei J, Wu J, Jing J, Ma W, Zhang J, Gui JF (2015) A comprehensive transcriptome provides candidate genes for sex determination/differentiation and SSR/SNP markers in yellow catfish. Mar Biotechnol 17(2):190–198

Chen L, Au DWT, Hu C et al (2017) Linking genomic responses of gonads with reproductive impairment in marine medaka (*Oryzias melastigma*) exposed chronically to the chemopreventive and antifouling agent, 3,3-diindolylmethane (DIM). Aquat Toxicol 183:135–143

Corradetti B, Stronati A, Tosti L et al (2013) Bis-(2-ethylexhyl) phthalate impairs spermatogenesis in zebrafish (*Danio rerio*). Reprod Biol 13:195–202

Cuvillier-Hot V, Lenoir A (2020) Invertebrates facing environmental contamination by endocrine disruptors: novel evidences and recent insights. Mol Cell Endocrinol 504:110712

Dar SA, Najar AM, Balkhi MH, Rather MA (2012) Histological observations of ovarian development in endemic snow trout, *Schizopyge Esocinus* (Heckel, 1838). Indian Vet J 89(8):110

De Wolf H, De Coen W, Backeljau T, Blust R (2001) Intersex and sterility in the periwinkle Littorina littorea (Mollusca: Gastropoda) along the Western Scheldt estuary, The Netherlands. Mar Environ Res 52(3):249–255

Devlin RH, Nagahama Y (2002) Sex determination and sex differentiation in fish: an overview of genetic, physiological, and environmental influences. Aquaculture 208:191–364

Dominguez GA, Bisesi JH, Kroll KJ, Denslow ND, Sabo-Attwood T (2014) Control of transcriptional repression of the Vitellogenin receptor gene in largemouth bass (*Micropterus salmoides*) by select Estrogen receptors isotypes. Toxicol Sci 141(2):423–431

Duffy TA, Iwanowicz LR, McCormick SD (2014) Comparative responses to endocrine disrupting compounds in early life stages of Atlantic salmon, *Salmo salar*. Aquat Toxicol 152:1–10

EPA Environmental Protection Agency (EPA) (2014) Endocrine disruptor research. http://www.epa.gov/research/endocrine disruption/faq.htm. Accessed 5 Dec 2014

Fenske M, Segner H (2004) Aromatase modulation alters gonadal differentiation in developing zebrafish (*Danio rerio*). Aquat Toxicol 67(2):105–126

Fenske M, Maack G, Schäfers C, Segner H (2005) An environmentally relevant concentration of ethinylestradiol induces arrest of male gonad development in zebrafish, *Danio rerio*. Environ Toxicol Chem 24:1088–1098

Giari L, Vincenzi F, Badini S et al (2016) Common carp Cyprinus carpio responses to sub-chronic exposure to perfluorooctanoic acid. Environ Sci Pollut Res Int 23:15321–15330

Godfrey A, Hooser B, Abdelmoneim A, Horzmann KA, Freemanc JL, Sepúlveda MS (2017) Thyroid disrupting effects of halogenated and next generation chemicals on the swim bladder development of zebrafish. Aquat Toxicol 193:228–235

Goksøyr A (2006) Endocrine disruptors in the marine environment: mechanisms of toxicity and their influence on reproductive processes in fish. J Toxic Environ Health A 69:175–184

Gomes T, Gonzalez-Rey M, Bebianno MJ (2009) Incidence of intersex in male clams Scrobicularia plana in the Guadiana estuary (Portugal). Ecotoxicology 18(8):1104–1109

Grilo TF, Rosa R (2017) Intersexuality in aquatic invertebrates: prevalence and causes. Sci Total Environ 592:714–728

Guo J, Mo J, Qi Q, Peng J, Qi G, Kanerva M, Iwata H, Li Q (2021) Prediction of adverse effects of effluents containing phenolic compounds in the Ba river on the ovary of fish (*Hemiculter leucisculus*) using transcriptomic and metabolomic analyses. Sci Total Environ 801:149554

Harding LB, Schultz IR, Goetz GW, Luckenbach JA, Young G, Goetz FW et al (2013) High-throughput sequencing and pathway analysis reveal alteration of the pituitary transcriptome by 17alpha-ethynylestradiol (EE2) in female coho salmon, *Oncorhynchus kisutch*. Aquat Toxicol 142–143:146–163

Hatef A, Alavi SMH, Abdulfatah A et al (2012) Adverse effects of bisphenol a on reproductive physiology in male goldfish at environmentally relevant concentrations. Ecotoxicol Environ Saf 76:56–62

Henley DV, Korach KS (2006) Endocrine-disrupting chemicals use distinct mechanisms of action to modulate endocrine system function. Endocrinology 147:s25–s32

Hoffmann JL, Torontali SP, Thomason RG, Lee DM, Brill JL, Price BB et al (2006) Hepatic gene expression profiling using gene chips in zebrafish exposed to 17alpha-thynylestradiol. Aquat Toxicol 79(3):233–246

Kidd K, Blanchfield P, Mills K, Palace V, Evan R, Lazorchak J, Flick R (2007) Collapse of a fish population after exposure to synthetic estrogen. Proc Natl Acad Sci U S A 104:8897–8901

Kwak HI, Bae MO, Lee MH, Lee YS, Lee BJ, Kang KS et al (2001) Effects of nonylphenol, bisphenol a, and their mixture on the viviparous swordtail fish (*Xiphophorus helleri*). Environ Toxicol Chem 20:787–795

Leet JK, Richter CA, Cornman RS, Berninger JP, Bhandari RK, Nicks DK, Zajicek JL, Blazer VS, Tillitt DE (2020) Effects of early life stage exposure of largemouth bass to atrazine or a model Estrogen (17α-Ethinylestradiol). PeerJ 8:e9614

Legrand E, Forget-Leray J, Duflflot A, Olivier S, Thomé J-P, Danger J-M, Boulangé-Lecomte C (2016) Transcriptome analysis of the copepod *Eurytemora affinis* upon exposure to endocrine disruptor pesticides: focus on reproduction and development. Aquat Toxicol 176:64–75

Lin R, Wang L, Zhao Y, Gao J, Chen Z (2017) Gonad transcriptome of discus fish (*Symphysodon haraldi*) and discovery of sex-related genes. Aquac Res 48(12):5993–6000

Ma Y, Han J, Guo Y, Lam PKS, Wu RSS, Giesy JP, Zhang X, Zhou B (2012) Disruption of endocrine function in in vitro H295R cell-based and in in vivo assay in zebrafish by 2,4-dichlorophenol. Aquat Toxicol 106–107:173–181

Mankidy R, Wiseman S, Ma H, Giesy JP (2013) Biological impact of phthalates. Toxicol Lett 217(1):50–58

Matsuda M, Nagahama Y, Shinomiya A, Sato T, Matsuda C, Kobayashi T, Hori H (2002) DMY is a Y-specific DM-domain gene required for male development in the medaka fish. Nature 417(6888):559–563

Mohapatra S, Kumar R, Sundaray JK, Patnaik ST, Mishra CSK, Rather MA (2021) Structural damage in liver, gonads, and reduction in spawning performance and alteration in the haematological parameter of *Anabas testudineus* by glyphosate-a herbicide. Aquac Res 52(3): 1150–1159

Morales M, Martínez-Paz P, Martín R, Planelló R, Urien J, Martínez-Guitarte JL, Morcillo G (2014) Transcriptional changes induced by in vivo exposure to pentachlorophenol (PCP) in Chironomus riparius (Diptera) aquatic larvae. Aquat Toxicol 157:1–9

Oehlmann J, Bauer B, Minchin D, Schulte-Oehlmann U, Fioroni P, Markert B (1998) Hydrobiologia 378(1/3):199–213

Okeke S et al (2022) Transcriptomic profiling and differential analysis revealed the neurodevelopmental toxicity and mechanisms of zebrafish (*Danio rerio*) larvae in response to tetrabromobisphenol a bis (2–hydroxyethyl) ether (TBBPA-DHEE) exposure. Comp Biochem Physiol C Toxicol Pharmacol 259:109382

Orn S, Yamani S, Norrgren L (2006) Comparison of vitellogenin induction, sex ratio, and gonad morphology between zebrafish and Japanese medaka after exposure to 17αethinylestradiol and 17β-trenbolone. Arch Environ Contam Toxicol 51(2):237–243

Puy-Azurmendi E, Olivares A, Vallejo A, Ortiz-Zarragoitia M, Pina B et al (2014) Estrogenic effects of nonylphenol and octylphenol isomers in vitro by recombinant yeast assay (RYA) and in vivo with early life stages of zebrafish. Sci Total Environ 466:1–10

Qiu W, Zhao Y, Yang M, Farajzadeh M, Pan C, Wayne NL (2016) Actions of Bisphenol a and Bisphenol S on the reproductive neuroendocrine system during early development in zebrafish. Endocrinology 157(2):636–647

Qiu W et al (2020) Transcriptomic responses of bisphenol S predict involvement of immune function in the cardiotoxicity of early life-stage zebrafish (*Danio rerio*). Environ Sci Technol 54:2869–2877

Rathor PK, Bhat IA, Rather MA, Gireesh-Babu P, Kumar K, Purayil SBP, Sharma R (2017) Steroidogenic acute regulatory protein (StAR) gene expression construct: development, nanodelivery and effect on reproduction in air-breathing catfish, Clarias batrachus. Int J Biol Macromol 104:1082–1090

Renaud L, Agarwal N, Richards DJ, Falcinelli S, Hazard ES, Carnevali O, Hyde J, Hardiman G (2019) Transcriptomic analysis of short-term 17α-Ethynylestradiol exposure in two Californian sentinel fish species sardine (*Sardinops sagax*) and mackerel (Scomber japonicus). Environ Pollut 244:926–937

Rotchell JM, Ostrander GK (2003) Molecular markers of endocrine disruption in aquatic organisms. J Toxicol Environ Health B 6:453–495

Saaristo M, Wong BB, Mincarelli L, Craig A, Johnstone CP, Allinson M, Lindstrom K, Craft JA (2017) Characterisation of the transcriptome of male and female wild-type guppy brains with RNA-Seq and consequences of Exposureto the pharmaceutical pollutant, 17alpha-Ethinyl Estradiol. Aquat Toxicol 186:28–39

Santos EM, Paull GC, Van Look KJW, Workman VL, Holt WV et al (2007) Gonadal transcriptome responses and physiological consequences of exposure to oestrogen in breeding zebrafish (*Danio rerio*). Aquat Toxicol 83:134–142

Schäfers C, Teigeler M, Wenzel A, Maack G, Fenske M, Segner H (2007) Concentration- and time-dependent effects of the synthetic estrogen, 17αethynylestradiol, on reproductive capabilities of the zebrafish, *Danio rerio*. J Toxicol Environ Health A 70:768–779

Scholz S, Mayer I (2008) Molecular biomarkers of endocrine disruption in small model fish. Mol Cell Endocrinol 293:57–70

Schulz RW, de França LR, Lareyre J-J, LeGac F, Chiarini-Garcia H, Nobrega RH, Miura T (2010) Spermatogenesis in fish. Gen Comp Endocrinol 165(3):390–411

Shanmugam G, Ramasamy K, Selvaraj KK, Sampath S, Ramaswamy BR (2014) Triclosan in fresh water fish Gibelion Catla from the Kaveri River, India, and its consumption risk assessment. Environ Forensic 15(3):207–212

Sun F, Liu S, Gao X, Jiang Y, Perera D, Wang X, Dunham R (2013) Male-biased genes in catfish as revealed by RNA-Seq analysis of the testis transcriptome. PLoS One 8(7):e68452

Sundaray JK, Rather MA, Kumar S, Agarwal D (eds) (2021) Recent updates in molecular Endocrinology and Reproductive Physiology of Fish. https://doi.org/10.1007/978-981-15-8369-8

Takehana Y, Matsuda M, Myosho T, Suster ML, Kawakami K, Shin T, Hamaguchi S (2014) Co-option of Sox3 as the male-determining factor on the Y chromosome in the fish *Oryzias dancena*. Nat Commun 5(1):1–10

Tao W, Chen J, Tan D, Yang J, Sun L, Wei J, Wang D (2018) Transcriptome display during tilapia sex determination and differentiation as revealed by RNA-Seq analysis. BMC Genomics 19(1): 363

Tokumoto T, Tokumoto M, Nagahama Y (2005) Induction and inhibition of oocyte maturation by EDCs in zebrafish. Reprod Biol Endocrin 3(69):1–9

Uren Webster TM, Laing LV, Florance H, Santos EM (2014) Effects of glyphosate and its formulation, roundup, on reproduction in zebrafish (*Danio rerio*). Environ Sci Technol 48: 1271–1279

Van Der Kraak G, Lister A (2011) The inhibitory control of oocyte maturation in the zebrafish (*Danio rerio*): the role of the G protein-coupled estrogen receptor and epidermal growth factor. Biol Reprod 85:6–8

Vetillard A, Bailhache T (2006) Effects of 4-n-nonylphenol and tamoxifen on salmon gonadotropin-releasing hormone, estrogen receptor, and vitellogenin gene expression in juvenile rainbow trout. Toxicol Sci 92(2):537–544

Wang P, Du Z, Gao S et al (2016) Impairment of reproduction of adult zebrafish (*Danio rerio*) by binary mixtures of environmentally relevant concentrations of triclocarban and inorganic mercury. Ecotoxicol Environ Saf 134:124–132

Wang J, Zhou J, Yang Q, Wang W, Liu Q, Liu W, Liu S (2020) Effects of 17 α Methyltestosterone on the transcriptome, gonadal histology and sex steroid hormones in Pseudorasbora Parva. Theriogenology 155:88–97

Weber GJ, Sepulveda MS, Peterson SM, Lewis SS, Freeman JL (2013) Transcriptome alterations following developmental atrazine exposure in zebrafish are associated with disruption of neuroendocrine and reproductive system function, cell cycle, and carcinogenesis. Toxicol Sci 132:458–466

Wiegand C, Krause E, Steinberg C, Pflugmacher S (2001) Toxicokinetics of atrazine in embryos of the zebrafish (Danio rerio). Ecotoxicol Environ Saf 49:199–205

Wilson EO (1994) Naturalist. Island Press/Shearwater Books, Washington, DC

Xu H, Yang J, Wang Y, Jiang Q, Chen H, Song H (2008) Exposure to 17αethynylestradiol impairs reproductive functions of both male and female zebrafish (*Danio rerio*). Aquat Toxicol 88(1): 1–8

Xu H, Lam SH, Shen Y, Gong Z (2013) Genome-wide identification of molecular pathways and biomarkers in response to arsenic exposure in zebrafish liver. PLoS One 8:e68737

Xu W, Yan W, Huang W, Miao L, Zhong L (2014) Endocrine disrupting chemicals in the pearl river delta and coastal environment: sources, transfer, and implications. Environ Geochem Health 36(6):1095–1104

Yano A, Guyomard R, Nicol B, Jouanno E, Quillet E, Klopp C, Guiguen Y (2012) An immune-related gene evolved into the master sex-determining gene in rainbow trout, *Oncorhynchus mykiss*. Curr Biol 22(15):1423–1428

Yilmaz O, Prat F, Ibañez AJ, Amano H, Koksoy S, Sullivan CV (2015) Estrogen-induced yolk precursors in European sea bass, Dicentrarchus labrax: status and perspectives on multiplicity and functioning of vitellogenins. Gen Comp Endocrinol 221:16–22

Yin P, Li YW, Chen QL, Liu ZH (2017) Diethylstilbestrol, flutamide and their combination impaired the spermatogenesis of male adult zebrafish through disrupting HPG axis, meiosis and apoptosis. Aquat Toxicol 185:129–137

Yokoi H, Kobayashi T, Tanaka M, Nagahama Y, Wakamatsu Y, Takeda H, Ozato K (2002) Sox9 in a teleost fish, medaka (*Oryzias latipes*): evidence for diversified function of Sox9 in gonad differentiation. Mol Reprod Dev 63(1):5–16

Zhang Y, Zhang S, Lu H, Zhang L, Zhang W (2014) Genes encoding aromatases in teleosts: evolution and expression regulation. Gen Comp Endocrinol 205:151–158

Zhang X, Shi J, Sun Y, Zhu Y, Zhang Z, Wang Y (2019) Transcriptome analysis provides insights into differentially expressed genes and long noncoding RNAs involved in sex related differences in Amur sturgeon (*Acipenser schrenckii*). Mol Reprod Dev 86(2):132–144

Correction to: Cypermethrin-Induced Reproductive Toxicity in Zebrafish: Biochemical and Molecular Perspective

Sana Aziz, Mumaiza Mumraiz, Fariha Latif, and Muhammad Sarfraz Ali

Correction to:
Chapter 4 in: M. A. Rather et al. (eds.), *Xenobiotics in Aquatic Animals*, https://doi.org/10.1007/978-981-99-1214-8_4

The book was inadvertently published with the fourth author name as Muhammad Sarfraz, which has been corrected as Muhammad Sarfraz Ali.

The updated original version for this chapter can be found at
https://doi.org/10.1007/978-981-99-1214-8_4

M. A. Rather et al. (eds.), *Xenobiotics in Aquatic Animals*,
https://doi.org/10.1007/978-981-99-1214-8_22

MIX
Papier aus verantwortungsvollen Quellen
Paper from responsible sources
FSC® C105338

If you have any concerns about our products,
you can contact us on
ProductSafety@springernature.com

In case Publisher is established outside the EU,
the EU authorized representative is:
Springer Nature Customer Service Center GmbH
Europaplatz 3, 69115 Heidelberg, Germany

Printed by Libri Plureos GmbH
in Hamburg, Germany